PEARSON

ALWAYS LEARNING

Warren W. Esty

Precalculus

Sixth Edition

Pearson Learning Solutions, 501 Boylston Street, Suite 900, Boston, MA 02116
A Pearson Education Company
www.pearsoned.com

Printed in the United States of America

 7 8 9 10 V0ZN 17 16 15 14

000200010271309885

JH

 ISBN 10: 1-256-67120-7
ISBN 13: 978-1-256-67120-6

CONTENTS

PREFACE

The goal of this course is to develop your ability to read, write, think, and do mathematics and to give you command of the facts and methods of algebra and trigonometry. It emphasizes everything you need to understand to be prepared for calculus.

This course is designed for students who want to become good at math. A unique feature is its **emphasis on developing fluency in the abstract and symbolic language of algebra.** That "foreign" language is extremely important because it is the most effective and efficient language in which to learn, understand, recall, and think mathematical thoughts.

THIS CLASS IS DIFFERENT

Calculators and computers are changing the face of mathematics. What mathematics remains essential now that "Calculators can do it all"? What should be in a state-of-the-art class? This text is my answer.

I am convinced that, even with calculators, you still need to learn a great deal about mathematics. Furthermore, I think that graphing calculators, properly used, can greatly facilitate *learning* math, not just *doing* math. In the homework you will see that it is still possible to test your learning and not just test the power of your calculator.

This course, like other math courses, teaches you how to calculate answers to problems. But that is not all. In addition, this text emphasizes mathematical concepts that you must develop and retain in order to advance in mathematics. This additional emphasis makes this class different from any other math class you have taken. You will notice some of the homework is different. It is different because it does not just train you to do the simplest sorts of problems of each type, but also trains you to think mathematically. It teaches you to use algebra the way it is used in higher-level mathematics like calculus, not just the way it is used in high school.

This course will help you

1. Learn the methods and facts of algebra so well that you have them at your command (even without recent review)
2. Learn the methods and facts of trigonometry so well that you have them at your command (even without recent review)
3. Learn how and when to use calculators and graphing technology
4. Become good at word problems
5. Understand what you do
6. Learn to read symbolic mathematics fluently (and, in the process, learn how to learn math by reading it)
7. Learn to work abstractly with symbols and functions (as comfortably as you now work with numbers)
8. Learn to explain (symbolically, and in English) key general results
9. Learn to illustrate (with illuminating pictures) key general results
10. Remember (with the help of symbols, English explanations, and pictures) key general results [What good does it do to have "taken" math if you don't remember it?]

PREREQUISITE

This is a "Precalculus" course which is intended to serve as a bridge from high-school Algebra II and trigonometry to college calculus. This course builds on what you already know about algebra. The apparent content (algebra and trig) may be familiar, but this course asks you to learn it, and mathematical language as well, in a new way that should raise you to a far higher level of mathematical ability. **A graphing calculator is required**.

HOMEWORK

Homework is categorized into "A" and "B" problems. The "A" problems are less complex. They require you to use a given formula, recall a given fact, or to solve a basic equation.

The "B" problems emphasize concepts, interpretations, explanations, visual images, notation, and applications of methods to more difficult problems.

> **Your goal should be to be able to answer the "B" problems.**

The symbol ☺ marks problems that are very short and could be done aloud in class. By asking many of these in a short amount of time, your instructor can make sure you and the class are on the right track.

Some problems ask about important concepts that are worth knowing by heart. These are starred (with an asterisk, *).

Some Answers in the Text

Many problems in the text have answers [in brackets] right alongside the problem. They are given with **two** significant digits. You can see right away if you did the problem right. However, when you hand in your homework, you must report your answers rounded to **three or more significant digits**. That way, we can check your work. If you get the third significant digit right, we can be confident you used the proper method.

PACE

The material has not been artificially subdivided into single-day chunks. On the contrary, material which is mathematically connected is grouped together and the connections are emphasized, not severed. Some sections are two or three days long. Here is a pace I use in a four-credit one-semester course (with 50-minute classes) at Montana State University. There is time for half-day weekly quizzes in this schedule, but exam days would be extra.

Section and numbers of days (excluding exams):

Chapter 1	Chapter 2	Chapter 3	Chapter 4	Chapter 5
1.1: 1-	2.1: 1½	3.1: 2	4.1: 1	5.1: 2
1.2: 1	2:2: 1½	3.2: 2	4.2: 2	5.2: 2
1.3: 1	2.3: 1	3.3: 1	4.3: 1	5.3: 1+
1.4: 1½		3.4: 1	4.4: 2	5.4: omitted
1.5: 1½		3.5: 2	4.5: 1½	
1.6: 1		3.6: 1	4.6: 2	
review 1		review ½	review ½	
----------	----------	----------	----------	----------
8	5	9½	10	5+

Chapter 6	Chapter 7
6.1: 1-	7.1: 1
6.2: 2	7.2: 2
6.3: 2	7.3: 2
6.4: 2	7.4: 2
-----------	-----------
7	7

Chapters 1-7 constitute a one-semester four-credit course at Montana State University.

SUPPLEMENTS

For students there is a solutions manual with solutions (not just answers) to half the problems. In it answers are rounded to two significant digits. The steps and rounded numbers in the manual should make it easy for students to duplicate the correct work. Nevertheless, they cannot simply copy the manual because solutions with three or more significant digits are required. Therefore, they must at least retrace the proper steps and do the computations themselves.

Instructors may obtain a complete solutions manual and an instructor's manual with section-by-section comments and suggestions for instructors, as well as a potential syllabus including daily homework assignments. Sample exams and quizzes are also available.

ACKNOWLEDGEMENTS

A previous version of this text was published by Prentice Hall under the title *Precalculus Concepts*.

Some good ideas take years to bear fruit. Anyone developing such an idea appreciates encouragement and moral support. I got a lot from my wife, Najaria, and my daughter, Norah. In addition, I appreciate the contributions and faith of Anne Teppo, Jean Schmittau, Ken Tiahrt, John Lund, Rufus Hurst, Sarah Boone, and Ken Bowers. I thank the anonymous reviewers who loved the manuscript and encouraged me to produce this text. I also appreciate the comments of dozens of Graduate Teaching Assistants who have taught from the book. Finally, I express my thanks to my colleagues in the Department of Mathematical Sciences at Montana State University for making it a great place to work.

Warren Esty
westy@math.montana.edu
Department of Mathematical Sciences
Montana State University
Bozeman, MT 59717

This text is designed to produce a deep understanding of algebra and trigonometry so that students will be comfortable with their next math. Students will be well-prepared for calculus. The content includes the usual precalculus material (functions, powers, polynomials, logs, exponentials, trig, etc.). Graphing calculators play an important role.

Why Use This Text? This text is unlike others in several important ways. Of course, the presentation of most topics resembles that of other precalculus texts. So does the organization, at least after Chapter 1 (which is unique). But it is particularly effective because of its numerous distinguishing features:

- It does not just use calculators to do old-style problems, but actually incorporates **calculators as a learning tool** and not just a "doing" tool. (Dr. Esty has given numerous conference talks about **learning** with the help of calculators.) Calculators can help in two ways: They can concentrate attention on essential points and they can increase the rate at which students gather experience with the subject.

- Emphasis on **learning how the symbolic language of math is used**. This is probably the most distinguishing feature of the text. It is an explicit goal of Chapter 1 that students learn how thoughts about methods are written in modern mathematical notation. This text has explicit reading lessons! They will increase students' ability to recognize patterns, generalize properly, and **retain** what they learn, as well as increase students' ability to learn outside of class.

- Most Precalculus texts simply teach Algebra II the same way again (an approach which seems bound to fail if the goal is to advance students to a higher level.) This text **emphasizes using the same algebra but at a higher level**. For example, it teaches how to solve for y in "$y^2 + 3xy + 2x^2 - 5x = 17$" in the section on quadratics. Most students who "know" the Quadratic Formula have only solved for "x" and cannot solve for "y" until taught how. (This is closely related to understanding the symbolic language.)

- Illuminating **homework to promote retention** of key ideas, in addition to the usual type of calculation problems for practice. (For example, many "B1" problems ask for an illustration, or explanation, rather than a computation.) Asking students to recreate memorable visual illustrations helps them generate and retain correct concepts. Asking students to reproduce theorems helps them learn to read theorems and use symbols properly.

- Emphasis on mathematical **concepts** that will not become obsolete when the next generation of calculators arrives.

- Emphasis on **how to do word problems**, in general (with two sections devoted to how to do them).

- Instructor-friendly. This text does not require new teaching techniques or classroom experiments. It has numerous problems marked, with ☺, as especially suitable to be asked aloud in class.

- Excellent for teaching oneself. It is hard to learn math outside of class. In contrast to all other texts, this one has lessons in Chapter 1 on how math is written and how to read it. This helps students get the most of the text, even those who are learning at home.
- Does not have hundreds of pages on topics that are not necessary to advance in mathematics.

Reasons to Not Use This Text. There are some reasons to not use this text.

- You might prefer a text with only low-level problems that can be graded by computer. In this text not all problems can be computer graded. Of course, there is a solution manual.
- Some texts are simpler because all the lessons are algorithmic and so is all the homework. You might prefer one of the many texts like that. In this text students are asked to think and understand, not just calculate.
- This text omits topics that are not necessary to advance to calculus, and some of those omitted topics might be topics you want to cover. For example, this text has nothing on matrices.

1

Reading Mathematics

Section 1-1 ALGEBRA

This chapter discusses what you need to know in order to do mathematics at the level of algebra and calculus. In arithmetic, the emphasis is on numbers. Algebra extends arithmetic by also emphasizing operations and order.

Learn what we mean by "operations and order." The following examples show that the operations can be more important than the numbers.

EXAMPLE 1A   Solve "$2(x + 5) = 24$."

It is easy: $x + 5 = \dfrac{24}{2} = 12$.

$$x = 12 - 5 = 7, \text{ so } x = 7.$$

EXAMPLE 1B   Solve "$2(x + 5) = 100$."

This one is hardly different. The number 100 is different from 24, but the order of operations on the left is the same. Do the same steps: Divide by 2 and subtract 5.

$$x + 5 = \dfrac{100}{2} = 50.$$
$$x = 50 - 5 = 45.$$

The method does not depend upon the number, 24 or 100. It depends upon the **sequence of operations** expressed by $2(x + 5)$: "Add 5" and then "Multiply by 2." To solve, we "undo" them using "Divide by 2" and then "Subtract 5." The key to solving it is the **sequence of operations.** ▬

EXAMPLE 2 Compare the two problems:

PROBLEM 1: Solve "$2(x + 5) = 24$."

PROBLEM 2: Solve "$2(x + 5) + x = 24$."

Problem 2 is harder because the order of operations is different. We need to "multiply out" the left side and "consolidate like terms."

Given $2(x + 5) + x = 24$ First change the operations and order. Use the identity "$2(x + 5) = 2x + 10$."

$\qquad\qquad 2x + 10 + x = 24$ Now change the operations and order again.

"Consolidate like terms." Use the identity "$2x + x = 3x$" to get:

$$3x + 10 = 24$$
$$3x = 24 - 10 = 14.$$
$$x = \frac{14}{3}.$$

To solve Problem 1 we did not need to multiply it out. In Problem 2 we do. The difference is the order of operations.

The solution process used the distributive property twice to change the order of operations. ▬

THE DISTRIBUTIVE PROPERTY 1-1-1

For all a, b, and c, $c(a + b) = ca + cb$.

This relates the operations of multiplication and addition. In Example 2 we used it to change $2(x + 5)$ into $2x + 10$, and to change $2x + x$ into $3x$.

The Distributive Property is a generalization stated with "For all a, b, and c." Often we omit that part and just state the rest as an identity.

DEFINITION 1-1-2

A <u>generalization</u> is a sentence that asserts that something is always true. An <u>identity</u> is an equation that has a variable and which is true for all values of the variable. We regard identities as abbreviated generalizations.

Identities express alternative sequences of operations.
[You can substitute either order (either side) for the other.]

EXAMPLE 3 "$2(x + 5) = 2x + 10$" is an identity. What does this equation say about the number x?

Nothing! It's not about x.

Equations are sentences in the language of mathematics. Equations have subjects—they convey information about something. In this example, the information is **not** about a number called x. The equation is true for **all** numbers, so it does not distinguish one number from another; it does not say anything about x, the number.

It says something about the operations (functions) "Add 5" and "Multiply by 2," and the order in which they are executed. It says that, regardless of the number you start with, you may instead "Multiply by 2 and then add 10." The identity gives an alternative sequence of operations.　　━━

Identities expressed using the letter "x" are not <u>about</u> "x". Identities are about operations and order.

In contrast, in Example 1 the equation "$2(x + 5) = 24$" tells you about the number named x: $x = 7$.

Nevertheless, to solve "$2(x + 5) = 24$" you must see the sequence of operations.

■ Focus on the Operations

Properties of functions are given by identities.

EXAMPLE 4 $\left(\dfrac{x}{2}\right)^2 = \dfrac{x^2}{4}$, for all x.

This is a "property" of squaring and division. It tells you nothing about the number x.

This identity gives alternative sequences of operations. You may divide by 2 and then square, or you may square and then divide by 4. The result will be the same either way (regardless of the number x).　　━━

■ The Language of Mathematics

The symbolic language of mathematics is designed for algebra. One goal of this text is to have you become so fluent in it that you will be able to learn new math easily, just by reading it, even without a teacher.

DEFINITION 1-1-3　　<u>Expressions</u> are the nouns and pronouns of the language. Expressions include numbers such as "7", "$5(3^2)$", and "sin 2". The expressions "x", "$2(x + 5)$", and "sin x" include variables (letters).

Expressions with the numerical variable "x" simultaneously represent two distinct things:

1. a *number* (which depends upon the value of x), and
2. a *sequence of operations* (which does not depend upon the value of x).

   [In the expression "$2(x + 5)$" the sequence of operations is "Add 5 and then multiply by 2."]

EXAMPLE 5   Here are two equations. What are they about?

(A) $(x + 1)^2 = 16$ You can solve this. It is about the value of x, a number.

(B) $(x + 1)^2 = x^2 + 2x + 1$ This is an identity. It is about operations and order-addition and squaring. It says that "Add 1 and then square" is equivalent to "Square, add two times the original, and add 1." ▬

EXAMPLE 6   Is the equation about a number, or about operations and order?

(A) $x^2 - 1 = (x - 1)(x + 1)$ You recognize this as an identity. It is therefore about operations and order.

(B) $x^2 - 1 = 17.$ You can solve this. It is about the number x. ▬

You have already seen that

1. You solve equations by dealing with the operations and order (Examples 1 and 2),
2. Identities give alternative operations in a different order (Example 3),
3. Properties of functions are about operations and order (Example 4), and
4. Identities, which are about operations and order, look similar to equations about numbers. Learn to tell the difference (Examples 5 and 6).

The next examples show that

5. Word problems use symbols (such as "x") to express operations and order.

■ Formulas and Word Problems

Formulas tell you what to do by stating the appropriate operations and order.

EXAMPLE 7   "$C = \pi d$" is the formula for the c̲ircumference of a circle.

"$A = \pi r^2$" is the formula for the a̲rea of a circle.

The letters are chosen to remind you of what they mean: d is for d̲iameter; r is for r̲adius. The formulas tell you which operations and order to use. ▬

Algebraic word problems emphasize thoughts about operations, as opposed to thoughts about numbers.

EXAMPLE 8A   A field is in the shape of a square with a semicircular cap on one side (Figure 1). The side is 100 meters. What is the perimeter of the field? ("Perimeter" means the distance around three sides and the semicircle.)

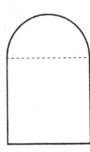

FIGURE 1 A field in the shape of a square with a semicircular cap

SOLUTION   The three sides of the square contribute 300 (meters). The semicircle contributes half the circumference of a circle of diameter 100. So it contributes $\dfrac{100\pi}{2}$.

$$\text{Total length: } 300 + \frac{100\pi}{2} = 457.08 \text{ (meters)}.$$

This is arithmetic, not algebra. Computing the perimeter does not require algebra. The next problem uses similar words, but the focus is not on numbers because it uses algebra. ▬

Convention: In this text, numbers (such as 457.08) expressed with three or more significant digits may be approximations and not exact.

EXAMPLE 8B   A field is in the shape of a square with a semicircular cap on one side (Figure 1, again). Its perimeter is 400 meters. How long is a side of the square?

We cannot simply plug in to a formula, because the formula for this shape is not given. First we need to create the formula.

SOLUTION   The same *operations* used in Example 8A will work for any length side. Call the side "x". This is where algebraic notation comes in. We use an algebraic expression to write about these operations, regardless of the unknown number x.

To find the perimeter, take the side (x) and *multiply by three* (to get "$3x$").

Separately, take the side and *multiply by* π (πx) and then *divide by two* $\left(\dfrac{\pi x}{2}\right)$.

Add those two separate contributions $\left(3x + \dfrac{\pi x}{2}\right)$.

This gives the formula for the perimeter P:

$$P = 3x + \frac{\pi x}{2}$$

Now the word-problem sentence about "400 meters" translates to the equation

$$3x + \frac{\pi x}{2} = 400.$$

To solve this equation, simplify the left side. You must continue to think about operations and order. The idea is to "isolate" x or "Consolidate like terms." Use the Distributive Property to rewrite the expression so it expresses a different sequence of operations.

$$\left(3 + \frac{\pi}{2}\right)x = 400.$$

Now $3 + \dfrac{\pi}{2}$ is just a number, so we can find x by dividing by it.

$$x = \frac{400}{3 + \dfrac{\pi}{2}} = 87.51 \text{ (meters)}.$$

The solution *process* (but not the solution) would be identical if the perimeter were some other number besides 400. The particular number 400 has nothing to do with creating the formula. Also, it has nothing to do with the change of

"$3x + \dfrac{\pi x}{2}$" to "$\left(3 + \dfrac{\pi}{2}\right)x$." ▬

> **Almost all of this problem is about operations and the order in which they occur, not about numbers.**

Finding the perimeter from the side (Example 8A) is arithmetic because it is "direct." In contrast, finding the side from the perimeter (Example 8B) is much harder because it is "indirect."

DEFINITION 1-1-4 A problem is <u>direct</u> when the given words, symbols, or basic formulas express the operations you actually do to solve the problem. A problem is <u>indirect</u> when the given words, symbols, or basic formulas suggest operations that you are *not* supposed to actually *do*. To do indirect problems you *write about* the operations (in symbolic notation) and then manipulate the operations (instead of just manipulating numbers).

> **Writing about operations we don't actually do is a characteristic of algebra.**

Here is another example to make the point that word-problems are mostly about operations, not numbers.

> Algebra problems are indirect. The symbolic language is used to *write about* operations you don't actually do.

EXAMPLE 9 Suppose ABCopy charges 7 cents each for the first 20 copies and only 4 cents each for copies after the first 20. You made enough copies so they average 4.5 cents each. How many do you make?

This problem is indirect. You cannot simply calculate the answer by doing the given operations. Instead, you *write about them* in symbols (and end up doing completely different operations).

Let x be any number of copies.

The formula for the cost, in cents, for more than 20 copies is

$$C = 7(20) + 4(x - 20).$$

The first term represents the cost of the first 20 copies: 7(20). The second term expresses the number of copies beyond 20("$x - 20$") and their cost at 4 cents each ("$4(x - 20)$").

We could simplify (change the order of operations) to:

$$C = 140 + 4x - 80 = 60 + 4x.$$

The average cost will be the cost divided by the number of copies:

$$\text{Average cost} = \frac{C}{x} = \frac{60 + 4x}{x} = \left(\frac{60}{x}\right) + 4.$$

Now set this equal to the given average cost, 4.5, and solve (Problem A19).

The key is to *write about the operations* and manipulate the operations. Your focus is on the operations, not the numbers. The given average number (4.5) does not have much to do with the real work of this algebra problem. ▬

Operations and order are what algebraic notation is designed to express.

C O N C L U S I O N   In algebra and calculus the emphasis is beyond numbers. Many algebra problems require you to focus on operations and order.

To solve an equation you must identify the operations and order.

Identities and properties of functions are about operations and order. They give alternative sequences of operations. Even when they are expressed in terms of x, they are, nevertheless, not *about x*.

To do algebraic word problems you use symbols to write about operations and order.

Algebra applies to numbers, but, when you do algebra, you must think about operations and order. Algebra problems are indirect. You write about operations you don't actually do.

Therefore, you must learn to read and write about operations and order. You must learn how the symbolic language works. Reading and writing is the subject of all the sections through Section 1-4.

Terms: operations and order, sequence of operations, identity, expression, direct, indirect.

Exercise 1-1

A ☺ *State the formula for*

A1. (A) The area of a circle.
(B) The circumference of a circle.

A2. (A) The area of a square.
(B) The perimeter of a square.

A3. (A) The area of a rectangle.
(B) The perimeter of a rectangle.

A4. State the formula for the area of a triangle.

A5. State the Pythagorean Theorem.

A6. State the formula for the volume of a (rectangular) box.

Word Problems

A7. In a rectangle the base, b, is 2 units longer than the height, h.
(A) Find a formula for the perimeter in terms of h alone.
(B) The perimeter is 32. Find the height.

A8. The number, A, of bacteria of the first type is 4 times the number, B, of the second type.
(A) Find a formula for the total number of bacteria in terms of the number of the second type.
(B) If the total number of bacteria is 22,000, how many are of the second type?

A9. There are 15 times as many students as professors at the meeting.
(A) Find a formula for the total number of students and professors at the meeting.
(B) If the total number of students and professors at the meeting is 80, how many are professors?

A10. There are 12 times as many coyotes as wolves. If the total number of coyotes and wolves is 117, how many are wolves?

A11. Now River C has half the pollutant it had 5 years ago. Now it has 38 ppm. How many ppm did it have then?

A12. A wire 100 inches long is cut into two pieces. One is three times as long as the other. How long is the other?

A13. A wire 200 inches long is cut into three pieces, two of equal length, each twice as long as the third. How long is the third piece?

A14. Mike and John split the cost of a $270 car trip. Mike pays $20 more than John. How much did John pay?

☺ *Decide if the following is an expression or an equation.*

A15. $3(x + 5)$ **A16.** $3(x + 5) = 17$

A17. $5x - 2 = 0$ **A18.** $5x - 2$

A19. Answer the question in Example 9. How many copies do you buy?

A20. (A) In Example 9, how much would 50 copies cost?
(B) What would the average cost per copy be for 50 copies?

A21. In Example 8A, what would the perimeter be if the side were 150 meters?

A22. In Example 2 we used the Distributive Property (1-1-1) to change $2(x + 5)$ into $2x + 10$. What is the "a" in 1-1-1? The "b"? The "c"?

A23. In Example 2 we used the Distributive Property (1-1-1) to change $2x + x$ into $3x$. What is the "a" in 1-1-1? The "b"? The "c"?

B

B1.* (A) Is algebra only about numbers?
(B) What else is algebra about?
(C) What is algebraic notation with variables designed to express?
(D) An expression with a variable simultaneously represents two distinct things. Which two?

B2.* Algebra concerns indirect problems, as opposed to direct problems. What makes a problem "indirect"?

B3.* In Example 8B, if the perimeter were given as 550 meters instead of 400 meters, what part of the work would remain the same?

B4.* ☺ (A) When an identity is stated in terms of "x", what does it say about the number x?
(B) (In general) What are identities about?

☺ *These problems can be solved in two steps, regardless of whether the number on the right is 97 or some other number. **State, as two operations (commands)**, the two steps, in the right order, which will solve the problem. Do not solve it! Just state the two steps in order.*

B5. $3(x - 5) = 97$ **B6.** $4(x + 5) = 97$

B7. $6x + 7 = 97$ **B8.** $8x - 9 = 97$

B9. $\dfrac{x - 10}{11} = 97$ **B10.** $\dfrac{x + 12}{13} = 97$

B11. $\dfrac{x}{15} - 16 = 97$ **B12.** $\dfrac{x}{3} + 2 = 97$

B13. $5.7(x + 4) = 19$ **B14.** $\dfrac{(x - 7)}{8} = 3.4$

B15. $9x + 4 = 12$

B16. In which of Problems B5–B15 would changing the number on the right change the two steps?

☺ *To solve some of these problems algebraically it is best to "multiply out" the product or square. In others, it is not necessary. Is it best to multiply out the expression? (Answer "Yes" or "No". Do not solve it.)*

B17. $3(x + 2) = 19$

B18. $5(x - 7) = 19$

B19. $5(x + 2) - 8 = 100$

B20. $5(x + 2) - x = 100$

B21. $7(x - 3) + x = 200$

B22. $7(x - 3) + 12 = 200$

B23. $(x + 1)^2 = 100$

B24. $(x + 1)^2 + 1 = 100$

B25. $(x + 1)^2 + x = 500$

B26. $(x + 1)^2 - x = 600$

B27. $(x + 1)^2 + 20 = 700$

B28. $(x + 1)^2 + 12 = 800$

☺ *Decide if the subject of the equation is a number (N) or operations and order (OO). Do not solve! Just answer "N" or "OO".*

B29. $(x + 1)^2 = x^2 + 2x + 1$

B30. $x + 3 = 2x - 6$

B31. $(x + 1)^2 = 16$

B32. $x + 3 = \dfrac{(2x + 6)}{2}$

B33. $3(x + 6) = 3x + 18$

B34. $3(x + 6) = 2x + 20$

B35. $2(x + 1) = 2x + 2$

B36. $2x + 6 = 2(x + 3)$

B37. $\left(\dfrac{x}{3}\right)^2 = \dfrac{x}{16}$

B38. $\left(\dfrac{x}{3}\right)^2 = \dfrac{x^2}{9}$

B39. $x(a - b) = ax - bx$

B40. $x(a - b) = a + b$

Write the formula which expresses the given relationship.

B41. See Figure 1 (a square with a semicircular cap). Create the formula for the area of the figure in terms of the side of the square. [Use A and x.]

B42. Let taxes be T and income be x. Taxes are 10 percent of her income in excess of $12,000. [Assume $x > 12,000$.]

B43. The cost (C) of photocopies is 6 cents each for the first 30 and 4 cents for each one after the first 30. [Use x for the total number of photocopies. Assume $x > 30$.]

B44. The rental car cost $40 a day for up to 150 miles plus 35 cents per mile for all miles after the first 150. [Use C and m for the total number of miles per day. Assume $m > 150$.]

B45. A right triangle has two equal legs. Give the relationship between the leg, x, and the hypotenuse, c.

B46. A right triangle has two equal legs. Give the relationship between the leg, x, and the perimeter, P.

B47. A right triangle has one leg twice the other. Give the relationship between the shorter leg, x, and the hypotenuse, c.

B48. A right triangle has one leg twice the other. Give the relationship between the shorter leg, x, and the perimeter, P.

B49. The cost of mailing a letter is 37 cents for the first ounce and 23 cents for each additional ounce. [Use C and x = ounces.]

B50. The cost of n copies is $50 + 3n$ cents. What is the average cost per copy? [Use A and n.]

B51. The cost of b books is $10 + 13b$ dollars. What is the average cost per book? [Use A and b.]

B52. Abe has twice as much money as Mike, and Sal has $30 more than Abe. How much money do the three of them have together? [Use only M and T, for total.]

B53. One side of a rectangle is 3 inches longer than the other, x. Find the formula for its area.

B54. One side of a rectangle is twice as long as the other, x. Find the formula for its area.

B55. The <u>Distributive Property</u> states: $a(b + c) = ab + ac$. Write a similar identity about multiplication and subtraction.

B56. The <u>Distributive Property</u> states: $a(b + c) = ab + ac$. Write a similar identity about division and addition.

B57. An expression such as "$2x + 5$" can be regarded as simultaneously representing two things. What are its two interpretations?

B58. An expression such as "$x^2 - 1$" can be regarded as simultaneously representing two things. What are its two interpretations?

B59. The numbers which solve "$5(x + 3) = 34$" and "$5(x + 3) = 95.6$" are different. But something about solving them is the same for both. What?

B60. The numbers which solve "$5x + 2 = 90$" and "$5x + 2 = 3.19$" are different. But something about solving them is the same for both. What?

B61. In Example 9, if the average cost were given as 4.7 cents per copy instead of 4.5, what part of the work would remain the same?

Section 1-2 ORDER MATTERS!

Algebraic expressions such as "$\pi + 3x$" often indicate more than one operation. Then the order in which the operations are carried out usually makes an important difference. You must distinguish

1. the left-to-right order in which it is written,
2. the order in which the operations are to be executed, and
3. the order of keystrokes you need to use to make *your* calculator evaluate the expression correctly.

Order Matters!

Algebraic notation is a language. Rules called the "algebraic conventions" tell us how to read it.

Convention 1
Operations enclosed in parentheses are executed first.

There are other grouping symbols, such as the horizontal fraction bar and the extended square root symbol (discussed below), that serve the purpose of parentheses.

Convention 2
Squaring and other operations with exponents are evaluated before multiplication or division, unless parentheses indicate otherwise.

Convention 3
Multiplication and division are performed before addition or subtraction, unless parentheses indicate otherwise.

> Here are two ways to remember the order conventions:
> PEMA: Parentheses, Exponents (such as squaring), Multiplication (and division), Addition (and subtraction).
> or
> "Please Excuse My Dear Aunt Sally": Parentheses, Exponents, Multiplication, Division, Addition, Subtraction.

EXAMPLE 1

Expression	First operation	Comment
$1 + 2x$	Multiply by 2	$= 1 + (2x)$, not $(1 + 2)x$
$3(x + 5)$	Add 5	by Convention 1
$3x + 5$	Multiply by 3	by Convention 3
$x - \dfrac{6}{2}$	Divide 6 by 2	by Convention 3
$\dfrac{x - 6}{2}$	Subtract 6	by Convention 1

Convention 4
$-x^2 = -(x^2)$, not $(-x)^2$. Squaring is done **before** the negative is attached.

EXAMPLE 2 $-7^2 = -49$. If you want to square "-7", do not write "-7^2". Use parentheses and write $(-7)^2$. $(-7)^2 = 49$.

Let $a = 3x$. Then $a^2 = (3x)^2$ [not $3x^2$].

Let $b = x + 1$. Then $\dfrac{1}{b} = \dfrac{1}{x + 1}$ $\left[\text{not } \dfrac{1}{x} + 1 \right]$.

Let $c = -x$. Then $c^2 = (-x)^2$ [not $-x^2$]. $-x^2 = -(x^2)$, by Convention 2.

Order matters!

■ Your Calculator

Learn how to make *your* calculator do the calculations.

CALCULATOR EXERCISE 1 Let $c = -1$. Find $-2^2 + c^2$.

$$-2^2 + c^2 = -4 + (-1)^2 = -4 + 1 = -3 \quad \text{[not 5]}.$$

Let $b = -\pi$. Find b^2. [Find and use your π key.]

$$b^2 = (-\pi)^2 = 9.87 \quad \text{[not } -\pi^2 = -9.87\text{]}.$$

Convention of This Text: Numbers expressed with three or more significant digits may be approximations and are not necessarily exact.

■ Significant Digits

You must report your homework answers with at least three significant digits. Answers are often given in the text right alongside the problems, but with only **two** significant digits. For example, the answer to Problem B8 in this section is given as "$-.35$." With greater accuracy, the answer is "$-.3472609163$." When you report the answer, give at least three significant digits: "$-.347$".

You can check that you did it right by noting that your answer would round off to "$-.35$" with two significant digits. Your instructor can check if you did it right by seeing if you got the third digit right.

Here are some numbers rounded off to two and three significant digits:

Exact number	With 3 digits	With two significant digits	
23417	23400	23000	Round down because it is closer.
53678	53700	54000	Round up because it is closer.
1.2268	1.23	1.2	Round to the closer number.
1.253	1.25	1.3	Round to the closer number.
.0024312	.00243	.0024	The two 0's are not "significant." Significant digits begin with the leftmost *non-zero* digit.
.012345	.0123	.012	The 0 on the left is not "significant."
8403	8400	8400	In some cases you cannot tell 3 significant digits from 2. Fortunately, this hardly ever matters.

In some cases the exact number has only one or two digits. Then, just give it exactly. In some cases the exact number is half way between two three-digit numbers. For example, 745 is half way between 740 and 750. Then the text answer will give all three digits.

Convention 5
Functions with special names (which have function keys, such as trig functions, logs, and exponentials) are executed before addition and subtraction.

EXAMPLE 3

$\log x + 2 = (\log x) + 2$ [not $\log(x + 2)$].

$\sin x - 5 = (\sin x) - 5$ [not $\sin(x - 5)$].

CALCULATOR EXERCISE 2

Reproduce these results on your calculator. Learn to use parentheses properly.

$$\log(\pi + 2) = .711.$$

$$\log \pi + 2 = 2.497.$$

$$\frac{\log 297}{2} = 1.236.$$

$$\log\left(\frac{297}{2}\right) = 2.17.$$

EXAMPLE 4

Do not handwrite "sin $x/2$" with a slanting division bar. It could mean "(sin x)/2" or "sin ($x/2$)," which are different. The precise position of the fraction bar makes a difference.

$\sin \dfrac{x}{2}$ indicates division is first; $\dfrac{\sin x}{2}$ indicates that sine is applied first.

■ Important Notation for Grouping

There are two common written grouping symbols that do not have keys on many calculators:

1. the long fraction bar and
2. the extended overhead line of the square root symbol.

   Calculators need parentheses to indicate the grouping that these symbols can indicate without parentheses.
3. Superscripts (exponents) are a third, less common, way to indicate grouping without parentheses.

Fraction Bars
The fraction bar is a grouping symbol.

CALCULATOR EXERCISE 3

Evaluate these to check your use of your calculator.

$$\frac{4 + \pi}{7} = \frac{(4 + \pi)}{7} = 1.02 \quad \left[\text{not } 4.45, \text{ which is } 4 + \frac{\pi}{7}\right].$$

$$\frac{19}{2 + \pi} = \frac{19}{(2 + \pi)} = 3.695 \quad \left[\text{not } 12.64, \text{ which is } \frac{19}{2} + \pi\right].$$

CALCULATOR EXERCISE 4

Different calculator models do this one differently. Learn what yours does. Find out how to "store" a number as, say, "A". Store 3 in the memory as A. Now, compute this:

$$\frac{12}{2A} = 2.$$ Does your calculator get 2 if you type in "12/2A"? Some get 18

and some get 2. You can avoid the problem by using parentheses: $12/(2A) = 2$.

■ Square Roots

The extended square root symbol is a grouping symbol which many calculators cannot use.

$$\sqrt{a + b} = \sqrt{(a + b)} = \sqrt{(a + b)}, \text{ not } \sqrt{a} + b.$$

Grouping symbols are hard to pronounce.

$\sqrt{a + b}$ is "The square root of the quantity a plus b."

$\sqrt{a} + b$ is "The square root of a [pause], plus b."

Only the phrase "the quantity" distinguishes these two very different expressions.

CALCULATOR EXERCISE 5

Check the use of your calculator by reproducing these results.

$$\sqrt{\pi + 9} = 3.48 \quad \left[\text{not } 10.77, \text{ which is } \sqrt{\pi} + 9.\right]$$

Let $b = -3.7$ and $d = 20$. $\sqrt{b^2 + d} = 5.80$.

$$\frac{1 + \sqrt{5.4 + 1.85}}{2} = 1.846 \quad \left[\text{not } 2.346, \text{ which is } 1 + \frac{\sqrt{5.4 + 1.85}}{2}.\right]$$

Superscripts
Superscripts are generally written in a smaller size. The intended order is given by the size and position.

$$2^{m+n} = 2^{(m+n)}, \text{ not } 2^m + n.$$

$$5^{m/n} = 5^{(m/n)}, \text{ not } \frac{5^m}{n}.$$

CALCULATOR EXERCISE 6

Learn how your calculator works by reproducing these results.

$$2^5 = 32.$$

$$8^{2/3} = 4 \quad \left[\text{not } 21.333, \text{ which is } \frac{8^2}{3}.\right]$$

$$2^{1.2+\pi} = 20.27 \quad [\text{not } 5.439, \text{ which is } 2^{1.2} + \pi].$$

$$2^{1.2(3.4)} = 16.9 \quad [\text{not } 7.81, \text{ which is } 2^{1.2}(3.4)].$$

■ Parentheses For the Quadratic Formula

The quadratic formula is usually written with an extended fraction bar and an extended square root bar as grouping symbols. For your calculator, you will need extra pairs of parentheses.

THE QUADRATIC FORMULA 1-2-1

If $a \neq 0$, $ax^2 + bx + c = 0$, is equivalent to

$$x = \frac{-b \pm \sqrt{b^2 - 4ac}}{2a}.$$

"Negative b plus or minus the square root of (the quantity) b squared minus four a c, all over two a."

The entire top is grouped by the long fraction bar and so is the "$2a$" in the denominator. Also, the interior of the square root is grouped, by the long square root bar.

This is tricky to enter into your calculator correctly because you need extra grouping symbols. Try these examples with your calculator.

EXAMPLE 5

Solve $3.67x^2 + 20x + 5 = 0$.

$a = 3.67$, $b = 20$, and $c = 5$.

Plugging in, and using * for multiplication, we get

$$x = \frac{-20 + \sqrt{20^2 - 4 * 3.67 * 5}}{2 * 3.67}$$ for the part with the plus sign. But your calculator may need this entered as if the whole formula were on a single line:

$$(-b + \sqrt{(b^2 - 4 * a * c)})/(2 * a).$$

Extra grouping symbols: ((   )) (   ).   ▬

Extra grouping symbols are necessary for the one-line version.

CALCULATOR EXERCISE 7

Learn how your calculator works by reproducing these results. Try to enter the whole formula at once, rather than breaking it up into separate calculations. Do the "plus" term and use $a = 3.67$, $b = 20$, and $c = 5$.

$$\frac{-b + \sqrt{b^2 - 4ac}}{2a} = -.2627.$$

The next one is a bit trickier.

CALCULATOR EXERCISE 8

Let $a = 4.69$, $b = -5.3$, and $c = -7$. Evaluate the "plus" term of the Quadratic Formula:

$$\frac{-b + \sqrt{b^2 - 4ac}}{2a} = 1.91.$$

In the quadratic formula, be careful with the final "divided by two times a." The grouping matters.

$12 \div (2 * 3) = 2$

$12 \div 2 * 3 = 18$. They are not the same.

■ Negatives and Squaring

Let $b = -2.3$. Find b^2.

Do not type in "-2.3^2". That is wrong.

$-2.3^2 = -(2.3^2) = -5.29$.

$b^2 = (-2.3)^2 = 2.3^2 = 5.29$.

To square a negative number, simply omit the negative sign when you type it in. $(-a)^2 = a^2$, for all a, the shortcut of just using 2.3^2 instead of $(-2.3)^2$ saves keystrokes and avoids the problem.

The next three exercises are easy to get wrong. Do them yourself. If you don't get them right the first time, think about how you inserted (or failed to insert) parentheses.

CALCULATOR EXERCISE 9

Let $b = -3.9$. Evaluate $\sqrt{b^2 - 4(-7.7)}$.

The answer is 6.78. This is easiest to evaluate by simply omitting the negative sign on b, because $(-3.9)^2$ is the same as 3.9^2.

CALCULATOR EXERCISE 10

Let $c = -9.1$, $d = 4.7$, and $k = 3.6$. Evaluate $\dfrac{\sqrt{c^2 - 2k}}{3d}$.

The value of the expression is .617, with three significant digits. The textbook answer with two would be .62, so you can check your work. However, you must give at least three significant digits.

CALCULATOR EXERCISE 11

Evaluate $\dfrac{\pi - \sqrt{2(3.47 - 3.1)}}{1.2 + 1.875}$.

The value of the expression is .742, with three significant digits. With only two significant digits it is .74. This one needs extra parentheses to close the square root and to close the numerator and denominator.

Guide to Pronunciation of Mathematical Expressions

-5	negative five [or] minus five
$5 - (-2)$	five minus negative two [or] five minus minus two
$5x$	five x [avoid "five times x"]
$\dfrac{x}{2}$	x over two [or] x divided by two
$3x + 2$	three x plus two [avoid "three times x plus two"]
$3(x + 2)$	three times the quantity x plus two [compare with the previous one] [or] three times (pause) x plus two

[In spoken Mathematics, parentheses are hard to say aloud. "The quantity" is a phrase used to alert the listener to parentheses. Also, pauses can be used to indicate parentheses.]

$3 + \dfrac{x}{2}$	three plus x over two [or] three plus x divided by two
$\dfrac{3 + x}{2}$	three plus x, the quantity, over two [compare with the previous one]
$3x^2$	three x squared
$(3x)^2$	three x, all squared [compare with the previous one] [or] three x, the quantity, squared [or] the quantity three x (pause) squared

$\sqrt{2}$	the square root of two [or "root two" or "radical two"]
$\sqrt{x+2}$	the square root of the quantity x plus two

[Sometimes the awkward phrase, "the quantity," is omitted, in which case this and the next expression are not easily distinguished aloud.]

$\sqrt{x}+2$	the square root of x (pause) plus two
$\|x\|$	the absolute value of x [some say "absolute x"]
$\|x-2\|$	the absolute value of the quantity x minus two [or] the distance between x and 2
$\|x\|-2$	the absolute value of x (pause) minus two
x^3	x cubed [or] x to the third [or] x to the third power
x^p	x to the pth power [or] x to the p
$=$	equals [or] is equal to [not "equals to"]
$<$	is less than [avoid "is smaller than"]
$\leq$	is less than or equal to
$>$	is greater than [avoid "is bigger than"]
$\geq$	is greater than or equal to
$-2 < x < 2$	negative two is less than x is less than two [or] minus two is less than x is less than two [or] negative two is less than x and x is less than two

CONCLUSION **Order matters!** The "algebraic conventions" determine the order of operations in expressions. To get your calculator to evaluate expressions correctly, you may need to use parentheses that are not in the original expression. You must be very careful with grouping and you must practice with *your* model of calculator.

Term: algebraic conventions.

Exercise 1-2

Note: Answer with <u>three or more significant digits</u>. An approximate value with two significant digits is often given in brackets. If your answer does not round off to the given answer, check your work. If the answer is "323", it will be given as "320", which has only two significant digits. If the answer is actually ".0258712886", you should report at least ".0259", which has three correct significant digits. The number ".026" has only two significant digits.

A

A1.* What is the main lesson of this section?

☺ *Rewrite these with exactly three significant digits.*

A2. (A) 1234 (B) 1237 (C) .01234 (D) 1.237

A3. (A) 35987 (B) 3.592 (C) .003592 (D) .03598

A4. (A) 3.14159 (B) 314.2 (C) .03142 (D) 45778

A5. (A) .058247 (B) 132456 (C) 697812 (D) .0001849

A6. (A) 12.851 (B) 1.2849 (C) 92.126 (D) .09187

☺ *Let $a = 2, b = 3$, and $c = -4$. Give*

A7. (A) ab^2 (B) c^2 (C) $-a^2$

A8. (A) ca^2 (B) $a + b^2$ (C) $2b^2$

A9. ☺ *Let $b = -2, c = 5$, and $d = 3$. Give*

(A) b^2 (B) $-c^2$ (C) $c + \dfrac{d}{4}$

☺ *Let $b = 2, c = 3$, and $d = 4$. Evaluate*

A10. (A) bc/d (B) $b + c/d$ (C) $d/b + d$

A11. (A) $b/c + d$ (B) bc^2 (C) $-c^2$

In all the following calculator exercises evaluate the expressions to <u>three or more significant digits</u>. To help you check your keystroke sequence, a two-significant-digit answer is given in brackets.

A12. 3.4567^4 [140]   **A13.** $\dfrac{3 + \pi}{5}$ [1.2]

A14. $\dfrac{3}{4 + \pi}$ [.42]   **A15.** $\sqrt{15 + \pi}$ [4.3]

A16. $1.2^{1+\pi}$ [2.1]   **A17.** $4^{2/3}$ [2.5]

A18. $\dfrac{2^3}{5}$ [1.6]   **A19.** $3^{2/5}$ [1.6]

A20. $\dfrac{17}{2 + \pi}$ [3.3]   **A21.** $\sqrt{51 + 6\pi}$ [8.4]

A22. $\sqrt{30} + \pi$ [8.6]   **A23.** $\log \pi + 3$ [3.5]

A24. $\log(\pi + 3)$ [.79]   **A25.** $\log 2 + \pi$ [3.4]

A26. $\log(2 + \pi)$ [.71]   **A27.** $10^{2+\pi}$ [140,000]

A28. $10^2 + \pi$ [100]   **A29.** $\dfrac{5 + \pi}{7}$ [1.2]

A30. $\dfrac{27}{\pi + 5}$ [3.3]   **A31.** $\dfrac{3 + \pi}{\pi + 6}$ [.67]

A32. $\sqrt{10 - 2\pi}$ [1.9]

Evaluate the Quadratic Formula with

A33. $a = 3.7, b = -1.2,$ and $c = -7.8.$ [1.6, −1.3]

A34. $a = 14.4, b = 19,$ and $c = -95.$ [2.0, −3.3]

A35. $a = 3, b = -2.1,$ and $c = -7.$ [−1.2, 1.9]

Reading and Writing Mathematics

☺ *Pronunciation*. *Write out, in words, how you would say these aloud.*

A36. (A) $x - 2 = 12$ (B) $\dfrac{x}{3} = 21$

A37. (A) $x + 7 = 19$ (B) $\dfrac{x}{5} = 20$

A38. (A) $5 - 12 = -7$ (B) $3x = 12$

A39. (A) $2x - 3 = -7$ (B) $5x = 50$

A40. (A) $x^2 = c$ (B) $x = \pm\sqrt{c}$

A41. (A) $x^2 = 16$ (B) $x = \pm\sqrt{16}$

A42. (A) $x < 3$ (B) $x > 4$

A43. (A) $3x$ (B) $\dfrac{x}{5}$

A44. (A) bc (B) $x + 3$

A45. (A) $x - 2$ (B) ax

A46. (A) $5x + 2$ (B) $\dfrac{c}{3}$

A47. (A) $x \le 5$ (B) $x \ge 2$

A48. (A) $-2 < x < 3$ (B) $x - 3 = 5$

A49. (A) $x - 12 = -5$ (B) $2 \le x < 6$

A50. (A) $3x^2$ (B) $(5x)^2$

A51. (A) -3^2 (B) $(-3)^2$

A52. (A) $a + \dfrac{b}{2}$ (B) $\dfrac{a + b}{2}$

A53. (A) $a - \dfrac{c}{d}$ (B) $\dfrac{a - c}{d}$

A54. (A) $\dfrac{\left(\dfrac{b}{c}\right)}{d}$ (B) $\dfrac{b}{\left(\dfrac{c}{d}\right)}$

A55. (A) $\sqrt{x + 2}$ (B) $\sqrt{x} + 2$

A56. (A) $a - (b - c)$ (B) $a - b - c$

A57. (A) $\dfrac{3}{x + 2}$ (B) $\dfrac{5}{x} + 2$

A58. (A) cd^2 (B) $(cd)^2$

A59. (A) $c + d^2$ (B) $(c + d)^2$

A60. (A) $\dfrac{a}{\left(\dfrac{b}{c}\right)}$ (B) $\dfrac{\left(\dfrac{a}{b}\right)}{c}$

A61. $\dfrac{a}{b} = c$ **A62.** $a(b + c) = ab + ac$

A63. (A) $\dfrac{3 + \pi}{5}$ (B) $3 + \dfrac{\pi}{5}$

A64. (A) $\sqrt{51 + 6\pi}$ (B) $\sqrt{30} + \pi$

A65. (A) $\sin 2 + \pi$ (B) $\sin(2 + \pi)$

☺ *Give an equivalent expression without parentheses:*

A66. $100 - (20 - \pi)$ **A67.** $\dfrac{100}{\left(\dfrac{20}{\pi}\right)}$

A68. $b - (c - d)$ **A69.** $\dfrac{b}{\left(\dfrac{c}{d}\right)}$

☺ *True or false?* *[No reason required.]*

A70. (A) $(x + 4)^2 = x^2 + 16$ (B) $(x - 3)^2 = x^2 - 9$

(C) $(x + 5)(x - 5) = x^2 - 25$

A71. (A) $\dfrac{1}{2} + \dfrac{1}{5} = \dfrac{1}{7}$ (B) $\left(\dfrac{1}{x}\right)^2 = \dfrac{1}{(x^2)}$

(C) $-x^2 = x^2$

A72. (A) $\sqrt{\dfrac{1}{x}} = \dfrac{1}{\sqrt{x}}$ (B) $\sqrt{x^2 + 25} = x + 5$

A73. (A) $(a + b) + c = a + (b + c)$
(B) $(a - b) - c = a - (b - c)$
(C) $(ab)c = a(bc)$

A74. (A) $\dfrac{\left(\dfrac{a}{b}\right)}{c} = \dfrac{a}{\left(\dfrac{b}{c}\right)}$ (B) $a + b = b + a$

(C) $a - b = b - a$

A75. (A) $ab = ba$ (B) $\dfrac{a}{b} = \dfrac{b}{a}$

A76. (A) How do you (personally) remember the algebraic conventions?
(B) Give an acronym which can help you remember the algebraic conventions.

A77.* In algebra we avoid the symbol "×".
(A) Why?
(B) How can we write "6 × 5" without using "×", still leaving it in factored form?

B

B1.* Name three ways to group expressions that are available in written mathematics that are not available on most calculators.

☺ *In all the following calculator exercises, evaluate the expressions to three or more significant digits. To help you check your keystroke sequence, a two-significant-digit answer is given in brackets.*

B2. Evaluate $\dfrac{3.98 + \pi}{1.2 + 3.5(4.17)}$. [.45]

B3. Evaluate $\dfrac{3.28 + 1.23(2.17)}{1.4 - 3(.42)}$. [42]

B4. Evaluate $\dfrac{2.68 + \sqrt{3.22 + 4.59}}{4.68 - 3.97}$. [7.7]

B5. Let $c = -3.97$.
Evaluate $\dfrac{3.22 - \sqrt{c^2 + 6(3.57)}}{3(5.96)}$. [-.16]

B6. Let $b = -3.11$. Evaluate $\dfrac{201 - 3.1(60.7)}{2 + \sqrt{b^2 + 10}}$. [2.0]

B7. Let $b = 23.4, c = -4.5,$ and $d = 6$.
Evaluate $\dfrac{-d + \sqrt{c^2 + bd}}{5bc}$. [-.013]

B8. Let $a = -4.1, b = 2.34,$ and $c = -5.6$.
Evaluate $\dfrac{-b - \sqrt{a^2 + c}}{7b}$. [-.35]

B9. Let $a = 1.4$ and $b = 7.8$. Evaluate $\dfrac{3^{a/b} - 3}{b - a}$. [-.28]

B10. Let $b = 2.3, c = 4.1$ and $k = -6$.
Evaluate $\dfrac{\sqrt{c^2 - k}}{5b}$. [.42]

B11. Let $c = 3.1, d = -5.6,$ and $k = 2.3$.
Evaluate $\dfrac{k + \sqrt{d^2 + c}}{cd}$. [-.47]

B12. Let $d = -2.6, b = 3.1,$ and $c = 9.7$.
Evaluate $\dfrac{d + c}{b + \sqrt{d^2 + c}}$. [.99]

B13. Let $a = 5$ and $b = -6$.
Evaluate $\dfrac{4 + \sqrt{b^2 + 12a}}{3.27a}$. [.84]

B14. Let $a = -3.7, x = 2.3,$ and $b = 5.12$.
Evaluate $\dfrac{b + \sqrt{b^2 - a^2}}{b + x}$. [1.2]

B15. Let $a = 1.45, c = 3.12,$ and $d = -1.67$.
Evaluate $\dfrac{a + \sqrt{d^2 + 5c}}{3.27\pi}$. [.56]

B16. Let $a = 10.18, b = -2.92,$ and $c = -2$.
Evaluate $\dfrac{a + 3c}{a + \sqrt{b^2 + a}}$. [.29]

Use the Quadratic Formula and your calculator to evaluate the "plus" solution. [Don't use a program, rather, check your own abilities with the given two-digit answers.]

B17. $3.45x^2 + 2x - 15 = 0$ [1.8]

B18. $6.54x^2 - 4x - 12.3 = 0$ [1.7]

B19. $-2.3x^2 + 6x - 1.5 = 0$ [.28]

B20. $1.23x^2 + 3x = 2 - 5x$ [.24]

B21. Learn how to program your calculator. Program your calculator with the Quadratic Formula.
(A) What are the solutions if $a = 2.3, b = 4.56,$ and $c = -7.8$? [1.1 or . . .]
(B) Solve for b if $3^2 = 7^2 + b^2 - 2b(7)\cos 25°$. [$b = 6.8$ or . . .]

B22. Learn how to program your calculator. Program your calculator with the Quadratic Formula. Use it to solve for y: $3y^2 + 6.7y - 42 = 12$. [$y = 3.3$ or . . .]

True or false? [No reason required.]

B23. (A) $(x + a)^2 = x^2 + a^2$ (B) $(x - a)^2 = x^2 - a^2$

(C) $(x + a)(x - a) = x^2 - a^2$

B24. (A) $\dfrac{x}{a} + \dfrac{x}{b} = \dfrac{x}{a + b}$ (B) $\left(\dfrac{a}{x}\right)^2 = \dfrac{a^2}{x^2}$

(C) $-b^2 = b^2$

B25. (A) $\sqrt{x^2 + a^2} = x + a$

(B) $\sqrt{\dfrac{x^2}{a^2}} = \dfrac{x}{a}$, if x and a are positive

B26. (A) $\sqrt{x^2 - 16} = x - 4$

(B) $\sqrt{16x^2} = 4x$, if $x > 0$.

Determine if these assertions are identities. If not, change the right hand expression to the correct expression. Note if there is no common alternative equivalent expression that can be used on the right side. Assume $a \geq 0$ and $b \geq 0$.

B27. (A) $(a + b)^2 = a^2 + b^2$ (B) $(ab)^2 = a^2b^2$

B28. (A) $\dfrac{a}{b} + \dfrac{c}{d} = \dfrac{a + c}{b}$ (B) $\dfrac{a}{b} + \dfrac{a}{c} = \dfrac{a}{b + c}$

B29. (A) $\sqrt{a^2 + b^2} = a + b$ (B) $\sqrt{ab} = \sqrt{a}\sqrt{b}$

B30. (A) $a - (b - c) = a - b - c$

(B) $\left(\dfrac{a}{b}\right)\left(\dfrac{c}{b}\right) = \dfrac{ac}{b}$

Reading and Writing Mathematics

B31. Using a calculator it is easy to make a mistake by grouping incorrectly. Think of a calculator mistake that might be promoted by poor pronunciation (pronunciation which does not express the proper grouping).

B32. Identify two expressions which are not equivalent, but which might be pronounced similarly.

B33. Create an expression without parentheses in which multiplication is encountered first but its execution is delayed until two other operations have been executed.

B34. "$3x$" means 3 times x. Explain what "37", "$3(7)$", "$3\frac{1}{2}$", and "$3\left(\frac{1}{2}\right)$" mean. Compare and contrast the interpretation of putting two symbols side-by-side in the five cases.

Section 1-3   FUNCTIONS AND NOTATION

The expression "$7(x + 3)$" expresses a rule for obtaining its calculated value (the value is called the "image") given an original number ("x", the "argument"). The rule is "Add 3 and then multiply by 7." In the terminology of this section, we identify the *function*, "Add 3 and then multiply by 7."

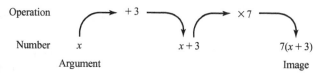

FIGURE 1 Whatever number "x" represents, add 3 to it and then multiply the sum by 7, to obtain $7(x + 3)$

A single expression such as "$7(x + 3)$" simultaneously represents two different types of mathematical objects:

1. a *number* (which depends upon the value of "x"), and

2. a *sequence of operations* (which does not depend upon the value of "x").

EXAMPLE 1   The sentence "For all x, $7(x + 3) = 7x + 21$," has no information about the number, x. It has information about the sequence of operations, "Add 3 and then multiply by 7." It says that sequence is equivalent to "Multiply by 7 and then add 21." The subject is the sequence of operations.

A sequence of operations is called a *function*. To solve an equation you must focus on the operations.

EXAMPLE 2 Suppose this is the problem: "Solve $7(x + 3) = 63$." What will you *do*, and *why*?

$$7(x + 3) = 63$$

Step 1: $x + 3 = \dfrac{63}{7} = 9$ [Undo the "Multiply by 7" part]

Step 2: $x = 9 - 3 = 6$. [Undo the "Add 3" part]

To solve the equation, divide by 7 and then subtract 3. The steps are determined by the function expressed in "$7(x + 3)$," not by its numerical value.

To solve "$7(x + 3) = 4233$" the same solution *process* applies. The number is different, but you still "Divide by 7 and then subtract 3." This process can be illustrated with a "function-loop" diagram:

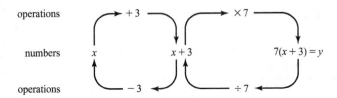

FIGURE 2 A "function-loop" diagram. The top displays the operations to go from x to $7(x + 3)$. The bottom displays the operations to go from $7(x + 3)$ back to x. That is, the bottom displays the solution process for the equation "$7(x + 3) = y$" ▬

Algebra is about operations and order. The term we have that concentrates attention on operations and order is *function*.

EXAMPLE 3 Give a mathematical expression expressing the rule: "Subtract 5 and then square."

The answer in terms of "x" is "$(x - 5)^2$." This expression expresses the function. ▬

DEFINITION 1-3-1 A <u>function</u> is a rule that relates exactly one number to each number in a given set.

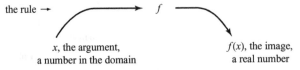

FIGURE 3 A function diagram. The function f takes numbers in the given set, called the domain (which will be all real numbers unless specified otherwise), and yields real numbers as images
Note the order: argument—function—image

TERMINOLOGY 1-3-2 Functions are often denoted by f or g. If a function, f, takes a number, x, and yields the number $f(x)$ [pronounced "f of x"—these parentheses do not symbolize multiplication], then the number x is called the <u>argument</u> and $f(x)$ is called the <u>image,</u> or the <u>value of f at x</u>. Using computer terminology, the argument is the "input" and the image is the "output." The "given set" of all arguments to which the rule applies is called the <u>domain</u>. The domain will often be the set of all real numbers.

The definition of *function* requires that each argument yields exactly one image. The argument *determines* the image. A function is a *rule*, not a number.[1]

When you see "*f(x)*"
you should see three things: the argument, *x*, the function (rule), *f*, and the image,
***f(x)*, in that order** (as in Figure 3).

EXAMPLE 4   The sentence "Let $f(x) = x^2$, for all x," defines the function f, "Square it!" by showing what it does to arguments. "Let" is a word used to introduce definitions. The operation "Square it!" is described by the pattern; the letter x merely holds places for any number in a pattern about operations. Therefore,

$$f(3) = 3^2.$$
$$f(a) = a^2.$$
$$f(2x + 1) = (2x + 1)^2.$$
$$f(x^2) = (x^2)^2.$$

DEFINITION 1-3-3   A letter (such as "*x*" in Example 4 above) is said to be a <u>placeholder</u> (also commonly known as a <u>dummy variable</u>) when it is used to hold places where any expression could go.

Placeholders are used to express thoughts about operations and order (not thoughts about numbers). Since the particular numbers involved do not matter, placeholders may be replaced by other expressions if they are consistently replaced throughout the whole sentence.

Functions are defined using placeholders.

EXAMPLE 4   Let $f(x) = x^2$. [This time we omit the "for all x," but it is understood anyway.]
CONTINUED   The argument, x, is a placeholder. That is, it holds two places and it may be replaced by any numerical expression, as long as it is replaced by the same expression in both places.

$$f(z) = z^2. \qquad \text{["x" is replaced by "z", twice]}$$
$$f(x + h) = (x + h)^2. \quad \text{["x" is replaced by "$x + h$", twice].}$$ ▬

DEFINITION 1-3-4   A letter is said to be an <u>unknown</u> when you are supposed to solve for its numerical value.

EXAMPLE 5   If the problem is to solve the equation "$x^2 = 16$", then the letter x is an unknown. The equation gives information about the number x.   ▬

---

[1]The functional rule creates a set of argument-image pairs. Defining a function to be a rule is the best way to begin, but for graphing we will emphasize the set of argument-image pairs (in Definition 2-1-2).

Unknowns and placeholders are much different.

1. Unknowns are used to express thoughts about numbers. [e.g. $x^2 = 16$]
2. Placeholders are used to express thoughts about operations and order. [e.g. $f(x) = x^2$]   ▬

Formulas

You are familiar with functions already, but you may know them as "formulas."

EXAMPLE 6   The formula for the area, A, of a circle of radius r is given by $A = \pi r^2$. The area is a function of the radius, "Square and then multiply by π."

We might write "$A(r) = \pi r^2$" to emphasize the dependence of the area on the radius. The argument is "r" and the image is "$A(r)$".

The "domain" of a function is the set of arguments. Here, the argument is the radius of a circle, which is never negative, so the domain is the set of all $r > 0$.   ▬

Functional versus Formula Notation

The formula notation "$A = \pi r^2$" corresponds to the functional notation "$A(r) = \pi r^2$". There are advantages to functional notation and other advantages to formula notation.

Formula notation is shorter and therefore more convenient. But all calculations have the rule "Square and then multiply by π" in common. In the functional notation "$A(r) = \pi r^2$," the area has the value $A(r)$ and rule has the name A.

Functional notation gives the rule a name of its own.

The function "Multiply by two" may be expressed by the equation "$y = 2x$." Then "x" is called the <u>independent variable</u> (the argument of the function) and "y" is called the <u>dependent variable</u> (the image). We emphasize the "$f(x)$" notation because it highlights the essential difference between a number and a rule (a process). The equation "$y = 2x$" creates many pairs of numbers. What do they have in common? The rule, "Multiply by 2!" Writing it using "$f(x) = 2x$" emphasizes the rule, f.

Functions of Two Variables

Some formulas relate three or more numbers to each other and determine functions of two or more variables.

EXAMPLE 7   Consider the formula "Distance equals rate times time." The formula, $d = rt$, determines the distance, d, traveled by an object moving at a constant rate, r, for a given amount of time, t. In this case we could think of distance as a function of two arguments, rate and time: $d(r, t) = rt$.   ▬

■ "*f*" versus "*f(x)*"

When you see the expression "$f(x)$" you should see three (yes, 3) things. The argument, x, the function, f, and the image, $f(x)$.

EXAMPLE 8   Give the function defined by the statement, "Let $f(x) = 4x + 3$."
Answer: Multiply by 4 and then add 3.

Do *not* describe the function as "Multiply x by four and then add three." The reference to "x" should be omitted.

The function has nothing to do with "x". Only the *notation* has to do with "x". The function will take *anything* and "Multiply by four and add three." In our notation we use "x" to represent *anything*. In our notation, "x" is merely a placeholder. The letter "x" could be replaced by "z" as long as it is replaced in both places: "Let $f(z) = 4z + 3$" defines exactly the same function as "Let $f(x) = 4x + 3$."   ▬

EXAMPLE 9   Define a function which would square any given number and then subtract 7 from the square.

"Let $f(x) = x^2 - 7$."

EXAMPLE 9
CONTINUED   Let $f(x) = x^2 - 7$. Determine $f(y)$.

This can fool beginners. They sometimes ask, "How can I tell? You only told me f of x." But, the point is, the function is "Square and then subtract 7" and has nothing to do with the name of the variable. Any number can go in the *place* of x.

$$f(x) = x^2 - 7.$$
$$f(3) = 3^2 - 7 \quad [= 2].$$
$$f(y) = y^2 - 7.$$
$$f(2x) = (2x)^2 - 7.$$
$$f(x + h) = (x + h)^2 - 7.$$
$$f\left(\frac{x}{2}\right) = \left(\frac{x}{2}\right)^2 - 7.$$ ▬

Placeholders
The fact that "x" need not be "x" is remarkable. The notation could use a black box as a placeholder instead.

EXAMPLE 10   Define f by $f(\blacksquare) = \blacksquare(\blacksquare + 3)$.
Find $f(5), f(17), f(x),$ and $f\left(\sqrt{x}\right)$.

$$f(\blacksquare) = \blacksquare(\blacksquare + 3).$$
$$f(5) = 5(5 + 3)$$
$$f(x) = x(x + 3).$$
$$f\left(\sqrt{x}\right) = \sqrt{x}\left(\sqrt{x} + 3\right).$$
$$f(x + h) = (x + h)(x + h + 3)$$

The *places* the symbol holds are important; the particular numerical expressions in those places are not.   ▬

Functional Notation and Order
The notation for functions is not executed left-to-right.

Functional notation is often understood <u>right-to-left</u>.

The argument (whatever is in the parentheses) comes first and *then* the operation indicated by f is applied to it. Thus, in "$f(x^2)$" [read "f of x squared"], x is squared first (on the right) and then f is applied.

If we want to apply f first and square last, we want $[f(x)]^2$.

In a very real sense the notation for functions is backwards. When read left-to-right like English, sentences like "$y = f(x)$" display the image y first, then the function f, and then the argument x. Completely backwards! Be sure to learn to think in the proper "argument—function—image" sequence.

■ *"f"* and *"f(x)"*

In this section we take pains to distinguish between "f" and "$f(x)$." Most texts do not. Why is the difference worth emphasizing?

First of all, f and $f(x)$ are conceptually quite different. The letter f stands for a function and, technically, $f(x)$ is a number, not a function. We do algebra by thinking about functions.

EXAMPLE 11   Suppose the problem is to "Solve $6x + 4 = 87.5$." What will you *do*, and *why*?

You will "Subtract 4" and then "Divide by 6." These steps are triggered by the function, not by the number 87.5.   ■

Algebra is about operations and order, not about numbers. This is a very strong reason for abstracting operations and order into a single concept. **The one term we have that concentrates attention on operations and order is *function*.**

There is no law that says functions have to be simple. You can define a function to be as complicated as you wish; the only qualification is that each argument in the domain must have exactly one image.

EXAMPLE 12   Suppose cloth costs $14 per yard for up to 10 yards, but only $12 per yard for each yard after the first 10 yards. Express the cost as a function of the number of yards.

Let the cost of x yards be $f(x)$.

$$f(x) = 14x \text{ if } 0 \le x \le 10, \text{ and}$$
$$f(x) = 140 + 12(x - 10) \text{ if } x > 10.$$

A particular function is *one* thing. Even this function f gives only *one* rule. You may, at first, think that it gives two rules, and in some sense that is so, but, from the point of view of mathematics, these are merely two parts of the *one* rule, f, which is defined "piecewise" (with two pieces).   ■

DEFINITION 1-3-5   The <u>natural domain</u> of a function is the set of all argument values such that the defining expression yields a real number.

The two most common operations which may limit the natural domain are

1. taking a square root
   (we do not take the square root of a negative number since the square root of a negative number is not a real number) and

2. dividing by 0 (we do not permit division by 0 since division by 0 is undefined).

EXAMPLE 13   Let $f(x) = \sqrt{x - 2}$. Find the natural domain.

To be able to take the square root, $x - 2$ must be greater than or equal to zero, because the square root of a negative number is not a real number. This requires $x \ge 2$. The natural domain is $[2, \infty)$.   ■

CALCULATOR
EXERCISE 1

Evaluate $\sqrt{-3}$.

You will get an error message, or a complex number if your calculator is sophisticated enough. Negative numbers are not in the natural domain of the square root function.

CALCULATOR
EXERCISE 2

Use your graphing calculator to graph $y = \sqrt{x}$.

There are no y-values for negative values of x. Use the *trace* key to display the x- and y-values. If the x-value is negative, there will be no y-value because negative numbers are not in the natural domain of the square root function.

EXAMPLE 14

Let $h(x) = \dfrac{x - 6}{x - 2}$. Find the natural domain.

When $x = 2$, the denominator is 0. Division is not defined when the denominator is 0. Therefore, the natural domain of h includes all real numbers *except 2*. ▬

DEFINITION 1-3-6

The <u>domain</u> of a function is the set of all the argument values that we choose to use. Sometimes the domain can be chosen arbitrarily, and sometimes it is motivated by physical considerations (Example 15, next).

The domain may be different from the "natural domain" because we might not choose to use all the argument values that are mathematically possible (in the sense that their image is a real number).

EXAMPLE 15

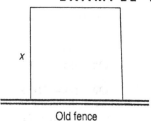

Old fence

FIGURE 4 A pen made from 100 feet of new fence on three sides and an old fence on the fourth side

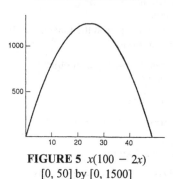

FIGURE 5 $x(100 - 2x)$
[0, 50] by [0, 1500]

Suppose a farmer wants to make a rectangular animal pen with an existing long straight fence for one side and 100 feet of new fencing and gates for the other three sides (Figure 4). If we write the area in terms of x in the picture, what is the domain? What dimensions of the pen would yield the maximum possible area of the pen?

To maximize the area, we need an expression for the area. Let one side of the pen be "x" (Figure 4).

If the pen is x feet deep, $2x$ feet are used for the two sides pictured vertically, and $100 - 2x$ feet of fence remain for the other side. The area is length times width, so the area function is

$$A(x) = x(100 - 2x).$$

What is the domain?

There is no point in using negative x's. Sides are not negative. Mathematically, negative x's make sense in the expression "$x(100 - 2x)$," but physically they do not. Similarly, if $x > 50$, there is not enough fence for the sides. The domain is, by definition, the set of x's we chose to use, which is not necessarily the "natural" domain. So, in this problem, we would regard the domain as [0, 50], even though the "natural" domain of the expression is all real numbers.

CALCULATOR
EXERCISE 3

Graph this [$y = x(100 - 2x)$] on your graphing calculator. To get a good picture, you must select the right domain. The standard window with x-interval from -10 to 10 is not appropriate. Figure 5 uses the domain [0, 50]. Furthermore, you will also have to select an appropriate "y" interval. Figure 5 uses [0, 1500]. ▬

CONCLUSION A function is a rule—an operation or sequence of operations. Distinguish the *function, f* (as a command, for example "Square it!"), from its image, $f(x)$, which is a number. The notation may mention "$f(x)$", for example, "Let $f(x) = x^2$," but, in such definitions, the letter "x" is a only placeholder. Placeholders may be replaced by any numerical expression, as long as they are replaced consistently everywhere they occur.

Terms: Function, argument, image, value of f at x, domain, function-loop diagram, placeholder (*dummy variable* is a synonym), natural domain.

Exercise 1-3

A

A1.* ☺ When you see the expression "$f(x)$" you should distinguish three things. Which three? In which order?

A2.* (A) Define "natural domain."
(B) Which are the two most commonly encountered operations that may cause the natural domain of a function to be restricted?

☺ *Give the functions as commands [do not mention "x"]:*

A3. $f(x) = 3x$

A4. $f(x) = x - 9$

A5. $f(x) = \dfrac{x}{5}$

A6. $f(x) = x + 4$

A7. $f(x) = 2(x + 1)$

A8. $f(x) = \dfrac{x}{2} - 5$

A9. $f(x) = \log(x + 1)$

A10. $f(x) = (x + 2)^2$

A11. $A = \pi r^2$

A12. $V = \left(\dfrac{4}{3}\right)\pi r^3$

A13. $A = s^2$

A14. $P = 4s$

☺ *Give the standard "f(x)" definitions of the functions:*

A15. "Add 20 and then divide by 3"

A16. "Divide by 2 and then subtract 1."

A17. "Subtract one and then take the absolute value."

A18. "Square and then add one."

A19. "Add 5 and then square."

A20. "Subtract 9 and then divide by 5."

Give the natural domain of f. In some cases, it may help to inspect a graph.

A21. (A) $f(x) = \dfrac{1}{x - 2}$ (B) $f(x) = x^2 - 3$
(C) $f(x) = \sqrt{x - 5}$

A22. (A) $f(x) = \dfrac{1}{x^2 - 4}$ (B) $f(x) = \dfrac{1}{x^2 + 3}$
(C) $f(x) = \sqrt{x - 9}$

Grammar

☺ *Which parentheses denote multiplication, and which denote the application of a function? Assume c is a constant.*

A23. (A) $7(x - 2)$ (B) $f(x - 2)$

A24. (A) $g(x - 2)$ (B) $5(x - 2)$

A25. (A) $f(x + 1)$ (B) $c(x + 1)$

A26. (A) $(x - 1)(x + 1)$ (B) $c(x + 1)$

☺ *Write out the pronunciation of these sentences.*

A27. $f(x) = 2x$ **A28.** $f(x) = \dfrac{x}{3}$

A29. $f(x) = \sqrt{x}$ **A30.** $g(x) = 2(x + 4)$

A31. $g(x) = x(x + 1)$ **A32.** $f(g(x)) = 2x + 1.$

B

B1.* (A) What is a function?
(B) Give an example.
(C) In your example, distinguish the function itself from the notation for its image.

B2.* Distinguish between the two types of objects denoted by "f" and "$f(x)$."

B3. ☺ Let $f(x) = x^2$. Express
(A) $f(4)$ (B) $f(z)$ (C) $f(x + 1)$ (D) $f\left(\dfrac{x}{2}\right)$

B4. ☺ Let $f(x) = 3x + 2$. Express
(A) $f(4)$ (B) $f(z)$ (C) $f(x + 1)$ (D) $f\left(\dfrac{x}{2}\right)$

B5. ☺ Let $f(x) = \dfrac{1}{x}$. Express
(A) $f(2x)$ (B) $f(x + h)$ (C) $f\left(\dfrac{1}{x}\right).$

B6. ☺ Let $f(x) = x(x - 2)$. Express

(A) $f(3)$ (B) $f(x - 1)$ (C) $f(x^2)$ (D) $[f(x)]^2$.

(E) Why are the last two different?

B7. ☺ Let $f(x) = \dfrac{2}{x}$. Express

(A) $f(5)$ (B) $f(x - 4)$ (C) $f\left(\dfrac{1}{x}\right)$

(D) $f(x^2)$ (E) $f(c)$

B8. ☺ Let $f(x) = (x + 1)^2$. Express

(A) $f(3)$ (B) $f(3 + h)$ (C) $f(x + h)$

*B9–B12: ☺ Each problem has two parts. For each f,
(A) Give the rule [Answer as a command in English], and
(B) What would you do to solve "$f(x) = c$" for x? [Answer
as a command in English.]*

B9. $f(x) = 3x + 7$.   **B10.** $f(x) = \dfrac{x}{5} - 8$.

B11. $f(x) = \dfrac{x - 4}{27}$.   **B12.** $f(x) = 9(x - 5)$.

B13. The cost of photocopying a single original is 6 cents each for the first 10, and 5 cents each for each copy after the first 10. Express the function which gives the cost of n copies.

B14. The cost of printing many copies of a single page is 50 cents for the setup charge, plus 3 cents per copy for the first 200. Each copy after the first 200 costs only 2 cents. Express the function which gives the cost of n copies.

B15. (After B13 and B14) When is the printing method in B14 cheaper than the photocopy method in B13?

B16. Suppose you apply the function "Multiply by 2" and then the function "Multiply by 10" to the image. This is equivalent to applying one function to the original argument. Which function? [Express your answer in mathematical notation using "f"].

B17. Suppose you apply the function "Add 12" and then the function "Subtract 9" to the image. This is equivalent to applying one function to the original argument. Which function? [Express your answer in mathematical notation using "f"].

B18. Suppose you apply the function "Multiply by 25" and then the function "Divide by 5" to the image. This is equivalent to applying one function to the original argument. Which function? [Express your answer in mathematical notation using "f"].

B19. Suppose you apply the function "Add 7" and then the function "Subtract 7" to the image. This is equivalent

to applying one function to the original argument. Which function? [Express your answer in mathematical notation using "f"].

B20. Suppose you apply the function "Add 12" and then the function "Subtract 3" to the image. This is equivalent to applying one function to the original argument. Which function? [Express your answer in mathematical notation using "f"].

Reading and Writing Mathematics

B21. (A) Very clearly distinguish, in general, between "$f(x) + h$" and "$f(x + h)$."

(B) Give a simple example to show that they are not the same.

B22.* In what sense is functional notation "backwards"?

B23.* (Comparing functional notation and formula notation.) (A) In functional notation there are letters for which mental objects? (B) In formula notation there are letters for which mental objects? (C) Which one has a letter for the rule?

B24.* Explain and illustrate with an example the meaning of "placeholder."

B25. Suppose we "Let $y = 3x$" When $x = 4, y = 12$. When $x = 5, y = 15$. When x changes the numbers change, but something remains the same. What?

B26. Suppose we "Let $y = 2x^2$." When $x = 1, y = 2$. When $x = 10, y = 200$. When x changes, the numbers change, but something remains the same. What?

B27. There are two commonly-encountered operations that may cause the natural domain of a function to be restricted: square roots and division by zero. Give at least one more function with a naturally restricted domain.

B28. (Short essay) Why are functions interesting? Why don't we just talk about expressions, or equations such as "$y = 3x$"?

B29. (A) What advantages does "formula notation" for functions have over "functional notation"?

(B) What advantages does "functional notation" for functions have over "formula notation"?

B30. Why might a mathematician have a technical objection to the label "log x" on a calculator key?

B31. ☺ "$5x$" means 5 times x. Explain what "56", "5(6)", "$5\frac{1}{2}$", "$5\left(\dfrac{1}{2}\right)$", and "$f(6)$" mean. Compare and contrast the interpretation of putting two symbols side-by-side in the six cases.

Section 1-4 READING AND WRITING MATHEMATICS

You already know how to *use* many mathematical methods. You know how to find the area of a circle, how to simplify "$\dfrac{(3/5)}{7}$", and how to solve many types of equations. In mathematical notation, methods can be written as facts. In this section learn *how to write* methods you already know and how to *read* new ones.

VARIABLES 1-4-1 In the language of mathematics, two similar sentences may use letters such as "x", "a", and "b" in radically different ways. Letters can be used to express

1. thoughts about numbers (the letters are unknowns), or
2. thoughts about operations and order (the letters are placeholders).

EXAMPLE 1 The equation "$3(x + 2) = 18$" is about the number x. The letter x is an unknown. Solving it, we discover that $x = 4$. ▬

EXAMPLE 2 The generalization "For all x, $3(x + 2) = 3x + 6$," has no information about the number x. The variable "x" is not an unknown; it is a placeholder that *holds places* so that operations (functions) can be discussed. The equation says that the function "Add 2 and then multiply by 3" is equivalent to the function "Multiply by 3 and then add 6."

The two places held by "x" could be occupied by other letters instead. The statement "For all a, $3(a + 2) = 3a + 6$" has exactly the same information. The letter is a placeholder and it does not matter which letter is used (as long as certain conventions are not violated. For example, in this text we do not use "f" for numbers because we use it for functions).

We often omit the "For all x" part of "For all x, $3(x + 2) = 3x + 6$" and just write the identity "$3(x + 2) = 3x + 6$." You are supposed to recognize when an equation is always true. ▬

DEFINITION 1-4-2 An equation with a variable is an <u>identity</u> when it is true for all values of the variable. We interpret identities as abbreviated generalizations. The variables are placeholders. The same information may be expressed with other letters.

EXAMPLES 1 AND 2 CONTINUED The identity "$3(x + 2) = 3x + 6$." is a sentence and its subject is a function, not a number. In order to comprehend it, you must understand functions.

The algebraic expression "$3(x + 2)$" has two different interpretations.

1. It represents a *number* (which depends upon the number represented by "x").
2. It also represents *operations in a particular order* ("Add 2" and "Multiply by 3," in that order).

Is "$3(x + 2)$" a number, or a sequence of operations?

It's both. You must recognize both interpretations, and switch back and forth between them.

In the problem "Solve the equation '$3(x + 2) = 18$'," the expression "$3(x + 2)$" obviously represents the number 18. Nevertheless, to solve it, the operations and order are more important. This is why we emphasize functions in this course.

Variables. Numerical variables can be used in different ways. The variable "x" may

1. *represent* particular numbers (in sentences about numbers), or
2. *hold places* where general numbers can go (in sentences about operations and order).

EXAMPLE 3   In the equation "$2x = 8$" the letter represents the particular number 4. The letter "x" is an unknown.   ▬

EXAMPLE 4   Consider this definition of f: "Let $f(x) = (x + 1)^2$, for all x."

Here the letter x is not intended to represent a particular number. It holds places so that a sequence of operations can be discussed. It defines the function f, which is "Add 1 and then square."

$$\text{If} \quad f(x) = (x + 1)^2, \text{ then}$$
$$f(5) = (5 + 1)^2,$$
$$f(2x) = (2x + 1)^2,$$
$$f(c) = (c + 1)^2, \text{ and}$$
$$f(x + h) = (x + h + 1)^2.$$

Note the importance of *place*. In the original sentence "Let $f(x) = (x + 1)^2$," the letter x appears in two places. You can put any numerical expression in those places, as long as you put the same expression in both places. The letter x is a placeholder. The choice of number does not matter; this definition of f is not about numbers!

Unknowns are not placeholders. Changing the letter would change the meaning of the sentence.   ▬

EXAMPLE 5   Do the equations "$3x = 18$" and "$3q = 18$" have the same meaning?

No. One is about the value of x, the other is about the value of q. Knowing that $3x = 18$ does not tell you that $3q = 18$.

Do the equations "$3x = 2x + x$" and "$3q = 2q + q$" have the same meaning?

Yes. Both say that multiplying by 3 is equivalent to multiplying by two and adding the original number. Neither equation tells you the value of x or q.   ▬

EXAMPLE 6   Decide if "x" is an unknown or a placeholder, and if the subject is a number or operations and order (O&O).

sentence	type of variable	subject
(A) $(x + 1)^2 = 16$	unknown	x, a number
(B) $(x + 1)^2 = x^2 + 2x + 1$	placeholder	O&O [Add 1 and then square]
(C) Let $f(x) = (x + 1)^2$	placeholder	O&O [This defines f.]
(D) $x + 1 = (x + 1)^2$	unknown	x, a number

▬

■ How to Write Methods

In mathematics, methods can be written symbolically using placeholders. You already know the methods discussed in this section. Now learn how to read and write them symbolically.

DEFINITION 1-4-3   A <u>method</u> is a set of instructions for doing some type of problem.

In algebra, numerous problems can be done by the same method. In mathematics, the *method* itself is a mathematical object. It can be written as a *fact*. Here are three primary ways of expressing methods as facts:

1. **formulas**,
2. **identities**, and
3. **relations between equations**.

■ Methods as Formulas

Formulas express operations and methods. Formulas are facts that can tell you what to do.

EXAMPLE 7   How do you find the area of a circle?

Take the radius, square it, and then multiply by π. This can be expressed with a formula: $A = \pi r^2$.

By itself, the formula does not tell you the radius, or the area. It tells you *how to compute* the area given the radius. It tells the operations and their order. ■

■ Methods as Identities

Identities are a second way to express methods as facts.

Identities are commonly interpreted as abbreviated generalizations, with the "For all . . ." suppressed. Identities express alternative sequences of operations. For example, when you first learned about negative numbers, you learned to do operations with negative numbers by actually doing different operations with positive numbers.

EXAMPLE 8   State the method for doing arithmetic problems such as "$5 - (-2)$" and "$19 - (-6)$" by working with positive numbers.

Here is the method: "$a - (-b) = a + b$." That is, "$a - (-b) = a + b$, for all a and b."

The method is expressed by an identity or its corresponding generalization.

Instead of actually doing the given operations in "$5 - (-2)$", which has a negative and subtraction, we may do "$5 + 2$", which uses only positive numbers. Similarly, instead of actually doing "$19 - (-6)$" we do "$19 + 6$".

The purpose of identities is to provide alternative sequences of operations for evaluating expressions.

One side of the identity expresses a problem-pattern ("$b - (-c)$") and the other expresses a solution-pattern ("$b + c$"), with different (simpler) operations in a different order.

> **Identities are about operations and order. Identities are written with placeholders. Identities express equivalence of expressions.**

The equation "$b - (-c) = b + c$" is an identity because the two expressions "$b - (-c)$" and "$b + c$" are not just equal, they are equivalent.   ▬

DEFINITION 1-4-4   Two *expressions* are <u>equivalent</u> when they always express the same numerical value (although usually with a different order of operations).

Perhaps unfortunately, we use the same symbol, "$=$" (equals), to express both equivalence and equality.

The Use of "$=$". The most common verb in Mathematics is "$=$" ["equals"]. The symbol "$=$" is used two significantly different ways. It may denote

1. *equality* in order to express facts about numbers, such as "$3 + 4 = 7$" or "$x + 5 = 9$" (which tells you $x = 4$). It may also denote

2. *equivalence of expressions* in order to express facts about operations and order, such as "$b - (-c) = b + c$, for all b and c" and "Let $f(x) = x^2$, for all x."

Learn to grasp the difference.

EXAMPLE 9   "$3 + 4x = 6x$." The equal sign denotes equality, but not equivalence because the two sides are not always equal. The equation has information about the value of x that makes the two sides equal.

"$3x + 4x = 7x$, for all x," which is often written simply "$3x + 4x = 7x$." The equal sign denotes equivalence of expressions. They are always equal, regardless of the value of x.

"Let $f(x) = 2x + 1$, for all x," which is often written "Let $f(x) = 2x + 1$." The equal sign denotes equivalence of expressions. They are equivalent *by definition*. Either side can replace the other (until we pick a new meaning for f).   ▬

The misuse of "$=$". Sometimes the symbol "$=$" is misused. Many students erroneously use "$=$" to mean "I'm about to do the next step." For example, they may solve the equation "$2x = 30$" with the work:

$$\text{"}2x = 30 = x = 15.\text{"}$$

$$\uparrow \qquad \uparrow \qquad \uparrow$$

OK wrong! OK

This says "$30 = x$" which is wrong. The student **did** the right thing and **wrote** the wrong thing.

Most people don't write any symbol between the equations "$2x = 30$" and "$x = 15$." However, they are related (they are equivalent **equations**), but not by "equals", which is a verb that only relates **expressions**, never equations. If you choose to write something between them, write "is equivalent to" or use a symbol such as " $\Leftrightarrow$ " or " iff" (an abbreviation for "if and only if") for equivalent **equations**:

$$\text{``}2x = 30 \quad \text{iff} \quad x = 15,\text{''}$$

$$\uparrow \qquad \uparrow \qquad \uparrow$$

$$\text{OK} \qquad \text{OK!} \quad \text{OK}$$

■ Methods of Arithmetic

Many methods of arithmetic can be written as identities.

EXAMPLE 10   Explain how to divide a fraction by a fraction.
First look at two examples.

Instances: $\dfrac{\left(\dfrac{3}{4}\right)}{\left(\dfrac{5}{8}\right)} = \left(\dfrac{3}{4}\right)\left(\dfrac{8}{5}\right) = \dfrac{3 \times 8}{4 \times 5}. \qquad \dfrac{\left(\dfrac{2}{3}\right)}{\left(\dfrac{7}{2}\right)} = \left(\dfrac{2}{3}\right)\left(\dfrac{2}{7}\right) = \dfrac{2 \times 2}{3 \times 7}.$

The process can be expressed in words with the command, "Invert and multiply." Here is how that method is written. Look at the pattern. Look at the *places* where the numbers go.

THEOREM   For all b, c, and d that are not zero,

$$\dfrac{\left(\dfrac{a}{b}\right)}{\left(\dfrac{c}{d}\right)} = \left(\dfrac{a}{b}\right)\left(\dfrac{d}{c}\right) = \dfrac{ad}{bc}.$$

The theorem tells you that, rather than actually doing the three operations of division expressed on the left, you may do, instead, the two multiplications ("*ad*" and "*bc*") and one division (the quotient of those two) expressed on the right. The problem-pattern is on the left (it represents all similar problems) and the solution-pattern is on the right.   ▄▄

Math literacy largely depends upon pattern recognition.
Facts written in symbolic patterns can tell you what to do and when to do it.
The places in the patterns are held by letters called *placeholders*.

■ Methods as Relations Between Equations

To solve an equation we operate on it and replace it with a new, simpler, equation. Theorems can express how this works. They state the problem-pattern-equation and the pattern of the equation resulting from the operation, and how they are related.

EXAMPLE 11   State the method for the first step in solving equations such as "$x^2 - 7 = 32$" and "$3x - 19 = 57.3$."
The **first step** is to add something (7 or 19) to both sides. This type of operation takes an equation and yields a new, related, equation. It is an operation *on an equation*. The method could be stated several ways.   ▄▄

THEOREM 1-4-5A   "$x - a = b$" is equivalent to "$x = b + a$".

THEOREM 1-4-5B   "$a = b$" is equivalent to "$a + c = b + c$".

The placeholders x, a, and b are used to generalize the problems into a problem-pattern. The placeholder "x" holds the place where the problems in Example 11 have "x^2" and "$3x$." This is because the theorem expresses only the first step, which has nothing to do with what is in the place of "x". Anything can be in that place.

DEFINITION 1-4-6   Two equations are said to be <u>equivalent</u> if they are true for the same values of the variables.

Equivalent equations with unknowns have the same solutions. Equivalence of equations is a relation between equations. The phrase "is equivalent to" may be abbreviated by "$\Leftrightarrow$" or "iff", which is itself an abbreviation for "if and only if."

To solve an equation you operate on it to convert it to a new equation. The method is expressed with a theorem that gives

1. the pattern of the problem (the *problem-pattern*), and
2. the corresponding pattern of the new equation (the *solution-pattern*).

In the next theorem the two patterns are related by "is equivalent to."

E X A M P L E   1 2   Write the method for solving equations such as "$x^2 = 39$."

THEOREM 1-4-7   "$x^2 = c$" is equivalent to "$x = \pm\sqrt{c}$."
This says you may replace "$x^2 = 39$" with its solution "$x = \pm\sqrt{39}$." ▬

E X A M P L E   1 3   Write the theorem which expresses the method for the **first step** in solving for k in the equation "$(3k + 1)^2 = 90$."

Write Theorem 1-4-7, just as it is. The first step addresses only the squaring. The "x" in the theorem holds the place occupied by "$3k + 1$" in the problem. Theorems are written with placeholders that hold places where any expression can go. ▬

**A common problem-pattern is exhibited in one half of a theorem
and a useful replacement solution-pattern is exhibited in the other half.**

**Theorems are written with placeholders (= dummy variables).
Theorems are about methods, not about numbers.**

Distinguishing the Three Ways
What is the difference between methods given as 1) formulas, 2) identities, and 3) relations between equations?

To use a formula, you must know the context and what the letters mean. For example, the meaning of "$A = \pi r^2$" is not self-contained. You must know the context (circles) and the meaning of the letters "A" and "r". The left side of the equation, A, expresses a problem-pattern ("What is the area?") and the right, πr^2, gives the solution-pattern.

In contrast, identities give their own context. You do not need to know what the letters mean to use an identity. For example, the left side of "$a - (-b) = a + b$" gives the problem-pattern. The right side gives the alternative solution-pattern.

Relations between equations tell you how to change one equation into another. To solve an equation, you take the original equation and change it into a simpler equation. Relations between two equations tell you what to do and when to do it. One of the two equations expresses the problem-pattern. The other gives the solution-pattern replacement equation. For example:

THEOREM 1-4-8 (THE QUADRATIC THEOREM)

If $a \neq 0$,

$$ax^2 + bx + c = 0 \text{ is equivalent to } x = \frac{-b \pm \sqrt{b^2 - 4ac}}{2a}.$$

The two equations in this theorem are related; they are equivalent, if $a \neq 0$. The solution-pattern by itself is known as the "Quadratic Formula." The formula by itself is useless if you don't know the problem-pattern to which it applies.

EXAMPLE 14

Use the Quadratic Formula to solve "$4x - 3 + 2x^2 = 8$."

To use the Quadratic Formula you must first identify "a", "b", and "c" of the problem-pattern. You must make the problem fit the problem-pattern.

The equation is equivalent to this equation:

$$2x^2 + 4x - 11 = 0,$$

which does fit the problem-pattern,

$$ax^2 + bx + c = 0.$$

Now, read the problem-pattern to find $a = 2$, $b = 4$, and $c = -11$. Plug these into the solution-pattern (the Quadratic Formula) to get the answer (Problem A23).

Again, note the importance of *place*. You must identify the places held by a, b, and c in the sequence of operations. The letters a, b, and c, are parameters used to determine a member of the family of quadratic equations. ▬▬

DEFINITION 1-4-9

Parameters are arbitrary constants. They are letters used to identify and distinguish members of a family of equations or expressions.

■ Why Do We Care about Parameters?

Parameters are used to specify a member of a family. The number of parameters tells us the number of pieces of information needed to determine which member it is.

For example, the Quadratic Theorem has **three** parameters. To specify a particular quadratic you need **three** values. **Three** points on the graph will determine which quadratic it is. Two points are not enough.

EXAMPLE 15

"$y = mx + b$" describes a family of lines. "m" and "b" are **two** parameters that distinguish different members of the family. "x" and "y" are not parameters. x is the argument and y is the image. In geometry you learned a line is determined by **two** points. Values for the **two** parameters m and b determine a line. ▬▬

EXAMPLE 16

Which of these sentences use "x" as a placeholder?

sentence	type of variable	subject
(A) $2x + 5x = 7x$	placeholder	operations and order
(B) $2x + 5x = 70$	unknown	x, a number

	sentence	type of variable	subject
(C)	Let $f(x) = 2x$	placeholder	operations and order

[This defines f as "Multiply by 2."]

(D) $x + b = c$ unknown (b and c are parameters) x, a number

(E) $x + b = c$ iff $x = c - b$ placeholder (b and c are parameters)

A method (not "a number")

The theorem in part (E) is about a method, not about the numbers. The places are what is important; the theorem applies even if "$2x$" or "x^2" or "z" are in the two places of "x". Part (D) is much different. If you put "$2x$" in for "x", you get a completely different equation.   ▬

■ Reading Mathematics

Theorems and definitions are written with placeholders. The letters in a theorem may represent numbers, and the letters may be switched for other letters. Look for letter-switching in the next example. It is about how to read theorems, not just about logarithms. Focus on the places held by the letters.

> For Example 17 you do not need to already know about logarithms. In English you can read and understand sentences about subjects you have never studied before. **This section and this example are intended to help you learn to read Mathematics**—even new mathematics you have never seen before.

EXAMPLE 17
THEOREM 1-4-10

$10^a = b$ is equivalent to $a = \log b$.
   This theorem tells us how to solve certain equations.

EXAMPLE 17
LEFT-TO-RIGHT

Solve "$10^x = 70$."
   This fits the problem-pattern, "$10^a = b$," on the left of Theorem 1-4-10:
   $10^a = b$ is equivalent to $a = \log b$.

	problem-pattern	solution-pattern
Theorem	$10^a = b$	$a = \log b$
Problem	$10^x = 70$	$x = \log 70 \; (=1.845)$

Note that *place* is important, the choice of letter is not.

EXAMPLE 17
RIGHT-TO-LEFT

Solve "$\log x = 2.13$."
   This fits the pattern on the right, "$a = \log b$" (with its sides switched).

	problem-pattern	solution-pattern
Theorem	$\log b = a$	$b = 10^a$
Problem	$\log x = 2.13$	$x = 10^{2.13} \; (= 134.9)$

CALCULATOR
EXERCISE 1

Learn the keystroke sequences for evaluating "$\log x$" and "10^x." Reproduce these results:

$\log 70 = 1.845$.

$10^{2.13} = 134.9$.

$10^{-1.2} = .063$.

Many texts would write our sample theorem as two theorems, one to emphasize "a" as the unknown, and another to emphasize "b" as the unknown. Pay attention to how the letters switch in these versions.

THEOREM 1-4-10B $10^x = y$ is equivalent to $x = \log y$.

THEOREM 1-4-10C $\log x = y$ is equivalent to $x = 10^y$.

These say nothing that was not already said in the first version, but they do it with x, which some people prefer.

EXAMPLE 17 CONTINUED   Solve 10^{x^2}

$$10^{x^2} = 4 \ \text{iff} \ x^2 = \log 4.$$

The theorem takes you this far. Now use Theorem 1-4-7 to handle the squaring.
$$x = \pm\sqrt{(\log 4)} = \pm.776.$$

This example makes the point that

Theorems are often about one operation at a time, and are applicable when that operation is *last*.

MORE ABOUT EXAMPLE 17   Solve "$\log(x - 3) = 1.4$."
  Because the last operation is "Take the log!", Theorem 1-4-10C applies.
  $\log(x - 3) = 1.4$. $x - 3 = 10^{1.4}(= 25.12)$. $x = 28.12$.

STILL MORE ABOUT EXAMPLE 17   Solve "$\log x - 3 = 1.4$."
  Now subtraction is last. It is not yet ready for the log theorem.
  $\log x = 1.4 + 3$, by Theorem 1-4-4. Now undo the log by using Theorem 1-4-10C: $x = 10^{4.4}$.   ▬

■ Placeholders in Definitions

Suppose you read a definition you have never seen before. Could you use it?

EXAMPLE 18   Here is a definition: $\begin{vmatrix} a & b \\ c & d \end{vmatrix} = ad - bc$. It defines the determinant in matrix theory. Read the definition and use it to do these two problems.

Find $\begin{vmatrix} 3 & 5 \\ 2 & 9 \end{vmatrix}$.

Solve $\begin{vmatrix} 6 & x \\ 4 & 5 \end{vmatrix} = 18$.

Use this example to learn to read Mathematics. You are *not* expected to memorize this definition and notation. Although important in matrix theory, here it is just an example of the use of placeholders.

You may be wondering what those tall vertical lines outside the groups of four numbers mean. But, that is precisely what the definition tells you. It has four placeholders that hold four places.

$$\begin{vmatrix} 3 & 5 \\ 2 & 9 \end{vmatrix} = 3(9) - 5(2) = 17.$$

Inspect the places to see why this is so.

To solve $\begin{vmatrix} 6 & x \\ 4 & 5 \end{vmatrix} = 18$, replace the left side, using its definition.

$$\begin{vmatrix} 6 & x \\ 4 & 5 \end{vmatrix} = 6(5) - x(4) = 30 - 4x. \text{ So, } 30 - 4x = 18. \ 4x = 12. \ x = 3.$$

EXAMPLE 19   Let $f(x) = 3x - 2$. Solve for x: $f(x) = 2f(4)$.

Use the definition of f.

Replace $f(x)$ with $3x - 2$:
$$3x - 2 = 2(3(4) - 2)$$
$$3x - 2 = 20$$
$$3x = 22.$$
$$x = \frac{22}{3}.$$

EXAMPLE 20   Let $\boxtimes$ be defined by this definition: $x \boxtimes y = 2x - 3y$, for all x and y.
Find $x \boxtimes 7$.

Read the definition. $x \boxtimes 7$ is equivalent to $2x - 3(7)$, or $2x - 21$.
Find $a \boxtimes b$.

It is $2a - 3b$. (It is the left and right places that matter. The definition applies regardless of the expression in those places.)
Find $5 \boxtimes x$.

Now the symbol x is on the right, not the left. So it is in the place of "y" in the definition, and "5" is in the place of "x". So, $5 \boxtimes x = 2(5) - 3(x)$. (The definition says to subtract three times the thing on the right of the symbol, even if the thing on the right is "x" instead of "y".)
Solve for x in this equation: $4 \boxtimes x = x \boxtimes 7$.

Now both sides must be replaced by an equivalent expression without the $\boxtimes$ symbol.

On the left, $4 \boxtimes x$ is equivalent to $8 - 3x$. On the right, $x \boxtimes 7$ is equivalent to $2x - 21$.

So, the equation is equivalent to: $8 - 3x = 2x - 21$. $29 = 5x$. $x = \frac{29}{5}$.

CONCLUSION   You are learning to read and write mathematics. Much of algebra is about operations, order, and relations (Equations are relations between expressions). These are not objects from ordinary life. Therefore, communication about them requires a special language with special features. English is not good enough.

If a mathematical definition, identity, or theorem expresses a thought about operations or methods, the variable is a placeholder. Placeholders hold places in sentences about operations so that the operations themselves can be discussed. The places they hold are important; the choice of letters to put in those places is not.

In contrast, unknowns are used in sentences about numbers.

Algebraic methods can be written in symbolic sentences with problem-patterns and solution-patterns.

Algebraic notation is well-designed to express mathematical methods as facts. Three of the primary ways are:

1. **formulas** (To use formulas you must know the context and what the letters mean)
2. **identities** (The meaning is self-contained), and
3. **relations between equations** (They explain when and how to change one equation into another).

Terms: unknown, placeholder, identity, equivalent expressions, equivalent equations, parameter.

Exercise 1-4

A

A1. In previous math courses did you (personally) learn primarily by *reading* your text or by *doing* what your teacher showed you how to do? [Or some other way? There is no "correct" answer to this question.]

A2.* ☺ (A) Which type of variable is used to state theorems?
(B) Which type of variable is used to state identities?
(C) When you are asked to solve for x, which type of variable is "x"?

A3.* (A) If "x" is used in a sentence about an operation, what type of variable is "x"?
(B) If "x" is used in a sentence about a number, what type of variable is "x"?

☺ *For A4–6 answer using mathematical symbolism.*

A4. $1 \times 5 = 5$. $1 \times 97 = 97$. State the corresponding abstract fact about multiplication by 1.

A5. $\dfrac{17}{1} = 17$. $\dfrac{324}{1} = 324$. State the corresponding abstract fact about division by 1.

A6. $99 + 0 = 99$. $47 + 0 = 47$. State the corresponding abstract fact about addition of zero.

True or False

A7. ☺ Methods can be expressed as facts.

A8.* ☺ "x" is always a placeholder.

A9. ☺ Simplify "$\dfrac{a}{(b/c)}$".

A10. ☺ Simplify "$\dfrac{(a/b)}{c}$".

Use Theorem 1-4-10 to solve these equations. [You can use the given solution with two significant digits in brackets to check your method. Report your solution with at least three significant digits.]

A11. $10^x = 20$ [1.3]   **A12.** $10^x = 500$ [2.7]

A13. $10^{2x} = 3000$ [1.7]   **A14.** $10^{x-5} = .01$ [3.0]

A15. $\log x = 2.5$ [320]   **A16.** $\log x = -1.8$ [.016]

A17. $\log(4x) = 3$ [250]   **A18.** $\log(x - 2) = 1.3$ [22]

☺ *Restate the theorem with some other letters.*

A19. Theorem 1-4-5A   **A20.** Theorem 1-4-5B

A21. Theorem 1-4-7   **A22.** Theorem 1-4-10

A23. Solve the equation in Example 14.

B

B1.* (A) What types of thoughts are expressed by sentences with unknowns?
(B) What types of thoughts are expressed by sentences with placeholders?

B2.* Name three ways in which methods can be expressed as facts and give one new example of each.

B3.* (A) Which types of things can be related by the symbol "="?
(B) What are its two uses in algebra? [Answer with more than just the word "equals".]
(C) What is the meaning sometimes erroneously given it by students who work only with commands?

B4.* (A) If we want to assert two *expressions* are equivalent, which symbol is usually used?
(B) Which symbol or phrase can be used to assert two *equations* are equivalent?
(C) Distinguish "equivalent" from "equal" expressions. Also, distinguish the types of variables they use.

☺ *Is the subject of the sentence a number (N) or operations and order (OO)?*

B5. (A) $x(x + 1) = 3x$.　(B) $x(x + 1) = x^2 + x$.

　　(C) $x + 0 = x$.　(D) $2x = c$ iff $x = \dfrac{c}{2}$.

B6. (A) $x^2 \geq 0$.

　　(B) $xy = 0$ iff $x = 0$ or $y = 0$.

　　(C) $(x + a)(x - a) = 0$.

　　(D) $(x + a)(x - a) = x^2 - a^2$.

B7. (A) $x - a = b$ iff $x = b + a$.

　　(B) Let $f(x) = x^2$.

　　(C) $3x - 15 = 50$.

　　(D) $x - a = -(a - x)$.

B8. (A) $\dfrac{x}{5} = c$ iff $x = 5c$.　(B) $x - a = b$.

　　(C) Let $f(x) = 3x - 5$.　(D) $3x - 5 = 20$.

☺ *Is "x" a placeholder in the given sentence?*

B9. $5x = 60$　　　**B10.** $5x = 2x + 3x$

B11. $2x > b$ iff $x > \dfrac{b}{2}$　**B12.** $4(x + 2x^2) = 4x + 8x^2$

B13. $4(x + 2x^2) = 16$　**B14.** $x + 9 = 42$

B15. Let $f(x) = \dfrac{(x + 4)}{3}$　**B16.** $\dfrac{x + 4}{3} = 19$

B17. $\dfrac{x + 4}{3} = \dfrac{x}{3} + \dfrac{4}{3}$　**B18.** $(x + 1)^2 = 16$

B19. $(x + 1)^2 = x^2 + 2x + 1$

B20. Let $f(x) = (x + 1)^2$

☺ *Give the problem-pattern (just the problem-pattern).*

B21. The Quadratic Theorem (1-4-8)

B22. Theorem 1-4-5A

B23. Theorem 1-4-7

B24. Theorem 1-4-10

☺ *Give the solution-pattern (just the solution-pattern).*

B25. The Quadratic Theorem (1-4-8)

B26. Theorem 1-4-5A

B27. Theorem 1-4-7

B28. Theorem 1-4-10

☺ *State, in symbolic notation, the theorem that expresses the method for the first step in solving these. Give both a problem-pattern and a solution-pattern, and relate the two patterns properly.*

B29. Both "$x + 22 = 45$" and "$3x + 5 = 92$."

B30. Both "$x^2 = 19$" and "$(2x + 1)^2 = 150$."

B31. Both "$\log x = -.12$" and "$\log(3x) = 2.34$".

B32. Both "$x - 13 = 45$" and "$x^2 - 13 = 39$."

B33. Both "$2x = 56$" and "$2(x - 19) = 100$."
[Be careful about dividing by zero.]

B34. Both "$\dfrac{x}{3} = 20$" and "$\dfrac{(6x + 1)}{5} = 1.92$."
[Be careful about multiplying by zero.]

B35. Both "$x^2 = \log 4$" and "$(\log x)^2 = 2.37$."

B36. Both "$\sqrt{x} = 20$" and "$\sqrt{x} = x - 10$."
[Be careful about negative numbers.]

☺ *State the algebraic formulation of the method used to evaluate these expressions in terms of simpler operations (as in Examples 8 and 10). [That is, state an identity with the appropriate problem-pattern and solution-pattern. The solution pattern should express simpler operations with positive integers. In the solution pattern, subtraction, if used, should express subtraction of smaller positive integers from larger positive integers.]*

B37. $19 - 40$　　**B38.** $\dfrac{(3/4)}{5}$

B39. $(-3)7$　　**B40.** $\dfrac{4}{(-6)}$

B41. $12 - (-8)$　　**B42.** $-6 - 5$

B43. $(-5)(-3)$　　**B44.** $\dfrac{2}{(3/4)}$

B45. $\dfrac{1}{3} + \dfrac{2}{5}$　　**B46.** $-34 + 12$

B47. $\dfrac{5/6}{7/8}$　　**B48.** $\dfrac{-3}{-6}$

B49. $\dfrac{(4/5)}{7}$　　**B50.** $12 - 43$

B51. $\dfrac{3}{(-5)}$　　**B52.** $23 - (-4)$

B53. $5(-8)$　　**B54.** $-4 - 9$

B55. $\dfrac{(2/3)}{4}$　　**B56.** $\dfrac{1}{2} + \dfrac{3}{5}$

B57. $\dfrac{3}{(5/7)}$

B58. $\left(\dfrac{5}{7}\right)\left(\dfrac{3}{11}\right)$

B59. $-54 + 19$

B60. $\left(\dfrac{-52}{-13}\right)$

B61. The expression "$3x + 9$" may be correctly interpreted in two radically different ways.
  (A) What are the two interpretations?
  (B) If you want to solve "$3x + 9 = 17.3$," which interpretation determines the steps?

B62. The expression "$\dfrac{(x + 2)}{5}$," may be correctly interpreted in two radically different ways.
  (A) What are the two interpretations?
  (B) If you want to solve "$\dfrac{(x + 2)}{5} = 89.6$", which interpretation determines the steps?

B63. An expression, $f(x)$, may be interpreted in two distinct ways.
  (A) Which two?
  (B) Which interpretation is more important for determining the steps to solve the equation "$f(x) = c$".

B64.* (A) What are identities about?
  (B) What is the purpose of identities?

Use Example 18 to:

B65. (A) Find $\begin{vmatrix} 3 & 9 \\ 2 & 7 \end{vmatrix}$ (B) Solve: $\begin{vmatrix} x & 7 \\ 2 & 3 \end{vmatrix} = 13$

B66. (A) Find $\begin{vmatrix} 2 & -3 \\ 5 & 12 \end{vmatrix}$ (B) Solve: $\begin{vmatrix} x & 2x \\ 2 & 9 \end{vmatrix} = 30$

B67.* Distinguish between placeholders and unknowns. Which is a "placeholder," and what does "placeholder" mean?

B68. Rewrite the identity in Example 8 with different letters.

B69. (A) In the equation "$mx + b = c$," which letter is the unknown?
  (B) How can you tell?
  (c) In the equation "$\dfrac{a}{b} = \dfrac{c}{d}$," which letter is the unknown? Note that this is ambiguous.

B70. (A) In the equation "$ax^3 + bx = d$" which letter is the unknown?
  (B) How can you tell?
  (C) In the equation "$ab + cd = 1$," which letter is the unknown? Note that this is ambiguous.

Reading and Using Definitions and Theorems

Here is a theorem about sums of consecutive positive integers:

"$1 + 2 + 3 + \ldots + n = \dfrac{n(n + 1)}{2}$."

Read it and use it to rewrite the value of the given expression.

B71. ☺ (A) $1 + 2 + 3 + \ldots + 100$
  (B) $1 + 2 + 3 + \ldots + k$

B72. ☺ (A) $1 + 2 + 3 + \ldots + 250$
  (B) $1 + 2 + 3 + \ldots + j$

B73. ☺ $1 + 2 + 3 + \ldots + 2n$

B74. ☺ $1 + 2 + 3 + \ldots + n + (n + 1) + \ldots + (n + 5)$

B75. $101 + 102 + 103 + \ldots + 500$

B76. $2 + 4 + 6 + \ldots + 2n$

Here is a theorem: $1 + 3 + 5 + \ldots + (2n - 1) = n^2$. [The letter "n" holds the place of an integer.] Read it and use it to rewrite these sums:

B77. (A) $1 + 3 + 5 + \ldots + 99$
  (B) $1 + 3 + 5 + 7 + \ldots + (2k - 1)$
  (C) $1 + 3 + 5 + \ldots + (4k - 1)$

B78. (A) $1 + 3 + 5 + \ldots + 59$
  (B) $1 + 3 + 5 + \ldots + n$, where n is odd.

B79. ☺ Here are two theorems:
  (T1) $|x| < c$ if and only if $-c < x < c$.
  (T2) $a < b$ if and only if $a + c < b + c$.

  (A) What is the problem-pattern of T1?
  (B) What is its solution pattern?
  (C) Which would you use first to solve the inequality "$|x - 3| < 12$"?
  (D) Solve it.
  (E) Which would you use first to solve the inequality "$|x| - 3 < 12$"?
  (F) Solve it.

☺ *Rewrite these expressions using the identity "$10^a 10^b = 10^{a + b}$."*

B80. $10^5\left(10^3\right)$

B81. $10^3\left(10^{1.2}\right)$

B82. $10^4\left(10^{2.7}\right)$

B83. $10^x\left(10^2\right)$

B84. $10^x \left(10^5\right)$

B85. $10^x \left(10^{3x}\right)$

B86. $10^x \left(10^{-1}\right)$

☺ *Here is a property of powers:* $\dfrac{a^p}{a^r} = a^{p-r}$. *Read it and use it to rewrite the given expressions.*

B87. (A) $\dfrac{x^3}{x^2}$ (B) $\dfrac{x^a}{x^b}$

B88. (A) $\dfrac{a^5}{a^2}$ (B) $\dfrac{a^b}{a^c}$

B89. $\dfrac{x^2}{x^{1/3}}$ **B90.** $\dfrac{x^3}{x^{1/2}}$

B91. $\dfrac{x^{1/2}}{x^{1/3}}$ **B92.** $\dfrac{x^{2/3}}{x^{1/2}}$

B93. $\dfrac{x^2}{x^{-1}}$ **B94.** $\dfrac{x^4}{x^{-3}}$

B95. ☺ Here is a sample theorem about the natural logarithm function:

"$e^a = b$ iff $a = \ln b$." ["e" is the particular number 2.718 . . . and not a placeholder.]

(A) Rewrite it to emphasize problems of the form "$e^x = c$."

(B) Rewrite it to emphasize problems of the form "$\ln x = c$."

Read the theorem in B95 and use it to solve the following equations. [You can use the given solution with two significant digits in brackets to check your method. Report your solution with at least three significant digits.]

B96. $e^x = 5$ [1.6] **B97.** $e^x = 100$ [4.6]

B98. $\ln x = 7$ [1100] **B99.** $\ln x = -3$ [.050]

B100. $e^{x-4} = 150$ [9.0] **B101.** $e^{3x} = .34$ [−.36]

B102. $\ln(2x) = 5$ [74] **B103.** $\ln(x + 4) = 1.1$ [−1.0]

☺ *Here is a sample identity:* "$sin(2\theta) = 2(sin\,\theta)(cos\,\theta)$." *Read it and use it to rewrite the given expression.*

B104. (A) $sin(2a)$ (B) $sin(40°)$

B105. (A) $sin(2x)$ (B) $sin(100°)$

B106. $sin(2(x + h))$ **B107.** $sin(2(\theta + 1))$

B108. $sin(4\theta)$ **B109.** $sin(4x)$

B110. Here is a theorem from trigonometry: In the interval $0° \le x \le 180°$, "$sin\,x = c$" is equivalent to "$x = sin^{-1} c$ or $x = 180° - sin^{-1}c$."

Read it and use it to solve: $sin\,x = .4$ for x in the interval $0° \le x \le 180°$. [Make sure your calculator is in "degree mode" so, for example, $sin\,30° = .5$].

B111. ☺ Here is a sample theorem: "If $n \ne 0$ and $f(x) = x^n$, then $f'(x) = nx^{n-1}$."
Read it and use it to find $f'(x)$ if $f(x)$ is
(A) x^3 (B) x^{-1} (C) x^p

B112. For positive integers denoted by n, let $n! = n(n - 1)(n - 2)\ldots(2)1$.
(A) Find 4! (B) Find 6!

(C) Solve $5! = x(3!)$ (D) Simplify $\dfrac{n(n!)}{((n - 1)!)}$

B113. Here is a sample theorem: For $a > 0$ and $b > 0$, $\log(ab) = \log a + \log b$.

(A) Use it to rewrite "$\log 5x$".

(B) Use it to rewrite "$\log(x^2 + x)$".

B114. How do we add a negative number to a negative number?
(A) Express your answer in English.
(B) Express your answer in mathematical notation.

B115. How do we multiply fractions?
(A) Express your answer in English.
(B) Express your answer in mathematical notation.

B116. How do we divide a fraction by a number?
(A) Express your answer in English.
(B) Express your answer in mathematical notation.

B117. (A) What would you *do* to solve the equation "$x - 34 = 457$"? [Write your answer as a command.]
(B) State an abstract *fact* which corresponds to your method.

B118. (A subtle difference) In the problem-pattern of the Quadratic Theorem, "x" is an unknown. However, in the theorem as a whole, it is a placeholder. Explain how this can be, given that unknowns are not placeholders.

☺ *Here is a theorem:* $1 + 2 + 4 + 8 + \ldots + 2^n = 2^{n+1} - 1$. *Read it and use it to find these.*

B119. $1 + 2 + 4 + \ldots + 2^{10}$

B120. $1 + 2 + 4 + \ldots + 2^{20}$

B121. $1 + 2 + 4 + \ldots + 2^k + 2^{k+1}$

B122. $1 + 2 + 4 + 8 + \ldots + 2^{2n}$

B123. Let $f(x) = 2x - 1$. Solve for x: $f(x) = 3f(5)$.

B124. Let $f(x) = 2x - 1$. Solve for x: $f(x) = f(5)f(4)$.

B125. Let $f(x) = (x - 2)^2$. Solve for x: $f(x) = 9f(7)$.

B126. Let $f(x) = (x - 2)^2$. Solve for x: $f(x) = f(4)f(7)$.

B127. Let $f(x) = 3x - 1$. Solve for x: $f(x) = 2f(4)$.

B128. Let $f(x) = 4x - 2$. Solve for x: $f(x + 1) = 16$.

B129. Let $f(x) = 3x - 1$. Solve for x: $f(x + 1) = 2f(4)$.

B130. Here is a definition: $(a, b) \cdot (c, d) = ac + bd$. Read it and use it to
(A) Find $(3, 5) \cdot (2, 9)$
(B) solve $(3, x) \cdot (2x, 5) = 20$.

B131. Here is a definition: $x\#y = 2x + 3y$. Read it and use it to
(A) Find $4\#7$
(B) Find $c\#x$
(C) Solve for x: $(x + 1)\#x = 23$.

B132. Here is a definition: $a@b = 2b - a^2$, for all a and b. Read it and use it to
(A) Find $3@7$
(B) Find $5@a$
(C) Solve for x: $x@4 = 7@x$.

B133. Here is a definition: Let $x \square y = 5y - 2x$, for all x and y. Read it and use it to
(A) Find $7 \square 4$.
(B) Find $a \square b$
(C) Solve for x: $23 \square x = 14$.

B134. Define ☆ this way: $x☆ = x^2 - 2$. Solve for x in this equation: $x☆ = (4☆)(3☆)$.

Section 1-5 GRAPHS

The setup for a graph (technically, a "rectangular coordinate graph") begins with two perpendicular number lines.

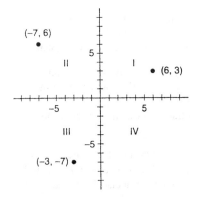

FIGURE 1A An axis system

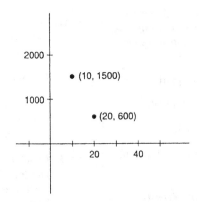

FIGURE 1B Another axis system, with a different scale

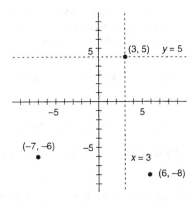

FIGURE 2 The locations of $(3, 5)$ and some other points (ordered pairs)

Each number line is called an <u>axis</u> and exhibits a <u>scale</u>, which is an indication of which locations correspond to which numbers. The place where both number lines have zeros is the <u>origin</u>. The combination of the axes and their scales is called the <u>axis system</u> (Figures 1A and 1B). The two axes naturally divide the plane into four regions called <u>quadrants</u>, which are sometimes designated by Roman numerals as in Figure 1A.

The equation "$x = 3$" has two different meanings, depending upon the context. In one dimension, "$x = 3$" describes a point on a number line. In two dimensions, the equation "$x = 3$" means "$x = 3$ and $y =$ anything." In two dimensions "$x = 3$" is the equation of a vertical line. Similarly, "$y = 5$" is a horizontal line.

> vertical line: $x = c$, for some c.
> horizontal line: $y = c$, for some c.
> x-axis: $y = 0$.
> y-axis: $x = 0$.

Points are located at the intersection of lines. The lines $x = 3$ and $y = 5$ intersect at the point $(3, 5)$ (Figure 2).

The notation "(x, y)" serves for both points and ordered pairs. The number in the position of "x" is called the <u>first-coordinate</u> (or "x-coordinate") and the number in the position of y is called the <u>second-coordinate</u> (or "y-coordinate").

■ Expressions and Graphs

One sequence of operations makes *many* number pairs—ordered pairs.

EXAMPLE 1   The expression "$2x + 1$" creates many argument-image pairs, including $(0, 1)$, $(1, 3)$ and $(5, 11)$. The set of all such ordered pairs is equivalent to the function f given by $f(x) = 2x + 1$.   ▬

DEFINITION 1-5-1 (GRAPH)   The <u>graph</u> of f is the set of all points (x, y) where $y = f(x)$. It is the set of all points of the form $(x, f(x))$. It is the graph of the equation "$y = f(x)$"

Commonly the term <u>graph</u> is used to mean a *picture* of the set of those points. In practice, because the size of a picture is limited, a graph actually consists of a picture of all the pairs that fit in the <u>window</u> (or <u>viewing rectangle</u>), which is the region shown.

EXAMPLE 2   To graph "$y = 2x + 1$" mark all the pairs of the form $(x, 2x + 1)$ that fit in the window (Figure 3).

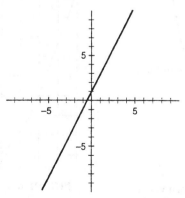

FIGURE 3 $y = 2x + 1$. The window is $-10 \le x \le 10$ and $-10 \le y \le 10$

Graphs show (x, y) pairs. An expression can be evaluated from its graph, even if no algebraic representation is given. ▬

EXAMPLE 3

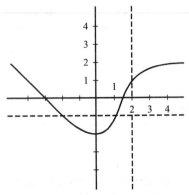

FIGURE 4 A graph of $f(x)$ with no algebraic representation given

Figure 4 gives the graph of $y = f(x)$, without giving its algebraic representation. Tick marks are one unit apart.

Find $f(2)$.

The graph gives $f(x) = y$. $f(2)$ is the y-value of the point where $x = 2$. Make the vertical line $x = 2$. Find the point of intersection. Read its y-value by looking across to the scale on the y-axis. $f(2) = 1$.

Solve $f(x) = -1$.

The graph gives $f(x) = y$. You want $y = -1$. So, graph the horizontal line $y = -1$ and see where the two graphs intersect. Then read the x-values of those points.

The solution is $x = -2$ or $x = 1$.

Solve $f(x) = 0$.

The graph gives $f(x) = y$. $y = 0$ is the x-axis. So $f(x) = 0$ when the graph touches the x-axis: $x = -3$ or $x = 1.5$.

Solve $f(x) > 0$.

The graph gives $f(x) = y$. $f(x)$ is *greater* than zero when $y > 0$, that is, the corresponding point is *above* the x-axis (which is $y = 0$). Give the x-values of all points above the x-axis. The solution is: $x < -3$ or $x > 1.5$. ▬

■ The Intersection Method

Graphs can be used to solve equations. Graph both sides and find where they intersect.

EXAMPLE 4

Solve $x^2 = x + 6$ graphically.

Graph both $y = x^2$ (the left side) and $y = x + 6$ (the right side) (Figure 5). The two graphs intersect where $x^2 = y = x + 6$. The solutions are the x-values of those points.

$x = -2$ or $x = 3$. ▬

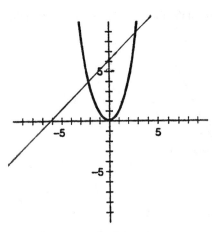

FIGURE 5 $y = x^2$ and $y = x + 6$
$[-10, 10]$ by $[-10, 10]$

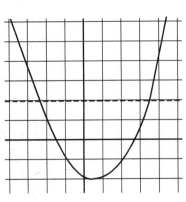

FIGURE 6 $y = f(x)$. Also, $y = 2$

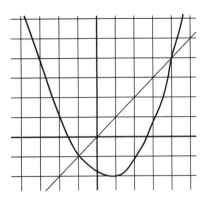

FIGURE 7 $y = f(x)$ Also, $y = x$

EXAMPLE 5   Figure 6 gives the graph of $y = f(x)$. Grid lines are one unit apart. Solve $f(x) = 2$.
Graph the horizontal line $y = 2$. Then the graphs intersect where $f(x) = y = 2$.
The solution is the x-values of the points of intersection.
   $x = -2.4$ or $x = 3.4$ (approximately).
Now solve $f(x) = x$.
Graph $y = x$ (Figure 7). Read where the two graphs intersect.
   $x = -1$ or $x = 4$. ▬

■ Graphing with Calculators

Learn how to graph expressions with *your* calculator.

**CALCULATOR
EXERCISE 1**   Learn how to

1. enter expressions to graph,
2. select the desired window,
3. draw the graph, and
4. read the coordinates of points on the graph.

We use the terms in this table.

FIGURE 8 $y = x^2$.
$[-10, 10]$ by $[-10, 10]$

Term	Meaning
standard window	We use $-10 \leq x \leq 10$ and $-10 \leq y \leq 10$.
trace	Identify points on a graph and display their x- and y-values

Now, graph "$y = x^2$."
Make the window our "standard" window (Figure 8).
Now use the *trace* feature to display the (x, y) pair at the cursor. You can move the cursor to the left or right to exhibit different (x, y) pairs.

> We express windows using interval notation.
> "$-10 \leq x \leq 10$ and $-10 \leq y \leq 10$"
> is written
> "$[-10, 10]$ by $[-10, 10]$."

■ Selecting a Window

Most graphs do not illustrate the whole function. For example, when you graph the $y = x^2$ (Figure 8), the point $(7, 49)$ will not appear in the standard window because 49 is too large. But the standard window is often fine for creating a "representative" graph.

DEFINITION 1-5-2   A <u>representative</u> graph is one that is not misleading.
   A representative graph might not include all the detail you want, but it does indicate the general behavior of the expression, even for points not in the window.

For example, the graph of x^2 in Figure 8 is representative. It gives the right impression about where the points outside the window would be.

CALCULATOR
EXERCISE 2

Graph $10x^2 + 100$.

The picture on the standard window is blank. Try it and see for yourself. You need a different window. How can you select a "good" window?

There is not just one "good" window. Many different choices are fine. First find one that is reasonable, and later modify it to an even better one if you want.

The standard window is blank because the y-values are off the screen. Determine some y-values and adjust the y-interval so you can see them.

When $x = 0$, $y = 100$. When $x = 10$, $y = 1100$. So a y-interval that brackets 100 and 1100 will show some action. Overall, you might use, say, $0 \le y \le 1500$, and put tick marks every 500 units (Figure 9).

You can use the *trace* feature to do these calculations. If nothing appears on your screen, use the *trace* feature to find y-values. Then adjust the window to display them.

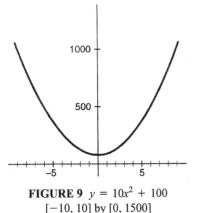

FIGURE 9 $y = 10x^2 + 100$
$[-10, 10]$ by $[0, 1500]$

A way to select a window

1. Graph it in the standard window.
   If the graph looks good, quit. If not,

2. Find relevant y-values (use the *trace* feature), adjust the y-interval, and redo the graph. If it looks good, quit. If not

3. Decide if the x-interval should be changed. If so, change it and redo the graph.

4. Repeat steps 2 and 3 until you are satisfied [Your window does not have to be "right" the first time—it is easy to change].

CALCULATOR
EXERCISE 3

Graph $200\sqrt{x}$.

The picture in the standard window is virtually blank. Try it and see for yourself. Which windows will provide illuminating pictures?

First find relevant y-values. Use the *trace* key. When $x = 0$, $y = 0$. At the right edge, when $x = 10$, $y = 200\sqrt{x} = 200\sqrt{10} = 632.5$. So, pick a y-interval which includes y-values from 0 to 632.5, say 0 to 1000. Negative x's are not in the domain, so make the x-interval begin at 0. All positive values of x are possible, but you cannot graph them all. So pick a convenient maximum x-value, say $x = 10$ (Figure 10).

■ Limitations of Calculator Graphs

Graph any simple graph and look closely at the picture. Graphing calculators are wonderful, but they are not perfect. They have some important limitations. By studying even a simple graph such as the graphs of "x^2" you can learn a lot about how they work. Use your calculator to graph x^2 (Figure 11). Figures 8 and 11 graph the same expression with the same window. Why does Figure 11 appear ragged?

A graphing calculator does not display all the points in the window. The calculator has a screen which is made up of discrete dots. Each dot, called a <u>pixel</u>,

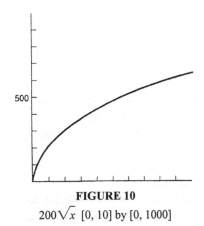

FIGURE 10
$200\sqrt{x}$ $[0, 10]$ by $[0, 1000]$

can be illuminated or not. Pixels have a small, but non-zero, width. If you look closely at your calculator screen you can see the pixels arranged in vertical columns and horizontal rows. Each column corresponds to an exact horizontal position—an exact x-value. Some calculators have 95 columns of pixels. To graph an expression the calculator selects 95 x-values evenly spaced across the window and evaluates the expression at each of the 95 values. Points with other x-values in between those 95 x-values are omitted.

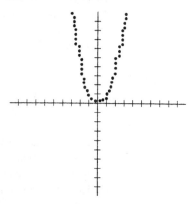

FIGURE 11 A screen-like picture of the graph of x^2, illustrating the discrete pixels

CALCULATOR EXERCISE 4

Graph x^2 on the standard scale. Put the cursor on the graph using the *trace* key. Move the cursor to the left and right slowly and note how the x-value changes. Each column of pixels corresponds to a particular x-value. Neighboring columns correspond to x-values some fixed distance apart (about .2 units on some calculators). Other x-values between those x-values do not yield points.

The picture is composed of discrete dots, which explains why curves which should be smooth may appear ragged on your screen. Consider two neighboring columns of pixels. Suppose the graph is steep and the left column illuminates the pixel in row 21 and the right column illuminates the pixel in row 25 (Figure 12, left). Should your calculator leave a gap between those two dots? Or should it "fill in" between them to visually connect them? The true graph would have points for the many x-values between the two x-values labeling the two columns. Most of the time the true graph would have y-values between the two given y-values and the graph would be connected (Figure 12, right).

Most of the time it is right to "fill in" between neighboring columns of pixels, which is why graphing calculators usually do fill in between dots. Nevertheless, sometimes dots are connected which should not be connected.

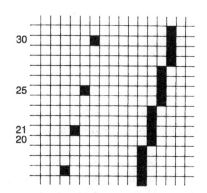

FIGURE 12 Pixels in neighboring columns not connected (left), and connected (right)

DEFINITION 1-5-3

A part of a calculator's graph that appears to be incorrect because of the way the dots are, or are not, connected is said to be an <u>artifact</u> of the calculator's programming.

CALCULATOR EXERCISE 5

Try these graphs to see if your calculator produces artifacts.

Graph $y = 1/(x - 1.5)$. Your picture should not show a vertical line through $x = 1.5$. If there is a vertical line there, it is an artifact.

Graph $y = \sqrt{83 - x^2}$. The picture should show the top half of a circle centered at the origin. However, many calculators show a gap between the curve and the x-axis. The gap should not be there. It is a result of the limited number of columns of pixels used to evaluate the expression. The x-value where $y = 0$ was not used, so the calculator did not connect the dots to the x-axis (Problem B48).

■ Guess-and-Check

To <u>solve</u> the equation "$f(x) = c$," find the values of x such that the expression "$f(x)$" has the value c. If the expression $f(x)$ is simple enough, we can use algebra to solve for x. There is another approach that avoids algebra. It is simply to use guess-and-check.

EXAMPLE 6 Solve $x^3 + x = 5$.

This equation is not easy to solve. The expression on the left is a cubic (third power) polynomial. There is a cubic formula, somewhat like the quadratic formula, but it is so complex that hardly anyone knows it or uses it. So how does a mathematician solve that problem?

Mathematicians are likely to use guess-and-check. They use a calculator or computer to evaluate the expression "$x^3 + x$" for various values of x. When they find an x for which the value is close to 5, they may quit, figuring close is good enough. How many decimal places of accuracy do you want, anyway?

Guess-and-check is effectively what calculators do with their equation-solvers when algebraic methods fail to work.

The method is to *evaluate the expression* "$f(x)$" for various values of x to try to *solve the equation* "$f(x) = c$." It is generally easier to evaluate an expression than to solve an equation.

> **Expressions tell you what to do to evaluate themselves,**
> **but equations do not tell you what to do to solve themselves.**

EXAMPLE 6 CONTINUED The *expression* "$x^3 + x$" tells you how to find its value when you are given x. It says to "Cube the number and add that to the number." If $x = 2$, its value is $2^3 + 2 = 10$. That computation is <u>direct</u> because the process for doing it is given in the problem itself.

In contrast, the *equation* "$x^3 + x = 5$" does not tell you how to find x. The process given by the expression "$x^3 + x$" is *not* the solution process. When the process you need to do is *not* indicated by the problem, we call the problem <u>indirect</u>. Solving equations is more difficult than evaluating expressions.

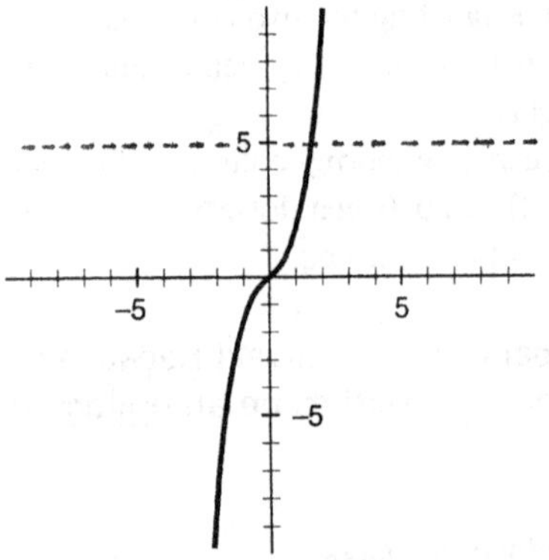

> If symbols, words, or a basic formula give the process you should do, the problem is said to be <u>direct</u>. Evaluating an expression is a direct process. If the process for obtaining a desired number is not given, the problem is <u>indirect</u>. Algebra provides indirect methods for solving equations.

To solve "$x^3 + x = 5$" you may graph both "$y = x^3 + x$" and "$y = 5$" (Figure 13).

Look for where the two graphs intersect. The x-values of those points are the solutions. In this example, there is only one solution: $x = 1.52$. ■

FIGURE 13 A representative graph of $y = x^3 + x$. Also, the graph of $y = 5$. $[-10, 10]$ by $[-10, 10]$

**CALCULATOR
EXERCISE 6**

Graph "$y = x^3 + x$" and use the *trace* feature to try to solve "$x^3 + x = 5$" (Figure 13).

Trace until you find a y-value close to 5. Read the corresponding x-value as an approximate solution. The exact x-value which solves the equation probably will not even be tried. However, you can find an x-value where the y-value is slightly less than 5, and a neighboring x-value where the y-value is slightly greater than 5 (Figure 13). The solution is somewhere between those two x-values.

This is a solution by guess-and-check. The calculator evaluates the expression at many x-values and displays the results graphically. It tries many guesses, directly calculates the images, and displays the results for you to read.

To get more accuracy you could change the scale—magnify the window about the key point so that neighboring columns of pixels have x-values which are closer together so the solution can be bracketed more closely. Figure 14A illustrates the original standard picture, a point with y-value close to 5, and the region to be magnified. Figure 14B then illustrates the picture magnified by four using the *zoom* feature.

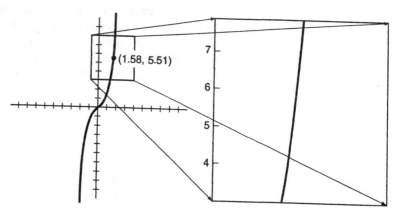

FIGURE 14A A standard graph $x^3 + x$ and a region to be magnified by four

FIGURE 14B The region magnified by four

Learn to "zoom in" on a point of a graph. Learn to do it at least two ways. One is to use the *zoom* feature. Another is to use the *window* feature to manually select a window.

In the magnified picture (Figure 14B) the x's the calculator tries are closer together. Using this procedure repeatedly, you can bracket the solution with narrower and narrower intervals and obtain the solution to any desired degree of accuracy. The solution is, to three significant digits, $x = 1.52$.

Some calculators have a *solve* key for solving equations. The *solve* key uses an automatic guess-and-check procedure to find a solution. The method is very similar in spirit to this example. Be sure you understand the guess-and-check process, even if your calculator has a *solve* key and you do not have to manually zoom in yourself.

**EXAMPLE 6
REVISITED**

Again, solve "$x^3 + x = 5$".

To solve an equation it is common to "Put everything on the left." The equation is equivalent to

$$x^3 + x - 5 = 0.$$

Now, you can graph "$y = x^3 + x - 5$" and solve the equation by finding the x-value where y is zero. "$y = 0$" is the x-axis. Points on the graph near the x-axis yield x-values near the solution.

Guess-and-check is the approach of many graphing calculators with a "solver." They evaluate the expression at numerous x-values and "zero in" on the x-value that yields a zero. Again, $x = 1.52$. ▬

**CALCULATOR
EXERCISE 7**

Solve "$10 \ln x - x = 0$" by the guess-and-check method using a graph.

Figure 15 graphs "$y = 10 \ln x - x$" on the standard scale. From the graph you expect a solution near $x = 1$. Zoom in and you can find it: $x = 1.12$, to three significant digits.

This answer is right, as far as it goes, but it does not go far enough. This is why we introduced the concept of a "representative" graph. A representative graph is one which is not misleading. The graph in Figure 15 is not representative. Zoom out to the interval $0 \le x \le 40$ (Figure 16) and you will see that the strong upward trend toward the right of Figure 15 is misleading. The curve turns around and comes back down. The second solution is $x = 35.77$.

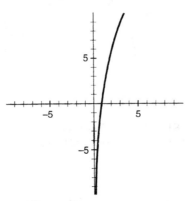

FIGURE 15 $y = 10 \ln x - x$.
$[-10, 10]$ by $[-10, 10]$

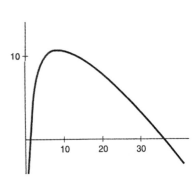

FIGURE 16 $y = 10 \ln x - x$.
$[0, 40]$ by $[-5, 15]$

■ Maximizing or Minimizing an Expression

A major type of calculus problem is to maximize or minimize the value of an expression. Graphically, it is easy to find the x-value which yields the maximum of an expression. Just graph the expression and look for the highest point on the graph.

**CALCULATOR
EXERCISE 8**

Use your calculator to find the maximum value of $x^2(5 - 2x)$ for $x > 0$, and find the corresponding x-value.

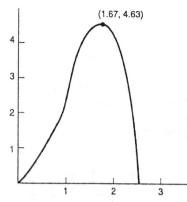

FIGURE 17 $x^2(5 - 2x)$. [0, 3] by [0, 5]. The highest point is identified

Graph the expression and identify the maximum y-value and corresponding x-value with the *trace* key (Figure 17). Zoom in until you are satisfied with the accuracy of your approximate solution.

The maximum value for $x > 0$ appears to be about $y = 4.63$ when $x = 1.67$. The exact answer, $y = \dfrac{125}{27}$ (about 4.6296) at $x = \dfrac{5}{3}$ (about 1.6667), could be obtained using calculus.

Calculators that have a "solver" have a "maximizer" as well. It uses guess-and-check, speeded up by the calculator.

Graphing Terminology
You do not graph the same sort of equation that you solve.

To "solve" the equation with **one** variable, "$x^5 - x - 7 = 0$" you do not graph it. You graph the related equation "$y = x^5 - x - 7$" with **two** variables. Use the graph to find the x's that yield y-value zero. The "y" variable and the graph record the results of many guesses for x.

To make a graph, most calculators require equations to be in "$y = f(x)$" form. For other equations with two variables, you will need to solve for y.

EXAMPLE 7   "$x^2 + y^2 = 1$" is the equation of the important "unit circle" of radius 1 centered at the origin.

Solving this for y, we obtain:

$$y = \sqrt{1 - x^2} \text{ or } y = -\sqrt{1 - x^2}$$

These two graphs plotted together on one axis system will make the picture (Figure 18).   ▬

CONCLUSION   In two dimensions, the point (c, d) is at the intersection of the vertical line $x = c$ and the horizontal line $y = d$. The graph of f has points (x, y) where $y = f(x)$. To find $f(c)$, find the y-value of the point on the graph that is also on the vertical line $x = c$. To solve "$f(x) = d$," find the x-values of all the points on the graph of $y = f(x)$ that are also on the horizontal line $y = d$.

To solve the equation "$f(x) = g(x)$" you may graph both "$y = f(x)$" and "$y = g(x)$" and find where the graphs intersect. The x-values of the points of intersection are the solutions.

Calculators can be used to solve equations with guess-and-check. You may solve the equation "$f(x) = 0$" by graphing "$y = f(x)$" and reading the graph, "zooming in" to get more accuracy. This is effectively what a automatic calculator "solver" does.

Graphing calculators are wonderful, but they are not perfect. They may display artifacts (parts of the graph that look wrong) and are sometimes misleading. Graphs that are not misleading are said to be representative.

Terms: graph, origin, axis, scale, quadrant, ordered pair, point, coordinate, window, standard window, trace, representative graph, pixel, artifact, guess-and-check method.

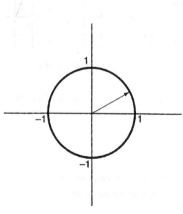

FIGURE 18 Graph of the unit circle, $x^2 + y^2 = 1$. [−2, 2] by [−2, 2]

Exercise 1-5

A

A1.* ☺ In this text, what is the "standard window"?

A2.* ☺ What does the "trace" feature on a calculator do?

A3.* ☺ What makes a graph "representative"?

A4.* ☺ What is an "artifact" (in the context of a graph produced by a graphing calculator)?

A5.* ☺ What is a "pixel"?

A6.* ☺ Give the equation of the a) x-axis. b) y-axis.

☺ *Give the equation of the*

A7. (A) vertical line through (a, b).
(B) horizontal line through (a, b).

A8. (A) vertical line through $(2, \pi)$.
(B) horizontal line through $(-1, -4)$.

A9. (A) vertical line through $(7, -2)$.
(B) horizontal line through $(3, 5)$

☺ *Which number quadrant is*

A10.* (A) the upper left?
(B) the lower left?

A11.* (A) the upper right?
(B) the lower right?

A12. (A) What model of calculator do you use?
(B) Use your trace key to find out, on the standard scale, the difference in x-values of neighboring columns of pixels.
(C) How many columns of pixels does your calculator have? You can find out by dividing the width of the standard interval by the difference from part (b), and then adding 1. Or, you can count them.

A13. ☺ To solve the equation "$f(x) = 0$" graphically, which equation would you graph?
[Hint: You do not graph "$f(x) = 0$".]

B

B1.* Given a representative graph of an expression, $f(x)$, explain how to (approximately)
(A) Evaluate $f(a)$.
(B) Solve $f(x) = c$.

In your answers, be sure to distinguish points on graphs from their coordinates.

B2. Explain, in English, what a calculator's "equation solver" does to solve an equation when the equation cannot be solved algebraically.

☺ *B3–B8. Grid lines are one unit apart. Approximate answers are good enough.*

B3. The figure gives a representative graph of $f(x)$ without giving its algebraic expression.
(A) Find y for $x = 2$.
(B) Find x such that $y = 3$.

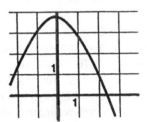

FIGURE for B3

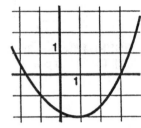

FIGURE for B4

B4. The figure gives a representative graph of $f(x)$ without giving its algebraic expression.
(A) Find y for $x = 1$.
(B) Find x such that $y = -1$.

B5. The figure gives a representative graph of $f(x)$ without giving its algebraic expression.
(A) Find y for $x = 3$.
(B) Find x such that $y = 1$.

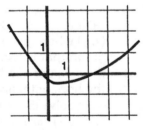

FIGURE for B5

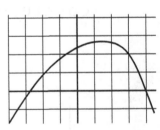

FIGURE for B6

B6. The figure gives a representative graph of $f(x)$ without giving its algebraic expression.
(A) Find $f(3)$.
(B) Solve $f(x) = 0$.
(C) Solve $f(x) \geq 0$

B7. The figure gives a representative graph of $f(x)$ without giving its algebraic expression.
(A) Find $f(-2)$.
(B) Solve $f(x) = 0$.
(C) Solve $f(x) < 0$.

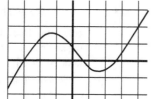

FIGURE for B7   **FIGURE** for B8

B8. The figure gives a representative graph of $f(x)$ without giving its algebraic expression.
(A) Find $f(4)$.
(B) Solve $f(x) = 0$.
(C) Solve $f(x) \geq 0$.

☺ *In the figures for B9–16 grid lines are one unit apart and the graphs are representative.*

B9. Solve $f(x) = g(x)$ **B10.** Solve $g(x) < f(x)$

B11. Solve $f(x) = g(x)$ **B12.** Solve $g(x) < f(x)$

B13. Solve $f(x) = x$ **B14.** Solve $f(x) = -x$.

B15. Solve $f(x) = f(2)$ **B16.** Solve $f(x) = f(4)$.

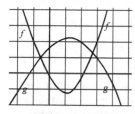

FIGURE for B9–10   **FIGURE** for B11–12

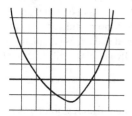

FIGURE for B13–16

Problems B17–B25 give graphs in certain windows with $-10 \leq x \leq 10$. Find the y-intervals of the given windows.

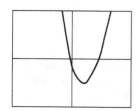   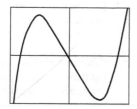

B17. $10x(x - 5)$ **B18.** $x(x - 8)(x + 8)$

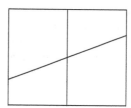   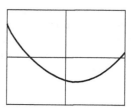

B19. $x + 50$ **B20.** $(x + 6)(x - 9)$

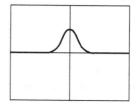   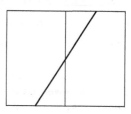

B21. $\dfrac{1}{(x^2 + 1)}$ **B22.** $\left(\dfrac{x}{5}\right) + 1$

Problems B23–25 give graphs in certain windows. Find the x- and y-intervals of the given windows.

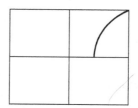   

B23. $\sqrt{x - 1}$ **B24.** $\sqrt{x + 20}$

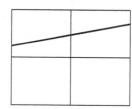

B25. $x + 1$

B26. The figure locates (a, b) and (c, d) on an axis system. Locate (b, a) and (d, c).

B27. The figure locates (a, b) and (c, d) on an axis system. Locate $(-a, b)$ and $(c, -d)$.

B28. ☺ The figure gives (a, b) and (c, d). Find the coordinates of the point P.

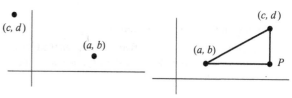

FIGURE for B26 and B27   **FIGURE** for B28

Sometimes the "standard" scale is not the one you want.
(A) Sketch, in a window you draw, a "representative" graph of the equation. (B) Give the x and y-intervals you used. (C) find the minimum value of y for x > 0 in your window. [Great accuracy is not required, just get it close enough to show you know how to do it.]

B29. $y = (x - 20)(x - 50)$

B30. $y = (2x - 10)(x - 100)$

B31. $y = x^3 - 20x^2 + 150$

B32. $y = x^3 - 50x + 300$

*B33–B42. Solve these equations. **Note**: This text occasionally supplies solutions with two significant digits in brackets (For example, Problems B35–39). These are given so you can immediately determine whether you used the correct solution process. For all such homework problems, prove that you know the correct process by giving a more accurate answer rounded to three or more significant digits.*

B33. $10\sqrt{x + 7} - x - 20 = 0$

B34. $x^3 + 26x^2 - 132x - 360 = 0$

B35. $x^3 - 5x - 6 = 0$ [2.7]

B36. $8\sqrt{x} + x^2 = 12$ [1.5]

B37. $x + \sin x = 1$ (use radians)   [.51]

B38. $x + \ln x = 2$ [1.6]

B39. $x + e^x = 2$ [.44]

B40. $x + \tan x = 5 \left(\text{use radians}, 0 < x < \dfrac{\pi}{2} \right)$ [1.3].

B41. $x^3 - 3x < 1$ [$x < -1.5$ or ....].

B42. $\sin x > x^2$ (use radians)   [$0 < x$ and ....]

B43. (A) Solve $8\sqrt{x} - x = 12$ graphically. Give at least one decimal place of accuracy.
(B) What mistake would be easy to make in part (a)?

B44. (A) Solve $100x + 50\sqrt{x} - x^2 = 500$ graphically.
(B) What mistake would be easy to make in part (a)?

Find the x-value that yields the maximum y-value.

B45. $y = x^2 + 3x - x^4$ [1.1]

B46. $y = 6x + \sqrt{x} - x^2$ [3.1]

Reading and Writing Mathematics

B47. There is an important difference between the types of equations you "solve" and the types you "graph." What is it?

B48. (A) Look at the graph of $\dfrac{1}{(x - 1.5)}$ on your graphing calculator. Does it produce an artifact? What is the artifact, if any? [Mention the model of calculator you use.]
(B) Look at the graph of $\sqrt{83 - x^2}$ on your graphing calculator. Does it produce an artifact? What is the artifact, if any?

B49. Find an expression and a window such that your graphing calculator produces an artifact different from the ones given in the previous problem. Mention the model of calculator you use.

B50. Here is a graph that may produce an <u>artifact</u>.
Let $f(x) = \dfrac{(1 - \cos(x^4))}{x^8}$.
Use radian mode (not degree mode). The domain does not include $x = 0$.
(A) Graph it to approximate the value for x near, but not equal to, zero. Zoom in at most once to find an approximation.
(B) Now change to the scale $[-.1, .1]$ by $[-.1, .6]$. Does your graph produce nearly a horizontal line at $y = \dfrac{1}{2}$? It should, but some calculators show the height dropping to zero for $|x| < .01$. This is a mistake—an artifact.

B51. Here is a graph that may produce an artifact. First graph "sin x" on your calculator's standard so-called "trigonometric scale." We want to compare the next graph to it. Now graph "sin(50x)" or may be "sin(65x)." They should have 50 and 65 times as many bumps, respectively. Most calculators cannot display so many bumps, and make these graphs look far different than they really are. (We will study such graphs in Chapter 2.)

☺ *When you sketch a graph with these features, what order would they come in?*

B52.* (A) the point (a, b)
(B) the line $x = a$.
(C) the x-axis.

B53. (A) the point (a, b)
(B) the line $y = b$.
(C) the y-axis.

B54. (Short essay) Graphing calculators are wonderful, but they are not perfect. Discuss two things that are not perfect about calculator graphs.

Section 1-6   FOUR WAYS TO SOLVE EQUATIONS

Given an equation to solve, how should you solve it? The answer depends upon the order and operations in the equation. The purpose of this section is to show you that, when all is said and done, there are only four basic ways to solve an equation. Therefore, you should always aim for one of the four basic formats that allow you to finish off easily.

Here are the four primary methods of solving equations:

1. Do the inverse operations in the reverse order,
2. Use the Zero Product Rule,
3. Use the Quadratic Formula, and
4. Guess-and-check, also known as trial-and-error.
(This includes graphical methods).

The first three methods are traditional algebraic methods. They are <u>indirect</u> because you do not actually do the operations represented in the problem. The fourth method is direct; it is not algebraic.

Sometimes equations must be rearranged into equivalent equations before the algebraic methods apply.

To solve an equation its component expressions must fit, or be rearranged to fit, certain patterns which express operations in convenient orders.

Look for the importance of order in the following discussions of the four equation-solving methods.

■ The "Inverse-Reverse" Method

The "inverse-reverse" method is also known as "doing and undoing." The name "inverse-reverse" is short for "inverse operations in the reverse order."

EXAMPLE 1   Solve $3x + 5 = 26$.

First, look at the expression on the left and identify the operations and the order in which they are to be executed on x. There is "Multiply by 3" and then "Add 5", in that order. Thus, to solve it you may *subtract* 5 and then *divide* by 3.

$$\frac{(26 - 5)}{3} = 7. \text{ The solution is } x = 7.$$

In the solution process, subtraction was used *first* because the expression "$3x + 5$" expresses addition *last*—illustrating the "reverse order" part of the "inverse-reverse" name. The *addition* in the expression caused *subtraction* (the inverse of addition) in the solution process—illustrating the "inverse operations" part of the "inverse-reverse" method name.

The process can be illustrated by a "function-loop" in which the top shows the operations applied to "x" to obtain "$3x + 5$" and the bottom shows the functions which "undo" those operations.

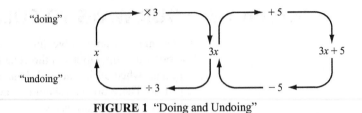

FIGURE 1 "Doing and Undoing"

The inverse-reverse method uses a certain format.

Inverse-Reverse Format: The equation has **only one appearance of the unknown** (often "*x*"). We can use inverse-reverse if the equation can be rearranged into that format.

Some equations do not have only one appearance of x, but are easily rearranged into inverse-reverse form.

EXAMPLE 2   Solve $\pi x = 3 + x$.
     This one is solved with the inverse-reverse method even though there are two appearances of "*x*". You just need to "consolidate like terms" (rearrange it) so that there is only one.

$$\pi x = 3 + x \text{ iff } \pi x - x = 3 \qquad [\text{subtracting } x \text{ from both sides}]$$
$$\text{iff } (\pi - 1)x = 3 \qquad [\text{Distributive Property}]$$

[Now it *is* in inverse-reverse form.]

$$\text{iff } x = \frac{3}{\pi - 1} = 1.40 \qquad [\text{inverse-reverse}]$$

This problem illustrates the purpose of identities such as the Distributive Property, which is used here to change the order of operations.

EXAMPLE 3   Solve $\dfrac{x^2 - 1}{3} = 15$.
     There are several operations expressed here, but only one appearance of "*x*". Use inverse-reverse.

$$\frac{x^2 - 1}{3} = 15 \text{ iff } x^2 - 1 = 45$$
$$\text{iff } x^2 = 46$$
$$\text{iff } x = \pm 6.78.$$

The square-root function on calculators gives the square root of 46, which is positive. *You* must remember that the negative of the square root is another solution (Theorem 1-4-7). The square root symbol refers only to the non-negative solution of $x^2 = c$, so $\sqrt{c}$, is never negative. The solution to $x^2 = 25$ is not just $\sqrt{25}$. It is $\pm\sqrt{25}$ (that is, 5 or -5).

In summary, one method of equation-solving which works when the unknown appears only once is to

Do the inverse operations in the reverse order.

■ The Zero Product Rule Method

The Zero Product Rule method consists of replacing one equation with two related, but simpler, equations according to the Zero Product Rule.

THEOREM 1-6-1 THE ZERO PRODUCT RULE

$ab = 0$ iff $a = 0$ or $b = 0$.

To use this rule, the equation must fit, or be altered to fit, a certain format. The name of this rule describes the two keys.

Zero Product Rule Format: One side of the equation is *zero* and the other is a *product* (that is, multiplication is last). You can use the Zero Product Rule if the equation can be rearranged into that form.

EXAMPLE 4

Solve $(x - 5)(x^2 - 16) = 0$.

The left side is a product and the right side is zero. The problem-pattern is "$ab = 0$." It fits the Zero Product Rule. This equation is equivalent to

$$x - 5 = 0 \text{ or } x^2 - 16 = 0.$$

These are equivalent to

$$x = 5 \text{ or } x = 4 \text{ or } x = -4,$$

which exhibit the solution. ▬

One key to the Zero Product Rule is that it uses a product of two factors. Therefore, factoring may be necessary.

EXAMPLE 5

Solve $x^2 + 5x - 6 = 0$.

For the Zero Product Rule you need multiplication *last*. The identity "$x^2 + 5x - 6 = (x + 6)(x - 1)$," reorders the operations so the equation can be rewritten in Zero Product Rule format:

$$(x + 6)(x - 1) = 0.$$

The purpose of algebraic identities is to give alternative orders of operations.

By the Zero Product Rule,

$x + 6 = 0$ or $x - 1 = 0$. The solution is $x = -6$ or $x = 1$

This solution also could have been obtained using the Quadratic Formula, which we will discuss later in this section. ▬

The Zero Product Rule is the primary motivation for factoring.

The Zero Product Rule converts a longer equation into two shorter equations. You still have to be able to solve the shorter equations.

EXAMPLE 6

Solve $(x - 5)(\ln x + 2) = 0$.

"ln" is the symbol for the "natural logarithm" function which is very important in calculus.

To solve "$(x - 5)(\ln x + 2) = 0$" note that the left side is a product and the right side is 0. The form is "$ab = 0$." Therefore, the Zero Product Rule can be used to get

$$x - 5 = 0 \text{ or } \ln x + 2 = 0.$$

The solution to "$x - 5 = 0$" is obviously "$x = 5$." The solution to the second equation, "$\ln x + 2 = 0$" is less obvious. By the inverse-reverse method you subtract 2 to find the equivalent equation, "$\ln x = -2$," but now you need the inverse of the natural logarithm function.

THEOREM 1-6-2 (THE INVERSE OF *LN*) $\ln x = c$ iff $x = e^c$.

$\ln x = -2$ iff $x = e^{-2} = .135$. Therefore, the solution to the original problem in Example 6 is

$$x = 5 \text{ or } x = .135.$$ ▬

CALCULATOR EXERCISE 1 Reproduce these results:

$$e^{-2} = .135. \quad e^3 = 20.09 \quad \ln 10 = 2.30.$$

You will study the properties of logarithms and exponentials later. Right now your responsibility is merely to note that

---

To solve an equation you may need to know the inverses of the component functions expressed.

EXAMPLE 7 Solve $(x - 3)(x - 5) = 10$.

The left side is factored. Do you like that?

You spent a great deal of time in school learning to factor, so your first reaction might be to like this factored form. But *why* do we like factored form?

We like factored form because of the Zero Product Rule, which deals only with *zero* products. This product is not zero; the product is useless. We must get rid of the factors. To solve this equation algebraically, multiply out ("unfactor") the left, subtract 10 from both sides to obtain a zero on the right, and use the "$ax^2 + bx + c = 0$" form in the upcoming Quadratic Theorem (Problem A48). ▬

■ The Quadratic Theorem

The third primary way to solve an equation is to use the famous "Quadratic Formula." Equations which are in the form

$$ax^2 + bx + c = 0,$$

or which can be rearranged into this form, where a, b, and c are constants and a is not zero, are called "quadratic" equations. All quadratic equations have either zero, one, or two real-valued solutions. They can be found easily using the Quadratic Formula.

THEOREM 1-6-3 (THE QUADRATIC THEOREM) For $a \neq 0$, the solutions to

$$ax^2 + bx + c = 0$$

are given by

$$x = \frac{-b \pm \sqrt{b^2 - 4ac}}{2a},$$

where the symbol "$\pm$" (read "plus or minus") indicates two distinct solutions when the square root term is positive. The Quadratic Formula is the solution-pattern.

Quadratic Theorem Format: The equation is in the form

$$\text{``}ax^2 + bx + c = 0\text{''}$$

where a is not zero, and "x" is the unknown. You can use the Quadratic Formula if the equation can be rearranged into that form.

Theorems express their own "formats." One half of this theorem expresses a format in abstract algebraic notation using placeholders. It is the form or pattern of operations that matters, not the letters used to express the pattern. The other half expresses the alternative form into which the original may be changed.

EXAMPLE 8   Solve $7x^2 - 5x = 3$.

The first step is to convert the given equation into the standard quadratic form with all terms on the left. It is equivalent to

$$7x^2 - 5x - 3 = 0.$$

This does not factor easily. $a = 7$, $b = -5$, and $c = -3$. Using the Quadratic Formula:

$$x = \frac{-(-5) \pm \sqrt{(-5)^2 - 4(7)(-3)}}{2(7)} = \frac{5 \pm \sqrt{109}}{14}.$$

$$x = 1.103 \text{ or } x = -0.389.$$

To solve a quadratic equation it is *sometimes* possible to use the Zero Product Rule, but it is *always* possible to use the Quadratic Formula.   ▬

EXAMPLE 5
REVISITED   Solve $x^2 + 5x - 6 = 0$.

In Example 5 we factored to get "$(x + 6)(x - 1) = 0$", which has solution $x = -6$ or $x = 1$. Of course, we could use the Quadratic Formula instead we would get the same answer. Here $a = 1$, $b = 5$, and $c = -6$. Therefore

$$x = \frac{-5 \pm \sqrt{5^2 - 4(1)(-6)}}{2(1)} = \frac{-5 \pm \sqrt{49}}{2} = \frac{-5 \pm 7}{2} = -6 \text{ or } 1,$$

as before.   ▬

■ Guess-and-Check

The fourth basic method of equation-solving is guess-and-check. Another common name for it is "trial-and-error." Guess-and-check is the basic method behind solving solutions graphically. Calculators can rapidly evaluate an expression for many values of x and then display the results with a graph.

CALCULATOR
EXERCISE 2   Solve $xe^x = 7$.

It is common to "Put everything on the left." This step yields "$xe^x - 7 = 0$."

None of the three algebraic methods work. Figure 2 graphs "$y = xe^x - 7$." From the picture it looks as if the x-value which yields height 0 is somewhere between 1 and 2. A three-significant-digit approximation is $x = 1.52$.

Guess-and-check can work without a graph. You could try many values and zero in on the solution. That is what a calculator's "solver" does.

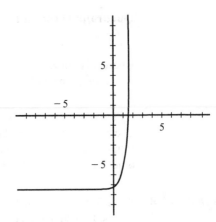

FIGURE 2 $y = xe^x - 7$. $[-10, 10]$ by $[-10, 10]$

Guess-and-Check Format Requirements: None. (However, it is easier to use if one side is zero or a constant.)

Many problems require guess-and-check. Examples include:

$$x^5 + x = 1, x + \sin x = 3, x + \log x = 3, \text{ and } 10^x = 2 + x.$$

The guess-and-check method is a part of the important subject area of mathematics called "numerical analysis."

■ Format

What do you do if the format of the original equation is not right? Change it! Rearrange it to satisfy one of the formats.

EXAMPLE 9   Solve algebraically: $(x + 4)(x - 1)^2 + (x - 1)(x + 4)^2 = 0$.
  The word "algebraically" tells you to use one of the three traditional methods. It's not a quadratic and inverse-reverse will not work because there are too many appearances of x. Aim for the Zero Product Rule.
  Now it is a sum (addition is last). Use the Distributive Property to turn the sum into a product by factoring out common terms.

$$(x + 4)(x - 1)^2 + (x - 1)(x + 4)^2$$
$$= (x + 4)(x - 1)[(x - 1) + (x + 4)] \quad \text{[by the Distributive Property]}$$
$$= (x + 4)(x - 1)[2x + 3]$$

A product!
  Now, $(x + 4)(x - 1)[2x + 3] = 0$ if and only if

$$x + 4 = 0 \text{ or } x - 1 = 0 \text{ or } 2x + 3 = 0.$$

Therefore the solution is: $x = -4$ or $x = 1$ or $x = -\dfrac{3}{2}$.

By the way, this type of problem arises frequently in calculus.   ▬

■ Choosing a Method

Because the guess-and-check method is not algebraic, it is ruled out when the point of a problem is to see if you understand algebraic methods.

The other three methods (inverse-reverse, Zero Product Rule, and the Quadratic Formula) are algebraic methods. They are indirect because you do *not* do the operations represented in the problem.

> When we ask you to "solve algebraically," we mean that you should use one of the three traditional, algebraic, indirect methods. We do not mean that you must report the exact symbolic solution which the method is capable of yielding. Just give the solution in decimal form rounded off to three or more significant digits.

EXAMPLE 10 For each of the following name the **algebraic method best suited** to solving the equation. If and only if none of the three algebraic methods work, pick guess-and-check.

<u>method</u>

(10a) $3x + 17 = 87.4$ inverse-reverse (There is only one appearance of "x")

(10b) $(x - 1)(x^2 - 20) = 0$ Zero Product Rule (problem pattern $ab = 0$)

(10c) $x^2 + \sqrt{x} - 6 = 0$ Guess-and-check

(10d) $(x - 1)(x + 3) = 4.6$ Quadratic Formula (after multiplying it out and "consolidating like terms")   ▬

EXAMPLE 11 To solve equations algebraically you sometimes want to "multiply out" expressions. Is it best to multiply these out?

(11a) $3(x + 7) = 60$ No. To solve it, just undo the given operations.

(11b) $3(x + 7) + x = 60$ Yes. Multiply it out because there are two appearances of x. Consolidate like terms. Then use inverse-reverse.   ▬

EXAMPLE 12 To solve these algebraically is it best to multiply them out?

(12a) $(x + 2)^2 = 20$ No. Just undo the given operations.

(12b) $(x + 2)^2 + x = 20$ Yes, you must multiply it out and consolidate like terms into the Quadratic Theorem format.   ▬

EXAMPLE 13 To solve these algebraically is it best to multiply them out?

(13a) $(x + 1)(x + 3) + 2(x + 1)(x - 7) = 0$ No. You can factor out "$x + 1$" and use the Zero Product Rule.

(13b) $(x + 1)(x + 3) + 2(x + 2)(x - 9) = 0$ Yes, you must multiply it out. There is no common factor as in the first part. Consolidate like terms and use the Quadratic Formula.   ▬

EXAMPLE 14 For each of the following name the **algebraic method best suited** to solving the equation. If and only if none of the three algebraic methods work, pick guess-and-check.

(14a) $3x + 5 = x \log 5000$

Inverse-reverse
(consolidate like terms)

(14b) $\dfrac{4 + x^2}{5} = 12$

Inverse-reverse

(14c) $(5x + 3)(x - 2)^2 + 3(x - 2)(5x + 3)^2 = 0$

Zero Product Rule
(factor it)

(14d) $x = 7 + x^3$

Guess-and-check ▬

CONCLUSION To solve an equation algebraically it must fit, or be rearranged to fit, certain patterns of operations. The patterns are expressed by the formats of the inverse-reverse method, the Zero Product Rule, and the Quadratic Formula. **If an equation does not fit any of those formats, first use identities and theorems to reorder the operations so that the formats are satisfied.**

The guess-and-check method is very important, but it is not algebraic. If an equation cannot be reorganized to fit any of the three algebraic methods, guess-and-check will work because it has no format requirement.

Terms: inverse-reverse, Zero Product Rule, Quadratic Theorem, Quadratic Formula, guess-and-check, direct, indirect, solve algebraically.

Exercise 1-6

A

A1.* ☺ (A) The name "inverse-reverse" is an abbreviation of what?
(B) What is the format for the inverse-reverse method?

A2.* ☺ (A) State (with variables) the Zero Product Rule.
(B) Give, in English, the format for the Zero Product Rule method.

A3.* ☺ What is the format for the Quadratic Theorem?

A4.* ☺ What is the format requirement for the guess-and-check method?

☺ *The formats of which methods are already met by the following equations?*

A5. $(x - 5)(\ln x + 7) = 5$

A6. $xe^x = 0$

A7. $5 - \ln x^2 = 0$

A8. $x(\sin x + 1) = 7$

A9. $(x - 2)(x^2 + 3x - 10) = 0$

A10. $x^2 + 5x - 10 = 0$

A11. ☺ (A) Why can't you use the inverse-reverse method on "$5.4x + 3.2x = 76$" as it stands?
(B) Solve it algebraically.

A12. ☺ (A) Why can't you use the Zero Product Rule on "$(x - 5)(x + 3) + 1 = 0$" as it stands?
(B) Solve it algebraically. [4.9, ...]

A13. ☺ (A) Why can't you use the Zero Product Rule on "$(x - 5)(x + 3) = 1$" as it stands?
(B) Solve it algebraically. [5.1, ...]

A14. ☺ (A) Why can't you use the Zero Product Rule on "$(x - 5)(x + 3) = 4x$" as it stands?
(B) Solve it algebraically. [7.9, ...]

☺ *To solve some of these problems you would want to "multiply out" the product or square. In others, it is not necessary. To solve these algebraically, is it **best** to multiply out the expression? (Answer "Yes" or "No". Do not solve it.)*

A15. $3(x + 2) = 19$

A16. $5(x - 7) = 19$

A17. $5(x + 2) - 8 = 100$

A18. $5(x + 2) - x = 100$

A19. $7(x - 3) + x = 200$

A20. $7(x - 3) + 12 = 200$

A21. $(x + 1)^2 = 100$

A22. $(x + 1)^2 + 1 = 100$

A23. $(x + 1)^2 + x = 500$

A24. $(x + 1)^2 - x = 600$

A25. $(x + 1)^2 + 20 = 700$

A26. $(x + 1)^2 + 12 = 800$

☺ *The purpose of identities is to provide alternative sequences of operations. Sometimes you want to "multiply out" products, other times you do not. Do **not** solve these. Just say if, to solve them algebraically, it is best to multiply out the product. (Yes or No).*

A27. $(x + 1)^2 = 90$

A28. $(x + 1)^2 = x$

A29. $(x + 1)^2 - x = 90$

A30. $3(x + 5) = 36$

A31. $3(x + 5) + 7 = 36$

A32. $3(x + 5) - x = 26$

A33. $(x + 3)(2x - 7)(x + 1) = 0$

A34. $(x + 3)(2x - 7) - 5 = 0$

A35. $(x + 3)(2x - 7) = 0$

A36. $(x + 3)(2x - 7) = 5x$

A37. $(2x - 3)(x - 8) = 3(x - 8)$

A38. $3(x + 5) = 7x - 9$

A39. $3(x + 5) - 2(x - 7) = 0$

A40. $3(x + 5) - 2(x - 7) = 9$

A41. $(x - 5)(x - 3) = 0$

A42. $(x - 5)(x - 3) = 2$

A43. $(x - 5)(x - 3) - 6 = 0$

A44. $-.35 = .43(x - 6.2)$

A45. $4(x - 6) = 73$.

A46. $4(x - 6) = 2.6x$.

A47. Suppose a student begins to solve the equation in Example 9 by multiplying out the terms.
(A) Are any of the four methods easier to use after the terms have been multiplied out?
(B) What does this teach you about multiplying out long expressions?

A48. Solve $(x - 3)(x - 5) = 10$

Reading and Writing Mathematics

A49. State the Distributive Property using "x", "y", and "z".

A50. State the Zero Product Rule (Theorem 1-6-1) using some other letters.

A51.* (A) What are identities about?
(B) What is their use?

A52.* Which property is commonly used to turn a sum or difference into a product?

B

B1.* Name the four ways to solve an equation and their corresponding formats.

B2.* Suppose you are to solve an equation which does not fit any of the three indirect-method formats as it stands and you are forbidden to use guess-and-check. What can you do?

☺ *There is a big difference in how the two equations usually would be solved, in spite of superficial similarities. Identify the different approaches (do not solve them).*

B3. "$x^2 + 7x = 0$" and "$x^2 + 7x - 2 = 0$."

B4. "$x^2 - x = 7$" and "$x^3 - x = 7$."

B5. "$x^2 + \sqrt{x} - 3 = 0$" and "$x^2 + x - 3 = 0$."

B6. "$x(x + 3) = 0$" and "$x(x + 3) = 1.4$."

B7. "$(x - 2)(x - 5) = 0$" and "$(x - 2)(x - 5) = 1$."

B8. "$(x - 4)^2 = 16$" and "$(x - 4)^2 = x$."

☺ *For each of the following name the **algebraic method best suited** to solving the equation. If and only if none of the three algebraic methods work, pick guess-and-check. Do NOT solve these! Name the method (abbreviations are ok)!*

B9. $2x + 5 = 23$

B10. $3 + 5x = x^2$

B11. $(x - 1)(\log x - 2.4) = 0$

B12. $\dfrac{(x-5)}{4} = 3x + 7$

B13. $x^2 + x^3 = 20$

B14. $x^2 + \sqrt{x} = 12$

B15. $3x + 9 = 2x - 7$

B16. $(x-3)^3 = 5$

B17. $(x-3)^3(x-1) = 0$

B18. $x^2 + 5 = 4x + 19$

B19. $x^3 + 5 = 90$

B20. $x + (x-1)(x+2) = 0$

B21. $(x-2)(x+3) = 1$

B22. $\log(2x+1) - 5 = 5.3$

B23. $3x + 6 = 5x - 7$

B24. $\dfrac{x}{2} = 5 + \dfrac{x}{3}$

☺ *The purpose of identities is to provide alternative sequences of operations. Sometimes you want to "multiply out" products, other times you do not. Do **not** solve these. Just say if, to solve them algebraically, it is best to multiply out the product. (Yes or No).*

B25. $(x+3)(x+5) - 2(x+3) = 0$

B26. $(x+3)(2x+5) - (x+3)^2 = 0$

B27. $(x+3)(x+5) - 2(x+3)(x-7) = 0$

B28. $(x+3)(x+5) - 2(x+2)(x-7) = 0$

B29. $(x+3)(x+5) - 2(x+2)(x-4) = 0$

B30. $(x+3)(x+5) - 2(x+3)(x-7) = 1$

B31. $(x+3)^2 = 40$

B32. $(x+3)^2 + x = 40$

B33. $\dfrac{(x+3)^2}{12} - 19 = 40$

B34. $(x+3)^2 = x + 17$

B35. $(x-3)(x-2) - (x-3)(x-5) = 0$

B36. $(2x-3)(x-2) - (x-3)(3x-5) = 0$

B37. $4(x-5) + 7 = 3x + 12$

B38. $4(x-5) + 2(x-5) = 92$

B39. $(x-4)^2 + 13 = 56$

B40. $(x-4)^2 + 13 = 56x$

B41. $(x-3)(x+5) + (x-3)^2 = 0$

B42. $(x-5)(x-2) + (x-3)^2 = 0$

☺ *For each of the following name the **algebraic method best suited** to solving the equation. If and only if none of the three algebraic methods work, pick guess-and-check. Do NOT solve these! Name the method (abbreviations are ok)!*

B43. $(x+2)(x-3) + 3(x-3) = 0$

B44. $(x+2)(x-1) - 5.9 = 0$

B45. $(x+2)(x^2+2x-6.2) = 5.7$

B46. $(x-4)(x+2)^2 + 4(x-4)^2(x+2) = 0$

B47. $x^2 - x - \sqrt{x} - 12 = 0$

B48. $(x+1)^2 + x = 90$

B49. $x^2 + \sqrt{x} - 7 = 0$

B50. $(x-3)^2(x+6) - 3(x-3)(x+6)^2 = 0$

B51. $(x-3)(x+4) + 2(x-1)(x+2) = 0$

B52. $(x-3)^2(x+6) - 3(x-4)(x+2)^2 = 0$

B53. $x(x+6) = 12.3$

B54. $\sin(2x) = .73$

B55. $x(x+6)^2 = 12.3$

B56. $\ln(2x-7) = 3.7$

B57. $x^2 + \sqrt{x} - 16 = 0$

B58. $(x-3)(x+1) = 5$

B59. $x(x^2-3) = 5$

B60. $(x-3)(x-5)^2 + 4(x-5)(x-3)^2 = 0$

B61. $(x-3)(x+4)^2 - (x-2)(2x+1) = 0.$

B62. $(x-2)(x+4)^2 - 3(x-3)(x+5)^2 = 0$

B63. $3\left(\dfrac{x}{2} + 5\right) = 17$

B64. $(x^2-3)^2 + 17 = 42$

B65. $\dfrac{(x-3)}{(x-1)} = 7$

B66. $\dfrac{(x-3)}{(x-1)} = 2x$

B67. $(x-1)(x+2)^2 - (x-3)(x+5.2) = 0$

B68. $(x-1)(x+2)^2 - (x-1)(x+3)(x+5) = 0$

Solve algebraically:

B69. $(x-5)(x^2-9) + 2x(x^2-9) = 0$

B70. $4x(x^2-5) + (x^2-5)^2 = 0$

B71. $(x-3)^2(x+5) - 3(x-3)(x+5)^2 = 0$

B72. $4x^3(1-x)^3 - 3(1-x)^2x^4 = 0$

B73. $(3x+1)(x-2)^2 - 4(x+7)(x-2)^2 = 0$

B74. $(x-1)(x+2)^2 - (x-1)(x+3)(x+5) = 0$

Note: This text occasionally supplies solutions with two significant digits in brackets. These are given so you can immediately determine whether you used the correct solution process. On the homework, prove that you know the correct process by giving a more accurate answer rounded to three or more significant digits.

B75. Solve: $10 \ln x = x$ [Hint: Did you find a representative graph?].

B76. Solve: $x + 2 = e^x$ [1.1 and a second solution].

B77. Solve $x + \ln x = 3$ to three significant digits on your calculator.   [2.2]

Graph these by algebraically solving for y and then using a graphing calculator:

B78. $3x + xy + 7 = 5y$.

B79. $x^2 + 2x - 3xy + y^2 - 17 = 0$.

Reading and Writing Mathematics

B80. Theorem 1-6-2 is stated with letters chosen to emphasize its role in solving problems in which the natural logarithm function is applied to the unknown.
  (A) Restate it, using its placeholders differently, to emphasize problems in which the exponential function is applied to the unknown.
  (B) Restate it in a neutral manner which does not use "*x*" to suggest which position holds the unknown.

State an identity which allows you to regard

B81. "$5 - x$" as beginning with x (instead of beginning with 5).

B82. "$\dfrac{7}{x}$" as beginning with x (instead of beginning with 7).

B83. "$\dfrac{120}{x}$" as beginning with x (instead of beginning with 120).

B84. "$99 - x$" as beginning with x (instead of beginning with 99).

B85. (A)* Theorems express their own "formats." How?
  (B) Use the Zero Product Rule to illustrate your answer to part (a).

B86. State the version of the Distributive Property which applies to
  (A) multiplication and subtraction
  (B) division and addition

B87. State the version of the Distributive Property which applies to division and subtraction.

CHAPTER

2

Functions and Graphs

Section 2-1   FUNCTIONS AND GRAPHS

Functions can be represented by graphs and tables. This section emphasizes the importance of the window of a graph.

■ Graphs

Technically, by definition, the <u>graph</u> of the function f is the set of all ordered pairs (x, y) where $y = f(x)$, for all x in the domain. But, the graphs we draw usually do not exhibit *all* the pairs because the size of the window is limited. Furthermore, different windows can make the same function look different.

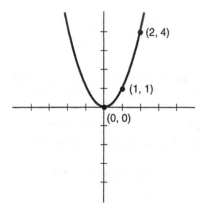
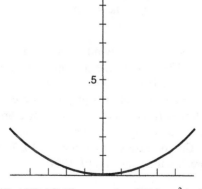
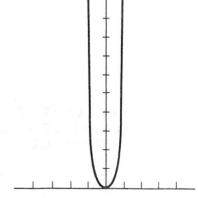

FIGURES 1A, 1B, AND 1C   Three graphs of $f(x) = x^2$ in different windows

$[-5, 5]$ by $[-5, 5]$ $\left[\dfrac{-1}{2}, \dfrac{1}{2}\right]$ by $[0, 1]$ $[-50, 50]$ by $[0, 100]$

Obviously, f given by $f(x) = x^2$ has more than one picture. The window strongly affects the appearance.

CALCULATOR EXERCISE 1 Graph x^2 on the standard scale. Then change the x-interval to make it look wider. Which x-interval, $[-20, 20]$ or $[-5, 5]$, makes it look wider? Why? (Problem A28.)

DEFINITION 2-1-1 The scale of a graph is <u>square</u> when one unit in the x-direction is the same distance as one unit in the y-direction.

Figure 1A had a square scale. Changing the scale made the graph look different.

■ Windows

How can you make the same function look different? It is easy–just change the window.

CALCULATOR EXERCISE 2 Use the window $[-10, 10]$ by $[0, 10]$ to graph $f(x) = x^2$ on your calculator (Figure 2A). Now, adjust only the x-interval to make the graph go through the upper right corner of the window (Figure 2B).

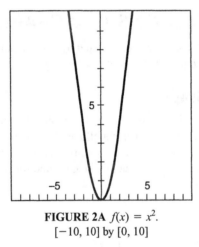

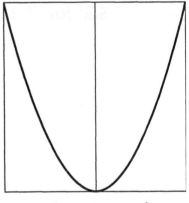

FIGURE 2A $f(x) = x^2$.
$[-10, 10]$ by $[0, 10]$

FIGURE 2B $f(x) = x^2$.
What is the window?

The top of the window is $y = 10$. So, make the largest x-value such that $x^2 = 10$. Change the x-interval to $[-\sqrt{10}, \sqrt{10}]$ which is approximately $[-3.16, 3.16]$.

CALCULATOR EXERCISE 2, CONTINUED Return to the original graph: x^2 in the window $[-10, 10]$ by $[0, 10]$ (Figure 2A). Adjust only the y-interval so that the graph goes through the upper right corner of the window (Figure 2B, again).

The right edge is the line $x = 10$. The corresponding y-value is 100. Therefore, use the y-interval $[0, 100]$.

CALCULATOR EXERCISE 3 Figure 3A is a graph in the standard window of $f(x) = 4x - x^3$. If you want to make the bumps the same width, but look taller in the window, how could you do it?

The width is determined by the x-values of the points, and the height by the y-values of the points. If we want a particular point to be higher in the window, we need to adjust the y-interval so that the same y-value looks higher. For example, the point $(1, 3)$ is on the graph of f and it is near the highest point in the first quadrant (Figure 3A). When the y-interval is $[-10, 10]$, that point is less than half way up in the first quadrant, since 3 is less than half of 10. However, in Figure 3B where the y-interval is $[-5, 5]$, the point $(1, 3)$ is more than half way up in the first quadrant, since 3 is more than half of 5. A shorter y-interval makes the same points look higher, since the same y-values are larger relative to the size of the interval.

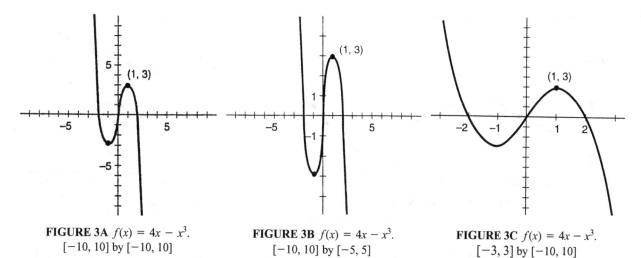

FIGURE 3A $f(x) = 4x - x^3$. $[-10, 10]$ by $[-10, 10]$

FIGURE 3B $f(x) = 4x - x^3$. $[-10, 10]$ by $[-5, 5]$

FIGURE 3C $f(x) = 4x - x^3$. $[-3, 3]$ by $[-10, 10]$

CALCULATOR EXERCISE 3 CONTINUED

Return to Figure 3A. If you want to make the bumps the same height, but look wider in the window, how could you do it?

Use a narrower x-interval, say $[-3, 3]$.

The point $(2, 0)$ is on the graph. When the x-interval is $[-10, 10]$, it is only two tenths the way from the center to the edge. The original graph is narrow (Figure 3A). If the x-interval were $[-3, 3]$, that point would be two-thirds the way to the edge. The bumps would look wider (Figure 3C).

> **The points on a graph of a function are determined by the function, but the apparent location of those points is determined by the window.**

Interval change	graph appears
x-interval: larger	narrower
smaller	wider
y-interval: larger	shorter
smaller	taller

EXAMPLE 1 Figure 4A plots the yearly average of the Dow Jones Industrial Stock price index for four years.

year	1990	1991	1992	1993
Dow average	2679	2929	3284	3754

The average gained 1075 points over those years, as the upward trend in the graph illustrates. How could you change the window to make the gain look more dramatic?

The vertical interval in Figure 4A is 2000 to 5000. The gain will appear more dramatic if the low is nearer the bottom of the window and the high nearer the top. Figure 4B uses the interval [2500, 4000].

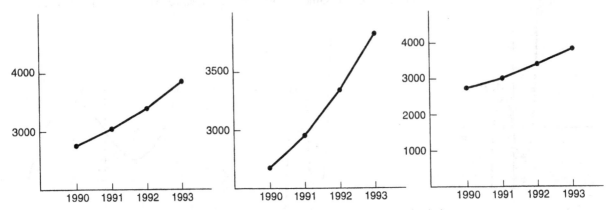

FIGURES 4A, B, AND C The Dow Jones Industrial Stock price index average, 1990 to 1993. Note how different vertical scales change the appearance

EXAMPLE 1 CONTINUED   Suppose you want to make the increase in the Dow Jones average appear less dramatic. How could you adjust the scale to do so?

One way is to use a longer y-interval, so that the change in y is a smaller fraction of the window. Figure 4C uses 0 to 5000. ▬

Sometimes we need to adjust the window so we can see the points of interest. For example, the minimum value might not be visible in the standard window.

CALCULATOR EXERCISE 4   Let $f(x) = 10x^2 - 300x + 10,000$. Use a graph to find x such that $f(x)$ is its minimum.

> Some calculators have a "minimizer" that will automatically solve this problem. It uses guess-and-check to determine the x-value by repeatedly evaluating the expression directly and comparing the values to each other.

Do not expect the standard scale to yield a representative graph. Try it and see. You need to select a better window.

You want a window that displays the minimum. The picture in the standard rectangle is blank because the y-values are off the screen. (When $x = 0$, $y = 10,000$, which is far off the screen. You may use the *trace* feature to find y-values.) To display $y = 10,000$, expand the y-interval to, say, [0, 20,000] as in Figure 5A. (Just select any y-interval that is possible and modify it later if you want.)

The minimum y-value is off the right edge of the graph and is certainly less than 10,000. So change the x-interval and, if you wish, the y-interval. Figure 5B uses the window [0, 30] by [0, 10,000]. Now the general location of the minimum is clear. It can be found by zooming in (Problem A27).

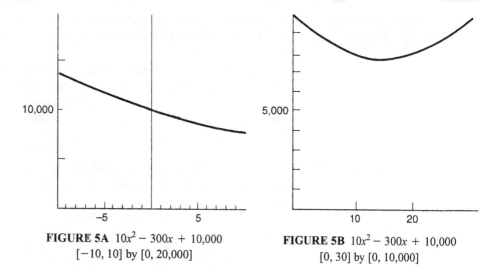

FIGURE 5A $10x^2 - 300x + 10,000$
[−10, 10] by [0, 20,000]

FIGURE 5B $10x^2 - 300x + 10,000$
[0, 30] by [0, 10,000]

Selecting a Custom Window
Step 1: Graph it in the standard window. If it looks good, quit. If not,
Step 2: Find y-values for those x-values (perhaps using the *trace* feature), adjust the y-interval, and redo the graph. If it looks good, quit. If not,
Step 3: Decide if the x-interval should be changed. If so, adjust it and redo the graph.
Step 4: Repeat Steps 2 and 3 until you are satisfied. [Your window does not have to be "right" the first time—it is easy to change.]

CALCULATOR EXERCISE 5

Figure 6 pictures the graph of $f(x) = x^4(1 - x)^6$ in a certain window. Find the window.

The picture in the standard window is nothing like this. Try it and see. The only place on the graph in the standard window that appears horizontal is between $x = 0$ and $x = 1$, so you might use the *zoom* feature to expand that region. That does not work either. Try it.

The idea that the region is between $x = 0$ and $x = 1$ is right. Use the *trace* feature to exhibit (x, y) pairs and see which y-values occur. For example, when $x = \dfrac{1}{2}$, $y = \left(\dfrac{1}{2}\right)^{10} = .00098$, so the y-interval needs to be very small to make small y-values look substantial. The window is [0, 1] by [−.002, .002].

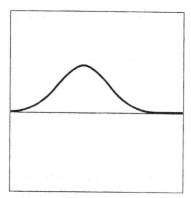

FIGURE 6 $f(x) = x^4(1 - x)^6$.
What is the window?

■ Functions, Ordered Pairs, and Tables

Now you can see why functions and graphs are so closely related. Functions relate first numbers to second numbers and points on graphs have first numbers and second numbers (coordinates). Furthermore, both functions and graphs take these many pairs and make them into one object.

DEFINITION 2-1-2
(OF "FUNCTION")

A <u>function</u> is a set of ordered pairs such that each first-coordinate value is paired with exactly one second-coordinate value.

This is compatible with the "rule" definition given as 1-3-1 where we used the "$f(x)$" notation, which defines a set of all ordered pairs of the form (x, y) that satisfy $y = f(x)$. The set of all first-coordinate values (x's, arguments) is the <u>domain</u>. The set of all second-coordinate values (y's, images) is the <u>range</u>. Therefore, the range of f is the set of all y's for which the equation "$f(x) = y$" has a solution for x.

EXAMPLE 2

Let $f(x) = x^2$. Its range is the set of all real numbers greater than or equal to zero. When you graph this, you see no points with y-values below zero (There are no points below the x-axis).

Let $g(x) = x + 3$. The range is the set of all real numbers. You can always solve for x in $x + 3 = y$.

Let $h(x) = x^2 + 100$. The range is the set of all real numbers greater than or equal to 100. If you want to see the graph, the y-interval of your window must include numbers above 100. ▬

If a function has a simple rule, the "rule" definition in Definition 1-3-1 is simplest. However, some numerical relationships cannot be expressed with convenient rules. Then we can regard the function as a set of ordered pairs, and graph the pairs or list them in a table.

You do not need to memorize anything about the particular example which follows; it is intended merely to illustrate how tables work.

EXAMPLE 3

Table 2-1-3 gives values of the so-called "Normal Distribution" function from statistics. It gives the probability that random "standard normal" numbers are less than or equal to z. Usually denoted by the symbol Φ (the upper case Greek letter phi), this function is fundamental to statistics, but has no convenient algebraic representation. Many calculators do not have a key for it. Consequently, statistics textbooks usually include a table at the back with a selection of argument-image pairs.

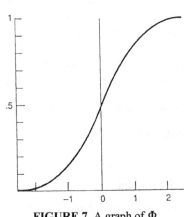

FIGURE 7 A graph of Φ.
[−2.5, 2.5] by [0, 1]

TABLE 2-1-3 The standard-normal cumulative-distribution function

z	$\Phi(z)$	z	$\Phi(z)$	z	$\Phi(z)$	z	$\Phi(z)$
−3	.0014	−1.0	.1587	.4	.6554	1.8	.9641
−2.5	.0062	−.8	.2119	.6	.7257	2	.9772
−2	.0228	−.6	.2743	.8	.7881	2.5	.9938
−1.8	.0359	−.4	.3446	1	.8413	3	.9986
−1.6	.0548	−.2	.4207	1.2	.8849		
−1.4	.0808	0	.5000	1.4	.9192		
−1.2	.1151	.2	.5793	1.6	.9452		

From the table we see the standard-normal cumulative-distribution function value associated with argument 0 is .5. $\Phi(0) = .5$. $\Phi(1) = .8413$. $\Phi(−2) = .0228$.

The value of Φ at unlisted arguments such as 1.32 can be approximated from the given entries. If more accuracy is desired, and occasionally it is, go find a better table! If you want to know the rule expressed as a command, the rule is, "Look it up in Table 2-1-3!"

To solve the equation "$\Phi(x) = c$" means to find the value of x which yields the desired y value, c. For example, to solve "$\Phi(x) = .8849$," find the image .8849 in the "$\Phi(x)$" column of the table and report the corresponding x value. The solution is $x = 1.2$.

> **To evaluate the expression "$f(x)$" is to go from x to y.**
> **To solve the equation "$f(x) = y$" is to go from y to x.**

EXAMPLE 4

Figure 8 is a representative graph of f. Grid lines are one unit apart.
(A) Find $f(2)$. (B) Solve $f(x) = 2$. (C) Solve $f(x) < 0$.

$f(2)$ refers to $x = 2$. Look for the y-value on the vertical line $x = 2$. $f(2) = 2.6$.

"Solve $f(x) = 2$" refers to $y = 2$. Look for the x-value(s) on the horizontal line $y = 2$.

The solution is $x = -4$ or .8 or 3.

"Solve $f(x) < 0$" refers to y-values less than 0—below the x-axis.

The solution is $-3 < x < -1$ or $x > 4$.

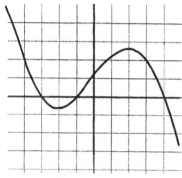

FIGURE 8 A representative graph of f

■ Describing a Function

A function can be described several ways:

1. as a rule
    (A) in English (as a command), or
    (B) in algebraic notation,

2. as a set of ordered pairs
    (A) described in algebraic notation, or
    (B) pictured as a graph, or
    (C) listed in a table.

CONCLUSION

The window strongly affects the appearance of a graph. For example, a narrower x-interval makes a graph appear wider. A shorter y-interval makes a graph appear taller. Learn how to find a window that yields a representative graph.

There are two ways to define the term *function*: as a rule, or as a set of ordered pairs. Each definition says that the image must be determined by the argument.

Terms: graph, window, function, domain, range.

Exercise 2-1

A

A1. ☺ The figure gives a representative graph without giving its algebraic expression.

(A) Evaluate $f(-1)$

(B) Approximate the solution to $f(x) = 2$.

A2. ☺ The figure gives a representative graph without giving its algebraic expression.

(A) Evaluate $f(-2)$

(B) Approximate the solution to $f(x) = 1$.

For A1

For A2

A3. The figure gives a representative graph without giving its algebraic expression.

(A) Find $f(2)$ (B) Solve $f(x) = 3$.

A4. The figure gives a representative graph without giving its algebraic expression.

(A) Find $f(2)$ (B) Solve $f(x) = 3$.

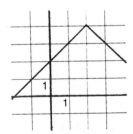

For A3

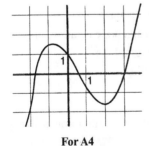

For A4

A5. The figure gives a representative graph of f, without giving its algebraic expression.

(A) Find $f(3)$ (B) Solve $f(x) = -1$.

A6. The figure gives a representative graph of f.

(A) Find $f(1)$ (B) Solve $f(x) = 1$.

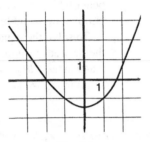

For A5   **For A6**

A7. The figure locates (a, b) and (c, d) on an axis system. Locate (b, a) and (d, c).

A8. The figure locates (a, b) and (c, d) on an axis system. Locate (b, a) and (d, c).

A9. The figure gives (a, b) and (c, d). Sketch the locations of (a, d) and (c, b).

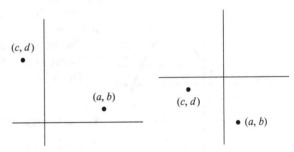

For A7   **For A8–9**

A10. The figure gives (a, b) and (c, d). Find the coordinates of the point in the picture

(A) P (B) Q.

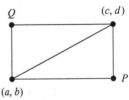

For A10

A11.* The origin is a point on which of these graphs?

(A) x^2 (B) x^3 (C) $\sqrt{x}$

(D) $\dfrac{1}{x}$ (E) $|x|$

A12.* Which of the expressions in the previous problem have the point $(1, 1)$ on their graphs?

The figure (shown below) marks some points on the standard scale. Make three sketches of where they would be in the three given windows (and note if they would be outside the window).

A13. $[-5, 5]$ by $[-10, 10]$; $[-20, 20]$ by $[-10, 10]$; $[-2.5, 2.5]$ by $[-2.5, 2.5]$.

A14. $[-10, 10]$ by $[-5, 5]$; $[-10, 10]$ by $[-20, 20]$; $[-40, 40]$ by $[-40, 40]$.

A15. $[-5, 5]$ by $[-10, 10]$; $[-20, 20]$ by $[-10, 10]$; $[-2.5, 2.5]$ by $[-2.5, 2.5]$.

A16. $[-10, 10]$ by $[-5, 5]$; $[-10, 10]$ by $[-20, 20]$; $[-40, 40]$ by $[-40, 40]$.

A17. $[-5, 5]$ by $[-10, 10]$; $[-20, 20]$ by $[-10, 10]$; $[-2.5, 2.5]$ by $[-2.5, 2.5]$.

A18. $[-10, 10]$ by $[-5, 5]$; $[-10, 10]$ by $[-20, 20]$; $[-40, 40]$ by $[-40, 40]$.

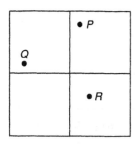

For A13–14

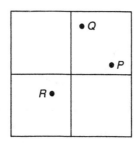

For A15–16

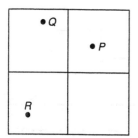

For A17-18

Give the range of f. (You might want to look at the graph.)

A19. (A) $f(x) = x^2$. (B) $f(x) = x^3$.
(C) $f(x) = |x|$. (D) $f(x) = \dfrac{1}{x}$.

A20. (A) $f(x) = x^2 - 7$. (B) $f(x) = \sqrt{x}$.
(C) $f(x) = |x - 3|$. (D) $f(x) = |x| + 6$.

Solve for x using Table 2-1-3.

A21. $\Phi(x) = .9772$. **A22.** $\Phi(x) = .6554$.

Let f be a set of ordered pairs. (A) Is the given f a function? (B) Give the domain as a set. (C) Give the range as a set. (D) Solve $f(x) = 5$.

A23. $f = \{(0, 10), (1, 5), (2, 4)\}$.

A24. $f = \{(1, 5), (2, 5), (2, 10)\}$.

A25. $f = \{(-1, 3), (0, 12), (1, 5)\}$.

A26. $f = \{(0, 4), (0, 6), (0, 5)\}$.

A27. Find the *x*-value that minimizes "$10x^2 - 300x + 10,000$" (from Calculator Exercise 4).

A28. Do Calculator Exercise 1.

Reading and Writing Mathematics

A29. We will call the scale <u>square</u> if the vertical scale and the horizontal scales are the same, that is, one unit is the same distance on each scale. Square scales are important in calculus.
(A) What model of calculator do you have?
(B) Check to see if your calculator has a keystroke sequence which yields a "square" scale. If it does, graph something on a scale which is not square and then try that sequence to see what it does. Does it change the vertical scale or the horizontal scale?

A30.* Which ordered pairs constitute the graph of the function *f*?

A31.* Define "representative" in the context of graphs.

A32.* The range of *f* is the set of all _____ .
Therefore, the range of *f* is the set of all *c* such that the equation _____ has a solution.

A33.* (A) In the context of functions, to evaluate an expression is to go from ____ to ____.
(B) In the equation "$f(x) = y$," to solve is to go from ____ to ____.

B

B1.* (A) Which *x*-interval, $[-10, 10]$ or $[-20, 20]$, makes graphs look wider?
(B) Which *y*-interval, $[-10, 10]$ or $[-5, 5]$, makes graphs look taller?

B2.* (A) Which *y*-interval, $[-10, 10]$ or $[-20, 20]$, makes graphs look shorter?
(B) Which *x*-interval, $[-10, 10]$ or $[-5, 5]$, makes graphs look narrower?

B3–B6. *Suppose you graph f (x) in the standard window ([−10, 10] by [−10, 10]) and the window seems wrong for the given reason. You have the best chance of getting what you want if you (pick one):*

    (A) Change the *y*-interval to [−20, 20]
    (B) Change the *y*-interval to [−5, 5]
    (C) Change the *x*-interval to [−20, 20]
    (D) Change the *x*-interval to [−5, 5].

B3. It seems too wide. You want it to appear narrower in the window.

B4. It seems too tall. You want it to appear shorter in the window.

B5. It seems too narrow. You want it to appear wider in the window.

B6. It seems too short. You want it to appear taller in the window.

B7. The figure gives a representative graph without giving its algebraic expression.
    (A) Evaluate $f(2) - f(0)$
    (B) Solve $f(x) < 0$.

B8. The figure gives a representative graph without giving its algebraic expression.
    (A) Evaluate $f(1) - f(-1)$
    (B) Solve $f(x) > 0$.

B9. The figure gives a representative graph without giving its algebraic expression.
    (A) Evaluate $f(2)$ (B) Solve $f(x) = 1$.
    (C) Solve $f(x) > 2$.

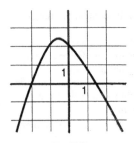

For B7   **For B8**

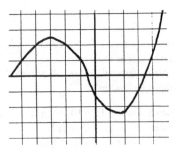

For B9

Problems B10–B24 picture the given graphs in a certain window.

    (A) Find the window (approximately).
    (B) Find the coordinates of *P*. [Great accuracy is not required, just get it close enough to show you know how to do it.]

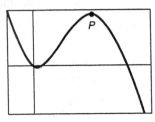

B10. $x^2(120 - 4x)$ **B11.** $x(x - 20)(200 - x)$

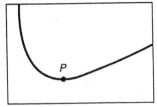

B12. $12x^2 + \dfrac{12{,}000}{x}$ **B13.** $x^2(1 - x)^4$

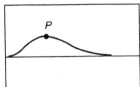

B14. $x^4(1 - x)^{12}$ **B15.** $x^3 - 20x^2 + 50x + 100$

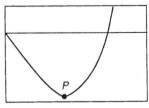

B16. $x(x - 20)(x - 40)$ **B17.** $\sin(3x)$ [in degrees]

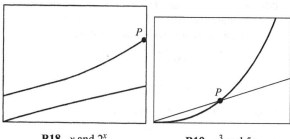

B18. x and 2^x

B19. x^3 and $5x$

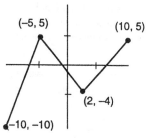

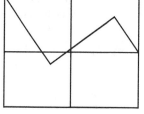

For B26, 28, and 30   **For B31 and 32**

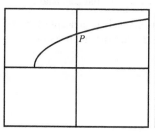

B20. $\log(x - 20)$

B21. $x^2 + 100x$

Sketch the graph (given above in [−10, 10] by [−10, 10])
in the requested window.

B25. In [−5, 5] by [−10, 10].

B26. In [−5, 5] by [−10, 10].

B27. In [−10, 10] by [−5, 5].

B28. In [−10, 10] by [−5, 5].

B29. In [−2.5, 2.5] by [−2.5, 2.5].

B30. In [−2.5, 2.5] by [−2.5, 2.5].

B31. In [5, 10] by [5, 10].

B32. In [−5, 0] by [−10, 0].

B33. Look at the graph of $f(x) = \sqrt{x}$ on the standard
scale.
  (A) Which new x-interval would make the graph
      exit the window at the upper right corner?
  (B) Which new y-interval would make the graph exit
      the window at the upper right corner?

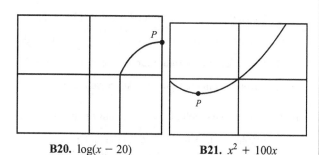

B22. $20\sqrt{x + 1}$

B23. $\dfrac{1}{x^2 - x}$

B34. Look at the graph of $f(x) = x^3$ on the standard
scale.
  (A) Which new x-interval would make the graph
      exit the window at the upper right corner?
  (B) Which new y-interval would make the graph exit
      the window at the upper right corner?

B35. Look at the graph of $y = x$ on the standard scale.
Then change the x-interval to make it look steeper.
  (A) Which x-interval, [−20, 20] or [−5, 5], makes it
      look steeper? Why?
  (B) Now return to the standard x-interval and
      change the y-interval to make it look steeper.
      Which y-interval, [−20, 20] or [−5, 5], makes it
      look steeper? Why?

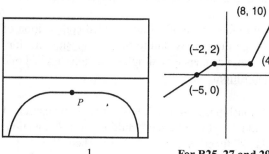

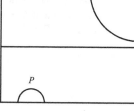

B24. $\dfrac{1}{(x - 10)(x - 50)}$

For B25, 27 and 29

B36. Look at the graph of $y = |x|$ (the absolute value
function) on the standard scale. Then change the
x-interval to make it look steeper.
  (A) Which x-interval, [−20, 20] or [−5, 5], makes it
      look steeper?

(B) Now return to the standard x-interval and change the y-interval to make it look steeper. Which y-interval, $[-20, 20]$ or $[-5, 5]$, makes it look steeper?

B37. The graph of $x^3 - 40x^2 - 165x + 1300$ crosses the x-axis in three places. Sketch a representative graph of it in a window that shows all three places and all the y-values between those zeros.

B38. During a time period the price of gold rose from $320 per ounce to $385 per ounce and in between it fluctuated up and down. Suppose you must graph this. (A) If you want to make the gain seem large, how should you pick the y-interval? [Suggest a specific interval.] (B) If you want to make the price seem relatively stable, how should you pick the y-interval?

B39. During a time period the price of silver rose from $4.19 per ounce to $4.87 per ounce and in between it fluctuated up and down. Suppose you must graph this. (A) If you want to make the gain seem large, how should you pick the y-interval? [Suggest a specific interval.] (B) If you want to make the price seem relatively stable, how should you pick the y-interval?

B40. (A) Sketch, in a rectangle you draw, a "representative" graph of $y = x^3 - 20x^2 + 150$.
(B) Give the x and y-intervals you used.
(C) For $x > 0$, find the minimum value of y [Great accuracy is not required, just get it close enough to show you know how to do it.]

B41. Suppose you wished to solve $f(x) = 4.32$. Inspect their graphs to determine which of the following functions would yield a unique solution.

(A) x^2 (B) x^3 (C) $\dfrac{1}{x}$

B42. In the figure for A2, solve for x: $f(x) = 2f(3)$.

B43. In the figure for A4, solve for x: $f(x) = 2f(0)$.

Use inverse-reverse to solve these for x.

B44. Given the f in the figure for A6, solve $f\left(\dfrac{x}{3}\right) = 1$.

B45. Given the f in the figure for A6, solve $f(2x) = 3$.

B46. Given the f in the figure for A6, solve $f(x + 1) = 3$.

B47. Given the f in the figure for A6, solve $f(x - 2) = 1$.

B48. Given the f in the figure for A6, solve $f(x + 1) = 1$.

B49.* Look at the graph of $\dfrac{1}{x - 1}$ on your calculator. Some graphics calculators include in the plot a vertical line near $x = 1$ when the graph does not have one. Why?

B50.* Look at the graph of $\dfrac{x + 1}{x - 3}$ on your calculator. Some graphics calculators include in the plot a vertical line near $x = 3$ when the graph does not have one. Why?

Section 2-2   COMPOSITION AND DECOMPOSITION

How many graphs do you know by heart? The purpose of this section is to greatly increase that number by including graphs obtained by applying functions one after another.

Applying a second function to the image of a first is called <u>composition</u> of functions. The resulting two-stage function is said to be a <u>composite</u> function. Rewriting a composite function as a sequence of simpler functions applied one after the other is called <u>decomposition</u>.

EXAMPLE 1   The Celsius-to-Fahrenheit relationship is $F = 1.8C + 32$. It is a composition of "Multiply by 1.8" and then "Add 32." Let g be the "Multiply by 1.8" function: $g(x) = 1.8x$. Let f be the "Add 32" function: $f(x) = x + 32$. Then

$$f(g(x)) = f(1.8x) = 1.8x + 32$$

Reading it, you see f first. Nevertheless, g is applied first because it is inside the parentheses.

Order matters. This relationship says to multiply by 1.8 *first* and *then* add 32. In the other order, $g(f(x)) = g(x + 32) = 1.8(x + 32) = 1.8x + 57.6$, which is not the Celsius-to-Fahrenheit relationship. Order matters. ▬

EXAMPLE 2   The formula for the area of a circle given its radius is $A = \pi r^2$. Define two functions, f and g, such that A is a composition of them: $A(r) = f(g(r))$.

Solution: To evaluate A, we "Square and then multiply by π." Write the two components, with g first. $g(r) = r^2$ (or $g(x) = x^2$, the variable is a placeholder). $f(x) = \pi x$. Then $f(g(r)) = \pi r^2$, the area function. ▬

A single *sequence* of operations can be regarded as one thing (for which the term *function* is appropriate), or as several things—several operations—one after another. Decomposition is essential for understanding complicated functions.

EXAMPLE 3   Let $h(x) = (x - 3)^5$. Find $f(x)$ and $g(x)$ such that $h(x) = f(g(x))$.

h is "Subtract 3 and then take the fifth power." In "$f(g(x))$", g is first. So, let $g(x) = x - 3$ [g is "Subtract 3"], and $f(x) = x^5$ [f is "Take the fifth power"]. Then $f(g(x)) = f(x - 3) = (x - 3)^5$, as desired. ▬

EXAMPLE 4   Decompose $k(x) = \sqrt{x^2 + 9}$ into as many components as are appropriate.

Think functionally. The rule is "Square, then add 9, then take the square root." Use three functions so $k(x) = f(g(h(x)))$. Let $h(x) = x^2$, $g(x) = x + 9$, and $f(x) = \sqrt{x}$. ▬

EXAMPLE 5   Let $f(x) = 2x + 1$ and $g(x) = x^2$. Find $f(g(x))$.

$f(g(x)) = f(x^2) = 2(x^2) + 1 = 2x^2 + 1$. This is not the product of the two expressions: $(2x + 1)(x^2)$. Composition is not multiplication.

Order matters, so the other order would yield a different result:

$$g(f(x)) = g(2x + 1) = (2x + 1)^2 = 4x^2 + 4x + 1.$$ ▬

EXAMPLE 6   Let $f(x) = \sqrt{x}$. Let $g(x) = x + 4$. Express $f(g(x))$.

f is "Take the square root." g is "Add 4." In the expression "$f(g(x))$," the first function is g, not f. $f(g(x)) = \sqrt{x + 4}$. The extended square root symbol does the grouping of "$x + 4$" before the square root is taken.

The parentheses in "$f(g(x))$" do *not* refer to multiplication. Similar notations can have different meanings. "$f(\ldots)$" signals the application of a function, but "$3(\ldots)$" signals multiplication. ▬

EXAMPLE 7   Analyze $h(x) = (x + 4)\sqrt{x}$.

Composition refers to *successive* operations, beginning with the argument. $h(x)$ is a product, not a composition. ▬

■ Composition for Calculus

Composition and decomposition are used almost every day in calculus. One of the important compositions occurs in the expression known as the "difference quotient" of f,

$$\frac{f(x + h) - f(x)}{h}.$$

This gives the slope of the line through two points on the graph of $f(x)$ (Figure 1). One point is arbitrary: $(x, f(x))$. The other point is also arbitrary: $(x + h, f(x + h))$. The numerator gives the difference of the y-values (the "rise"). The denominator gives the difference in x-values (the "run"). The quotient is, therefore, the "difference quotient."

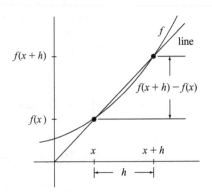

FIGURE 1 $f(x)$ and a line through two arbitrary points

EXAMPLE 8 Express and simplify the difference quotient if $f(x) = x^2$.

$$\frac{f(x + h) - f(x)}{h} = \frac{(x + h)^2 - x^2}{h}$$

$$= \frac{x^2 + 2xh + h^2 - x^2}{h}$$

$$= \frac{2xh + h^2}{h}$$

$$= 2x + h, \text{ if } h \neq 0.$$

In examples like this students sometimes make the mistake of treating $f(x + h)$ as $f(x) + h$. Don't! Order matters.

In calculus we then "Let h go to zero." As h goes to zero, the difference quotient, $2x + h$, goes to $2x$. (We may not just let h <u>be</u> zero in the original difference quotient, because the quotient would be the undefined expression "0/0".) In calculus, this process defines the "derivative" of f. The derivative of x^2 is $2x$. ▬

■ Composition and Graphs

Composition of a given function with addition, subtraction, multiplication, or division produces a simple change in its graph.

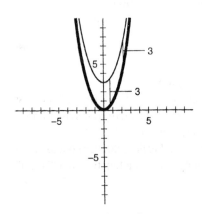

FIGURE 2 x^2 and $x^2 + 3$.
$[-10, 10]$ by $[-10, 10]$

EXAMPLE 9 The graph of $f(x) = x^2$ is well-known (Figure 2, bold). Give the graph of $h(x) = x^2 + 3$.

The graph is 3 units up.

The order in h has "Add 3" *after* the squaring function f creates its images. The "Add 3" function affects the images—the heights. Similarly, the graph of "$x^2 - 5$" would be moved down five units. ▬

> **Adding or subtracting *after* applying *f*
> changes the vertical location.**

EXAMPLE 10   Given the graph of $|x|$ (Figure 3, bold), graph $|x| - 6$.

The graph of "$|x| - 6$" is simply down 6 units, since "Subtract 6" applies to the images on the vertical scale (Figure 3).

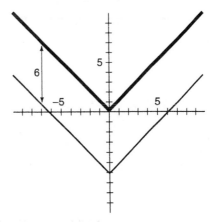

FIGURE 3 $|x|$ and $|x| - 6$. $[-10, 10]$ by $[-10, 10]$ ▬

EXAMPLE 11   Given the graph of $\sin x$ (Figure 4, bold), find the graph of $h(x) = 2 \sin x$.

The image of h is 2 times the image of sine, so the graph of h is magnified vertically by a factor of 2 (Figure 4).

This expands the graph away from the horizontal axis. This is not quite the same as making it "taller," since multiplication of negative numbers by 2 makes them even more negative (that is, even lower).

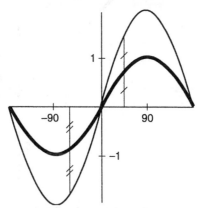

FIGURE 4 $\sin x$ [bold] and $2 \sin x$.
$[-180°, 180°]$ by $[-2, 2]$

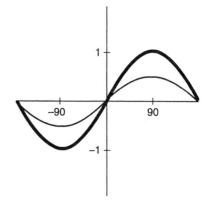

FIGURE 5 $\sin x$ [bold] and $\dfrac{\sin x}{2}$

$[-180°, 180°]$ by $[-2, 2]$ ▬

EXAMPLE 12   Define g by $g(x) = \dfrac{\sin x}{2}$. Graph g.

The images of "$\sin x$" are divided by 2. Points are closer to the horizontal axis (Figure 5).   ▬

Multiplying or dividing *after* applying *f* changes the vertical locations.

EXAMPLE 13   Let $f(x) = x^2$. Graph $g(x) = -f(x)$.

$g(x) = -x^2$. The sign of the images is changed, so that positive y-values become negative. The graph of x^2 is turned upside down (Figure 6). It is reflected through the x-axis.

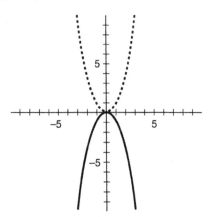

FIGURE 6 $f(x) = x^2$ [dashed] and $g(x) = -f(x) = -x^2$. $[-10, 10]$ by $[-10, 10]$ ▬

In the previous examples composition affected the image. Only the heights were changed. When composition affects the argument, the horizontal coordinate is affected and the relationship to the original graph may be surprising.

EXAMPLE 14   Given the graph of $|x|$, graph $|x - 6|$.

It is 6 units to the *right* of the graph of $|x|$ (Figure 7).

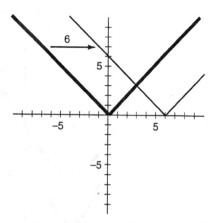

FIGURE 7 $|x|$ [bold] and $|x - 6|$ $[-10, 10]$ by $[-10, 10]$

Why would subtraction move a graph to the *right*? (Usually we think of sub-traction as moving numbers to the left.)

The graph of $|x|$ is well-known. It has a corner (vertex) at $(0, 0)$. When the argument is zero, the image is zero. Where is the vertex on the graph of $|x - 6|$?

The argument of this absolute-value function is "$x - 6$", which must be 0 at the vertex. Therefore, $x = 6$ at the vertex (instead of $x = 0$). The vertex is 6 units to the *right* of the vertex of $|x|$ (Figure 7). The fact that subtraction *before* apply-ing a function produces a shift *right* is surprising. ▬

CALCULATOR Graph "|x|" and "|x + 3|." Why is the vertex of |x + 3| where it is?
EXERCISE 1

Adding or subtracting *before* applying *f* changes the horizontal locations.

EXAMPLE 15   Let $f(x) = x^2$. Graph $f(x + 4) = (x + 4)^2$.
It is 4 units to the *left* (Figure 8). Why?
A key point on the graph of $f(x) = x^2$ is $(0, 0)$, the origin. $f(0) = 0^2 = 0$. In the expression "$f(x + 4)$" when will the argument of f be zero?
For "$x + 4$" to be 0, x must be -4. When $x = -4, f(x + 4) = (x + 4)^2 = (-4 + 4)^2 = 0$. The $(0, 0)$ point on x^2 occurs at $(-4, 0)$ on $(x + 4)^2$. Therefore the graph is shifted *left* 4 units (Figure 8). The fact that *addition* before applying *f* is associated with a shift *left* is surprising (since we usually think of addition as moving numbers to the right).

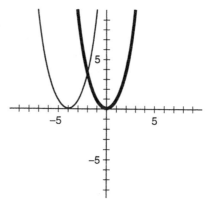

FIGURE 8 x^2 [bold] and $(x + 4)^2$. $[-10, 10]$ by $[-10, 10]$

EXAMPLE 16   Given the graph of sin x (Figure 9, bold), graph $g(x) = \sin(2x)$.
It is closer to the y-axis. Why?
"$\sin(2x)$" differs from "$\sin x$" only in its argument. Only the horizontal positions are changed. The images (heights) are the same, but they correspond to different x-values. For example, on the graph of sin(2x) the image "sin 180" will occur when $2x = 180$, that is, $x = 90$. The image "sin = 90" occurs when

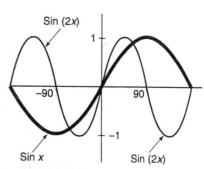

FIGURE 9 sin x [bold] and sin(2x). $-180° \le x \le 180°$

$2x = 90$, so $x = 45$. Every point on the $\sin x$ graph corresponds to a point on the new graph with half the old x-value. The $\sin x$ graph is compressed towards the y-axis to form the $\sin(2x)$ graph. ▬

Multiplication before applying f changes the horizontal scale.

EXAMPLE 17 Figure 10 (bold) graphs $f(x)$. Graph $f(-x)$.

The graph of $f(-x)$ is the graph of $f(x)$ reflected through the y-axis (Figure 10).

In $f(-x)$ the argument is multiplied by negative one before the function is applied. Therefore, on the graph of $f(-x)$ the image "$f(2)$" is not plotted above $x = 2$, but above $x = -2$. All the images normally associated with positive x-values become associated with negative x-values, and vice versa. Therefore, it is a mirror image of the original graph through the y-axis (Figure 10).

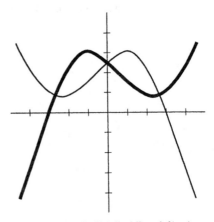

FIGURE 10 $f(x)$ [bold] and $f(-x)$ ▬

Composition With Two Arithmetic Functions
Some graphs combine two location shifts or two scale changes.

EXAMPLE 18 The graph in Figure 11 has the shape, but not the location, of the graph of x^2. Identify the expression graphed.

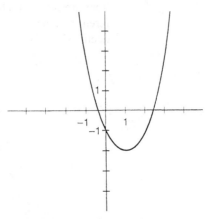

FIGURE 11 The graph of x^2, shifted: $(x - 1)^2 - 2$. $[-5, 5]$ by $[-5, 5]$

The graph is the graph of x^2 shifted to the right 1 and down 2. Its expression is $(x - 1)^2 - 2$. Beginning with "x", subtracting 1 before squaring shifts it to the right, and subtracting 2 after squaring moves it down 2.   ▬

We can also identify scale changes.

EXAMPLE 19   The graph in Figure 12 has the shape, but not the scale, of the graph of sin x. Identify the expression graphed.

The vertical scale is twice that of the sine function, so the images are multiplied by 2. Also, the horizontal scale is compressed by a factor of three (changes happen three times as fast), so the argument is multiplied by 3. The expression is "2 sin($3x$)."

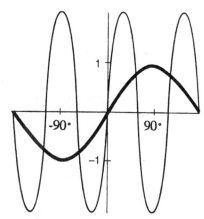

FIGURE 12 sin x [bold] and 2 sin($3x$). [$-180°$, $180°$] by [$-2, 2$]   ▬

EXAMPLE 20   Consider the graph of $f(x)$ given in Figure 13. How does the graph of $\dfrac{f(x + 4)}{2}$ compare to it?

For $\dfrac{f(x + 4)}{2}$, note that "Add 4" applies to the argument, so it causes a shift left of 4 units (Figure 14, dashed curve). After f is applied, "Divide by 2" is applied to the images. The graph is compressed vertically by a factor of 2 (Figure 14, solid curve).

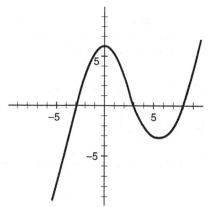

FIGURE 13 Graph of $f(x)$. [$-10, 10$] by [$-10, 10$]

FIGURE 14 Graphs of $f(x + 4)$ [dashed] and $\dfrac{f(x + 4)}{2}$.
[$-10, 10$] by [$-10, 10$]   ▬

EXAMPLE 21 Suppose we know $f(4) = 7$. So we know the point $(4, 7)$ is on the graph of $f(x)$. Find a point on the graphs of $f(x) + 2$, $-f(x)$, $f(x + 1)$, $f\left(\dfrac{x}{2}\right)$, and $f(-x)$.

function	point	
$f(x)$	$(4, 7)$	
$f(x) + 2$	$(4, 9)$	
$-f(x)$	$(4, -7)$	
$f(x + 1)$	$(3, 7)$	[The given argument of f is 4, so $x + 1 = 4$ and $x = 3$.]
$f\left(\dfrac{x}{2}\right)$	$(8, 7)$	[The given argument of f is 4, so $\dfrac{x}{2} = 4$ and $x = 8$.]
$f(-x)$	$(-4, 7)$	[For the argument of f to be 4, x must be -4.] ▬

EXAMPLE 22 The amount of any radioactive substance decays over time as in Figure 15. Decay is rapid in a nuclear reactor. Decay of Carbon 14, used to date archaeological materials, is slow. The *shape* in Figure 15 is constant, but the *scale* varies from example to example.

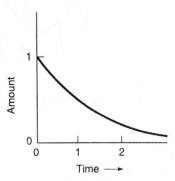

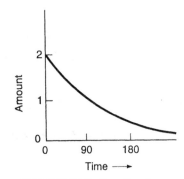

FIGURE 15 Radioactive decay. $y = \left(\dfrac{1}{2}\right)^{t}$

$A(t)$ is the amount remaining at time t

FIGURE 16 Radioactive decay.

$A(0) = 2$, and $A(90) = 1 = \left(\dfrac{1}{2}\right)A(0)$

How can we rescale the function in Figure 15 to make Figure 16, in which there are initially two (instead of 1) grams of radioactive substance and half the original amount will remain at time 90 (instead of at time 1)?

The basic shape is $f(t) = \left(\dfrac{1}{2}\right)^{t}$ and we want to change the scale. We want it twice as tall and 90 times as wide. Use $2f\left(\dfrac{t}{90}\right)$.

$A(t) = 2\left(\dfrac{1}{2}\right)^{t/90}$. When $t = 90$, the exponent will be $\dfrac{90}{90} = 1$. There will be a factor of $\left(\dfrac{1}{2}\right)^{1} = \dfrac{1}{2}$, as desired. ▬

The following tables summarize the effects of composition on graphs.

Notation $(c > 0)$	Change in graph	Operation applied to the . . .
$f(x) + c$	shift up c	
$f(x) - c$	shift down c	image of f (changes are vertical)
$cf(x)$ $(c > 1)$	expand vertically away from the x-axis by a factor of c	
$\dfrac{f(x)}{c}$ $(c > 1)$	contract vertically toward the x-axis by a factor of c	
$-f(x)$	flip upside down (reflect) through the x-axis	
$f(x + c)$	shift left c	
$f(x - c)$	shift right c	argument of f (changes are horizontal)
$f(cx)$ $(c > 1)$	*compress* horizontally toward the y-axis with a factor of c	
$f\left(\dfrac{x}{c}\right)$ $(c > 1)$	*expand* horizontally away from the y axis by a factor of c.	
$f(-x)$	reflect through the y-axis	

Type of arithmetic operation	Applied to the	Effect on the graph
add or subtract $c > 0$	argument	shift left or right
	image	shift up or down
multiply by $c > 1$	argument	*compress* horizontally
	image	expand vertically
divide by $c > 1$	argument	*expand* horizontally
	image	compress vertically

CONCLUSION Composite functions consist of simpler functions applied in sequence. The *argument-operation-image* order is critical. In "$f(g(x))$," g is applied first and then f. The image of the first function becomes the argument of the second. **Order Matters!**

Composition with addition, subtraction, multiplication, or division produces either a shift or a scale change. **Operations applied *after* f affect the vertical. Operations applied *before* f affect the horizontal in unexpected ways. "Add c" and "*left* c" are a surprising combination. Similarly, the graph of f(2x) is *half* as wide as the graph of f(x). While 2f(x) is expanded vertically, f(2x) is compressed horizontally.**

Terms: composition, composite, decompose, location, scale.

Exercise 2-2

A

A1.* ☺ The notation "5(. . .)" signals multiplication, but the notation "g(. . .)" signals something else. What?

A2. ☺ Let $f(x) = 2x$, Find $f(f(3))$.

A3. ☺ Let $f(x) = 3x + 1$. Find $f(f(2))$.

A4. ☺ Let $g(x) = x^2$. Find $g(g(3))$.

A5. ☺ Let $g(x) = 2x + 3$. Find $g(g(2))$.

A6. ☺ Let $h(x) = x^2 - 2$. Find $h(h(3))$.

A7. Let $f(x) = x^2$ and $g(x) = 2x + 7$.

　(A) Find $f(g(3))$　(B) Find $g(g(4))$　(C) Find $f(f(3))$

A8. Let $f(x) = 3x - 1$ and $g(x) = \dfrac{x + 1}{3}$.

　(A) Find $f(g(5))$　(B) Find $g(f(7))$　(C) Find $f(f(2))$

☺ *Give the vertex of the graph.*

A9. (A) $|x - 2|$　(B) $|x| + 2$　(C) $|x + 5|$
　　(D) $|x| - 7$

A10. (A) $x^2 - 9$　(B) $(x - 8)^2$　(C) $(x + 3)^2$
　　(D) $x^2 + 4$

☺ *Compare the graph to the graph of* $|x|$:

A11. (A) $|x| + 7$　(B) $|x + 3|$

A12. (A) $|x - 4|$　(B) $|x| - 2$

A13. Do Calculator Exercise 1.

B

☺ **Graphs.** *Compare the graphs of the given expression to the graph of* $f(x)$. *[Assume* $c > 1$*].*

B1.* $f(x) + c$　　　**B2.*** $cf(x)$

B3.* $f(x + c)$　　　**B4.*** $f(x) - c$.

B5.* $f(x - c)$　　　**B6.*** $\dfrac{f(x)}{c}$

B7.* $f(cx)$　　　　**B8.*** $f(-x)$

B9.* $-f(x)$　　　　**B10.*** $f\left(\dfrac{x}{c}\right)$

Notation

B11. ☺ (A) In the definition "Let $f(x) = x^2$," what is being defined?
　　(B) In that definition, what type of variable is "x"?
　　(C) In the composite function $f(g(x))$, which component function is applied to x first?

B12. ☺ (A) Distinguish clearly between "$f(x + 2)$" and "$f(x) + 2$".
　　(B) Give an illuminating example (your choice) of the difference.

Composition

B13. Let $f(x) = x + 5$. Find and simplify $f(f(x))$.

B14. Let $g(x) = 3 - 2x$. Find and simplify $g(g(x))$.

B15. Let $f(x) = 2x - 1$ and $g(x) = 3x + 2$. Find and simplify $f(g(x))$.

B16. Let $f(x) = 4x + 2$ and $g(x) = x + 3$. Find and simplify $f(g(x))$.

B17. Let $g(x) = 2x + 5$. Find and simplify $g(g(x))$.

B18. Let $f(x) = 3x - 1$. Find and simplify $f(f(x))$.

☺ *Express* $h(x) = f(g(x))$ *if*

B19. $f(x) = x^2$ and $g(x) = x + 1$

B20. $f(x) = 2x$ and $g(x) = x + 1$

B21. $f(x) = x^2$ and $g(x) = \ln x$

B22. $f(x) = e^x$ and $g(x) = 4x$

B23. $f(x) = \sin(2x)$ and $g(x) = x^2$

B24. $f(x) = x^2$ and $g(x) = x + 5$

☺ **Decompose** $h(x)$ *into* $f(g(x))$ *by identifying simpler* $f(x)$ *and* $g(x)$.

B25. $h(x) = |5x|$　　　**B26.** $h(x) = 3x + 2$

B27. $h(x) = \dfrac{x - 2}{3}$　　**B28.** $h(x) = \ln(2x)$

B29. $h(x) = 3e^x$　　　**B30.** $h(x) = (5x)^2$

B31. $h(x) = 2x^2$　　　**B32.** $h(x) = \sqrt{x + 7}$

In the following problems you need not simplify. Just give the correct composition.

B33. Let $f(x) = 2x + 1$. Give $f(3x - 5)$.

B34. Let $f(x) = 2(x + 4)$. Give $f(x + 1)$.

B35. Let $f(x) = 3x - 2$. Give $f(2x - 1)$.

B36. Let $f(x) = 2x + 5$. Give $f(f(x))$.

[Inverses] Express $f(g(x))$ *and then* $g(f(x))$ *if*

B37. $f(x) = 2x + 1$ and $g(x) = \dfrac{x - 1}{2}$

B38. $f(x) = \dfrac{x}{5} + 3$ and $g(x) = 5(x - 3)$

Find and simplify the "difference quotient." (See above Example 8.)

B39. Let $f(x) = 3x$.

B40. Let $f(x) = 2x + 4$.

B41. Let $f(x) = (x + 1)^2$.

B42. Let $f(x) = \dfrac{1}{x}$.

Graphs

☺ *Give the vertex of the graph.*

B43. $|x - 3| + 4$

B44. $|x + 2| - 5$

B45. $(x - 8)^2 - 1$

B46. $(x + 3)^2 + 2$

B47. The graph is the graph of $|x|$, shifted. Give the expressions graphed in Figures (A) and (B).

B48. The graph the graph of $|x|$, shifted. Give the expressions graphed in Figures (C) and (D).

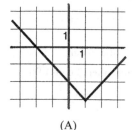

(A)

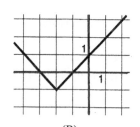

(B)

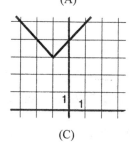

(C)

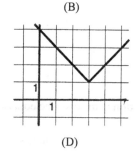

(D)

B49. Suppose $f(x)$ is graphed in Figure (A). Sketch the graph of $f(-x)$.

B50. Suppose $f(x)$ is graphed in Figure (B). Sketch the graph of $-f(x)$.

B51. Suppose $f(x)$ is graphed in Figure (C). Sketch the graph of $f(2x)$.

B52. Suppose $f(x)$ is graphed in Figure (D). Sketch the graph of $f\left(\dfrac{x}{2}\right)$.

B53. Suppose $f(x)$ is graphed in Figure (A). Sketch the graph of $f(x + 2)$.

B54. Suppose $f(x)$ is graphed in Figure (B). Sketch the graph of $f(x + 1) + 2$.

B55. Suppose $f(x)$ is graphed in Figure (C). Sketch the graph of $f(x - 2) - 3$.

B56. Suppose $f(x)$ is graphed in Figure (D). Sketch the graph of $f(x + 1) + 1$.

Here is a representative graph of $f(x)$ in the standard window: Reproduce it and then

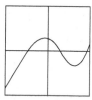

B57. On the same axis system, sketch the graph of $f(x) + 4$.

B58. On the same axis system, sketch the graph of $2f(x)$.

B59. On the same axis system, sketch the graph of $f(x - 3)$.

B60. On the same axis system, sketch the graph of $f\left(\dfrac{x}{2}\right)$.

B61. On the same axis system, sketch the graph of $f(2x)$.

B62. On the same axis system, sketch the graph of $f(x + 4)$.

☺ *Find the function with the graph with the given properties.*

B63. Twice as far from the x-axis as $y = \sin x$.

B64. Three times as far from the x-axis as $y = \log x$.

B65. Half as far from the x-axis as $y = 10^x$.

B66. One fifth as far from the x-axis as $y = |x + 2|$.

B67. Two units to the left of $y = x^2$.

B68. Three units to the right of $y = 3x - 5$.

B69. Twice as far from the y-axis as $y = 10^x$.

B70. Half as far from the y-axis as $y = \log x$.

☺ *How should we modify "$y = \sin x$" to produce a graph with*

B71. twice as many waves of the same height?

B72. half as many waves of the same height?

B73. twice as many waves, each three times as tall?

B74. half as many waves, each one quarter as tall?

☺ *How should we modify "$y = x^2$" to produce a graph which is shifted*

B75. left 5 units?

B76. up 3 units?

B77. down 2 units and to the left 5 units?

B78. left 4 units and turned upside down?

B79. Given the graph of $f(x)$, explain why the graph of $f(x + 4)$ is shifted 4 units *left*.

B80. Given the graph of $f(x)$, explain why the graph of $f(x - 5)$ is shifted 5 units *right*.

B81. Given the graph of $f(x)$, explain why the graph of $f(2x)$ is *half* as wide.

B82. Given the graph of $f(x)$, explain why the graph of $f\left(\dfrac{x}{2}\right)$ is *twice* as wide.

B83. Suppose $x = 3$ solves $f(x) = 0$. Find a solution to $f(2x) = 0$.

B84. Suppose $x = 5$ solves $f(x) = 0$. Find a solution to $f(x - 1) = 0$.

B85. ☺ If $f(x)$ is zero when $x = 7$, when is the value of the following function zero?

(A) $f(3x)$ (B) $f(x - 4)$ (C) $f\left(\dfrac{x}{2}\right)$ (D) $2f(x)$

B86. ☺ If $f(x)$ is zero when $x = -3$, when is the value of the following function zero?
(A) $f(x + 2)$ (B) $f(-x)$ (C) $f(2x)$ (D) $3f(x)$

B87. ☺ If $f(2) = 4$, find one point on the graph of

(A) $f(x + 3)$ (B) $f(x) + 2$ (C) $f(2x)$

B88. ☺ If $f(-2) = 5$, find one point on the graph of

(A) $f(x - 3)$ (B) $f(x) - 4$ (C) $f\left(\dfrac{x}{2}\right)$

B89. Let $f(x) = 3x - 5$. Solve for x in "$f(x + 1) = 2f(x)$."

B90. Let $f(x) = (x - 1)^2$. Solve for x in "$f(x + 2) = 4f(x)$."

Graphs

B91. (A) Suppose f is such that $f(x) = f(-x)$ for all x. Explain why the graph is symmetric about the y-axis.
(B) Find a different condition on f that makes the graph symmetric about the line $x = c$.

Find the function g(t) with graph that fits Figure 15, however with

B92. $g(0) = 3$ and half remaining after 6 units of time.

B93. $g(0) = .05$ and half remaining after .3 units of time.

Section 2-3   INVERSES

The problem in this section is to solve the equation "$f(x) = c$." What do you do to find x?

If the function f is not too complicated, you may solve for x by doing some steps. The steps determine the so-called "inverse." The function takes x and yields c. Its inverse takes c and yields x. The inverse solves the equation.

If there is exactly one solution for each value of c, then the inverse function gives it: $x = f^{-1}(c)$. Read this "x equals f inverse of c." The superscript "-1" is not a power when it is on a function—it is a special notation for *inverse*. If there is more than one solution, there are complications that we will discuss shortly.

EXAMPLE 1   To solve "$x + 5 = c$," subtract 5.

The inverse of "Add 5" is "Subtract 5."

The inverse of any f is denoted by f^{-1} [The notation "f^{-1}" is read "f inverse".] When $f(x) = x + 5, f^{-1}(x) = x - 5$.   ▬

EXAMPLE 2   How do you solve "$\dfrac{x}{2} + 3 = 17$" and "$\dfrac{x}{2} + 3 = 98.4$"?

By inverse-reverse thinking, the method is to "Subtract 3 and then multiply by 2." Therefore, the inverse of "Divide by 2 and add 3 " is "Subtract 3 and multiply by 2."

When $f(x) = \dfrac{x}{2} + 3, f^{-1}(x) = 2(x - 3)$.

Figure 1 illustrates why the *inverse-reverse* method is sometimes called "doing and undoing."

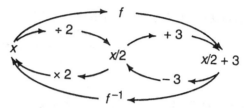

FIGURE 1 A "function loop" diagram for $f(x) = \dfrac{x}{2} + 3$ and its inverse. f takes x to $f(x)$ along the top. f^{-1} returns x from $f(x)$ along the bottom

The inverse undoes what the function does (Figure 2). The inverse expresses the steps for solving the equation "$f(x) = y$". ▬

To find f^{-1}, solve $f(x) = y$ for x. The solution is $f^{-1}(y)$.

(If you want $f^{-1}(x)$, simply switch letters when you are done.)

EXAMPLE 2 AGAIN

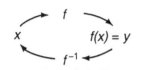

FIGURE 2 A function loop

Find the inverse of f when $f(x) = \dfrac{x}{2} + 3$.

To find the inverse, set $\dfrac{x}{2} + 3 = y$ and solve for x. The solution will be $f^{-1}(y)$.

$$\frac{x}{2} + 3 = y$$

$$\frac{x}{2} = y - 3$$

$$x = 2(y - 3) = f^{-1}(y).$$

This inverse function exhibits the steps. The letter y is simply a placeholder, so it could be replaced by x. $f^{-1}(x) = 2(x - 3)$. ▬

Some functions have inverses with special names.

EXAMPLE 3 Solve $x^2 = 20$.

$$x = \sqrt{20} = 4.472 \text{ or } x = -\sqrt{20} = -4.472.$$

The square root function is used, but it is not enough. The trick is that there are two solutions, and the square root function yields only one. ▬

THEOREM 2-3-1 (INVERSE SQUARE RELATION)

$x^2 = c$ iff $x = \sqrt{c}$ or $x = -\sqrt{c}$.

The square root symbol, $\sqrt{\ }$, denotes an inverse *function*, and functions, by definition, yield only one output. (If there is more than one output, as here where c yields two x's, the term *relation* applies instead of *function*. Any set of ordered pairs is a <u>relation</u>, but only those with unique images are also functions.) Calculator keys operate as functions – they give one answer. If there should be two, as when solving "$x^2 = 4$," the inverse key (the square-root key) gives only *one*: $\sqrt{4} = 2$. You must think of the other one yourself.

For an inverse of a function to be a function (and therefore yield only one answer), the original function must be "one-to-one."

DEFINITION 2-3-2 A function is said to be <u>one-to-one</u> if every image (*y*-value) occurs exactly once. [Different *x*-values always yield different *y*-values, so there can not be two solutions to "$f(x) = y$." Formally, "if $x_1 \neq x_2$, then $f(x_1) \neq f(x_2)$," or "if $f(x_1) = f(x_2)$, then $x_1 = x_2$."]

EXAMPLE 4 Let $f(x) = x^2$. The function f is not one-to-one because some images appear twice. For example, the equation "$x^2 = 4$" has two solutions because 4 is both $f(2)$ and $f(-2)$ (Figure 3).

Graphically it is easy to see if a function is one-to-one. $f(x) = x^2$ is not one-to-one because horizontal lines above the *x*-axis intersect the graph twice (Figure 3).

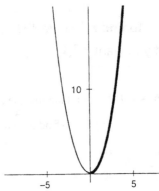

FIGURE 3 $y = x^2$ with the region where $x = \sqrt{y}$ emphasized.
[−10, 10] by [0, 20]

THE HORIZONTAL-LINE TEST 2-3-3 The function f is not one-to-one if and only if there is a horizontal line which intersects the graph of f more than once.

When a function (like squaring) is not one-to-one, there is a trick to its inverse. The inverse *function* must be defined only on a limited domain where the original function *is* one-to-one. The inverse function will find one solution *in that limited domain*, but it will not find any other solutions outside that domain. For example, the square root function does not find the negative solution to $x^2 = c$ because the inverse function (the square root function), uses only $x \geq 0$ for the domain of x^2 (so that x^2 is one-to-one there, Figure 3).

THEOREM 2-3-4 If a function f is one-to-one, then its inverse is a function and any solution to the equation "$f(x) = c$" is unique and is given by $x = f^{-1}(c)$.

If a function f is not one-to-one on its domain, but is one-to-one on domain *D*, then an inverse function can be defined for f on *D*. If there are solutions to "$f(x) = c$" that are not in *D*, $f^{-1}(c)$ will not give them.

THEOREM 2-3-5 If f is one-to-one on domain *D*, then, for *x* in *D*, $f^{-1}(f(x)) = x$. Also, if f^{-1} is defined on *B*, then, for *y* in *B*, $f(f^{-1}(y)) = y$.

EXAMPLE 5   On domain $x \geq 0$, x^2 is one-to-one. Then, $\sqrt{x^2} = x$. (But for $x < 0$, $\sqrt{x^2}$ is not x; it is $|x|$. Figure 3.)

For $y \geq 0$, $\left(\sqrt{y} \right)^2 = y$.   ▬

■ Complications When Functions Are Not One-to-One

Squaring is not one-to-one and its inverse function does not give all the solutions. Many other inverse functions do not give all the solutions. *Sine* is a well-known trigonometric function (Figure 4). Given the value of x, your calculator will find the unique value of sin x. But, given "sin $x = .5$," you cannot be sure of the value of x.

EXAMPLE 6   Solve "sin $x = .5$."

Apply the inverse to .5. One solution is $\sin^{-1}(.5)$ ["inverse sine of point five"]. $\sin^{-1}(.5) = 30°$ ("30 degrees"). The "inverse sine" function on your calculator gives *one* answer. But there are many more (Figure 4). How do you find the rest?

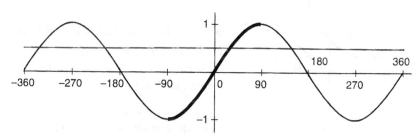

FIGURE 4 The graph of "$y = \sin x$" with the points where "$\sin^{-1} y = x$" emphasized. Also, $y = .5$ is graphed. The x-values of the points of intersection are the solutions to the equation "sin $x = .5$." The two graphs intersect in many places, but only one of them is called "$\sin^{-1}(.5)$."
Window: $-360° \leq x \leq 360°$, $-1 \leq \sin x \leq 1$

Here is the theorem about solving "sin $x = y$". (You are not expected to memorize it now.) It illustrates how theorems say that there is more than one solution.

▬

THEOREM 2-3-6 (THE INVERSE SINE RELATION)   sin $x = y$ iff

(A) $x = \sin^{-1} y$

(B) or $x = 180° - \sin^{-1} y$

(C) or $x = \sin^{-1} y \pm 360n°$, for any integer $n = 1, 2, 3, \ldots$

(D) or $x = 180° - \sin^{-1} y \pm 360n°$, for any integer $n = 1, 2, 3. \ldots$

Part A gives the only solution your calculator displays (Figure 4, bold region). However, Part B is almost as important. The number you want may not be the one your calculator provides; it may be one of the other alternatives. The solution to "sin $x = .5$" is

$x = 30°$ ($\sin^{-1}.5 = 30$ degrees, from Part A)

or $x = 150°$ ($= 180° - 30°$, from Part B),

or $x = 30° \pm 360n°$ (from Part C, which includes $390°$ when $n = 1$),

or $x = 150° \pm 360n°$ (from Part D, which includes $870°$ when $n = 2$).

Sine is not one-to-one if its domain is all real numbers. However, if we restrict the domain to $[-90°, 90°]$ it is (Figure 4, bold part), so we define its inverse function there.

The superscript "-1" in the notation for the inverse sine function is not a power. $\sin^{-1}x$ is certainly not $\dfrac{1}{\sin x}$. The superscript "-1" on a function indicates "inverse."

CALCULATOR EXERCISE 1   Set your calculator mode to degrees (instead of radians). Evaluate $\sin 30°$. You should obtain .5. Now find $\sin 150°$. Again the calculator displays .5. So $30°$ and $150°$ are *two* values of x related to the same value of $\sin x$ (Figure 4). The equation "$\sin x = .5$," has these two solutions, and many more. Take $\sin^{-1}(.5)$. It is $30°$. To find the additional solution $x = 150°$, use Part B.

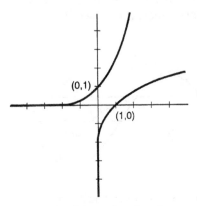

FIGURE 5 $y = e^x$, and $y = \ln x$ (lower). $[-10, 10]$ by $[-10, 10]$

■ Theorems About Inverses of Other Functions

The inverse-reverse method of solving equations requires knowing the inverses. For example, e^x is "undone" by the natural logarithm function, *ln* (pronounced "natural log," or just "log").

We will study e^x and *ln* more later. You are not expected to memorize this theorem now. The point is to learn to read theorems about how to solve equations using inverses.

CALCULATOR EXERCISE 2   Solve $e^x = 7$.

$e^x = 7$ iff $x = \ln 7$ ($= 1.9459$), according to the next theorem.

You may also think of e^x as applying a function to x to yield 7. To solve this, apply the inverse function, *ln*, to 7 to get back to x.

THEOREM 2-3-7 $e^x = y$ iff $x = \ln y$.

The letters in the theorem are placeholders. It can be rewritten with "x" for "y" and "y" for "x":

THEOREM 2-3-7B $\ln x = y$ iff $x = e^y$.

CALCULATOR EXERCISE 3   Solve $\ln x = 2.34$.

$\ln x = 2.34$ iff $x = e^{2.34}$ ($= 10.38$).

To solve an equation using the inverse-reverse method, you need to know the inverse functions. In addition, you must know if the inverse function finds *all* solutions.

If the original function is one-to-one, the inverse finds all solutions because there is only one. If the original function is not one-to-one, the inverse finds one and *you* need to find the others using a theorem.

EXAMPLE 7   Which of these equations may have more than one solution?
(If you look at the graph you can tell from the Horizontal Line Test.)

$5x = c$ $f(x) = 5x$ is one-to-one. There is exactly one solution.
$5x^2 = c$ $f(x) = 5x^2$ is not one-to-one. If $c > 0$ there are two solutions and the square root function finds only one.
$5x^3 = c$ $f(x) = 5x^3$ is one-to-one. There is exactly one solution.
$\ln x = c$ $f(x) = \ln x$ is one-to-one. There is exactly one solution.
$\sin x = c$ $f(x) = \sin x$ is not one-to-one. There are many solutions (if $-1 \le c \le 1$) and the inverse sine function finds only one.

The number of solutions to the equation "$x^n = c$" depends upon whether n is even or odd.

THEOREM 2-3-8   Let n be a positive integer ($n = 1, 2, 3, \ldots$).
If n is odd, $x^n = c$ if and only if $x = c^{1/n}$.
If n is even, $x^n = c$ if and only if $c \ge 0$ and $x = c^{1/n}$ or $x = -(c^{1/n})$.

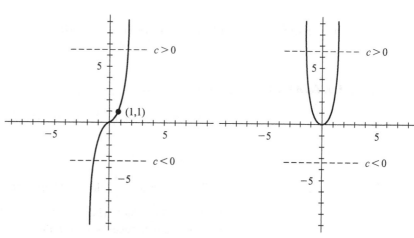

FIGURE 6 x^n, n odd, $n \ge 3$ **FIGURE 7** x^n, n even, $n \ge 2$

It makes a big difference whether the power is even or odd. Figure 6 illustrates that odd powers make the function one-to-one, but Figure 7 illustrates that even powers do not. Even powers are trickier.

EXAMPLE 8   Read this theorem and use it to solve $x^6 = 20$.

$x^6 = 20$ iff $x = \pm(20^{1/6}) = \pm 1.65$. There are two solutions because $n = 6$ is even. The inverse function (the $\dfrac{1}{6}$ power function) finds only one solution. Remembering that there is a second solution is your responsibility.

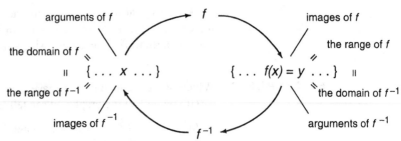

FIGURE 8 The relationship between f and f^{-1}

■ Subtraction and Division

To use the inverse-reverse method of solving "$f(x) = c$" we regard $f(x)$ as beginning with x and do the inverse operations in the reverse order. If $f(x)$ does not appear to begin with x, sometimes we can reorder the operations.

EXAMPLE 9   Solve "$5 - x = 17$."

One way is to subtract 5 and then change the sign. $-x = 17 - 5 = 12$. $x = -12$.

The expression "$5 - x$" does not appear to begin with x. However, "$5 - x$" is equivalent to "$(-x) + 5$" which does. Then the inverse is "Subtract 5 and change the sign." (Changing the sign is its own inverse). ■

THEOREM 2-3-9   (A) $c - x = (-x) + c = -(x - c)$
(B) $-x = c$ iff $x = -c$.

EXAMPLE 10   Solve "$\dfrac{3}{x} = 19$."

One way is to divide by 3 and then take the reciprocal: $\dfrac{1}{x} = \dfrac{19}{3}$.

$$x = \left(\frac{19}{3}\right)^{-1} = \frac{3}{19}.$$

The expression "$\dfrac{3}{x}$" does not appear to begin with x. However, it is equivalent to $(x^{-1})3$ which does. Then the inverse is "Divide by 3 and take the reciprocal." (Taking the reciprocal is its own inverse). ■

THEOREM 2-3-10   (A) $\dfrac{c}{x} = (x^{-1})c = \left(\dfrac{x}{c}\right)^{-1}$
(B) $\dfrac{1}{x} = c$ iff $x = \dfrac{1}{c}$.

■ Geometry and Inverses

According to the definition of "inverse," if (a, b) is on the graph of f, then (b, a) is on the graph of its inverse.

EXAMPLE 11   The point (2, 4) is on the graph of $f(x) = x^2$. Therefore, (4, 2) is on the graph of its inverse. $\sqrt{4} = 2$.

There is a simple geometric relationship between (a, b) and (b, a). They are mirror images through the line $y = x$. Figure 11 illustrates several pairs of such points and the line $y = x$.

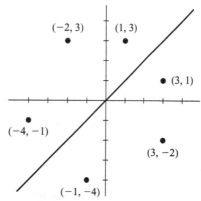

FIGURE 9 (a, b) and (b, a) are mirror images through the line $y = x$.
[$-5, 5$] by [$-5, 5$]

The graphs of *f* and *f* $^{-1}$ are mirror images of each other through the line *y* = *x*.

Figures 10 through 12 illustrate three pairs of inverse functions on a "square scale." A scale is <u>square</u> when one unit is the same distance both vertically and horizontally.

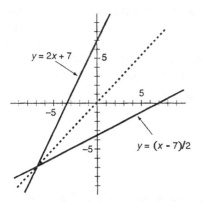

FIGURE 10 $f(x) = 2x + 7$
and $f^{-1}(x) = (x - 7)/2$.
[$-10, 10$] by [$-10, 10$]

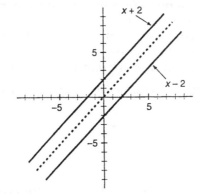

FIGURE 11 $f(x) = x + 2$
and $f^{-1}(x) = x - 2$.
[$-10, 10$] by [$-10, 10$]

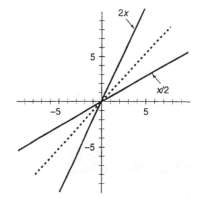

FIGURE 12 $f(x) = 2x$
and $f^{-1}(x) = x/2$.
[$-10, 10$] by [$-10, 10$]

CALCULATOR   Look at the graph of e^x and $\ln x$ on a square scale (Figure 5). Graph x^2 and $\sqrt{x}$.
EXERCISE 4   What is the relationship of the graphs (Problem B4)?
<u>For a proof that (a, b) and (b, a) are mirror images through the line $y = x$.</u>
Figure 13 has the lines and labels needed for a geometric proof. The first point

is labeled P and the second P^{-1}. A key idea is that angle 1 and angle 2 are congruent so angles 3 and 4 are also congruent. With a few more steps it can be shown that M is the midpoint of the line segment PP^{-1} which is perpendicular to the line $y = x$. Problem B65 asks for the details.

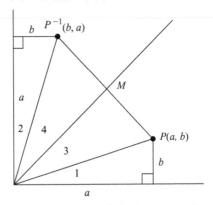

FIGURE 13 For a proof that (a, b) and (b, a) are mirror images of each other through the line $y = x$

CONCLUSION Solving equations uses inverses. To solve "$f(x) = c$" for x you apply the inverse to c.

In the expression "$f(x)$", "f" expresses the *evaluation* process.
In the equation "$f(x) = c$", "f^{-1}" expresses the *solution* process.
f^{-1} helps answer the problem: Solve "$f(x) = c$".

If there is only one solution, an inverse function finds it. However, sometimes there are two or more x's that yield the same c (as with squaring and *sine*, which are not one-to-one). Then the inverse *function* returns *one* of the x's. *You* must remember that the one solution your calculator gives may not be the only one you want.

Terms: Inverse, one-to-one, Horizontal Line Test.

Exercise 2-3

A

A1.* (A) What is another common name for the *inverse-reverse* method of solving equations?
(B) What is the format requirement for immediate application of the inverse-reverse method?

A2.* ☺ If (a, b) is on the graph of a function, then _____ is on the graph of its inverse.

A3.* (A) In the expression "$f(x)$", "f" expresses the _____ process.

(B) In the equation "$f(x) = y$", "f^{-1}" expresses the _____ process.

A4.* By definition, the <u>range</u> of f is the set of all its images. Therefore, the range of f is the set of all y such that the equation _____ has a solution.

Give $f^{-1}(x)$ for these. Write your answer in the form:
$f^{-1}(x) = \ldots$

A5. $f(x) = 3x$ **A6.** $f(x) = mx \ (m \neq 0)$

A7. $f(x) = x + 5$ **A8.** $f(x) = x + b$

A9. $f(x) = x/6$ **A10.** $f(x) = x/c \ (c \neq 0)$

A11. $f(x) = 1/x$.   **A12.** $f(x) = x$

A13. The first figure below locates (a, b) and (c, d) on an axis system. Locate (b, a) and (d, c).

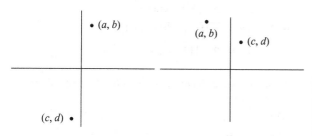

A14. The second figure above locates (a, b) and (c, d) on an axis system. Locate (b, a) and (d, c).

Make a function-loop diagram for f and its inverse.

A15. $f(x) = 5x + 7$ **A16.** $f(x) = 2(x + 1)$

A17. $f(x) = \dfrac{x - 3}{5}$

[You can use these problems and answers to learn the correct calculator keystrokes.] Evaluate these expressions with a calculator. Give 3 significant digits. A 2-digit answer is given.

A18. (A) $10^{1.3}$ [20]   (B) log 12   [1.1]

A19. (A) e^7 [1100]   (B) ln 23,000   [10]

A20. (A) e^{-5} [.0067]   (B) ln 12   [2.5]

A21. (A) $\sin(-35°)$ [−.57]

(B) $\sin^{-1}(.4)$ [24°]

A22. (A) $\sin(52°)$ [.79]   (B) $\sin^{-1}(.6)$ [37°]

A23. $\sin^{-1}(.1)$ [5.7°]   **A24.** $\sin^{-1}(.9)$ [64°]

Read Theorem 2-3-7 and use it to solve these equations algebraically.

A25. ln $x = -1.23$ [.29]   **A26.** ln $x = 5.6$ [270]

A27. $e^x = 20$ [3.0]   **A28.** $e^x = .1$ [−2.3]

A29. From Example 2, when

$$f(x) = \frac{x}{2} + 3, f^{-1}(x) = 2(x - 3).$$ Express and then simplify $f^{-1}(f(x))$.

A30. Let $f(x) = \dfrac{1}{x}$ and $g(x) = \dfrac{1}{x}$. Express and then simplify $f(g(x))$.

A31. Let $f(x) = 2x + 1$.

(A) Find f^{-1}.

(B) Express and then simplify $f^{-1}(f(x))$.

A32. Let $f(x) = 3(x - 2)$.

(A) Find f^{-1}.

(B) Express and then simplify $f^{-1}(f(x))$.

☺ *Look at a graph of these functions and decide if f appears to be one-to-one.*

A33. $f(x) = x^3$ **A34.** $f(x) = 10^x$

A35. $f(x) = x^2$ **A36.** $f(x) = (x - 4)(x - 6)$

A37. $f(x) = \dfrac{1}{x}$ **A38.** $f(x) = x^3 + x$

A39. $f(x) = x^3 - x$ **A40.** $f(x) = \ln x$

☺ *Give the range of **f** (the range is the set of all image values. You may wish to look at a graph.)*

A41. $f(x) = x^2$ **A42.** $f(x) = x^3$

A43. $f(x) = |x|$ **A44.** $f(x) = \dfrac{1}{x}$

A45. $f(x) = x^2 + 1$ **A46.** $f(x) = \sqrt{x + 5}$

A47. $f(x) = |x - 3|$ **A48.** $f(x) = x^2 - 5$

A49. If f has an inverse function, then the domain of its inverse is the _____ of f.

A50. If f has an inverse function, then the range of its inverse is the _____ of f.

Reading and Writing Mathematics

Read this theorem and use it to solve the next equations.
Theorem (about absolute values): $|x| = c$ iff $c \geq 0$ and $x = c$ or $x = -c$.

A51. $|x| = 6$.   **A52.** $|x| = -4$

A53. $|x| = 3.7$.   **A54.** $|x| = 16$.

A55. $|x| - 4 = 2.3$ **A56.** $|x - 3| = 1$

A57. $|x + 2| = 6$ **A58.** $|x - 3| = 2$.

State an identity giving an alternative way to

A59.* subtract.   **A60.*** divide.

A61.* Which condition guarantees that a function will have an inverse function?

B

B1.* (A) Given $f(x)$, explain how to find $f^{-1}(x)$, assuming it exists and is easy to find.

(B) What is the essential difference between algebraically solving "$f(x) = 5$" and finding f^{-1}?

B2.* Suppose f is one-to-one and f^{-1} is its inverse.

(A) Give $f^{-1}(f(x))$.

(B) $f(a) = b$, give $f^{-1}(b)$.

B3. Name two functions for which the usual inverse function does not find all the solutions to "$f(x) = c$."

B4. Sketch the graphs of x^2 and $\sqrt{x}$ for $x \geq 0$ together on a square scale. How is it graphically evident that they are inverse functions?

Find f^{-1} for the given f.

B5. $f(x) = \dfrac{x}{2} + 7$

B6. $f(x) = \dfrac{x - 5}{6}$

B7. $f(x) = 3(x + 4)$

B8. $f(x) = 5x - 2$

B9. $f(x) = 3\log x$

B10. $f(x) = e^{2x-1}$

B11. $f(x) = \log(10x) + 5$

B12. $f(x) = \dfrac{5}{x - \pi}$

B13. $f(x) = \dfrac{7}{9 - x}$

B14. $f(x) = \ln\left(\dfrac{5}{x}\right)$

B15. $f(x) = \dfrac{x}{x - 1}$

B16. $f(x) = \dfrac{2x}{x + 3}$

Read Theorem 2-3-6 to

B17. Find x such that $90° < x < 180°$ and $\sin x = .4$ [160°]

B18. Find x such that $90° < x < 180°$ and $\sin x = .6$ [140°]

B19. Find x such that $90° < x < 180°$ and $\sin x = .3$ [160°]

B20. Find x such that $90° < x < 180°$ and $\sin x = .9$ [120°]

B21. Find x such that $180° < x < 270°$ and $\sin x = -.4$ [200°]

B22. Find x such that $180° < x < 270°$ and $\sin x = -.2$ [200°]

B23. Find x such that $720° < x < 810°$ and $\sin x = 6$. [760°]

B24. Find x such that $-270° < x < -180°$ and $\sin x = .32$ [−200°]

Read Theorem 2-3-6 and use it to find three angles x (in degrees) such that

B25. $\sin x = .87$ [60° is one]

B26. $\sin x = .5$ [30° is one]

B27. $\sin x = -.18$ [−10° is one]

B28. $\sin x = .2$ [12° is one]

Read Theorem 2-3-8 and use it to solve these equations algebraically.

B29. $x^5 = 20$.

B30. $4.3x^5 = 19.5$

B31. $x^4 = 50$.

B32. $(3x)^4 = 200$

B33. $6x^3 = 19$

B34. $x^3 = .005$

B35. $x^8 = 1000$

B36. $(2x)^8 = 300$

From the given graph of f on a square scale, graph f^{-1}.

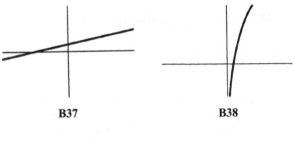

B37   **B38**

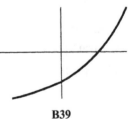

B39

How is it evident graphically that the given function is its own inverse?

B40. The reciprocal function.

B41. The "change sign" function.

B42. Find three functions which are their own inverses.

Solve $f(x) = 5$ when f is one-to-one and f^{-1} is as given.

B43. $f^{-1}(x) = x^3 + x$

B44. $f^{-1}(x) = \log x + x$

Reading and Writing Mathematics

Theorem: $|a| = |b|$ *iff* $a^2 = b^2$. *Use the theorem to solve*

B45. $|2x + 3| = |x - 1|$ **B46.** $|2x + 3| = |x|$

B47. $|2x| = |x + 1|$ **B48.** $|x - 2| = |2x|$

B49. Applying the the inverse function of squaring to c does not find all the solutions to "$x^2 = c$." Why not?

B50.* Explain the inverse-reverse method of solving equations.

B51.* If f is one-to-one, what is the relationship of the graphs of f and f^{-1} ?

B52.* If a function is not one-to-one, how can an inverse function be defined for it?

B53. (A) Sketch the graph of $\left(\sqrt{x}\right)^2$.

   (B) Graph $\sqrt{x^2}$.

   (C) Use the concepts of domain and range to explain why they are not the same.

   (D) "$\sqrt{(x^2)} = x$" is false as a generalization. Correct it.

B54. (A) Sketch the graph of $y = e^{\ln x}$.

   (B) Sketch the graph of $y = \ln(e^x)$.

   (C) Why are they not the same if the functions are inverses of one another?

B55. Rewrite Theorem 2-3-8 with different letters so it emphasizes solving equations similar to

   "$x^{1/3} = .42$" and "$x^{1/5} = 7$." [Odd roots only.]

B56. Give an example to show that this conjecture is false: $|x| = c$ iff $x = c$ or $x = -c$.

Relations

A <u>relation</u> is a set of ordered pairs. Functions are relations, but relations permit more than one image for an argument. Inverses are often relations rather than functions. Suppose the given function has its natural domain. Which inverses are not functions (but are relations)?

B57. The inverse of $f(x) = 3x + 5$

B58. The inverse of $f(x) = x^2$

B59. The inverse of $\sin x$

B60. The inverse of $f(x) = x^3$

B61. The inverse of $f(x) = x^4$

B62. The inverse of $f(x) = (x - 3)(x - 2)$

B63. [See the above definition of "relation".] The equation $x^2 + y^2 = 1$ defines a relation, but not a function. Why not?

B64. Compare and contrast the terms "function," "relation," and "graph" (the noun) in the context of ordered pairs of real numbers.

   (A) Do all functions have graphs?

   (B) Do all graphs correspond to functions?

   (C) Do all relations have graphs?

   (D) Do all graphs correspond to relations?

   (E) Are all functions relations?

   (F) Are all relations functions?

B65. Complete the geometric proof (begun in Figure 13) that (a, b) and (b, a) in the first quadrant are mirror images of each other across the line $y = x$.

B66. (A) Express "The inverse of 'Add c' is 'Subtract c'" in functional notation.

   (B) Express "The inverse of 'Multiply by c' is 'Divide by c,' if c is not 0," in functional notation.

B67. Restate Definition 2-3-2 of "one-to-one" using symbols. (Use the symbols "a", "b", "$f(a)$" and "$f(b)$" and the connective "if. . . , then. . .".)

B68. Calculators and Drawing Inverses.

   (A) What model of calculator do you have?

   (B) Does it have an automatic way to draw inverses? [Your calculator may have a "draw inverse" keystroke sequence. On a T1-83, graph, say, $y = 2x + 3$. Then go to the "DRAW" key menu. Select option 8 (DrawInv). Then use the VARS key. Arrow over to "Y-VARS" and select "Function" (with ENTER) and then "Y1" (with ENTER). Hit ENTER again. Bingo. It graphs the inverse. This is very easy for it to do because it takes each (a, b), reverses it to get (b, a), and graphs that. It does not have to solve for the inverse function algebraically.]

3

Fundamental Functions

Section 3-1 LINES

Lines are everywhere in the real world. This section connects lines to their equations.

Points are located on rectangular-coordinate-system graphs by the intersection of vertical and horizontal lines.

> In art, the term "lines" may include curves which are not straight, but, in Mathematics, lines are necessarily straight and geometric in nature.

EXAMPLE 1 $x = 3$ denotes the vertical line through the point on the x-axis labeled 3 (Figure 1). In the plane we can interpret "$x = 3$" as "$x = 3$ and $y =$ anything." Similarly, "$y = 2$" denotes the horizontal line through the point labeled "2" on the y-axis. The point (3, 2) is, by the definition of the axis system, at the intersection of these two perpendicular lines ($x = 3$ **and** $y = 2$) (Figure 1).

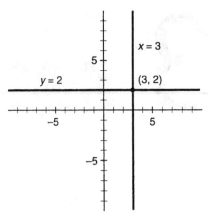

FIGURE 1 The lines $x = 3$ and $y = 2$ and the point $(3, 2)$

EXAMPLE 2   Find equations of the vertical and horizontal lines through the point $(\pi, 4.5)$.
That point is on the vertical line $x = \pi$ and the horizontal line $y = 4.5$.
The x-axis has equation $y = 0$, and the y-axis has equation $x = 0$. Be careful about this switch of x's and y's.

> horizontal line: $y = c$, for some c.
> vertical line: $x = c$, for some c.
> x-axis: $y = 0$.
> y-axis: $x = 0$.

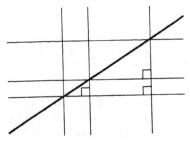

FIGURE 2 Three similar triangles formed by the intersection of a diagonal line with grid lines

■ Slope

When a line is neither vertical nor horizontal, all the horizontal and vertical lines of the rectangular coordinate system intersect it and create similar right triangles (Figure 2). From geometry, we know that the ratios of corresponding sides of similar triangles are equal. The ratio of the vertical side to the horizontal side is a characteristic of the line known as the slope and usually denoted by "m".

Figure 4 illustrates lines with various slopes. A line goes up from left to right (is "increasing") if and only if its slope is positive. A line goes down from left to right (is "decreasing") if and only if its slope is negative.

Line	Slope
horizontal	zero
upward to the right	positive
downward to the right	negative
vertical	does not exist

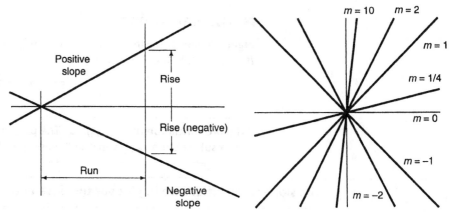

FIGURE 3 The slope is the rise over the run

FIGURE 4 Lines with various slopes, graphed on a square scale

The slope of any horizontal line is zero because its "rise" is zero.

The slope of any vertical line does not exist, since the denominator (the "run") is zero. The equation of a vertical line is "$x = c$" (and $y =$ anything), for some c.

■ The Two-Point and Point-Slope Formulas

The algebraic formula for a line follows from the similar triangles in Figure 5. Memorize the reasoning illustrated in Figure 5.

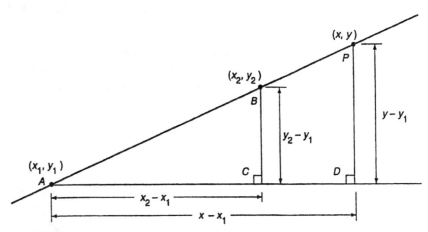

FIGURE 5 A line determined by two points, A and B, with another point, P, on it

The Two-Point Formula and Point-Slope Formula follow from the similar triangles in this figure. Memorize it.

■ The Reasoning

Figure 5 has two similar triangles, ABC and APD. Let $A = (x_1, y_1)$ and $B = (x_2, y_2)$. Using ABC the slope is given by

$$m = \frac{rise}{run} = \frac{y_2 - y_1}{x_2 - x_1} \tag{3-1-1}$$

> The pronunciation of "(x_1, y_1)" is "The point x sub one (pause) y sub one," or "x sub one (pause) y sub one," or, simply, "x one (pause) y one."

The slope is the difference in y's over the difference in x's.

Now, let P be any other point on the line. $P = (x, y)$. Using triangle APD, by similar triangles,

$$\frac{PD}{DA} = \frac{BC}{CA} = \text{the slope.}$$

$$\frac{y - y_1}{x - x_1} = \frac{y_2 - y_1}{x_2 - x_1} = m, \text{ the slope.} \tag{3-1-2}$$

Multiplying through by "$x - x_1$" yields both the "two-point" and "point-slope" formulas.

TWO-POINT FORMULA

$$y - y_1 = \left(\frac{y_2 - y_1}{x_2 - x_1}\right)(x - x_1)$$

$$\text{or } y = \left(\frac{y_2 - y_1}{x_2 - x_1}\right)(x - x_1) + y_1 \tag{3-1-3}$$

The second equation is simply a version of the first with "y_1" transferred to the right. This is essentially the "point-slope" formula for which the slope m and one point (x_1, y_1) are given (Figure 6).

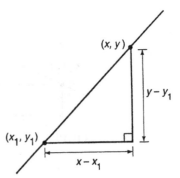

FIGURE 6 The point-slope formula: $y - y_1 = m(x - x_1)$. $\dfrac{y - y_1}{x - x_1} = m$

POINT-SLOPE
FORMULA

$$y - y_1 = m(x - x_1)$$
$$\text{or } y = m(x - x_1) + y_1.$$

(3-1-4)

> The point-slope form of a line is very important.
> Remember both formulas by knowing how they follow from similar triangles.

EXAMPLE 3   Find an equation of the line with slope $\frac{1}{2}$ through the point (3, 1) (Figure 7). Use the point-slope formula:

$$y - 1 = \left(\frac{1}{2}\right)(x - 3), \text{ or } y = \left(\frac{1}{2}\right)(x - 3) + 1.$$

These need not be "simplified." They nicely exhibit the point and slope.   ▬

The slope is easy to interpret. When the change in x values is 1, the ratio is simply the change in y values. We can think of the slope as the amount the line goes up for each unit it goes to the right (Figure 8).

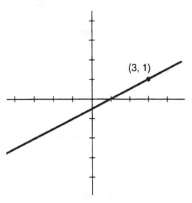

FIGURE 7   The line through (3, 1)
with slope $\frac{1}{2}$. $y - 1 = \left(\frac{1}{2}\right)(x - 3)$.
$[-5, 5]$ by $[-5, 5]$

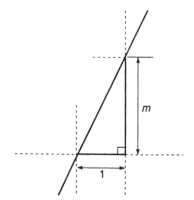

FIGURE 8   The slope of a line when
$x - x_1 = 1$

> The slope is the change in y-value corresponding to a change of 1 in x-value.

Because "slope = rise over run," we also have

(3-1-4, again)   rise = slope × run.

This is just the point-slope formula in words. Inspect the symbolic version and Figure 6 to see that this is really the same.

EXAMPLE 4 A line with slope 3 goes through (4, 2). What is the *y*-value when *x* = 4.1 (Figure 9)?

The *x*-value changes from 4 to 4.1, a change (run) of .1. Because the slope is 3, the *y*-value changes 3 times as much (.3), from 2 to 2.3. Therefore, *y* = 2.3.

What is the *y*-value when *x* = 3.8?

The *x*-value changes from 4 to 3.8, a change of −.2. The *y*-value changes 3 times as much (−.6), from 2 to 1.4. *y* = 1.4.

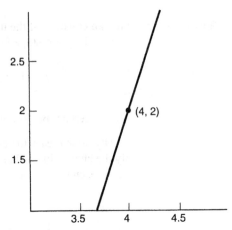

FIGURE 9 A line with slope 3 through the point (4, 2). [3, 5] by [1, 3] ▬

Slope in the Real World

The concept of slope as "rise over run" is abstract. In real life, the variable quantities represented by "*x*" and "*y*" would have units such as miles, hours, gallons, or dollars. Since slopes are quotients, particular examples of slopes would have the units of a quotient such as "miles per hour" or "dollars per gallon."

EXAMPLE 5 Suppose drilling a well costs $500 to set up and $20 for each foot drilled. Express the cost of drilling a well *x* feet deep.

The total cost is $500 plus ($20 per foot) × (the number of feet). Abstractly, *y* = 20*x* + 500, where "*y*" has the units of dollars and "*x*" has the units of feet. And the slope, 20, is really "20 dollars per foot." It has the units of the quotient formula for slope (3-1-1). ▬

Sometimes words can conceal the simple idea of putting a line through two points.

EXAMPLE 6 A plane flies due east at a constant speed. At 1:30 PM it is 100 miles west of Bozeman, and at 4:00 PM it is 250 miles east of Bozeman. Give the formula for the plane's location at any time (until the speed or direction changes). [Express time in hours after noon.]

Does this look like a line problem? It is. The "constant speed" in the problem corresponds to the constant slope of a line.

Let Bozeman be at location zero and east be the positive direction. The phrase "250 miles east" corresponds to +250, and "100 miles west" corresponds to −100. 1:30 PM corresponds to *t* = 1.5 (hours after noon) and 4:00 PM corresponds to *t* = 4. The formula for the plane's location is a line through two points, $(1.5, -100)$ and $(4, 250)$. Abstractly, according to Formula 3-1-3 (with "t" for "x"),

$$y = \frac{250 - (-100)}{4 - 1.5}(t - 1.5) - 100.$$

This simplifies to $y = 140t - 310$. The slope, 140, is really the speed, "140 miles per hour," which has the units of a quotient.

   The point is, we can study lines abstractly using "x" and "y", but they represent real formulas that have units that we omit in the abstract equations. ▬▬

EXAMPLE 7   There are 2.54 centimeters in an inch. To convert from inches to centimeters, multiply by 2.54 (centimeters per inch). The relationship is, abstractly, $y = 2.54x$. In reality, all of the terms in this equation have units. The number "x" has the units of inches. The number "y" has the units of centimeters. And "2.54," the slope, has the units of the quotient, "centimeters per inch." The slope gives the change in centimeters (change in y) corresponding to a change in <u>one</u> inch (change in x) (Figure 8). That is a real instance of "rise over run." ▬▬

■ Proportional

Two variable quantities are said to be <u>proportional</u> (or "to vary directly") if one is a constant multiple of the other. If one is denoted by x and the other by y, y is <u>proportional</u> to x if and only if there exists a number, $k \neq 0$, such that

$$y = kx. \tag{3-1-5}$$

The constant k is called the <u>constant of proportionality</u>. The two quantities are functionally related by the function, "Multiply by k."

EXAMPLE 7 REVISITED   Measurements in inches and centimeters are proportional. Measurements in centimeters are always 2.54 times measurements in inches. The constant of proportionality is $k = 2.54$ (centimeters per inch). ▬▬

EXAMPLE 5 REVISITED   In Example 5, the cost of a well is not proportional to its depth. With the formula "$y = 500 + 20x$," a one-foot well costs \$520 and a two-foot well costs \$540, not twice as much. The fixed set-up cost makes the total cost not proportional to the depth. ▬▬

   The graph of a proportional relationship will be a line *through the origin* with slope k; "k" plays the role of "m" in the usual notation. When y is proportional to x, x is also proportional to y, so they are proportional to each other (Problem B50).

EXAMPLE 8   Physics (Hooke's Law) tells us that the amount a spring stretches is proportional to the force applied to it. If a force of 5 pounds stretches a spring 3.2 inches, how far will a force of 7 pounds stretch it?

   This is often done without bothering to obtain the whole function, since only one number is unknown. Call the stretch s.

s is to 7 as 3.2 is to 5. $\dfrac{s}{7} = \dfrac{3.2}{5}$ $s = \dfrac{7(3.2)}{5} = 4.48$ (inches).

Find the relationship between force and stretch for the spring.

In calculus we often want the entire relationship, rather than just the particular stretch for a particular force. We can obtain a general formula for the stretch, s, in terms of the applied force, x.

$$s \text{ is to } x \text{ as } 3.2 \text{ is to } 5. \quad \frac{s}{x} = \frac{3.2}{5}. \quad s = \left(\frac{3.2}{5}\right)x. \quad s = .64x.$$

So the constant of proportionality, k, is .64.

Here is another way to find k. This way uses the definition, $s = kx$. Use the given (x, s) pair,

$$3.2 = k(5), \text{ so } k = \frac{3.2}{5} = .64, \text{ and } s = .64x, \text{ as before.}$$

Technically, the units of "s" are inches and the units of "x" are pounds. And, the units of ".64" the slope, are the units of the quotient 3-3-1, "inches per pound." ▬

EXAMPLE 9 Suppose photocopies cost 6 cents each. Is the total cost of n copies proportional to n?

Yes, the formula is "$C = 6n$." "C" has the units of cents. "n" has the units of photocopies. The slope (constant of proportionality) is a quotient, 6 (cents per photocopy). ▬

EXAMPLE 10 Is the area of a square proportional to its side?

No. The term "proportional" does not mean simply "increasing together." It means one is a *constant* times the other. The area formula is $A = x^2$ in which x is not multiplied by a constant. They are not proportional. ▬

■ Parameters

A <u>parameter</u> is an arbitrary constant. It is a letter used to describe members of a family of expressions or equations.

For example, the equation "$y = mx$" has one parameter, m. Different values of m yield different lines (Figure 7). The "m" plays a different role than the "x" and the "y". For a given equation we imagine m to be constant (say, 2.54, as in "$y = 2.54x$") and the x and y to vary. However, for different lines there are different values of m, so m may vary too.

A Line Through Two Points
Together, the formula for the slope and the point-slope formula are sufficient to fit a line through two given points.

EXAMPLE 11 Find an equation of the line through $(1, 2)$ and $(3, 8)$ (Figure 10).

The slope is $m = \dfrac{8 - 2}{3 - 1} = 3$. Now use either point in the point-slope formula:

$$y = 3(x - 1) + 2.$$

> Equations in two variables may have *graphs* which are lines. The equation itself is not a line. Nevertheless, it is common, convenient, and not very misleading to speak as if equations could be lines. For example, the sentence, "Consider the line $y = mx + b$," is technically incorrect, but unlikely to be misunderstood.

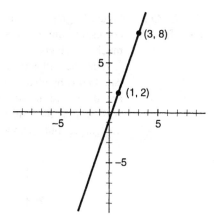

FIGURE 10 The line through $(1, 2)$ and $(3, 8)$. $y - 2 = 3(x - 1)$

The slope, 3, and the point used, $(1, 2)$, are clearly evident in this equation. The problem requested "an" equation of "the" line. This is because there are many, equivalent, equations. You could use the other point in the point-slope formula:

$$y = 3(x - 3) + 8.$$

This version clearly exhibits the slope and the second point, $(3, 8)$. If you simplified, a third equivalent equation in "slope-intercept" form would result:

$$y = 3x - 1.$$

When $x = 0$ (the y-axis), $y = -1$, which is the y-intercept. (The term is "intercept" with "cept," not "intersect," with "sect," as you might expect.) ▬

The slope-intercept form of a line emphasizes the slope and the y-intercept (Figure 11).

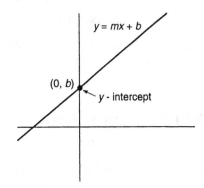

FIGURE 11 $y = mx + b$. b is the "y-intercept"

SLOPE-INTERCEPT $y = mx + b$.
FORM 3-1-6
This equation has two parameters: "m" for the slope and "b" for the y-intercept.

EXAMPLE 12 Find the slope and y-intercept of the line "$y = -11x + 7$."
The slope is -11 and the y-intercept is 7. ▬

The appearance of the slope, as reflected in the steepness of the line, is dependent upon the scale. By changing the window we can make the same line look flatter or steeper.

Figure 10 graphs the line "$y = 3x - 1$" in the standard window. The line is steep, corresponding to the slope, $m = 3$. Change the window so that the line looks much less steep.

The line will appear less steep if we make the change in y less impressive. We can do this by changing the window so that 2 or 5 or 10 times as many y-values are displayed in the same space. For example, change the y-interval to $-50 \le y \le 50$ (and still use $-10 \le x \le 10$). Then the slope of the line is unchanged, but its appearance is greatly changed (Figure 12).

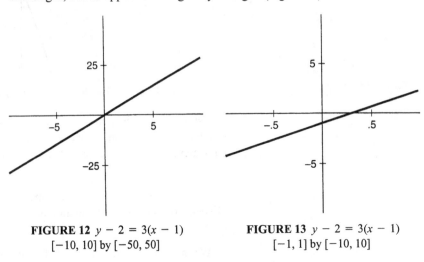

FIGURE 12 $y - 2 = 3(x - 1)$
$[-10, 10]$ by $[-50, 50]$

FIGURE 13 $y - 2 = 3(x - 1)$
$[-1, 1]$ by $[-10, 10]$

Another way to make a given line appear less steep is to change the x-scale, leaving the y-scale as it was. Small changes in x make for small changes in y, so we can create the appearance of a small change in y by making the window express only a small change in x. For example, let $-1 \le x \le 1$, as in Figure 13. Scale is important. ■

**DEFINITION 3-1-7
OF "SQUARE SCALE"**

A scale is <u>square</u> when one unit in the x-direction is the same distance as one unit in the y-direction.

If the scale is not square, the slope of a line not obvious visually.

EXAMPLE 13

Figure 14 graphs a line. Find its slope.

Note the scale, which is not square. Vertical grid lines are $\dfrac{20}{5} = 4$ units apart. Horizontal grid lines are $\dfrac{100}{5} = 20$ units apart. The line goes through $(8, 20)$ and $(16, 80)$. It goes up 60 units for each 8 horizontally. The slope is $\dfrac{60}{8} = 7.5$. ■

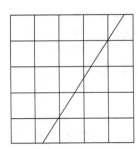

FIGURE 14 The window is $[0, 20]$ by $[0, 100]$. What is its slope of the line?

EXAMPLE 14   Find the line through the points on the graph of $f(x) = 1/x$ where $x = 2$ and $x = 3$ (Figure 15).

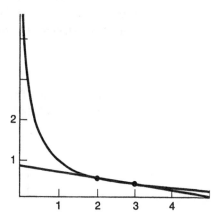

FIGURE 15 $f(x) = \dfrac{1}{x}$ and a line through

the points where $x = 2$ and $x = 3$.

[0, 5] by [0, 5]

Points on the graph are determined by their x values (2 and 3). The two points are $\left(2, \dfrac{1}{2} \right)$ and $\left(3, \dfrac{1}{3} \right)$. The slope is $\dfrac{1/3 - 1/2}{3 - 2} = -\dfrac{1}{6}$. An equation of the line in point-slope form is

$$y = -\left(\frac{1}{6} \right)(x - 2) + \left(\frac{1}{2} \right).$$

There is no need to "simplify" to slope-intercept form.   ▬

■ Applications

There are many cases where it is natural to put a line through two points, even if we know that a line is not quite right.

EXAMPLE 15   The standard-normal cumulative-distribution function, Φ, is very important in statistics, but difficult to compute. Use Table 2-1-3 to estimate $\Phi(1.65)$.

The table gives $\Phi(1.6) = .9452$ and $\Phi(1.8) = .9641$, but it does not give $\Phi(1.65)$. We need a better table. Or, we could try approximating the value on the curve by using the value on a line instead.

The two points are $(1.6, .9452)$ and $(1.8, .9641)$. According to the two-point formula (3-1-3), the line would be

$$y = \left(\frac{.9641 - .9452}{1.8 - 1.6} \right)(x - 1.6) + .9452$$

$\Phi(1.65)$ has $x = 1.65$:

$$\frac{.9641 - .9452}{1.8 - 1.6}(1.65 - 1.6) + .9452 = .9499$$

Figure 16 displays the relevant part of the graph of Φ.

On the line, the change in y is proportional to the change in x. The change in y between the two points we know is $.9641 - .9452$. The change in x for those two points is $1.8 - 1.6$. The change in x we want to use is $1.65 - 1.6$. The corresponding change in y will be proportional.

$$\frac{y - .9452}{1.65 - 1.6} = \frac{.9641 - .9452}{1.8 - 1.6}.$$

This is (3-1-2) again. The only difference is that "y" is computed for only the particular x-value 1.65, rather than for a general "x" value.

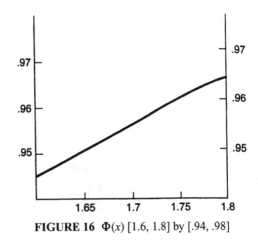

FIGURE 16 $\Phi(x)$ [1.6, 1.8] by [.94, .98]

Linear interpolation consists of putting a line through two points on a curve, and then using the line to approximate the curve.

EXAMPLE 16 Suppose market research shows The Bozeman Widget Company that it will probably sell about 2000 widgets per week if they are priced at $50, but only 1500 widgets per week if they are priced at $60. How many will they sell per week if widgets are priced at x dollars?

> In business examples, a <u>widget</u> is a mythical product manufactured by a mythical company. Widgets can serve to illustrate the relationship between price and sales.

Who knows? The true relationship between price and the number sold is some unknown curve. However, without a better idea, let's try to answer the question by fitting a straight line through the given points (50, 2000) and (60, 1500) (Figure 17).

The slope is

$$\frac{1500 - 2000}{60 - 50} = -50.$$

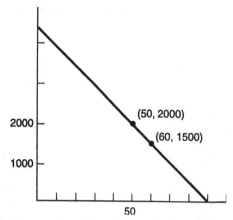

FIGURE 17 The line through (50, 2000) and (60, 1500).
$y - 2000 = -50(x - 50)$. [0, 100] by [0, 5000]

This says that for every dollar increase in price, 50 fewer widgets will be sold per week. Now use the point-slope formula with either point. With the point (50, 2000),

$$y = -50(x - 50) + 2000.$$

Now, for any price, x, you could estimate the sales, y.

You do not need to "simplify" this to Slope-Intercept form "$y = mx + b$." There is no point in emphasizing "b", which would be the number "sold" when the price is zero!

Also, if we foolishly extend our line far enough to the right, we obtain negative numbers for the number sold at high prices. Also nonsense. It is common and reasonable to fit a line over a region where you have good data, but it is not reasonable to extrapolate that line far beyond the data.

If we use a line to estimate a y-value *between* x-values for which the y-values are known, the process is called <u>interpolation</u>. If we use a line to estimate functional values for x's *outside* that interval, the process is called <u>extrapolation</u>. As in this example, extrapolation may produce ridiculous results. Avoid extrapolation.   ▬

Solving Equations

Linear interpolation may also be used to solve equations.

EXAMPLE 17   Suppose we want to solve "$f(x) = 0$," but $f(x)$ is hard to evaluate and we only know $f(.785) = -.00015$ (slightly too low) and $f(.786) = +.00071$ (slightly too high). Use linear interpolation to find a line and approximate x such that $f(x) = 0$ to four or more decimal places (Figure 18).

First fit a line through the two given points.

$$y - (-.00015) = \frac{.00071 - (-.00015)}{.786 - .785}(x - .785)$$

$$y - (-.00015) = .86(x - .785)$$

Set $y = 0$ and solve for x.

$$.00015 = .86(x - .785)$$

$$\frac{.00015}{.86} = x - .785$$

$$.78517 = x.$$

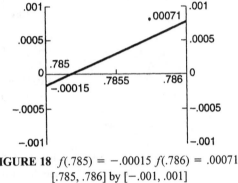

FIGURE 18 $f(.785) = -.00015 \ f(.786) = .00071.$
$[.785, .786]$ by $[-.001, .001]$

A fundamental idea of calculus is that most curves that are not straight lines can, nevertheless, be approximated by straight lines over small regions.

CALCULATOR EXERCISE 1 Use your calculator to graph any curved graph of your choice (for example, $y = x^2$). Zoom in several times about any point on the curve (for example, $(1, 1)$ on $y = x^2$) to obtain the picture in a very small window. Does the graph look like a straight line? Usually, but not always, it will (Problems B66 and B67).

EXAMPLE 18 Suppose that $f(x)$ is not a straight line, but can be approximated by a straight line. Suppose further that $f(35.4) = .0518$ and the slope of the tangent line to the graph at that point is $-.16$. Approximate the solution to $f(x) = 0$ (Figure 19).

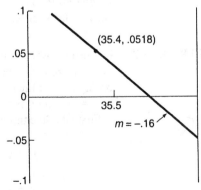

FIGURE 19 $f(35.4) = .0518.$ $m = -.16$ there. $[35, 36]$ by $[-.1, .1]$

The point is (35.4, .0518) and the slope is −.16. The line is
$$y - .0518 = -.16(x - 35.4).$$
Set $y = 0$ and find the corresponding x.
$$0 - .0518 = -.16(x - 35.4).$$
Dividing by −.16 and then adding 35.4,
$$x = .32 + 35.4 = 35.72.$$

The approach of Example 18 can be generalized (Problem B61). It is called "Newton's Method" after Sir Isaac Newton (the co-inventor, with Gottfried Wilhelm Leibnitz, of calculus).

■ Parallel and Perpendicular Lines

Two lines are parallel if they have equal slopes (Figure 20). They form congruent angles with the horizontal.

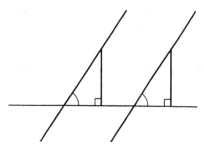

FIGURE 20 Parallel lines have the same slope

PARALLEL LINES 3-1-8   A line has slope *m* if and only if any line parallel to it has slope *m*.

Of course, all vertical lines are parallel. It is fair to say that two vertical lines have the same slope, in the sense that both slopes do not exist.

EXAMPLE 19   Find an equation of the line through (6, −5) parallel to the line $y = -3x + 1$.
Use point-slope form with $m = -3$.

$$y = -3(x - 6) - 5.$$

You may wish to "simplify" this, but it is fine as it is. It exhibits the slope and the point, which is all you need to know to graph the line.

In their initial geometric definition, lines are determined by any *two* points they go through. Given two points, the algebraic "two-point form" is appropriate.

Then, the "parallel postulate" states, "Given a line and one point not on it, there exists exactly one line through the point parallel to the line." So a line is also determined by *one* point in conjunction with some way of describing parallel lines. Slopes provide that way. Given the slope and *one* point, the algebraic "point-slope form" is appropriate.

Of course, vertical and horizontal lines are perpendicular. If a line is neither vertical nor horizontal, then the next theorem applies (Figure 21, Problem B65).

It says that two lines are perpendicular when the slope of one is the negative reciprocal of the slope of the other.

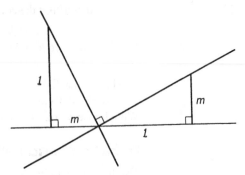

FIGURE 21 Slopes of perpendicular lines

PERPENDICULAR LINES 3-1-9   If a line has slope $m \neq 0$, then any line perpendicular to it has slope $\dfrac{-1}{m}$.

EXAMPLE 20   Find the line through $(7, 3)$ perpendicular to the line with equation $y = 2x - 9$.
Because the slope of the original line is 2, the slope of the perpendicular line is $\dfrac{-1}{2}$. Use the point-slope formula.

$$y = \left(\frac{-1}{2}\right)(x - 7) + 3.$$

This need not be "simplified." It emphasizes the key point as it is.   ▬

EXAMPLE 21   Find the line through $(3, 5)$ perpendicular to the line with equation $y = 4.3$.
The line $y = 4.3$ is horizontal and any line perpendicular to it is vertical. Vertical lines have the equation "$x = c$," for some constant c. Since the line we seek goes through $(3, 5)$, c must be 3. The equation is simply "$x = 3$."
There is a form for an equation of a line which handles all lines including vertical lines, which the formulas with slopes do not handle. The <u>general form</u> of a line is

$$ax + by + c = 0, \text{ where } a \text{ and } b \text{ are not both zero.} \qquad (3\text{-}1\text{-}10)$$

The case $b = 0$ handles vertical lines. The "general" form is not particularly important—it is just general. The point-slope and slope-intercept forms are far more important.   ▬

Six Forms for Lines	
Vertical:	$x = c$, for some c.
Horizontal:	$y = c$, for some c.
Two-Point Formula:	$y - y_1 = \left(\dfrac{y_2 - y_1}{x_2 - x_1}\right)(x - x_1).$
** Point-Slope Formula:	$y - y_1 = m(x - x_1)$ or
	$y = m(x - x_1) + y_1.$
Slope-Intercept Form:	$y = mx + b.$
General form:	$ax + by + c = 0.$

CONCLUSION Memorize Figure 5, which illustrates the two-point and point-slope forms of lines. They are simply symbolic versions of "Slope is rise over run." In calculus the "point-slope" form is more important than the famous "$y = mx + b$" form. Linear interpolation uses the idea of approximating a curve by a line through two points on the curve.

Terms: slope, two-point formula, point-slope formula, rise, run, proportional, parameter, y-intercept, slope-intercept form, widget, parallel, perpendicular, general form (of a line), linear interpolation, x-intercept.

Exercise 3-1

A

A1. ☺ (A) Vertical lines have equation
_____, for some c.
(B) Horizontal lines have equation
_____, for some c.
(C) The equation of the x-axis is
_____.
(D) The equation of the y-axis is
_____.

A2. ☺ (A) The slope of a horizontal line is _____.
(B) The slope of a line which goes upward to the right is _____.
(C) The slope of a line which goes downward to the right is _____.
(D) The slope of a vertical line
_____.

☺ *Find the equation of the*

A3. (A) vertical line through (6, 9).
(B) horizontal line through (123, 456).

A4. (A) vertical line through (−4, 123).
(B) horizontal line through (19, 21).

☺ *Give the slope of the line.*

A5. $y = 4x - 7$

A6. $y = -\left(\dfrac{1}{2}\right)x + 5$

A7. $y = 14(x - 6) - 2$

A8. $y = \dfrac{x - 3}{5} + 7$

A9. $2x + 3y = 18$

A10. $4y - 5x = 60$

Find an equation of the line through the given point with the given slope

A11. (5, 7), $m = 3$

A12. (0, 2), $m = -3$

A13. (−2, −6), $m = 4$

A14. (100, 60), $m = m = \dfrac{1}{2}$

Find an equation of the line through the two points.

A15. (3, 9) and (2, 5)

A16. (−1, 5) and (3, 7)

A17. (100, 50) and (300, 1000)

A18. (5, 7) and (25, 47)

The slope of a line in a window depends upon the scale of the graph. Estimate the slope of the given line after noting the scale.

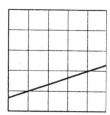

A19. [0, 10] by [0, 20]

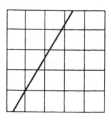

A20. [0, 100] by [0, 10]

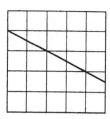

A21. [0, 10] by [0, 1000]

The slope of a line in a window depends upon the scale of the graph. Estimate the slope of the given line after noting the scale.

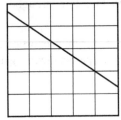 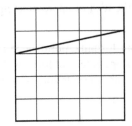

A22. [0, 1] by [0, 20]   **A23.** [0, 100] by [0, 5]

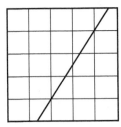

A24. [0, 10] by [0, 1]

☺ *Give the slope of a line perpendicular to the given line.*

A25. $y = 2x + 7$ **A26.** $y = 3(x - 5) + 9$

A27. $y = -3x + 12$ **A28.** $y = 10 - 5x$

Find an equation of the line through the given point and perpendicular to the given line.

A29. $(3, 5)$ $y = 2x + 10$ **A30.** $(1, 7)$ $y = \dfrac{x - 8}{3}$

A31. (A) Find an equation of the line through the point $(50, 210)$ and parallel to the line $y = 5x + 12$.
(B) Find the slope of a line perpendicular to that line.

A32. (A) Find an equation of the line through the point $(65, 12)$ and parallel to the line $y = \left(\dfrac{1}{4}\right)x + 987$.
(B) Find the slope of a line perpendicular to that line.

Linear Interpolation

A33.* ☺ True or False: The idea behind linear interpolation is basically putting a line through two points.

A34.* ☺ Pick one. Linear interpolation can be used for
(A) approximately evaluating $f(x)$
(B) approximately solving $f(x) = c$.
(C) both of (a) and (b)
(D) Neither of (a) and (b).

Proportionality

A35. Suppose x and y are proportional. When $x = 4$, $y = .2$. Find the formula for y in terms of x.

A36. Suppose distance and time are proportional. In one hour and 12 minutes the distance is 30 miles. Find the formula for distance in terms of time.

A37. A spring stretches 1.2 inches when a force of 24 pounds is applied. Find the formula for the stretch (y) given the force x.

A38. A spring stretches $\dfrac{1}{2}$ inch when a force of 8 ounces is applied. Find the formula for the stretch (y) given the force x.

A39. There are 45 calories in $\dfrac{1}{4}$ cup of a certain food.

(A) How many calories are in $\dfrac{1}{3}$ cup?
(B) Give the formula for the number of calories in any number of cups.

A40. There are 80 calories in $\dfrac{1}{3}$ cup of a certain food.

(A) How many calories are in $\dfrac{1}{8}$ cup?
(B) Give the formula for the number of calories in any number of cups.

Reading and Writing Mathematics

A41. ☺ True or false?
(A) If $y = 3x$, y is proportional to x.
(B) If $y = 5x - 1$, y is proportional to x.
(C) The area of a square is proportional to its side.
(D) The perimeter of a square is proportional to its side.

A42. ☺ True or false?
(A) If $C = 10n$, then C is proportional to n.
(B) If $y = 2(x + 3)$, then y is proportional to x.
(C) The area of a circle is proportional to its radius.
(D) The circumference of a circle is proportional to its radius.

A43. True or false?
(A) At a constant speed, the distance traveled is proportional to the time elapsed.
(B) At a constant speed, the time elapsed is proportional to the distance traveled.
(C) Over a fixed distance, the average speed is proportional to the time elapsed.
(D) Over a fixed distance, the time elapsed is proportional to the average speed.

A44. ☺ True or false?

(A) The volume of a cube is proportional to its side.

(B) For a constant height, the area of a triangle is proportional to its base.

(C) The area of a rectangle is proportional to its width.

A45. The figure locates (a, b) and (c, d) on a line, and forms the right triangle with those two points as vertices. Give the coordinates of the vertex at the right angle.

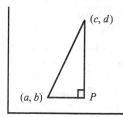

A46. ☺ List the parameters in the equation "$y = mx + b$."

A47. ☺ List the parameters in the Quadratic Formula.

Word Problems

A48. A car gets 26 miles per gallon and its tank holds 19 gallons. Give a formula for the number of gallons left in the tank after driving x miles.

A49. The cost of printing is 55 cents for the setup charge and then 3.5 cents for each copy. What is the cost of n copies?

B

B1.* (A) State the Two-Point Formula for a line.

(B) Sketch a (large) picture to illustrate *all* the points and distances in the formula.

(C) Use a bit of geometry to explain (or prove) why the formula in part (a) is, in fact, the right formula.

B2.* Suppose a line with slope m goes through the point (x_1, y_1).

(A) State the formula for its equation.

(B) Sketch a picture to illustrate the formula and label on it *all* the points and distances in the point-slope Formula.

(C) Use a bit of geometry to explain (or prove) why the formula in part (a) is, in fact, the right formula.

The graph of $y = f(x)$ is a line. Solve $f(x) = 0$ if

B3. $m = .3$ and $f(1.5) = .04$ [1.4]

B4. $m = -.5$ and $f(12) = .1$ [12]

B5. $m = .8$ and $f(4.6) = -.5$ [5.2]

B6. $m = -.7$ and $f(.95) = -.2$ [.66]

Use linear interpolation to a) find the relevant line, and b) approximate a solution to $f(x) = 0$

B7. $f(12.6) = .02$, $f(12.8) = -.01$

B8. $f(.78) = -.04$, $f(.81) = .017$

Use linear interpolation and Table 2-1-3 to a) find the relevant line, and b) approximate

B9. $\Phi(1.05)$ **B10.** $\Phi(.78)$

Use linear interpolation and Table 2-1-3 to a) find the relevant line, and b) approximate the solution to

B11. $\Phi(x) = 0.95$ **B12.** $\Phi(x) = .8$

B13. $\Phi(x) = .9$ **B14.** $\Phi(x) = .1$

Use linear interpolation to a) first find the relevant line (Be careful to use the two most relevant points), and b) then solve the equation.

B15. $f(.6) = .92$. $f(.7) = .74$. $f(.8) = .58$. Solve for an approximate solution: $f(x) = .67$

B16. $f(3.2) = 2.3$. $f(3.4) = 3.3$. $f(3.6) = 4.8$. Solve for an approximate solution: $f(x) = 3.5$.

Find the slope of the line through the points on the graph of

B17. $y = x^2$ where $x = 1$ and $x = 2$

B18. $y = \dfrac{1}{x}$ where $x = 1$ and $x = 2$

B19. $y = x^3$ where $x = 1$ and $x = 2$

B20. $y = \sqrt{x}$ where $x = 1$ and $x = 4$

B21. Consider the graph of $y = x^2$. Find the equation of the line through the points on the graph where $x = 2$ and where $x = 2.1$.

B22. Consider the graph of $y = \sqrt{x}$. Find the equation of the line through the points on the graph where $x = 2$ and where $x = 2.1$.

B23. Find and simplify the slope of the line through the points on the graph of x^2 where $x = 1$ and where $x = 1 + h$.

B24. Find and simplify the slope of the line through the points on the graph of x^2 where $x = a$ and $x = b$. [The slope simplifies nicely.]

B25. Find and simplify the slope of the line through the points on the graph of $\dfrac{1}{x}$ where $x = 1$ and $x = 1 + h$.

B26. Find and simplify the slope of the line through the points on the graph of $\dfrac{1}{x}$ where $x = a$ and $x = b$.

Word Problems

B27. If you want to rent a car and Firm A charges $40 a day plus 30 cents a mile and Firm B charges $30 a day and 34 cents a mile, when is Firm A better for you?

B28. Suppose you can rent a car from firm A at $30 a day plus 30 cents per mile. You can rent a comparable car from firm B for $40 a day plus 20 cents per mile. When is firm B the better deal for you?

B29. Suppose photocopies of a single original cost 6 cents each for the first 20 copies, and 5 cents each for additional copies after 20.
(A) Find the formula for the cost of n copies.
(B) Suppose copying a single original at QCopy costs 48 cents for the set up cost and then 4 cents per copy, no matter how many you make. Find a formula for the cost of n copies using QCopy.
(C) When is it cheaper to use QCopy? [Answer in terms of the number of copies you wish to make.]

B30. Suppose you can photocopy any number of copies of the same original for 6 cents each, or you can have the copies printed at 2.5 cents each after a setup charge of 50 cents. When is it cheaper to print copies than photocopy them?

B31. Suppose you want to make many copies of a flyer and Photocopy World charges 7 cents per copy and Print-King charges a $3 setup charge but after that only 4.5 cents per copy, when is PrintKing better for you?

B32. The president of Consolidated Widgets International thinks that, for widgets at any price, an increase in price of $5 per widget would decrease sales by 400 widgets per week. The price is now $65 and sales are 5000 per week. Find a straight line fit for the price (x) to sales (y) relationship.

B33. You want a cell phone, and the cell phone company offers two plans. Under Plan A you pay a $17.95 monthly access fee and all your calls cost you 15 cents per minute. Under Plan B you pay $29.95 per month, get 60 free minutes, and pay 10 cents per minute for all minutes after the first 60. Show all relevant formulas and then solve: When is Plan A cheaper? [< 120]

B34. Suppose you want to use mayonnaise with no more than 55 calories per tablespoon. You have both regular at 100 calories per tablespoon and Lite at 40 calories per tablespoon (but it's not as good).
(A) If you mix x tablespoons of Lite with 20 tablespoons of regular, how many calories, total, will be in the mix?
(B) How many tablespoons of Lite should you mix in with 20 tablespoons of regular to make a mix with 55 calories per tablespoon?

B35. A cup of lentils has 646 calories of which 19 are from fat. Sausage is 50 calories per ounce of which 27 calories come from fat.
(A) If you add x ounces of sausage to a cup of lentils, how many total calories will there be?
(B) How many calories from fat will there be?
(C) You want to add sausage to your lentil soup, but you do not want the calories from fat to be more than 15% of the calories. How many ounces of sausage can you add to a cup of lentils to have 15% of the calories from fat?

B36. Robin walks and runs a total of 6 miles. She walks at 4 miles per hour and runs at 8 miles per hour.
(A) If she runs x miles ($x < 6$) and walks the rest, how long will it take her to finish?
(B) If she wants to finish in exactly one hour, how far should she run?

Reading and Writing Mathematics

B37.* Give the formula for the slope of a line through points (a, b) and (c, d).

B38.* Give the formula for a line through (a, b) with slope c.

B39.* (A) How can you tell, just by looking at their equations, if two lines are parallel?
(B) How can you tell, just by looking at their equations, if two lines are perpendicular?

B40. Suppose a line goes through the points (a, b) and (c, d).
(A) State the formula for its slope.
(B) Sketch a picture to illustrate the formula and label on it *all* the points and distances in your slope formula from part (a).

B41.* Why might the phrase "*The* equation of a line" be misleading?

B42.* Which lines can *not* be written in the form $y = mx + b$?

B43. What can the "general" form of a line do that "point-slope form" cannot?

B44.* Explain how to find the equation of a line through a given point and parallel to a line with a given equation. Write your answer in the format of a theorem. Make appropriate use of symbols.

B45. Explain how to find the equation of a line through a given point and perpendicular to a line with a given equation. Write your answer in the format of a theorem. Make appropriate use of symbols.

B46.* Why might point-slope form be preferred to slope-intercept form?

B47.* Explain what "linear interpolation" is.

B48.* Define "proportional." Be sure to give the proper context.

B49.* (A) Define "parameter."
(B) Give an example of a parameter.

B50. Prove that, according to the definition, if y is proportional to x, then x is proportional to y. So we can say they are "proportional to each other."

B51. Prove that "$ca = cb$" is not equivalent to "$a = b$" by giving an example with three numbers where one equation is true and the other is not.

B52. Given the graph of $y = x$, interpret the graph of $y = mx + b$ as a scale change and then a location shift.

B53. Lines are *not* good examples with which to graphically illustrate the effects of shifts due to composition with addition or subtraction. This is because shifts left or right cannot be distinguished graphically from shifts up or down. Explain, mathematically, how horizontal shifts of the graph of $f(x) = 3x - 1$ can be interpreted as vertical shifts of the same graph.

B54. Conjecture: "$\dfrac{a}{b} = c$ is equivalent to $a = bc$." The conjecture is false. Show it is false by giving an example with three numbers where one equation is true and the other is not.

B55. Two variable quantities x and y are said to be <u>inversely proportional</u> (to <u>vary indirectly</u>) iff there is a constant k ($\neq 0$) such that $xy = k$. For example, for rectangles of fixed area k, the length and width are inversely proportional. Give another example of quantities that are inversely proportional.

B56. Prove: If the graph of f is a line with slope m, then the graph of $f(f(x))$ is also a line.

B57. Prove that the composition of any two linear functions is a linear function.

B58. $y - y_1 = \left(\dfrac{y_2 - y_1}{x_2 - x_1}\right)(x - x_1)$. Now set $y = 0$ and solve for x. [The result simplifies nicely. It is an important formula in the subject of "numerical analysis."]

B59. The <u>midpoint</u> of the line segment from (a, b) to (c, d) is given by $\left(\dfrac{a + c}{2}, \dfrac{b + d}{2}\right)$. Sketch a picture to illustrate why.

B60. Consider the line through points (a, b) and (c, d). Show that either point can be used in the point-slope form and equivalent equations result.

Advanced Line Problems

B61. (Newton's Method) Some equations are difficult to solve, even by guess-and-check because the function is difficult to evaluate. Example 18 illustrates the idea of "Newton's Method" for solving such equations. Basically, the idea is to pick a potential solution, x, and to evaluate both $f(x)$ and the slope of the curve at $(x, f(x))$. Then use the method in Example 18 to improve the pick. Solve $x^2 - 10 = 0$, beginning with first guess $x = 3$ and using slope $2x$. Generate one improved approximation to the solution. How accurate is it?

B62. Find and simplify the x-value of the point of intersection of the two lines
$y = m_1x + b_1$ and $y = m_2x + b_2$.

These are functional equations. In each, a function is unknown. Solve for f.

B63. $f(x + 1) = f(x) + 3$, for all x.

B64. $f(x + 5) = f(x) - 3$, for all x.

B65. Prove 3-1-9 for positive slope m. Let the given line be L_1 with slope m and let the perpendicular line, L_2 intersect it at point P. On the horizontal line through P mark point Q one unit to the right. The vertical line through Q will intersect L_1 m units up at, say, R. Locate point S m units to the left of P on the horizontal and then locate T at the intersection of L_2 and the vertical line through S. Show that point T is in fact one unit above S. Then from T to P on L_2 the rise is -1 and the run is m, so the slope of the perpendicular line is $\dfrac{-1}{m}$.

Tangent Lines

B66. Look at the graph of x^2. Zoom in about the origin until the window is very small and the curve looks almost straight. What line does the graph resemble near the origin? How did you determine the line?

B67. Look at the graph of x^2. Zoom in about the point $(1, 1)$ until the window is very small and the curve looks nearly straight. Which straight line does the graph resemble? How did you determine the line?

B68. Suppose $f(3) = 10$ and we know that the slope of the tangent line at any point on the graph is given by $2x$.
(A) Find the tangent line at $(3, 10)$. Then use the tangent line to approximate
(B) $f(3.1)$
(C) $f(2.8)$
(D) The true f is given by $f(x) = x^2 + 1$. How inaccurate are these approximations in (b) and (c)?

B69. Consider the graph of $\sin x$ (in radians) near $(0, 0)$. From calculus, we know the slope of the tangent line at $x = 0$ is 1.
(A) Find the tangent line at $(0, 0)$. Use the tangent line to approximate
(B) $\sin (.1)$
(C) $\sin (-.2)$
(D) How accurate are these approximations?

B70. We can approximate the curve $y = x^2$ near $x = 5$ by a line of slope 10.
  (A) Find the equation of that line. Approximate the value of the function by using the line, and find the error is so doing, at the points where
  (B) $x = 5$;   (C) $x = 5.1$;
  (D) $x = 4.95$;   (E) $x = 6$.
  (F) Comment on your findings.

Other

B71. (This type of problem occurs in the definition of integrals in calculus.) Suppose the interval [2, 5] is subdivided into n intervals of equal length. Number the subintervals from left to right.
  (A) What is the right endpoint of the 7th subinterval?
  (B) What is the right endpoint of the kth subinterval?

B72. (This type of problem occurs in the definition of integrals in calculus.) Suppose the interval $[a, b]$ is subdivided into n intervals of equal length. Number the subintervals from left to right.

  (A) What is the right endpoint of the 5th subinterval?
  (B) What is the right endpoint of the kth subinterval?

B73. In the subject of "differential equations," functions are described, not by giving an expression for $f(x)$ as usual, but by giving two pieces of information:
  (1) a point the graph goes through, and (2) an expression for the slope of the tangent line to the curve at any point. Here is the problem: Suppose we do not know the expression for $f(x)$, but we know $f(25) = 5$ and we know that the slope of the tangent line at any point on the curve is given by

$$\frac{1}{\left(2\sqrt{x}\right)}.$$

  (A) Use this information to approximate $f(26)$.
  (B) In calculus you will learn how to solve for the unknown function f with the given slope formula. It turns out to be $f(x) = \sqrt{x}$. How much error was there in approximating $f(26)$? [Note that the approximation was pretty good!]

Section 3-2 QUADRATICS

Linear functions are both the simplest and most important functions. Quadratic functions (with an "x^2" term) are perhaps the next most important. In physics, quadratic functions occur when an object is dropped or projected subject to the force of gravity. The shape of the graph of a quadratic function is called a parabola, and parabolic shapes have excellent reflective properties. Satellite dishes, microwave relay stations, and telescopes have parabolic reflectors to focus signals. In business, some simple profit functions are quadratics. And, of course, in mathematics, quadratic equations are very common, partly because they are simple enough to solve algebraically by using the Quadratic Formula.

DEFINITION 3-2-1   Any expression equivalent to "$ax^2 + bx + c$" for some a, b, and c, where $a \neq 0$, is called a <u>quadratic</u> in "x". It is "standard notation" to use "a" for the coefficient on "x^2", "b" for the coefficient on "x", and "c" for the constant term. The letters "a", "b", and "c" are the parameters of this family of expressions. The <u>leading</u> <u>coefficient</u> is a. Often, a is simply 1.

EXAMPLE 1   The basic quadratic is x^2 (Figure 1A). All the important properties of quadratics can be studied by studying x^2. The graph of x^2 has the shape known as a parabola. In an important sense, all parabolas are alike. For example, if a parabola seems "wide," it could be the graph of x^2 in a window which emphasizes points near the origin (Figure 1B, on [−.5, .5] by [0, 1]). "Wide" parabolas also could be the graph of x^2 with a large vertical interval (Figure 1C, on [−10, 10] by [0, 1000]).

If a parabola seems "narrow," it could be the graph of x^2 in a window which is wide (Figure 1D, on [−100, 100] by [0,100]). "Narrow" parabolas also could be the graph of x^2 with a small vertical interval (Figure 1E, on [−10, 10] by [0,4]).

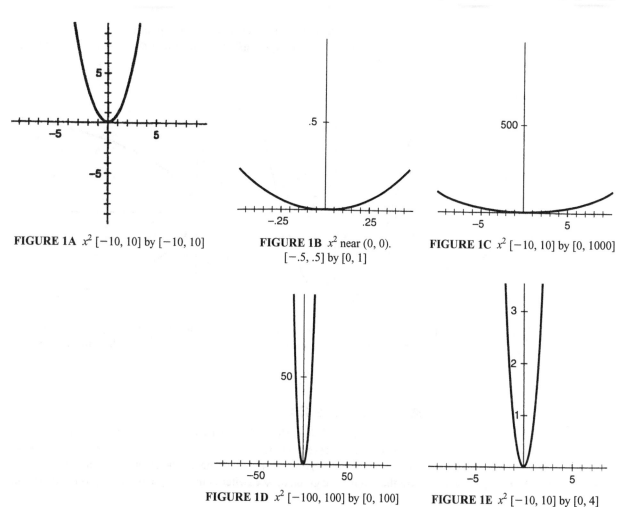

FIGURE 1A x^2 [−10, 10] by [−10, 10]

FIGURE 1B x^2 near (0, 0). [−.5, .5] by [0, 1]

FIGURE 1C x^2 [−10, 10] by [0, 1000]

FIGURE 1D x^2 [−100, 100] by [0, 100]

FIGURE 1E x^2 [−10, 10] by [0, 4]

Figures 1A through 1E show that scale changes can have a dramatic effect on the appearance of a graph.

The most important point on the graph of a quadratic is its <u>vertex</u>, the point corresponding to (0, 0) on the graph of x^2. If the graph <u>opens up</u> like the graph of x^2 does, the vertex is the lowest point on the graph. If the graph is upside down, it <u>opens down</u>, and the vertex is the highest point on the graph (Figure 8).

■ Symmetry

The graph of $f(x) = x^2$ is symmetric about its central axis, the y-axis. This is because the image of $−x$ is the same as the image of x:

$$f(-x) = (-x)^2$$
$$= x^2$$
$$= f(x).$$

Therefore the heights are the same on either side of the y-axis. The y-axis serves as an axis of symmetry.

If $c > 0$ there are two solutions to the equation "$x^2 = c$": $x = \sqrt{c}$ and $x = -\sqrt{c}$. They are at equal distances on either side of the axis of symmetry (Figure 2).

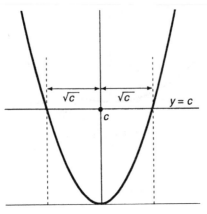 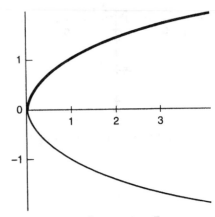

FIGURE 2 $y = x^2$ and $y = c$ intersect at distance $\sqrt{c}$ from the axis of symmetry

FIGURE 3 $y^2 = x. y = \sqrt{x}$ bold. $y = -\sqrt{x}$ on the bottom. [0, 4] by [−2, 2]

■ The Square Root

$y = \sqrt{x}$ implies $y^2 = x$. This is the reverse of the familiar "$x^2 = y$" relationship; we can graph it by reversing the axes. The graph of $y = \sqrt{x}$ (Figure 3) is the upper half (since $\sqrt{x} \geq 0$) of the horizontal version. The graph of x^2 is nearly horizontal near the origin. The graph of $\sqrt{x}$ switches the axes, so it is nearly vertical near the origin.

■ Location Changes

Consider shifting the graph left or right, and up or down. Shifts left or right shift the axis of symmetry.

EXAMPLE 2 $y = x^2 - 6x$ defines a quadratic function. Graph it.
  The Zero Product Rule yields the x-intercepts (where $y = 0$).

$$x^2 - 6x = x(x - 6) = 0 \text{ iff } x = 0 \text{ or } x = 6.$$

The graph crosses the x-axis at 0 and 6 (Figure 4). By the symmetry of the graph, the lowest point, the vertex, is half way between 0 and 6, at $x = 3$. The y-value there is $y = 3^2 - 6(3) = -9$.
  The shape of the graph is the shape of $y = x^2$, but its location is different. In some important ways all quadratics are alike, so by studying x^2 you can learn about them all. We will see that the coefficient "a" on the "ax^2" term can change the scale, but first we note that all quadratics with $a = 1$ (as in Figure 4) have exactly the same shape but not necessarily the same location.
  The lowest point on the graph of x^2 occurs when $x = 0$, since $x^2 \geq 0$ for all x. The lowest point on the graph of "$(x - h)^2 + k$" is when $x = h$ (so the argument

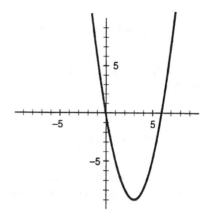

FIGURE 4 $y = x^2 - 6x$. $[-10, 10]$ by $[-10, 10]$

of the squaring function is zero). Then y is k. Therefore, the graph of $(x - h)^2 + k$ has its vertex at (h, k). That is:

SHIFTS 3-2-2 The graph of $(x - h)^2 + k$ differs from the graph of x^2 only in its location. It has the same shape. The <u>vertex</u> of the graph of "$(x - h)^2 + k$" is (h, k). The use of the letters "h" and "k" in this context is traditional.

This expression is a perfect square plus or minus some constant.

DEFINITION 3-2-3 (COMPLETE THE SQUARE) To <u>complete the square</u> of a quadratic expression "$x^2 + bx + c$" (for which $a = 1$) is to rewrite it as an equivalent expression in the form $(x - h)^2 + k$ or, if we use "d" in place of "$-h$": $(x + d)^2 + k$.

When the square is complete there is only one appearance of "x". Then

1. the location of the graph is easy to identify (Example 2), and
2. it fits the inverse-reverse format (Example 2, continued).

To complete the square we must discover h (or d) and k. We prefer to use the form with "h", since that is the x-value of the vertex. An expression such as "$(x + 5)^2$" is a perfect square and is fine to work with, but, in the end, we will treat it as "$(x - -5)^2$," so we can see the "h".

We need the expanded form of perfect squares.

THEOREM 3-2-4 $(x - h)^2 = x^2 - 2hx + h^2$, for all x and h.

$(x + d)^2 = x^2 + 2dx + d^2$, for all x and d.

EXAMPLE 2, AGAIN Complete the square of the quadratic $x^2 - 6x$.

The objective is to rewrite the given expression as a perfect square plus or minus some constant.

$$x^2 - 6x = x^2 - 6x + 9 - 9$$
$$= (x - 3)^2 - 9.$$

In the final expression the square is complete. $h = 3$ and $k = -9$. The vertex is $(3, -9)$ (Figure 4). There is only one appearance of "x".

In this example we added and subtracted 9. How did we know to use 9?

If $a = 1$, **to complete the square, take half the coefficient on x, and add and subtract its square.** Then rewrite the expression to exhibit the perfect square.

**THEOREM 3-2-5
(COMPLETING THE
SQUARE)**

$$x^2 + bx + c = x^2 + bx + \left(\frac{b}{2}\right)^2 - \left(\frac{b}{2}\right)^2 + c$$

$$= \left(x + \frac{b}{2}\right)^2 - \left(\frac{b}{2}\right)^2 + c.$$

The last expression "completes the square" of the first.

**EXAMPLE 2
RECONSIDERED**

Complete the square of $x^2 - 6x$.

Half the coefficient on "x" is -3. Add and subtract $(-3)^2 = 9$.

$$x^2 - 6x = x^2 - 6x + 9 - 9.$$

Now regroup, that is, reorder the operations to exhibit the perfect square.

$$x^2 - 6x = (x - 3)^2 - 9.$$

From this expression we can see that the vertex is $(3, -9)$ (Figure 5).　▬

EXAMPLE 3

Complete the square of $x^2 + 7x + 3$.

$$x^2 + 7x + 3 = x^2 + 7x + \left(\frac{7}{2}\right)^2 - \left(\frac{7}{2}\right)^2 + 3$$

$$= \left(x + \frac{7}{2}\right)^2 - \left(\frac{7}{2}\right)^2 + 3.$$

The vertex has x-value $\dfrac{-7}{2}$.　▬

EXAMPLE 4

Complete the square of $3x^2 + 5x - 1$.

$$3x^2 + 5x - 1 = 3\left(x^2 + \left(\frac{5}{3}\right)x\right) - 1$$

[The leading coefficient is not 1, so factor out a 3 so the squared term has coefficient 1, as in Theorem 3-2-5.]

$$= 3\left[x^2 + \left(\frac{5}{3}\right)x + \left(\frac{5}{6}\right)^2 - \left(\frac{5}{6}\right)^2\right] - 1$$

$$= 3\left[\left(x + \frac{5}{6}\right)^2 - \left(\frac{5}{6}\right)^2\right] - 1$$

$$= 3\left(x + \frac{5}{6}\right)^2 - 3\left(\frac{5}{6}\right)^2 - 1.$$

The vertex has x-value $\dfrac{-5}{6}$.　▬

**EXAMPLE 2
CONTINUED**

Solve $x^2 - 6x = 15$.

Pretend we do not know the Quadratic Formula. Note that "$x^2 - 6x$" has two appearances of "x", so it does not fit the inverse-reverse format requirement. But with the square completed it will have only one appearance of "x".

$$x^2 - 6x = 15 \quad \text{[Now, add 9 to both sides.]}$$
$$\text{iff } x^2 - 6x + 9 = 15 + 9 = 24$$

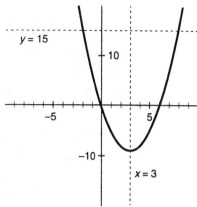

FIGURE 5 $y = x^2 - 6x$ and $y = 15$. $[-10, 10]$ by $[-20, 20]$

iff $(x - 3)^2 = 24$ [The square is "complete," this fits the inverse-reverse format]
iff $x - 3 = \pm\sqrt{24}$
iff $x = 3 \pm \sqrt{24}$.

The "3" is the "h", the x-coordinate of the vertex. Figure 5 shows that the two solutions are equal distances on either side of $x = 3$, which is the axis of symmetry.

■ The Quadratic Formula

The famous "Quadratic Formula" gives a one-step abbreviation of the process of first completing the square and then using the inverse-reverse method.

THEOREM 3-2-6 (THE QUADRATIC FORMULA)

If $a \neq 0$, $ax^2 + bx + c = 0$ is equivalent to

$$x = \frac{-b \pm \sqrt{b^2 - 4ac}}{2a}.$$

"Negative b plus or minus the square root of (the quantity) b squared minus four ac, all over two a."

The idea of the proof is to use Theorem 3-2-5 **to complete the square and then to use the inverse-reverse equation-solving method.** Because Theorem 3-2-5 applies only when $a = 1$, the first step is to divide through by a.

Proof: Because $a \neq 0$, $ax^2 + bx + c = 0$ is equivalent to

$$x^2 + \left(\frac{b}{a}\right)x + \frac{c}{a} = 0 \quad \text{[Next, put the constant on the right.]}$$

$$x^2 + \left(\frac{b}{a}\right)x = -\left(\frac{c}{a}\right) \quad \text{[Next, add the right amount to "complete the square"]}$$

$$x^2 + \left(\frac{b}{a}\right)x + \left(\frac{b}{2a}\right)^2 = -\left(\frac{c}{a}\right) + \left(\frac{b}{2a}\right)^2 \quad \text{[The left side is a square.]}$$

$$\left(x + \frac{b}{2a}\right)^2 = -\left(\frac{c}{a}\right) + \frac{b^2}{4a^2} \quad \text{[The square is "complete."]}$$

Now there is only one appearance of "x". The inverse-reverse format requirement is satisfied.

$$\left(x + \frac{b}{2a}\right)^2 = \frac{b^2 - 4ac}{4a^2} \quad \text{[}4a^2 \text{ is a common denominator for the right side]}$$

$$x + \frac{b}{2a} = \pm \sqrt{\frac{b^2 - 4ac}{4a^2}}$$

$$x = -\left(\frac{b}{2a}\right) \pm \frac{\sqrt{b^2 - 4ac}}{2a} = \frac{-b \pm \sqrt{b^2 - 4ac}}{2a}.$$

This is the famous Quadratic Formula.

EXAMPLE 5 Solve $4x + 2x^2 = 55$.

To use the Quadratic Formula, you must identify a, b, and c of the usual form, and this is not in the usual form. This equation is equivalent to

$2x^2 + 4x - 55 = 0$, in which $a = 2$, $b = 4$, and $c = -55$.

$$x = \frac{-4 \pm \sqrt{4^2 - 4(2)(-55)}}{2(2)} = 4.34 \text{ or } -6.34. \qquad \blacksquare$$

■ Symmetry and the Quadratic Formula

The axis of symmetry of the graph of $ax^2 + bx + c$ is apparent in the Quadratic Formula. The "negative b over $2a$" gives it (Figure 6). The axis of symmetry is just the first term in the Quadratic Formula. The two solutions lie at equal distances on either side of that line of symmetry ("plus or minus the square root of. . . ."). Also, the extreme value of y occurs on the axis of symmetry.

$$x = \frac{-b}{2a} \qquad \pm \qquad \frac{\sqrt{b^2 - 4ac}}{2a}$$

axis of symmetry   distance of solutions on either side.

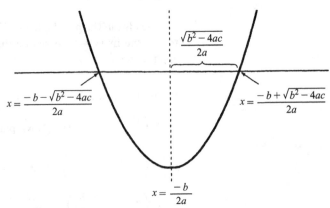

FIGURE 6 $ax^2 + bx + c$ for $a > 0$, labeled to show the terms of the Quadratic Formula

The graph of "$ax^2 + bx + c$" is similar to the graph of x^2 with a scale change and a shift. By completing the square we identify that shift, which is visible in Figure 6.

THEOREM 3-2-7 The vertex of $ax^2 + bx + c$ has

$$x = \frac{-b}{2a} \text{ (again) and } y = c - \frac{b^2}{4a} = -\left(\frac{b^2 - 4ac}{4a}\right)$$

The central axis (the axis of symmetry) is $x = \frac{-b}{2a}$.

If $a > 0$, the graph opens up and the minimum value occurs at $x = \frac{-b}{2a}$.

If $a < 0$, the graph opens down and the maximum value occurs at $x = \frac{-b}{2a}$.

Use the Quadratic Formula to help you remember the x-value of the vertex. Memorizing the y-value is not important; you can always find the y-value by substituting the x-value into the expression.

EXAMPLE 6 Graph $y = x^2 + 5x$. Find the x-value that makes $y = x^2 + 5x$ the minimum possible. Also, solve $x^2 + 5x = 3$ (Figure 7).

The minimum occurs at $x = \frac{-5}{2} [= \frac{-b}{2a}$, by Theorem 3-2-7]. (Since $a > 0$, the quadratic opens up and the vertex yields a minimum.)

Then, the solutions to "$x^2 + 5x = 3$" must be equidistant on either side of $x = -2.5$. The symmetry tells us this, and so does the Quadratic Formula. The solution is Problem A23.

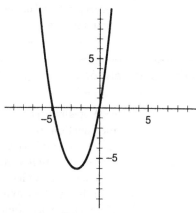

FIGURE 7 $x^2 + 5x$ $[-10, 10]$ by $[-10, 10]$

EXAMPLE 6 CONTINUED Suppose the problem had been slightly different: "For which values of k are there solutions to $x^2 + 5x = k$?"

This question determines the range of "$x^2 + 5x$." (The range is the set of all possible image values.)

Looking at the graph (Figure 7) and its vertex, we see that, for there to be solutions, k must be greater than or equal to the y-value of the vertex. Therefore, $k \geq\; = -6.25$. The range of "$x^2 + 5x$" is $[-6.25, \infty)$. ▬

EXAMPLE 7 Graph $5 + 2x - x^2$ by completing the square and noting the location shift.

Since Theorem 3-2-5 requires $a = 1$, factor out the minus sign.

$$5 + 2x - x^2 = -(x^2 - 2x - 5)$$
$$= -(x^2 - 2x + 1 - 1 - 5)$$
$$= -[(x - 1)^2 - 6]$$

We can graph this as the negative of the dashed curve in Figure 8, which is the graph of x^2 shifted right 1 and down 6. Then the "$-$" sign turns that graph upside down (Figure 8, solid curve).

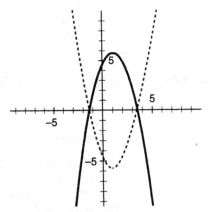

FIGURE 8 $5 + 2x - x^2$. $[-10, 10]$ by $[-10, 10]$ ▬

To complete the square of $ax^2 + bx + c$, you may factor out a ("-1" in Example 7) and use the method of Theorem 3-2-5. Be careful to distribute a and not just discard it.

If the coefficient on x^2 is negative, the graph opens downward.

A quadratic equation has either two, one, or no real-valued solutions. To "Solve the equation $ax^2 + bx + c = 0$" is also to "Find the zeros of the expression $ax^2 + bx + c$," which is to "Find the x-intercepts of the graph of $ax^2 + bx + c$." It is easy to see how many x-intercepts the graph has.

When the graph opens up ($a > 0$), there are three cases (Figure 9). If the vertex is below the x-axis, there will be two solutions. If the vertex is on the x-axis, there will be one solution. Finally, if the vertex is above the x-axis, there will be no real-valued solutions (but there will be two complex-valued solutions given by the Quadratic Formula). If the graph opens down ($a > 0$), similar results hold with the position of the vertex reflected through the x-axis.

The Quadratic Formula always gives the solutions. The number of real-valued solutions depends upon whether the part under the square-root sign is positive.

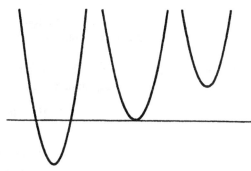

FIGURE 9 Quadratic equations have 2, 1, or no real-valued solutions. (This illustrates ($a > 0$), it opens up.)

THEOREM 3-2-8   If $b^2 - 4ac > 0$, there are two distinct real-valued solutions.

If $b^2 - 4ac = 0$, there is one real-valued solution.

If $b^2 - 4ac < 0$, there are no real-valued solutions (and two complex-valued solutions, because the square root of a negative number yields a complex number).

This is a direct result of the Quadratic Formula. This has nothing new to memorize. The expression "$b^2 - 4ac$" is called the <u>discriminant</u> because it discriminates between those three cases.

■ Complex Numbers

When the discriminant is less than zero, the Quadratic Formula expresses the square root of a negative number. The square root of a negative number cannot be a real number because $x^2 \geq 0$ for all real values of x. But it is possible to *define* a solution to $x^2 = -1$ and give it a name and properties. The solution will not be a "real" number, but, we can treat this "made up" solution as a number with properties in the same way that we treat "-3" as a number with properties. For many centuries mathematicians did not use negative numbers. It was "obvious" that you cannot take 10 coins from a pile of 7 coins. Physically, you cannot. But, now we know that you can, with the concepts of credit and debt. Of course, now we know negative numbers are "real" numbers, and very useful. Similarly, we now know that it is very useful to have the concept of an "imaginary" number, called "i" which satisfies "$i^2 = -1$." Unfortunately, the use and value of "imaginary" and "complex" numbers in electronics, physics, and mathematics is not easily explained in basic algebra. For now, we merely assert that they exist and are important.

DEFINITION 3-2-9   "*i*" is a number such that $i^2 = -1$. The number i can be added to or multiplied by a real number. For *a* and *b* real numbers, a number of the form "*a* + *bi*" is called a complex number, and "*a*" is its real part and b is its imaginary part. All the usual arithmetic operations applied to complex numbers yield complex numbers and all the usual rules of arithmetic operations (such as the commutative, associative, and distributive properties) apply to complex numbers. Also, for real-valued $c \geq 0$,

$$\sqrt{-c} = i\sqrt{c}.$$

With this definition the Quadratic Formula always yields solutions to a quadratic equation—but they might be complex-valued instead of real-valued.

EXAMPLE 8   Solve $x^2 + 2x + 5 = 0$.
The Quadratic Formula yields

$$x = \frac{-2 \pm \sqrt{2^2 - 4(1)(5)}}{2} = \frac{-2 \pm \sqrt{-16}}{2} = \frac{-2 \pm 4i}{2} = -1 \pm 2i.$$

For now, this is all you need to know about complex numbers. They exist. ▬▬

■ The Use of the Quadratic Formula

The Quadratic Formula applies to any quadratic equation, even if the letters are different.

EXAMPLE 9   Solve $x^2 + cx + 2b = 0$.
Now "c" and "b" do not play their usual roles. Read the Quadratic Theorem by position, not letter. $x = \dfrac{-c \pm \sqrt{c^2 - 4(1)(2b)}}{2}$.   ▬▬

EXAMPLE 10   Solve for *b* in the Law of Cosines: $c^2 = a^2 + b^2 - 2ab \cos C$.
Now *b*, not *x*, is the unknown. Here is the equation rearranged left-to-right:

$$b^2 - (2a \cos C)b + a^2 - c^2 = 0.$$

So, $b = \dfrac{2a \cos C \pm \sqrt{(-2a \cos C)^2 - 4(1)(a^2 - c^2)}}{2}$.

The Quadratic Theorem is about solving equations that express squaring, even if it is "*b*" squared instead of "*x*" squared. Also, the coefficients need not be numbers; they can be expressions.   ▬▬

CALCULATOR EXERCISE 1   Graph $x^2 + 3xy + y^2 - 14 = 0$.
Try it. This is not trivial.
Graphics calculators graph equations in the form "$y = \ldots$." *You* must solve for *y* yourself. Treat the unknown as *y* and solve for *y* (instead of *x*). Reorganize the equation to identify the coefficients on y^2 and *y*:

$$y^2 + 3xy + x^2 - 14 = 0$$
$$(1)y^2 + (3x)y + (x^2 - 14) = 0.$$
$$ay^2 + by + c = 0.$$

$$y = \frac{-3x \pm \sqrt{(3x)^2 - 4(1)(x^2 - 14)}}{2(1)}$$

Graph these two equations (Figure 10).

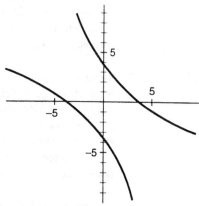

FIGURE 10 The graph of $x^2 + 3xy + y^2 - 14 = 0$. A hyperbola.
$[-10, 10]$ by $[-10, 10]$

> The variables in the Quadratic Formula are placeholders. The unknown does not have to be "x" and the coefficients do not have to be "a", "b", and "c". It can be used to solve any quadratic equation.

■ Applications

There are many applications of quadratics. Some arise in physics.

The formula for the vertical position, $s(t)$ (in feet), at time t (in seconds) of a projectile subject only to the force of gravity is

$$s(t) = -16t^2 + v_0 t + s(0), \tag{3-2-9}$$

where v_0 is the initial upward velocity in feet per second and $s(0)$ is the initial vertical position.

This is a quadratic "in t" instead of "in x." It opens downward, since the coefficient on t^2 is negative. The coefficient "-16" is determined by the force of gravity (and the units we use, which, here, are "feet per second squared").

EXAMPLE 11   If a rock is thrown upward from ground level with an initial velocity of 80 feet per second, how high will it go?

Let the ground level be 0 so $s(0) = 0$. Plug that and $v_0 = 80$ in to the formula.

$$s(t) = -16t^2 + 80t.$$

This is a parabola that opens downward. Its highest point occurs at

$$``\frac{-b}{2a}" \tag{3-2-7}$$

$$t_{max} = \frac{-80}{2(-16)} = \frac{80}{32}.$$

Substitute in to the formula to find the height at that time.

$$s(t_{max}) = -16\left(\frac{80}{32}\right)^2 + 80\left(\frac{80}{32}\right) = 100, \text{ the maximum height.}$$

When will the rock be 60 feet above the ground?

This merely asks you to solve the equation: $s(t) = -16t^2 + 80t = 60$.
There will be two solutions, one on the way up, and the other on the way down (Problem A24).   ▬

EXAMPLE 12   In business, <u>profit is revenue minus cost</u>. Suppose the cost of producing and selling each widget is $200. Suppose that market research tells the firm that, if they are priced at $300, 4000 will be sold. However, if they are priced at $400, only 2200 will be sold. Assume that the number sold at other prices would be on a straight line through those two points. What price will maximize profits?

If the price is higher, you make more profit per widget, but sell fewer widgets. If the price is lower, you sell more widgets, but make less profit per widget. Where is the optimum level?

Draw a graph that illustrates what is given (Figure 11). Let x be the price and y be the number sold at price x. The problem gives two points: (300, 4000) and (400, 2200) and says they are on a line.

$$y - 4000 = \left(\frac{2200 - 4000}{400 - 300}\right)(x - 300)$$

If we know how many will be sold, y, we can calculate the profit. For example, when the cost is $300 the profit per widget is $300 - $200 = $100. At that price the total profit would be $100(4000) = $400,000 because 4000 will be sold.

To find the price which maximizes profit, find the formula for profit in terms of price. When the price is x, call the profit $p(x)$.

$$p(x) = \text{(profit per widget) (number of widgets sold)}$$
$$= (x - 200)(y)$$

$$p(x) = (x - 200)\left[\left(\frac{2200 - 4000}{400 - 300}\right)(x - 300) + 4000\right]$$

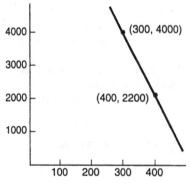

FIGURE 11 Price (x) to number sold (y) given the two points (300, 4000) and (400, 2200)

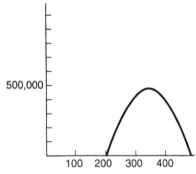

FIGURE 12 $-18p^2 + 13,000p - 1,880,000$
[0, 500] by [0, 1000000]

You can graph this to find the x-value (price) that yields the maximum profit. Or, because it is a quadratic, we can use

$$\text{``}x_{\text{max}} = \frac{-b}{2a}.\text{''}$$

$$p(x) = (x - 200)(-18x + 9400)$$
$$= -18x^2 + 13000x - 1{,}880{,}000.$$

The maximum profit occurs at $x = \dfrac{-b}{2a} = \dfrac{13{,}000}{36} = 361(\text{dollars})$. ▬

CALCULATOR EXERCISE 2

Graph the price-to-profit function.

The trick is to select an appropriate window. Graphing this in the "standard" window will show nothing.

The word problem itself mentions prices of $300 and $400, so the window should include those x-values. What about the y values? (When graphing $p(x)$, your calculator calls p "y".) We already calculated that profit could be as much as $400,000. So y must go up to at least that. Figure 12 graphs it on a window [0, 500] by [0, 1000000].

CONCLUSION

All quadratics are much alike; their graphs differ only by shifts and scale changes. The first term of the Quadratic Formula "$\dfrac{-b}{2a}$" exhibits the axis of symmetry and the x-coordinate of the extreme value of a quadratic. The vertex can be located by completing the square, which is a key step in the proof of the Quadratic Formula.

Terms: quadratic, coefficient, leading coefficient, standard form, complete the square, vertex, open up, open down, Quadratic Formula, axis of symmetry, discriminant.

Exercise 3-2

A

A1.* ☺ If you have the Quadratic Formula memorized, how can you use it to remember the location of the vertex and the axis of symmetry of the graph of a quadratic?

A2.* ☺ If $a > 0$, the minimum value of $ax^2 + bx + c$ occurs when $x =$ _____.

☺ Which x-value yields the minimum value of the expression?

A3. $x^2 - 9x + 3$ **A4.** $x^2 + 7x + 12$

A5. $5x + 3 + 2x^2$ **A6.** $3x + 14 + 4x^2$

☺ Which x-value yields the maximum value of the expression?

A7. $50 + 500x - x^2$ **A8.** $12 - 10x - x^2$

☺ Give the equation of the graph which has the same shape as the graph of x^2, but

A9. (A) 2 units to the left. (B) 4 units up.

A10. (A) 3 units to the right.
(B) 7 units down.

A11. (A) twice as tall.
(B) twice as wide (twice as far from the y-axis).

A12. (A) half as tall.
(B) half as wide (half as far from the y-axis).

Use the Quadratic Formula and your calculator to solve. As usual, give at least three significant digits.

A13. $x^2 - 3.2x + 1 = 0$ $[2.8, \ldots]$.

A14. $2.3x^2 - 1.6x - 4 = 0$ $[1.7, \ldots]$.

A15. $4 + 3.5x^2 = 12x$ $[3.1, \ldots]$.

A16. $-1.2x^2 + 10x - 2.3 = 0$ $[8.1, \ldots]$.

Complete the square of the expression.

A17. $x^2 - 2x$

A18. $x^2 + 7x + 2$

A19. $x^2 + 3x + 1$

A20. $x^2 - 10x + 31$

A21. Find a window $[0, 4]$ by $[0, d]$ such that the graph of x^2 goes through the upper right corner of the window.

A22. Find a window $[-10, 10]$ by $[0, d]$ such that the graph of x^2 exits the right side of the window half way up.

A23. Solve the equation in Example 6: $x^2 - 5x = 3$. $[5.5, \ldots]$.

A24. Solve $-16t^2 + 80t = 60$.

A25. Indicate on a sketch of $x^2 + 5x$ the solutions to $x^2 + 5x = 3$. Label the sketch with the two terms of the Quadratic Formula. Solve the equation.

A26. Indicate on a sketch of $2x^2 + 7x$ the solutions to $2x^2 + 7x = 8$. Label the sketch with the two terms of the Quadratic Formula. Solve the equation.

Use the Quadratic Formula to solve the following equations. Write your answer in "$a + bi$" form:

A27. $x^2 = -16$

A28. $x^2 + x + 1 = 0$

A29. $x^2 + 4x + 5 = 0$

A30. $3x^2 + 1 = 0$

A31. Simplify $\dfrac{f(x + h) - f(x)}{h}$ given $f(x) = x^2$.

A32. ☺ (A) Give the "discriminant" of a quadratic.
(B) What does it discriminate?
(C) Where is it in the Quadratic Formula?

A33.* What does it mean to "complete the square"?

B

B1.* If $a > 0$, illustrate the three possibilities for the number of real-valued solutions to a quadratic equation (Give three graphs and the corresponding number of solutions).

B2.* If $a < 0$, illustrate the three possibilities for the number of real-valued solutions to a quadratic equation (Give three graphs and the corresponding number of solutions).

B3.* Suppose "$ax^2 + bx + c = 0$" has two solutions and $a > 0$. Draw a sketch of the relevant graph and label it to illustrate the two major terms of the Quadratic Formula.

B4.* ☺ (A) How can you tell which quadratic expressions have minima and which have maxima?
(B) Where is the x-value at which the extreme value is attained?

B5.* ☺ How can you tell from the graph of a quadratic expression $f(x)$ how many complex-valued solutions the equation "$f(x) = 0$" has?

☺ *Find the x-value that yields the minimum of the expression.*

B6. $3x^2 + kx + d$

B7. $(x - a)(x - b)$

B8. $kx^2 - dx + 12$

Use the Quadratic Formula to solve

B9. $dx^2 + ex + f = 0$, for x.

B10. $x^2 + 2bx + a = 0$, for x.

B11. $3x^2 + cx = b$, for x.

B12. $y^2 - 2y - 5 = 0$, for y.

B13. $y^2 - 3xy - 20 = 0$, for y.

B14. $y^2 - 2xy + x^2 = 30$, for y.

B15. $bx^2 + 2ax + 3c = 0$ for x.

B16. $y^2 + 2by + a = 0$, for y.

B17. $cx^2 + bx + 1 = 0$, for x.

B18. $16^2 = 20^2 + b^2 - 40b \cos 39°$, for b. $[25, \ldots]$

B19. $3^2 = a^2 + 4^2 - 7a \cos 20°$, for a. $[5.2, \ldots]$

B20. Solve "$(\sin x)^2 + 3 \sin x = 2$" for x, with the help of Theorem 2-3-6 on the inverse sine function. [Give two solutions.]

Sketch the graph in the standard window. Use the Quadratic Formula first.

B21. $x^2 + xy + 2y^2 = 20$

B22. $x^2 + 3xy + y^2 = 17$

Give the equation of the graph which has the same shape as the graph of x^2, but

B23. 2 units down and 3 units to the right.

B24. 4 units to the left and 7 units up.

B25. 5 units down and 3 units to the left.

B26. 6 units to the right and 9 units up.

B27. 4 units to the right and half as wide.

B28. 2 units to the left and 3 times as wide.

B29. 10 times as tall with vertex at (2, 1).

B30. upside down with vertex at (5, 2).

In the picture the graph of x^2 is shifted. Identify the expression graphed.

B31. 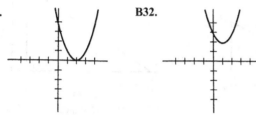   **B32.**

B33. 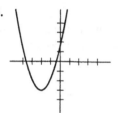   **B34.**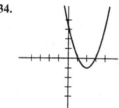

Complete the square.

B35. $2x^2 + 5x$ **B36.** $4x - 2x^2$

B37. $3x^2 + x + 1$ **B38.** $10x^2 + 12x + 7$

B39. Suppose the cost of producing and selling each widget is $100. Suppose that market research tells your firm that, if they are priced at $200, 50,000 will be sold. However, if they are priced at $175, 80,000 will be sold. Assume that the number sold at other prices would be on a straight line through those two points. What price will maximize profits?

B40. The Worldwide Widget Company calculates total profit as profit-per-widget-sold times the number-of-widgets-sold. If the profit-per-widget is $100, they sell 2000 widgets. If the profit-per-widget is $150, they sell only 1000 widgets. Assume a straight line fit for the profit-per-widget to number-of-widgets-sold relationship.
  (A) Find a formula for total profit in terms of the profit-per-widget.
  (B) What profit-per-widget should the company use to maximize total profit?

B41. Suppose a punt in football has a "hang time" of 3.6 seconds. About how high did it go?

B42. Suppose you drop a rock off a high bridge and count off the seconds until it hits the water below. Give a formula for the height of the bridge in terms of the elapsed time.

Reading and Writing Mathematics

B43. Write a theorem for solving all equations of the form "$x^2 < c$." [Be careful if c is negative.]

B44. Complete the theorem: If $x^2 + bx = (x + c)^2 - d$, then $c =$ _____ and $d =$ _____.

B45. Complete the theorem: The equation $x^2 = c$ has two real-valued solutions iff _____.

B46. Find the sum and product of the solutions to: $ax^2 + bx + c = 0$. [It simplifies nicely.]

B47.* The key ideas to deriving the Quadratic Formula were to first divide through by a to make the coefficient on x^2 be 1 and then to _____ and then to use the _____ method of solving equations.

B48. Prove the Quadratic Formula (without looking).

B49.* For the graph of x^2, a vertical scale change is exactly like a different horizontal scale change. Why?

Section 3-3 DISTANCE, CIRCLES, AND ELLIPSES

The Pythagorean Theorem can be used to find the distance between any two points in the plane.

**THEOREM 3-3-1
(THE PYTHAGOREAN
THEOREM)**

Let triangle ABC have sides a, b, and c opposite vertices A, B, and C, as in Figure 1.

$a^2 + b^2 = c^2$ if and only if the angle at vertex C is a right angle.

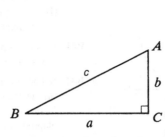

FIGURE 1 $a^2 + b^2 = c^2$ iff
angle ACB is a right angle

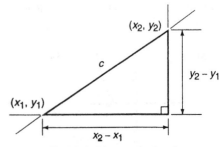

FIGURE 2 Distance in the plane.
$c^2 = (x_2 - x_1)^2 + (y_2 - y_1)^2$

The formula for the distance between any two points in the plane is derived using the Pythagorean Theorem (Figure 2). The distance between the two points is the length of the hypotenuse.

**THEOREM 3-3-2 (THE
DISTANCE FORMULA)**

The distance, d, between points (x_1, y_1) and (x_2, y_2) satisfies:

$$(x_2 - x_1)^2 + (y_2 - y_1)^2 = d^2.$$

$$\text{“}a^2 + b^2 = c^2\text{”}.$$

Therefore: $d = \sqrt{(x_2 - x_1)^2 + (y_2 - y_1)^2}$.

Use Figure 2 to remember this important formula. It is the Pythagorean Theorem.

"The difference in x's, squared, plus the difference in y's, squared, all in a square root."

EXAMPLE 1

Find the distance between the points $(5, -2)$ and $(-7, 3)$.
The distance is

$$\sqrt{(-7 - 5)^2 + (3 - -2)^2} = \sqrt{(-12)^2 + 5^2}$$
$$= \sqrt{144 + 25} = \sqrt{169} = 13.$$

EXAMPLE 2

Find the distance from (x, y) in the first quadrant to the y-axis.
The distance is simply x. The x-coordinate of any point is its directed distance from the y-axis. Similarly, the distance from (x, y) in the first quadrant to the x-axis is y. The distance formula 3-3-2 is not appropriate because it gives the distance between two points, not the distance from a point to a line.

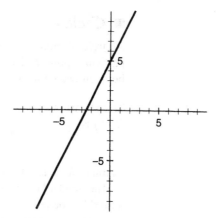

FIGURE 3 $y = 2x + 5.$ $[-10, 10]$ by $[-10, 10]$

EXAMPLE 3   Find the formula for the distances from points on the line $y = 2x + 5$ to the origin (Figure 3).

Meanwhile, the distance from any point (x, y) (not necessarily on the line) to the origin is

$$\sqrt{(x - 0)^2 + (y - 0)^2} = \sqrt{x^2 + y^2}.$$

We want the point to be on the line. Points on the line satisfy $y = 2x + 5$. Therefore,

$$d(x) = \sqrt{x^2 + (2x + 5)^2}.$$

This formula gives the distance from the point $(x, y) = (x, 2x + 5)$ on the line to the origin in terms of only the x-coordinate.

Use your graphics calculator to find the point on the line $y = 2x + 5$ closest to the origin.

The word "closest" tells us we are to minimize the distance. For that we need the distance formula that we just obtained. Graph the function (x horizontal and distance $d(x)$ vertical) and find the x-value of the lowest (smallest distance) point on the graph (Figure 4). It appears to be about $x = -2$. From the original equation, when $x = -2, y = 2x + 5 = 2(-2) + 5 = 1$. The closest point is $(-2, 1)$.

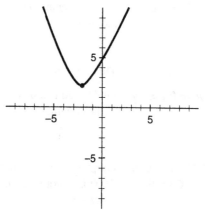

FIGURE 4 $d(x)$ from Example 3. The distance from the origin to $(x, 2x + 5)$ in terms of x. $[-10, 10]$ by $[-10, 10]$

■ Circles

A <u>circle</u> of radius r is, by definition, the set of all points at distance r from a particular point called the <u>center</u> of the circle. Formula 3-3-2 for the distance between two points yields the equations of circles.

CIRCLE 3-3-3 The <u>standard form</u> of the equation of the circle of radius r and center (h, k) is

$$(x - h)^2 + (y - k)^2 = r^2.$$

"x minus h, squared, plus y minus k, squared, equals r squared."

The standard form has three parameters—two for the location of the center and a third for the radius.

The standard form is just the square of the distance formula. The radius r is the distance between (h, k) and the general point (x, y) on the circle (Figure 5). In Figure 5 it is easy to see that

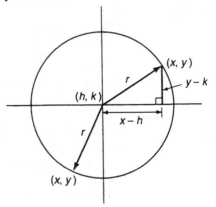

FIGURE 5 A circle. Center (h, k), radius r. $(x - h)^2 + (y - k)^2 = r^2$

The equation of a circle follows from the Pythagorean Theorem!

EXAMPLE 4 Find the equation of the <u>unit circle</u>, that is, the circle with center at the origin and radius 1.

$$(x - 0)^2 + (y - 0)^2 = 1^2, \text{ that is, } x^2 + y^2 = 1.$$

This is the most famous and important circle. ▬

EXAMPLE 5 Find the equation of the circle with radius 5 and center $(-4, 2)$. Plugging in to 3-3-3, the equation is

$$(x + 4)^2 + (y - 2)^2 = 25.$$ ▬

CALCULATOR EXERCISE 1 Graph the above equation on your calculator.

This is not in the usual "$y = \ldots$" form. You must solve for y. Use the inverse-reverse method.

$$(y - 2)^2 = 25 - (x + 4)^2.$$

$$y - 2 = \pm\sqrt{25 - (x + 4)^2}$$

$$y = 2 \pm \sqrt{25 - (x + 4)^2}$$

Graph these two equations to see the circle.

EXAMPLE 6 Identify the circle: $x^2 + 4x + y^2 - 6y = 5$.

A circle is identified by its center (h, k) and radius r, which are visible in the "standard form":

$$(x - h)^2 + (y - k)^2 = r^2.$$

To obtain that form, complete the square.

$$x^2 + 4x + y^2 - 6y = 5 \text{ iff}$$
$$x^2 + 4x + 4 + y^2 - 6y + 9 = 5 + 4 + 9 = 18,$$
$$\text{iff } (x + 2)^2 + (y - 3)^2 = 18 = (\sqrt{18})^2.$$

The center is $(-2, 3)$ and the radius is $\sqrt{18}$ (Figure 6).

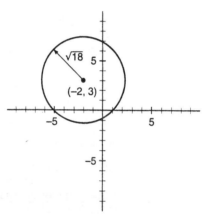

FIGURE 6 The circle with center $(-2, 3)$ and radius $\sqrt{18}$. $[-10, 10]$ by $[-10, 10]$

■ Ellipses

Equations of circles have terms with x squared and y squared. Scale changes convert circles to ellipses.

EXAMPLE 7 The equation of the unit circle is "$x^2 + y^2 = 1$." Graphed on a square scale, it looks like a circle, as it should. However, with a different window it can look different. Graph it with the window $[-1, 1]$ by $[-2, 2]$.

Your calculator wants y given functionally. So, solve for y.

$$y^2 = 1 - x^2,$$
$$y = \pm\sqrt{1 - x^2}.$$

Plot two graphs together, one for the plus sign and one for the minus sign.

Try different windows. Figure 7 is $[-1, 1]$ by $[-2, 2]$ and the shapes of the graph is an ellipse. Equations of circles yield circles *only on square scales*.

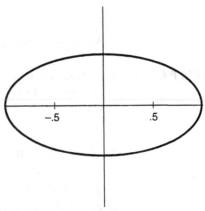

FIGURE 7 $x^2 + y^2 = 1$ on $[-1, 1]$ by $[-2, 2]$ ▬

EXAMPLE 8 Reconsider "$x^2 + y^2 = 1$." What would be the difference if "x" were replaced by "$\dfrac{x}{2}$"? Graph "$\left(\dfrac{x}{2}\right)^2 + y^2 = 1$" on a square scale.

Compare the two equations:
$$x^2 + y^2 = 1 \qquad \text{[unit circle]}$$
$$\left(\frac{x}{2}\right)^2 + y^2 = 1.$$

The only difference is that the x-value of the second is divided by two before it is squared. So $x = 1$ in the first corresponds to $x = 2$ in the second. $x = \dfrac{1}{2}$ in the first corresponds to $x = 1$ in the second. For a given y-value, the x-value of the second is doubled compared to the first. Therefore, the graph of the second is the graph of the first expanded horizontally by a factor of 2 (Figure 8). This shape is called an ellipse.

This fits the result from Section 2-2 about composition which states that dividing x by 2 before applying the function yields a graph which is expanded horizontally by a factor of 2.

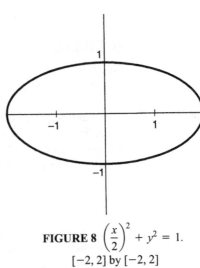

FIGURE 8 $\left(\dfrac{x}{2}\right)^2 + y^2 = 1.$
$[-2, 2]$ by $[-2, 2]$ ▬

EXAMPLE 9 Graph $x^2 + \left(\dfrac{y}{2}\right)^2 = 1$.

Compare this equation to the equation of the unit circle:

$$x^2 + y^2 = 1 \text{ [unit circle]}$$

$$x^2 + \left(\dfrac{y}{2}\right)^2 = 1.$$

For any given value of x, the y-value in the second equation would have to be twice the y-value in the first equation. The graph is the unit circle expanded vertically by a factor of two (Figure 9).

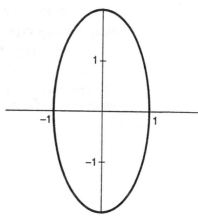

FIGURE 9 $x^2 + \left(\dfrac{y}{2}\right)^2 = 1. \ [-2, 2] \text{ by } [-2, 2]$ ▬

ELLIPSE 3-3-4 The equation $\dfrac{x^2}{a^2} + \dfrac{y^2}{b^2} = 1$ is the "standard form" of the equation of an ellipse centered at the origin. Its graph can be obtained from the graph of the unit circle by expanding the graph horizontally by a factor of a and vertically by a factor of b.

Of course, an ellipse centered at (h, k) can be represented by replacing "x" by "$x - h$" and "y" by "$y - k$," according to the ideas about location shifts in Section 2-2.

ELLIPSE 3-3-5 The equation $\dfrac{(x - h)^2}{a^2} + \dfrac{(y - k)^2}{b^2} = 1$ is the "standard form" of the equation of an ellipse centered at (h, k). It extends a units to the left and right of (h, k) and b units above and below (h, k). To describe this family of curves the standard form uses four parameters—two for the location of the center and two more for the size.

So ellipses are squashed or expanded circles. They have many important properties which will be studied later. If the equation is not in standard form, "Complete the square."

EXAMPLE 10 Graph the ellipse "$x^2 + 6x + 4y^2 - 32y + 37 = 0$" by putting the equation in standard form.

First, complete the squares on both "x" and "y"

$$x^2 + 6x + 4y^2 - 32y + 37 = 0$$

$$x^2 + 6x + 9 + 4y^2 - 32y + 37 = 0 + 9$$

$$(x + 3)^2 + 4y^2 - 32y = 9 - 37 = -28$$

$$(x + 3)^2 + 4[y^2 - 8y] = -28$$

$$(x + 3)^2 + 4[y^2 - 8y + 16 - 16] = -28$$

$$(x + 3)^2 + 4[(y - 4)^2 - 16] = -28$$

$$(x + 3)^2 + 4(y - 4)^2 = -28 + 64 = 36$$

Now, divide by 36 to obtain the "1" on the right required for standard form.

$$\frac{(x + 3)^2}{6^2} + \frac{(y - 4)^2}{3^2} = 1$$

The center is $(-3, 4)$. It stretches 6 units to the right and to the left and 3 units up and 3 units down (Figure 10).

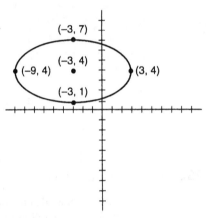

FIGURE 10 An ellipse centered at $(-3, 4)$ with $a = 6$ and $b = 3$

CONCLUSION Distance in the plane is given by the Pythagorean Theorem. Circles are determined as all points at a fixed distance (the radius) from the center. Ellipses are circles with the scale changed.

Terms: Pythagorean Theorem, distance, circle, standard form (of a circle), ellipse.

Exercise 3-3

A

A1.* ☺ The formula for the distance between two points in the plane is essentially just the _____ theorem from geometry.

Find the distance between

A2. (4, 6) and (2, −3). [9.2]

A3. (−1, 3) and (−9, 5). [8.2]

A4. (0, 0) and (3, 4).

A5. (5, 7) and (x, y).

A6. (x, 3) and (2x, 7).

☺ *Find the shortest distance from the given point to the y-axis.*

A7. (5, 6)   **A8.** $(9, 2\pi)$

A9. (−2, −5)   **A10.** (1, 7)

☺ *Find the shortest distance from the given point to the x-axis.*

A11. (3, 7)   **A12.** $(\pi, 14)$

A13. (−2, 4)   **A14.** (−41, −90)

A15. ☺ Find the shortest distance from the point (1, 2) to

(A) the line (x = −4)   (B) the line y = 10

A16. ☺ Find the shortest distance from the point (3, 8) to

(A) the line x = 10   (B) the line y = 6

Circles

A17.* ☺ (A) Give the "standard form" of the equation of a circle.

(B) How many parameters does it have?

A18. ☺ (A) Give the equation of the unit circle.

(B) How many parameters does it have?

☺ *Find the equation of the circle with given center and radius:*

A19. Center (2, 6) and radius 3.

A20. Center (−1, 4) and radius $\sqrt{5}$.

☺ *Identify the center and radius of the circle:*

A21. $x^2 + (y − 2)^2 = 16$.

A22. $(x − 1)^2 + (y + 5)^2 = 10$.

A23.* ☺ How are ellipses related to circles?

B

B1.* Sketch a picture to illustrate the distance formula and label on it *all* the relevant points and distances.

Note that the distance formula is just the Pythagorean Theorem from geometry.

B2.* (A) State the "standard form" of the equation of a circle.

(B) Draw a (large) picture to illustrate it, labeling *all* its variables and expressions on the picture.

(C) Explain, using some geometry, why the standard form is right.

B3. ☺ Give the shortest distance from the point (a, b) to the y-axis.

B4. ☺ Give the shortest distance from the point (c, d) to the x-axis.

Distance

B5. Find x such that the distance from (1, 2) to (x, 5) is 6 units. [6.2 or . . .]

B6. Find y such that the distance from (−5, −3) to (2, y) is 10 units. [4.1 or . . .]

B7. Find the point on the line y = 3x − 7 closest to the origin. [x = . . . and y = −.70]

B8. Find the point on the curve $y = x^2$ closest to the point (2, 1). [x = . . . and y = 1.4]

Give the formula for the distance from the given point to any point on the given line, in terms of the x-coordinate of the point on the line.

B9. (4, 7) to y = 3x − 1.

B10. (−1, 3) to y = 2x + 6.

B11. Use Problem B9 to find the x-value of the point on the line y = 3x − 1 closest to the point (4, 7). [x = . . ., distance = 1.3]

B12. Use B10 to find the x-value of the point on the line y = 2x + 6 closest to the point (−1, 3). [x = . . ., distance = .45]

B13. A pilot flies his fighter plane at low altitude from his base at (0, 0) into the first quadrant. His flight path has the equation y = x(8 − x). How close does he come to an enemy with a Stinger missile at location (5, 17)? [Distances are in miles.] [1.3]

B14. Find the distance from the point (3, 4) to the nearest point on the unit circle.

Circles

B15. Find the equation of the circle with center (−2, 1) that goes through (3, 5).

B16. Find the equation of the circle with center $(2, 5)$ that goes through $(-1, -2)$.

B17. (A) Identify and sketch the graph of
$$x^2 + (y + 2)^2 = 25.$$
(B) Is there a trick to getting your calculator to graph this? If so, what is it?

B18. (A) Identify and sketch the graph of
$$(x + 1)^2 + (y - 3)^2 = 4.$$
(B) Is there a trick to getting your calculator to graph this? If so, what is it?

Give the center and radius of the circle.

B19. $x^2 + 6x + y^2 - 10y = 0$

B20. $x^2 + y^2 + 4y = 6x - 12$

B21. $4x^2 + 6x + 4y^2 - 8y = 12$

B22. $x^2 - 5x + y^2 = 7$

B23. Find the equation of the circle with center $(2, 3)$ and tangent to the line $y = 7$.

B24. Find the equation of the circle with center $(-1, 2)$ and tangent to the line $x = 5$.

Consider the graph of $x^2 + y^2 = 1$ on $[-2, 2]$ by $[-2, 2]$.

B25. How could you choose a new y-interval to make it look half as high?

B26. How could you choose a new y-interval to make it look twice as high?

B27. How could you choose a new x-interval to make it look twice as wide?

B28. How could you choose a new x-interval to make it look one quarter as wide?

Complete the squares to find the standard form of the ellipse with the given equation.

B29. $x^2 - 2x + 3y^2 - 12y = 50$

B30. $4x^2 + 8x + y^2 - 6y = 90$

B31. The area of a circle is πr^2.

(A) Determine the area of the ellipse $\left(\dfrac{x}{2}\right)^2 + y^2 = 1$ by considering scale changes.

(B) Determine the area of the ellipse $\left(\dfrac{x}{a}\right)^2 + \left(\dfrac{y}{b}\right)^2 = 1$ by considering scale changes.

B32. Here is the graph of $x^2 - y^2 = 1$ (a hyperbola). Use it and a scale change to find the graphs of

(A) $\left(\dfrac{x}{2}\right)^2 - y^2 = 1.$

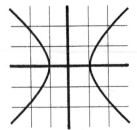

(B) $x^2 - \left(\dfrac{y}{2}\right)^2 = 1.$

B33. Suppose the outside of a circle with equation $x^2 + y^2 = 1$ is mirrored so it reflects light. Light from point $(2, 0)$ reflects off it through $(0, 4)$. Where does it reflect off the mirror? [According to the science of optics, it reflects so that the total distance the light travels from $(2, 0)$ to the point on the mirror to $(0, 4)$ is the minimum possible.]

B34. $x^2 + y^2 = 1$ is the equation of a circle.
(A) Use ideas from Section 2-2 to explain the shape of the graph of $(2x)^2 + y^2 = 1$.
(B) Use ideas from Section 2-2 to explain the shape of the graph of $x^2 + (2y)^2 = 1$.

B35. Find the equation of the line tangent to the unit circle $x^2 + y^2 = 1$ at the point in the first quadrant on the circle where $x = .2$.

B36. Find an equation of the line through $(5, 0)$ that is tangent to the unit circle in the first quadrant.

B37. Find the circle with center $(2, 1)$ tangent to the graph of $y = 4x + 1$.

B38. A baseball park has the outfield fence 325 feet from home plate down both foul lines and 400 feet from home plate in the middle of center field. The fence is an arc of a circle. Where is the center of the circle located?

B39. Redraw Figure 3 for finding the point on the line closest to the origin, as in the problem in "Example 3, continued." Note that the shortest distance will be measured perpendicular to the line. Use Theorems 3-1-9 (on the slopes of perpendicular lines) and 3-3-1 to determine the closest point.

B40. Redraw Figure 3 for finding the point on the line closest to the origin, as in the problem in "Example 3, continued." Note that the shortest distance will be measured perpendicular to the line. Use similar triangles and geometry to determine the distance to the closest point.

B41. The distance from the line $ax + by + c = 0$ to the point (x_0, y_0) is given by $\dfrac{|ax_0 + by_0 + c|}{\sqrt{a^2 + b^2}}$. Prove it.

[There are several ways. One parallels Problem B39, another parallels B40.]

B42. (A) Give the equation of the unit circle centered at the origin.

(B) Explain why $\left(\dfrac{x}{2}\right)^2 + y^2 = 1$ is twice as wide as the unit circle.

(C) Explain why $x^2 + \left(\dfrac{y}{2}\right)^2 = 1$ is twice as tall as the unit circle.

B43. Solve for the function f: $f(2x) = 4f(x)$, for all x.

Hyperbolas

B44. A basic hyperbola is given by the equation $x^2 - y^2 = 1$.

(A) Sketch its graph.

(B) What is the least positive x-value for which there is a y-value?

(C) According to Section 2-2, the graph of $\left(\dfrac{x}{2}\right)^2 - y^2 = 1$ is twice as wide (twice as far from the y-axis). What is the least positive x-value for which there is a y-value?

(D) Which similar equation has a graph that is a times as wide as the basic hyperbola?

(E) Reconsider the basic hyperbola. Which equation would yield a graph b times as far from the x-axis?

(F) Reconsider the basic hyperbola. Which equation would yield a graph that is both a times as far from the y-axis and b times as far from the y-axis?

(G) In all of the above, the graphs "open left and right." If you wanted a graph like that of the basic hyperbola but rotated 90 degrees so it opened up and down, give the new equation.

(H) In all of the above, the center is at $(0, 0)$. If you wanted the same graph as in part (f) shifted so its center was at (h, k), what would be the equation?

Section 3-4   GRAPHICAL FACTORING

To <u>factor</u> an expression means to write it as an equivalent product. To factor a quadratic expression means to write it as an equivalent product of linear terms, usually in one of the forms:

$$(x - b)(x - c) \qquad \text{[if the leading coefficient is 1] or}$$
$$k(x - b)(x - c) \text{ or } (kx - d)(x - c) \qquad \text{[the leading coefficient is } k\text{] or}$$
$$(ax - b)(cx - d) \qquad \text{[the leading coefficient is } ac\text{].}$$

The Zero Product Rule makes solving some quadratic equations extremely simple—if the expression is already factored.

EXAMPLE 1   Solve $(x - 2)(x + 3) = 0$.

This is trivial because it is already factored. The solution is: $x = 2$ or $x = -3$. Figure 1 graphs the expression. To solve the equation is to determine the x-values where the value of the expression is zero.

Factored form can be useful, however, there are many times when we do *not* want a quadratic expression factored.

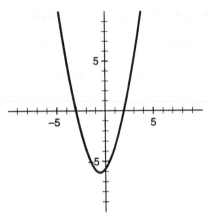

FIGURE 1 $(x - 2)(x + 3)$ $[-10, 10]$ by $[-10, 10]$

EXAMPLE 2 Solve $(x - 5)(3x + 4) = 5$.

In this equation factored form is useless. The Zero Product Rule only works for *zero* products, and this product is 5. You need to multiply it out and use a different method (Problem A11).

Of course, any quadratic equation can be solved with the Quadratic Formula, so you never need to use the Zero Product Rule to solve quadratic equations. Therefore, there is less reason to factor expressions after you understand the Quadratic Formula. Nevertheless, there are some other uses of factoring (see Section 4-3 on higher-degree polynomial equations and Section 4-5 on rational functions). Because factoring is the reverse of "multiplying out," it is appropriate to multiply out a few products first.

■ The Distributive Property

In previous sections we have had reasons to "consolidate like terms," "factor," and "multiply out" expressions. All of these processes rely on the Distributive Property (of multiplication over addition).

THE DISTRIBUTIVE PROPERTY 3-4-1 $a(b + c) = ab + ac$.

EXAMPLE 3 $3(102) = 3(100 + 2) = 3(100) + 3(2) = 306$.

You use the Distributive Property every time you multiply a two-digit number times a one-digit number by hand.

EXAMPLE 4

$$
\begin{array}{r@{\qquad}r@{\qquad}r}
24 & 20 + 4 & b + c \\
\times\, 3 & \times\, 3 & a \\
\hline
72 & 12 & ac \\
 & 60 & ab \\
\cline{2-3}
 & 72 & ab + ac \\
\end{array}
$$

You may have learned to write your work all on one line (as in the leftmost version) by "carrying." 4 times 3 equals 12, put down the 2 and carry the 1. Then 3 times 2 equals 6 plus the carried 1 is 7. Put down the 7.

Really, there is no "3 times 2" in this problem. There is "3 times 20." The algorithm (method) shortens the problem by dealing with the 20 as 2 tens by working with the tens digit. Also, there is no "1" to "carry"; it is really 10.

The Distributive Property also works for subtraction, which can be regarded as addition of the negative. ▬

EXAMPLE 5 $3(98) = 3(100 - 2) = 3(100) - 3(2) = 300 - 6 = 294.$

This subtraction version comes in handy when you are shopping. When you want to know how much 2 compact discs priced at $14.99 each cost, use the Distributive Property: $2(15) - 2(.01) = 30 - .02 = 29.98.$ ▬

The Extended Distributive Property deals with products of two sums.

THEOREM 3-4-2 (EXTENDED DISTRIBUTIVE PROPERTY) For all a, b, c, and d,

$$(a + b)(c + d) = ac + ad + bc + bd.$$

This theorem follows from using the Distributive Property three times.

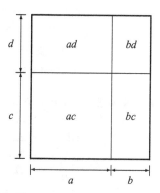

FIGURE 2 The extended Distributive Property

Proof:

$$(a + b)(c + d) = a(c + d) + b(c + d)$$

[using the Distributive Property with "$c + d$" as the factor and "$a + b$" as the sum]

$$= ac + ad + bc + bd$$

Many people remember this using the acronym FOIL: First, Outer, Inner, Last.

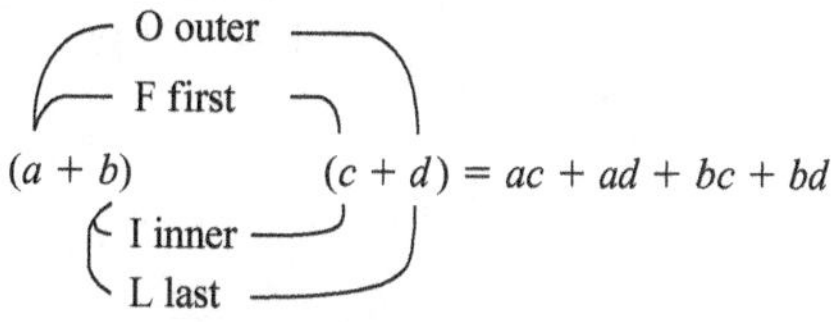

$$(a + b)(c + d) = ac + ad + bc + bd$$

- O outer
- F first
- I inner
- L last

FIGURE 3 FOIL

This pattern is exactly the same pattern we use to multiply two-digit numbers by each other:

EXAMPLE 6 Compute 27×34 by hand.

abbreviated	expanded	more expanded	with variables
34	34	30 + 4	$c + d$
27	27	20 + 7	$a + b$
238	28	7×4	bd
68	210	7×30	bc
918	80	20×4	ad
	600	20×30	ac
	918	918 total	$ac + ad + bc + bd$

FIGURE 4 Multiplication of two-digit numbers

The method you learned in school is used on the left of Figure 4. The "expanded" version in the second column shows the real steps without so much abbreviation. The role of the Distributive Property is emphasized in the "more expanded" version in the third column. The version with variables is on the right. But the idea is identical, whether we use numbers or variables. ■

By definition, a <u>corollary</u> to a theorem is another result so closely related to the first result that it needs little or no further proof.

COROLLARY 3-4-3
COMMON PRODUCTS

For all a, b, and x,
(A) $x^2 + (a + b)x + ab = (x + a)(x + b)$.
(B) $x^2 + 2ax + a^2 = (x + a)^2$.
(C) $x^2 - 2ax + a^2 = (x - a)^2$.
(D) $x^2 - a^2 = (x + a)(x - a)$.
(E) $x^2 + ax = x(x + a)$.

This theorem is used left-to-right for factoring. Used right-to-left, each product is "multiplied out" or "expanded."
Figure 5 illustrates Part A.

$$x + a$$
$$\times$$
$$x + b$$
$$\overline{bx + ab}$$
$$x^2 + ax$$
$$\overline{x^2 + (a + b)x + ab.}$$

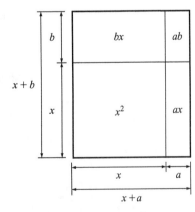

FIGURE 5 $(x + a)(x + b) = x^2 + ax + bx + ab$

Each term in each line multiplies each term in the other line. The "*bx*" and "*xa*" terms in the product consolidate into "$(a + b)x$" which is called the <u>cross product</u> term (perhaps from the cross in the image).

In 3-4-3 part B note that

The square of a sum is <u>not</u> the sum of the squares.

Order matters. Also, the square of a difference is not the difference of the squares (Part C). Order Matters. The difference of the squares is factored in part D. The expression "$x^2 + a^2$" does not appear. It does not factor using real numbers.

Usually students do not immediately learn to factor general quadratics, but only special cases that factor *using integers*.

EXAMPLE 7 Factor "$x^2 - 2x - 15$" using integers.

This can be done with guess-and-check. *If* it factors into $(x + a)(x + b)$, then, $ab = -15$ and $a + b = -2$ (Corollary 3-4-3, Part A). If a is to be integer-valued and $ab = -15$, then a must be 1, 3, 5 or 15, or the negative of one of these. By inspection, we see that $a = -5$ and $b = 3$ will work.

$$x^2 - 2x - 15 = (x - 5)(x + 3).$$

These factoring ideas can be summarized in a theorem. ▬

COROLLARY 3-4-4 ON FACTORING $x^2 + cx + d$ factors into $(x + a)(x + b)$ iff

$ab = d$ and $a + b = c$.

This follows immediately from Corollary 3-4-3A by matching the coefficients on "x" and the constants.

$$x^2 + cx + d$$
$$= (x + a)(x + b) = x^2 + (a + b)x + ab.$$

So $a + b = c$ and $ab = d$.

■ Advanced Factoring Methods

Sometimes we factor an expression so we can use the Zero Product Rule to solve an equation. The upcoming Factor Theorem reverses the process; we can solve an equation to factor an expression.

For this factoring method you must keep the distinction between expressions and equations in mind.

EXAMPLE 7 REVISITED "$x^2 - 2x - 15$" is a quadratic *expression*. Call it "$P(x)$."
"$x^2 - 2x - 15 = 0$" is a quadratic *equation*, "$P(x) = 0$."

Suppose we want to solve the equation "$P(x) = 0$." One way is to factor the expression to obtain the equivalent equation:

$$\text{"}(x + 3)(x - 5) = 0.\text{"}$$

Here the solutions "-3" and "5" are evident. Because "$x + 3$" is a factor, "-3" is a solution. Because "$x - 5$" is a factor, "5" is a solution.

The Factor Theorem allows us to reverse both the problem and the steps. It says that, because "-3" is a solution, "$x + 3$" is a factor. Because "5" is a solution, "$x - 5$" is a factor. The Factor Theorem says that we can factor a polynomial expression, "$P(x)$", by, instead, solving an equation, "$P(x) = 0$," whenever we can find some other way besides factoring to solve the equation. The Quadratic Formula and graphing provide other ways.

EXAMPLE 7 CONTINUED

Factor "$x^2 - 2x - 15$."

Pretend we don't see its factors. The upcoming Factor Theorem says solving "$x^2 - 2x - 15 = 0$" helps.

By the Quadratic Formula, $x = 5$ or $x = -3$. The factors are therefore "$x - 5$" and "$x - -3$."

$$x^2 - 2x - 15 = (x - 5)(x + 3).$$

This method does not use guess-and-check. **A quadratic expression can always be factored without any guessing** by using the Factor Theorem and the Quadratic Formula.

EXAMPLE 7 GRAPHICALLY

Factor "$x^2 - 2x - 15$."

We may graph "$y = x^2 - 2x = 15$" to solve the equation "$x^2 - 2x - 15 = 0$" (Figure 6). The solutions appear to be 5 and -3. Evaluating the expression at those values confirms they are solutions. Therefore, by the Factor Theorem, "$x - 5$" and "$x + 3$" are factors. $x^2 - 2x - 15 = (x - 5)(x + 3)$.

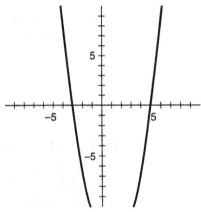

FIGURE 6 $x^2 - 2x - 15$. $[-10, 10]$ by $[-10, 10]$ ▬

THE FACTOR THEOREM 3-4-5

(A) "$x - c$" is a factor of the quadratic expression $P(x)$ if and only if c is a zero of $P(x)$.

[Zeros are solutions to "$P(x) = 0$." They are sometimes called "roots".]

(B) If the graph of $P(x)$ intersects the x-axis at c, then "$x - c$" is a factor of $P(x)$.

(C) If b and c are the two solutions to the quadratic equation "$P(x) = 0$," then $P(x) = k(x - b)(x - c)$, for some constant $k \neq 0$.

[*k* remains to be determined by some other fact. The Factor Theorem does *not* determine the leading coefficient, *k*, because the zeros of a polynomial do not determine its vertical scale.]

If *b* and *c* are the two solutions to the quadratic equation "*P(x)* = 0,"
then *P(x)* = *k(x* − *b)(x* − *c)*, for some constant *k* ≠ 0.

From the Zero Product Rule we already knew the reverse: factors of "*P(x)*" yield solutions to "*P(x)* = 0."

EXAMPLE 8   Factor $x^2 - 2x - 24$ using integers.

It factors into $k(x - b)(x - c)$, where we need to find $k, b,$ and c. By the Factor Theorem, solve

$$x^2 - 2x - 24 = 0.$$

The Quadratic Formula gives $x = 6$ or $x = -4$.
Therefore, by the Factor Theorem,

$$x^2 - 2x - 24 = k(x - 6)(x + 4).$$

Since the leading coefficient is 1, $k = 1$. Therefore,

$$x^2 - 2x - 24 = (x - 6)(x + 4).$$

If a quadratic expression has leading coefficient 1 and it factors in integers, the Factor Theorem and the Quadratic Formula or a graph make finding the factors **easy**.

EXAMPLE 8
GRAPHICALLY

Another way to factor "$x^2 - 2x - 24$" is to find its zeros using its graph (Figure 7). The zeros appear to be 6 and −4. Evaluating the expression at these values confirms they are solutions to the equation "$x^2 - 2x - 24 = 0$," so the Factor Theorem then gives the desired linear factors, "$x - 6$" and "$x + 4$."

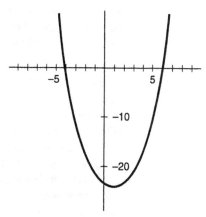

FIGURE 7 $x^2 - 2x - 24$.   [−10, 10] by [−30, 10]

EXAMPLE 9 Find the quadratic $P(x)$ given $P(4) = 0$, $P(-2) = 0$, and $P(1) = 7$ (Figure 8).

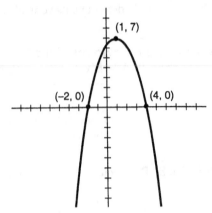

FIGURE 8 A quadratic with $P(4) = 0$, $P(-2) = 0$, and $P(1) = 7$.
$[-10, 10]$ by $[-10, 10]$

The zeros of P are given as 4 and -2. They tell us that $P(x)$ factors into

$$P(x) = k(x - 4)(x + 2),$$

where k remains to be determined. We can use the one other value, $P(1) = 7$, to determine k. Substituting 1 for x,

$$P(1) = k(1 - 4)(1 + 2) = -9k.$$

But $P(1) = 7$ was given.
Equating these two,

$$-9k = 7.$$

Therefore $k = \dfrac{-7}{9}$. $P(x) = \left(\dfrac{-7}{9}\right)(x - 4)(x + 2)$. ▬

A quadratic is determined by three parameters. In "standard" form they are the "a", "b", and "c" in "$ax^2 + bx + c$." In factored form they are the "k", "b", and "c" of "$k(x - b)(x - c)$." (The uses of the letters differ between the two forms). *Three* parameters can be determined by *three* facts. The *two* solutions to the equation in the Factor Theorem are not enough by themselves. The leading coefficient served as the third fact in Example 8, and "$P(1) = 7$" served as the third fact in Example 9.

Factoring using Real or Complex Numbers
In previous math classes, when you were asked to factor a quadratic, chances were good it would factor *in integers*. If a quadratic does not factor *in integers*, you usually don't want to factor it. This subsection considers those rare cases when you do.

There are many quadratics which do not factor using only integers, but which do factor if you use real numbers or complex numbers. So, when the instructions say, "Factor such-and-such a polynomial," you must first decide what the ground rules are. Probably the rules are to factor it using factors with *integer* coefficients. However, it is possible (and easy) to factor any quadratic into two linear factors using the Quadratic Formula, even if the factors do not have integer coefficients.

EXAMPLE 10   Factor $x^2 - 2x - 14$.

What are the ground rules? Are we to use integers? Or real numbers?

If it factors as in Corollary 3-4-4 using *integer-valued a* and *b*, then $ab = -14$ and *a* must be 1, 2, 7, or 14, or the negative of one of these. If you try every possibility, you discover that none work. This does not mean that it does not factor. It just means it does not factor *nicely* (that is, using integers). It does factor using real numbers, but you will not guess the factors easily. Use the Factor Theorem and the Quadratic Formula to find the factors.

The solutions to $x^2 - 2x - 14 = 0$ yield the factors. Using the Quadratic Formula we discover one solution is $1 + \sqrt{15}$ and the other is $1 - \sqrt{15}$. Treating these solutions as *a* and *b*, it factors into $(x - a)(x - b)$.

$$x^2 - 2x - 14 = \left[x - (1 + \sqrt{15})\right]\left[x - (1 - \sqrt{15})\right].$$ ▬

According to the Factor Theorem, quadratics can *always* be factored. That is not to say they always factor using integers. Real-valued solutions to the equation yield real-valued factors, and solutions that use complex numbers yield complex-valued factors.

EXAMPLE 11   Factor $x^2 - 2x + 5$ over the complex numbers.

By the Quadratic Formula, the solutions to $x^2 - 2x + 5 = 0$ are $1 \pm 2i$. Therefore, because the leading coefficient is 1, by the Factor Theorem,

$$x^2 - 2x + 5 = [x - (1 + 2i)][x - (1 - 2i)].$$ ▬

EXAMPLE 12   Does $x^2 + x - 1$ factor using integers?

This can be answered by trial-and-error using Corollary 3-4-4. Another approach is to graph it (Figure 9). If it factors into $(x - b)(x - c)$ using integers, *b* and *c* would have to be integer-valued zeros. Since the graph does not intersect the *x*-axis at integer values, it does not factor using integers. It will, however, factor using real numbers (Problem A32), as we can see from the fact the graph crosses the *x*-axis twice.

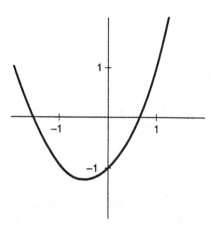

FIGURE 9 $x^2 + x - 1$. [−2, 2] by [−2, 2]   ▬

CONCLUSION The Zero Product Rule gives solutions from factors. The Factor Theorem gives factors from solutions. If b and c are the two solutions to the quadratic equation "$P(x) = 0$," then $P(x) = k(x - b)(x - c)$, for some constant $k \neq 0$. k is determined by some other point on the graph of $P(x)$.

The Quadratic Formula or graphing can be used with the Factor Theorem to determine factors of quadratics.

Terms: Distributive Property, factor (verb and noun), extended Distributive Property, cross product, Factor Theorem.

Exercise 3-4

A

A1.* ☺ Let $P(x)$ be a quadratic.
(A) The zeros of P are the solutions of

_____.

(B) If c is a zero of P, then _____ is a factor of $P(x)$.

A2.* ☺ (A) If a quadratic factors into "$(x - a)(x - b)$" using integers, what can we say about the graph of the quadratic?
(B) Which theorem tells you that?

☺ *The Factor Theorem does not determine constant factors. Give another quadratic expression, not equivalent to this one, which has the same zeros.*

A3. $(x - 3)(x - 5)$ **A4.** $(x + 2)(x - 7)$

A5. $3x^2 - 6x - 12$ **A6.** $2x^2 + 5x - 3$

A7–A10 factor using integers. Factor the expression by using a graph (and the Factor Theorem) to discover the factors.

A7. $x^2 - x - 12$ **A8.** $x^2 - 10x + 16$

A9. $2x^2 + 10x - 48$ **A10.** $3x^2 + 6x - 144$

A11–A14 do not factor using integers. Factor the expression by using the Quadratic Formula (and the Factor Theorem) to discover the factors.

A11. $x^2 + x - 1$ **A12.** $x^2 - 3x + 1$

A13. $2x^2 + x - 4$ **A14.** $5x^2 + x - 2$

A15. Solve $(x - 5)(3x + 4) = 5$.

A16. Solve $(x - 1)(x + 2) = 19$.

Factor the following expressions using traditional methods such as Corollaries 3-4-3 and 3-4-4.

A17. $x^2 - 4$

A18. $x^2 + 2x + 1$ **A19.** $x^2 + 6x + 8$

Multiply out ("expand") the following products:

A20. $(x - 3)(x - 2)$ **A21.** $(x - 3)^2$

A22. $\left(x - \sqrt{5}\right)^2$

A23. If "$3x - 5$" is a factor of a quadratic, $P(x)$, give a solution to $P(x) = 0$.

A24. If "$2x + 7$" is a factor of a quadratic, $P(x)$, give a solution to $P(x) = 0$.

A25.* True or false?
(A) $a^2 + b^2 = (a + b)^2$
(B) $a^2 - b^2 = (a - b)^2$
(C) $(a + b)(a - b) = a^2 - b^2$

A26. Can $x^2 + 4x - 13 = 0$ be solved
(A) by completing the square and using the inverse-reverse method?
(B) by the Quadratic Formula?
(C) by factoring using integers?
(D) by guess-and-check?

A27. Can "$x^2 - 5x - 14 = 0$" be solved
(A) by completing the square and using the inverse-reverse method?
(B) by the Quadratic Formula?
(C) by factoring using integers?
(D) by guess-and-check?

A28.* True or False?
(A) Any quadratic equation that can be solved can be solved using the Quadratic Formula.
(B) Any quadratic expression can be factored using integers.

(C) Factored form is always more useful than expanded (multiplied out) form.

(D) Expanded form is always more useful than factored form.

Consolidate like terms:

A29. $3x + x \tan 25 + 3\sqrt{x} + (\sqrt{x}) \ln 5$

A30. $\ln(x) + 3x + 2 \ln(x) - x \sin 36°$

A31. $x \ln 5 + 5 \ln(x) + 7x - \ln(x)$

A32. Solve $ab = -14$ and $a + b = -2$ (for Example 10).

Reading and Writing Mathematics

A33.* (A) What does the acronym "FOIL" mean?

(B) State, using variables, the theorem FOIL is intended to help you remember.

A34.* Restate the Distributive Property using some other letters.

A35. Restate Corollary 3-4-4 using some other letters.

A36. If $Q(x) = k(x - 5)(x + 3)$ and $Q(2) = 7$, find k.

A37. If $Q(x) = k(x - 2)(x + 5)$ and $Q(4) = 11$, find k.

Rewrite the expressions so all the numbers are integers:

A38. $12(x - 1.5)(x - 4)$

A39. $6(x - 3)(x + 2.5)$

A40. $24(x - 3.5)\left(x - \dfrac{4}{3}\right)$

B

B1.* Draw a figure to illustrate $(x + a)(x + b) = x^2 + (a + b)x + ab$. Label all the distances and areas mentioned in the identity.

B2.* Draw a figure to illustrate $(x + a)^2 = x^2 + 2ax + a^2$. Label all the distances and areas mentioned in the identity.

With the aid of a graph, factor these expressions using integers.

B3. $4x^2 + 4x - 3$ **B4.** $4x^2 - 5x - 6$

Solve algebraically:

B5. $(x - 2)^2(x + 1) + (x - 2)(x + 1)^2 = 0$

B6. $(x^2 - 4)(x + 3) + (x + 2)(x - 1)(x - 4) = 0$

B7. $3(x - 2)^5(x + 7)^2 - 5(x - 2)^4(x + 7)^3 = 0$

B8. $x^2(2x - 5)^3 - x^3(2x - 5)^2 = 0$

B9–B11. *Use the figure and the labeled points to find the expression graphed (factored form is recommended).*

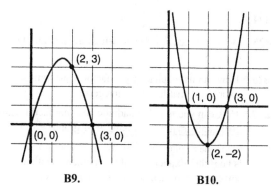

B9.   **B10.**

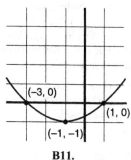

B11.

B12. Find a quadratic expression $P(x)$ with zeros at 2 and 5 such that $P(1) = 12$.

B13. Find a quadratic expression $P(x)$ with zeros at -3 and 4 such that $P(0) = 2$.

B14. Find a quadratic expression $P(x)$ with zeros at 1 and 3 such that $P(0) = -7$.

B15. Find a quadratic, $Q(x)$, such that $Q(3) = 0$, $Q(7) = 0$. and $Q(4) = 5$.

B16. Find a quadratic, $Q(x)$, such that $Q(2) = 0$, $Q(5) = 0$. and $Q(3) = 7$.

Give a linear factor of P(x) using integers.

B17. If $x = \dfrac{3}{2}$ solves $P(x) = 0$.

B18. If $x = \dfrac{-5}{7}$ solves $P(x) = 0$.

Factor

B19. $3x^2 - 5x - 1$ using real numbers.

B20. $2x^2 + x - 3$ using real numbers.

B21. $x^2 + x + 4$ using complex numbers.

B22. $3x^2 + 2x + 2$ using complex numbers.

B23. $3x^2 - 5x + 1$

B24. $4x^2 + 2x - 7$

B25. Factor as much as possible using only real numbers: $x^4 - 16$.

B26.* How can its graph help you factor a quadratic expression that factors in integers?

B27.* How can you tell if a quadratic does not factor using real numbers just by looking at its graph?

B28. Someone might say "$x^2 - 3x - 17$ does not factor." Is he right? Explain why this statement is either correct or incorrect, depending on his intention.

B29.* We do not always prefer quadratic expressions to be factored rather than multiplied out. Give a reason why we might rather have a quadratic multiplied out. Give an example with a factored quadratic that we would rather have multiplied out.

B30.* (A) Explain how the Quadratic Formula can be used to factor quadratics.
(B) Will it always work?

Reading and Writing Mathematics

B31.* Write (symbolically) how to do problems like "$3.98 times 4" and "$7.95 times 2" in your head.

B32.* Complete the theorem:
"$x^2 + kx + p = (x + b)(x + c)$, for all x, iff $k =$ _____ and $p =$ _____."

B33. Complete the theorem: Let $P(x)$ be a quadratic expression. If $P(b) = 0$, then _____ is a factor of _____.

B34. Complete the theorem: Let $P(x)$ be a quadratic expression. _____ is a factor of _____ iff _____ is a solution to _____.

B35.* Draw a figure to illustrate $(x - a)^2 = x^2 - 2ax + a^2$. Explain where the "$+a^2$" comes from.

When you multiply these by hand, there are four terms (sometimes written in only two lines). Write out, with all four terms, the product by hand. Draw and label a rectangle with these dimensions to illustrate the product as the area of the rectangle, and the four terms as smaller rectangles within the large rectangle (See Figure 2, but make your pictures fit the numbers given in B36 and B37).

B36. 23×37.

B37. 12×26.

B38. The "Remainder Theorem": Consider a polynomial $P(x)$ divided by another polynomial, $Q(x)$, of lower degree. If $P(x) = S(x)Q(x) + R(x)$, where $S(x)$ is a polynomial and $R(x)$ is of lower degree than $Q(x)$, then $R(x)$ is said to be the "remainder." [This is similar to the case with division of integers. When 17 is divided by 5 the remainder is 2 because $17 = 3(5) + 2$, where 2 is less than 5.] **The Remainder Theorem**: If a polynomial $P(x)$ is divided by $x - c$, then the remainder is $P(c)$.

(A) Let $P(x) = x^2 + 3x + 5$. Verify the Remainder Theorem for this polynomial and $c = 0$ by dividing this $P(x)$ by $x - 0$ and finding both the remainder and $P(0)$.
(B) Verify this for $c = 2$.

B39. (See the previous problem) The "Remainder Theorem": In the previous problem let $Q(x)$ be $x - c$. Prove: If a polynomial $P(x)$ is divided by $x - c$, then the remainder is $P(c)$.

B40. The Factor Theorem follows rapidly from the Remainder Theorem (see the previous problem). Prove it.

B41. Suppose $P(x) = cx^2 + dx + e$ is a quadratic with integer-valued c, d, and e. Suppose further that for some rational number, r, $P(r) = 0$. Write $r = \frac{n}{m}$ in lowest terms, where n and m are integers. Prove that the other solution also must be rational. What can we deduce about n and m in terms of c, d, and e?

Section 3-5 WORD PROBLEMS

Algebra is applicable to real-world problems, not just to homework problems with x's. Algebra helps design cell phones. Algebra helps analyze the controversial relationship between carbon dioxide and global warming. Real-world problems do not begin with x's. They begin with words. Real-world problems are "word problems."

In algebra, word problems are difficult because the words suggest operations that you are *not* supposed to *do*. On the contrary, you are supposed to *write* the operations in formulas.

■ Cue Words Indicate Operations

Words that suggest mathematical operations are called <u>cue</u> words (for example, "difference" may indicate subtraction). Word problems are said to be <u>direct</u> when the cue words indicate the action required to do the problem.

E X A M P L E   1   Sue sold 40 raffle tickets. John sold 25. How many fewer than Sue did John sell?

The phrase "fewer than" indicates subtraction and the problem gives the numbers in the right order. $40 - 25 = 15$. The problem is "direct." ▬

E X A M P L E   2   The theater has 20 rows of seats with 14 seats in each row. What is the total number of seats in the theater?

The words "each" and "total" indicate multiplication. The numbers are the right numbers to multiply $20 \times 14 = 280$. The problem is "direct." ▬

Direct problems are computational tasks expressed in words. *If* the cue words in a word problem indicate the required operations, the problem is easy. However, in algebra the cue words do not tell you what to do.

An <u>indirect</u> word problem is one in which the words suggest operations that are not the operations you actually do.

E X A M P L E   3   John sold 12 fewer tickets than Bill. John sold 35. How many did Bill sell?

The word "fewer" suggests subtraction, but the right operation to do is addition. The problem is indirect. The answer is $35 + 12 = 47$.

The words indicate the right formula. Let "J" denote the number John sold and "B" denote the number Bill sold. In mathematical notation,

"John sold 12 fewer tickets than Bill"

is   "$J = B - 12$."

Given this formula, the problem is easy. "John sold 35." Just plug in 35 for J to obtain

$$35 = B - 12. \quad B = 47.$$

In this example the cue word "fewer" does *not* suggest the right operation to *do*, but it does suggest the right formula $(J = B - 12)$. ▬

Indirect word problems use cue words to build formulas.

E X A M P L E   4   There are three weights. The second is twice the first, and the third is five times the first. The sum of the weights is 40 pounds. What is the first weight?

Build a formula for the sum of the weights in terms of the first weight.

Let the first weight be "x". Then the second is $2x$, and the third is $5x$. The formula for the total weight, W, is $W = x + 2x + 5x = 8x$.

Now set this sum equal to 40 and solve: $8x = 40$. $x = 5$. ▬

E X A M P L E   5   A field has the shape of a square with two semicircular caps on the ends. (Figure 1). Its area is 30. Find the side of the square.

Name the answer. Let the side of the square be, say, x. Label the picture.

Build your own formula for the area in terms of the side.

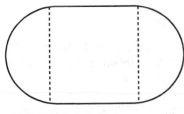

FIGURE 1 A field. A square with two semicircular caps

The field is a square plus two semicircles. **Write down relevant basic formulas.** The basic formulas are the area of a square and the area of a circle:

$$A_{\text{square}} = x^2. \; A_{\text{circle}} = \pi r^2.$$

The diameter is the side, so the radius is half the side: $r = \dfrac{x}{2}$. **Build the formula.**

So, the total area is

$$A(x) = x^2 + \pi \left(\frac{x}{2}\right)^2.$$

Now, set this equal to 30 and solve.

To do this problem you do not actually *do* the operations that would compute the area of the square or the circle. You *represent* them in a formula, without actually doing them. You end up doing completely different operations (Problem A2). ▬

■ Writing Formulas

The key to word problems is to "Build your own formula."

EXAMPLE 6   Phone calls cost 99 cents for all calls up to twenty minutes and 7 cents a minute for each minute after the first twenty. How much does a call cost?

The answer depends on the length of the call. The answer is a formula.

Let x be the length of the call. Let $C(x)$ be its cost.

$$C(x) = 99 \quad \text{if } x \le 20, \text{ and}$$

$$C(x) = 99 + 7(x - 20) \text{ if } x > 20. \quad [\text{``}x - 20\text{'' is the number of minutes after the first twenty.}]$$

If you don't see why this is the formula, compute the cost of any particular length call, say, 25 minutes. Its cost will be 99 cents for the first twenty minutes and 7 cents for each of the 5 minutes after $20(25 - 20 = 5)$:

$$C(25) = 99 + 7(25 - 20) = 134 \,(\text{cents}).$$

This displays the operations you do. Now, **write them** using x:

$$C(x) = 99 + 7(x - 20), \text{ if } x > 20. \quad ▬$$

EXAMPLE 7   Density is mass divided by volume. Lead has density 11.3 grams per cubic centimeter (cc). Copper has density 8.93 grams per cc. If you alloy (mix) copper with 200 cc's of lead, what will be the density of the alloy?

The density depends upon how much copper you add. The answer is a formula, not a number.

Let x be the number of cubic centimeters of copper added to the lead. The density is the total mass divided by the total volume.

The total mass will be the mass of the lead plus the mass of the copper.

$$M(x) = 11.3(200) + 8.93x$$

The total volume will be the volume of the lead plus the volume of the copper.

$$V(x) = 200 + x.$$

The formula for the density expresses mass divided by volume:

$$D(x) = \frac{11.3(200) + 8.93x}{200 + x}$$

▬

EXAMPLE 8 Consider a right triangle (Figure 2). One leg is twice the length of the other. The shorter leg is 6. What is its perimeter?

This is a calculation problem, not an algebra problem.

The longer leg is twice her shorter leg: $2(6) = 12$. The hypotenuse is given by the Pythagorean Theorem: $\sqrt{6^2 + 12^2}$. The perimeter is the sum:

$$P = 6 + 12 + \sqrt{6^2 + 12^2}$$

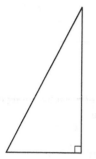

FIGURE 2 A right triangle with one leg twice the other

Consider a right triangle (Figure 2, again). One leg is twice the length of the other. What is its perimeter?

It depends. The answer is a formula.

Let x be the shorter leg.

Write down, in symbols, the same operations used to compute the perimeter:

$$P(x) = x + 2x + \sqrt{x^2 + (2x)^2}$$

This formula *represents* the steps in the calculation *without doing them*. This is what algebraic notation is good for.

EXAMPLE 8
THE WORD
PROBLEM

A right triangle has one leg that is twice the length of the other. The perimeter is 50. How long is the shorter leg?

Now that you have the formula, the problem is easy to set up:

$$x + 2x + \sqrt{x^2 + (2x)^2} = 50$$

Problem A3 asks for the solution.

▬

EXAMPLE 9 Suppose there are no taxes on the first $12,000 of income, but income in excess of $12,000 is taxed at a 15 percent rate. What fraction of income is paid in taxes?

It depends upon the income. The answer is a formula.

Let the income be x.

The formula for taxes is $T(x) = 0$ if $x \leq 12,000$
and $T(x) = .15(x - 12,000)$ if $x > 12,000$

which expresses (in symbols): "15 percent of the amount in excess of $12,000."

Now, reread the problem.

The "fraction of income paid in taxes" is 0 if $x \leq 12,000$, and

$$\frac{T(x)}{x} = \frac{.15(x - 12,000)}{x}, \text{ if } x > 12,000.$$

▬

■ Guess-and-Check

Guess-and-check can help you build formulas. Formulas express the operations in problems. You can see how to build a formula by guessing a number and doing the operations to it. Then write those same operations in terms of "x".

EXAMPLE 10 An open-topped box is formed from a 10 by 15 inch rectangular sheet of metal by cutting a square from each corner, folding up the sides, and sealing the seams. How big are the cutout squares which yield the maximum possible volume?

Reread the problem several times, if necessary. **Draw and label a picture** (Figure 3).

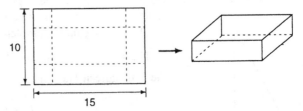

FIGURE 3 An open-topped box formed from a sheet of metal by cutting out the corners and folding up the sides

Name the answer. Let the side of the square be x.

You want the size that maximizes the volume, so you need a formula for the volume.

Write down any relevant basic formula:

$$V_{box} = bwh.$$

If you immediately see the formula in terms of x, fine. If not, **take a guess** at the answer. For example, suppose the cutout were 2 inches square. What would the volume of the box be?

Well, the base would be 6 inches by 11 inches (why?), and the depth would be 2 inches. Therefore, the volume would be $6 \times 11 \times 2 = 132$ (cubic inches).

The key is not the numbers. The key is to see *how* the numbers were obtained. Where did the "6", "11", "2" and "132" come from? It is the operations you want to discover. How did you find the short base side? The longer base side? The depth? The volume?

From the picture you can see that the sides are reduced by two times the amount folded up.

cutout size	shorter base side	longer base side	depth	volume
2	6	11	2	$6 \times 11 \times 2 = 132$
3	4	9	3	$4 \times 9 \times 3 = 108$
1	8	13	1	$8 \times 13 \times 1 = 104$
x	$10 - 2x$	$15 - 2x$	x	$(10 - 2x)(15 - 2x)x$

The last line expresses how the numbers were obtained. It expresses the evaluation process. It expresses the proper formula:

$$V = (10 - 2x)(15 - 2x)x.$$

Use this formula to find the cutout size that maximizes the volume. Graph the expression and locate the x-value which yields the maximum.

Consider the domain. What are reasonable values for x? Negative values are not possible. Also, you cannot cut out a square greater than 5 inches on a side (why?). So the appropriate domain is [0, 5]. Now choose the y-interval. The table shows y-values can be at least as large as 132. So, try a y-interval with larger y-value, say [0, 150] (Figure 4). It is easy to find the maximum given the formula (Problem A4).

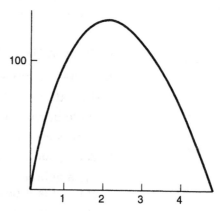

FIGURE 4 $V = (10 - 2x)(15 - 2x)x$ [0, 5] by [0, 150]

■ How to Solve Word Problems

The words in word problems suggest calculations. Sometimes you can just do the calculations. If so, do them. But, sometimes you cannot. Then the problem is indirect and algebraic notation is required.

EXAMPLE 11 A rectangle has one side three inches longer than the other. The total area is 75 square inches. How long are the sides?

What can you calculate?

Nothing yet. You know "one side is three inches longer than the other," but you can't compute the one side because you don't know the other side. You know "area equals base times height" but you don't know the base or the height. It's time for algebra! **Draw and label a picture** (Figure 5).

Name the answer. Let the shorter side be "x".

Write down what you know:

The other side is "$x + 3$."

Write down any relevant basic formula. The problem mentions area, so write down the area formula:

$$A = bh.$$

Substituting the sides:

$$A = x(x + 3).$$

This is the formula you need.

Now use "the area is 75 square inches."

$$A = x(x + 3) = 75.$$

Now the "word" part of the word problem is over (Problem A5).

The advice of this section is:

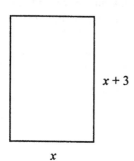

FIGURE 5 A rectangle with one side 3 inches longer than the other

Build your own formula.

■ How to do Word Problems (When You Don't Know What to Do)

0. Read the problem, several times if necessary.

    ***** Start *writing down, in algebraic notation,* what you do know! *****

1. Draw and label a picture, if possible.

2. Name the answer (perhaps "*x*").

Use letters to label any useful parts.

3. Write down any relevant basic formula.

If there is no well-known formula that fits perfectly,

4. Build your own formula

Use algebraic notation. The formula represents the relationships of the quantities in the problem.

Writing a formula for a problem illustrates the purpose of algebraic notation:

The purpose of algebraic notation is to represent operations and order, especially if the number to which the operations apply is unknown.

Here is another example of the guess-and-check approach to formula-writing.

EXAMPLE 12   A pair of animal pens are constructed from 100 linear feet of fence (including two gates) in the rectangular configuration in Figure 6. What is the area of the construction?

First read the problem, several times if necessary, until you understand it.

The area depends upon the shape, which is not given exactly. The answer is a formula.

Suppose you call the base x (label the picture). Find the area in terms of x.

The "guess-and-check" approach suggests you first try a few particular numbers. For example, if the width were 30 feet, what would the length and area be?

If the width were 30 feet, in the picture the top and bottom together would be 60 feet, and, out of the 100 total feet, the remaining 40 would be split among three equal lengths. So the length would be $\dfrac{40}{3}$ and the area $30 \times \dfrac{40}{3} = 400$ (Square feet).

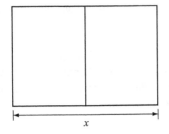

FIGURE 6 Animal pens constructed from 100 linear feet of fence

base	fencing for bases	height	area
30	60	$\dfrac{40}{3}$	$30 \times \dfrac{40}{3} = 400$
20	40	20	$20 \times 20 = 400$
40	80	$\dfrac{20}{3}$	$40 \times \dfrac{20}{3} = 266.67$
x	$2x$	$\dfrac{(100 - 2x)}{3}$	$\dfrac{x(100 - 2x)}{3}$

The formula symbolically represents the operations you did with particular numbers.

Suppose the total area of the construction is 384 square feet. How long is the base? We built the formula. The hard part is done.

$$A(x) = \frac{x(100 - 2x)}{3} = 384.$$

This is a quadratic which can be solved with the Quadratic Formula (Problem A6).

EXAMPLE 13 Find the point on the parabola $y = x^2$ closest to the point (3, 2).
Draw a picture (Figure 7). Label it.

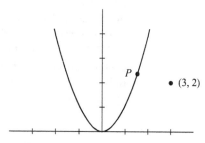

FIGURE 7 $y = x^2$, the point (3, 2), and a point on the curve

Name your answer. The answer is a point, so call it, say, P. Sketch it on your picture.
Use any relevant basic formula. What formula is relevant?
"Closest" refers to minimum distance. Use the distance formula.
The distance formula uses x- and y-values, so, on your picture, label the point $P = (x, y)$.
General Advice: Name and label both what you know and what you need to know. You need x and y to use the distance formula.
The distance from any point (x, y) to (3, 2) is

$$d = \sqrt{(x - 3)^2 + (y - 2)^2}$$

The point you want is not just "any point." It is on the parabola: $y = x^2$. Substituting:

$$d = \sqrt{(x - 3)^2 + (x^2 - 2)^2}$$

Now the formula is easy to minimize. Graph it to find the x-value of the closest point. The point's y-value will be x^2 (Problem A7).

EXAMPLE 14 An airplane flying due east at 120 miles per hour leaves Bozeman at noon. Another airplane leaves at 1:00 pm and flies due south at 150 miles per hour. When will they be 200 miles apart?

To do this problem, you do **not** need to know how to do this problem! **Start writing** what you do know, and the solution will develop.

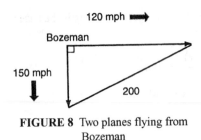

FIGURE 8 Two planes flying from Bozeman

Reread the problem. **Draw a picture** (Figure 8).

The question asks "When?" **Name your answer**, say, t (for time). The question gives a distance. Therefore you must relate time to distance. **Write down any relevant basic formula**.

$$d = rt.$$

"Distance equals rate times time."

This formula works for a single airplane, but this problem has two airplanes and requires a custom formula for distance between the two planes. **Build your own formula**.

If you don't see how to get the formula, guess-and-check can help. Guess an answer. Make it any convenient reasonable guess (say, two hours) and calculate the distance. Pay attention to the operations you use because you will use the same operations in the formula you build.

In two hours, the eastbound plane would have flown 240 miles ($=120$ miles per hour $\times$ 2 hours). The southbound plane would have flown 150 miles ($=150$ miles per hour $\times$ 1 hours). Their distance apart, d, would be given by the Pythagorean Theorem.

$$d^2 = 240^2 + 150^2.\, d = 283.0\text{ (miles)}.$$

You wanted to get "200" miles. The "guess" did not "check," but that does not matter. The guess helped expose the sequence of operations. To build the formula, write the sequence of operations.

Use "t" instead of "2" and use algebraic notation to express the guess-and-check process.

The eastbound plane is $120t$ miles east of Bozeman.

The southbound plane is $150(t - 1)$ miles from Bozeman. These express the legs of the triangle. The distance apart at time t is therefore,

$$d(t) = \sqrt{(120t)^2 + [150(t - 1)]^2}.$$

This is the formula you wanted to build. With the formula, the problem is straightforward. Now, and only now, you use the "200 miles" in the problem. "Plug in" the 200.

$$200 = \sqrt{(120t)^2 + [150(t - 1)]^2}.$$

Now solve for t (Problem A8).

CONCLUSION Formulas (not numbers) are the key to word problems in algebra. Algebraic problems are indirect because the words suggest operations you should write about, but do not actually do. You use the words to build your own formula.

Guess-and-check can be used to help discover the relevant formula because the formula expresses the guess-and-check process in mathematical notation. The purpose of algebraic notation is to express processes (operations and order), even if (*especially* if) the numbers to which they apply are not known.

To do a word problem you do not need to know how to do it! The solution will develop if you:

1. **Start writing** what you do know.

2. **Use letters** (such as "x") to represent parts it would help to know.

3. **Build your own formula.**

But, first you must **start writing!**

Here are more-detailed guidelines:

1. **Draw a picture**, if appropriate. Pictures can illustrate relationships between the components of the problem.
2. **Name the answer**, possibly "*x*". (Label the picture.)
3. **Write down relevant well-known formulas**. Use your "*x*" (and possibly other letters) to express the relationships in the problem.
4. **Build your own formula**.
   Cue words suggest operations to be expressed in algebraic notation in your custom-built formula. This is the stage where guess-and-check may help you find the formula. You may want to use additional letters to name other useful quantities.
5. Given the formula, plug in the known quantity and solve for the unknown.

Terms: cue word, indirect word problem.

Exercise 3-5

A

A1. (A)* Define "cue" word [in the context of a word problem].

   (B) Define "indirect" in the context of a word problem.

A2. Find the solution to Example 5.

A3. Find the solution to "Example 8, the Word Problem."

A4. Find the solution to Example 10.

A5. Find the solution to Example 11.

A6. Find the solution to "Example 12, continued."

A7. Find the point's *y*-value in Example 13.

A8. Find the solution to Example 14.

☺ *Use cue words and well-chosen letters to write a formula expressing the information.*

A9. Rose paid twice as much as Mike. [Use *R* and *M*.]

A10. Widgets cost $2 more from KingWidget than they do from NationalWidget. [Use *K* and *N*.]

A11. Widgets cost twice as much at KingWidget as they do at NationalWidget.

A12. There are three times as many widgets as dodads. [Hint: Do not answer $3W = D$ because that is wrong!]

A13. There are five fewer widgets than doodads. [Use *W* and *D*.]

A14. Taxes are 6 percent of income. [Use *T* and *x*.]

A15. There are twenty times as many students as professors. [Use *S* and *P*.]

A16. Taxes are one tenth income in excess of $15,000 (Give taxes, *T*, in terms of income, *x*. Assume income is in excess of $15,000).

A17. The cost is 2.5 cents each after paying a 40 cent setup charge. Give the total cost of *n*.

A18. The cost of the car is $40 plus 25 cents per mile. [Use *C* and *x*, for miles.]

Build the Formula

A19. A right triangle has one leg three times as long as the other. Write the formula for its area in terms of the shorter side.

A20. A right triangle has one leg twice as long as the other. Write the formula for its perimeter in terms of the shorter leg.

A21. A right triangle has one leg 2 units longer than the other. Write the formula for its perimeter in terms of the shorter leg.

A22. A right triangle has hypotenuse three times as long as one leg. Write the formula for the length of the other leg in terms of the one leg.

A23. A right triangle has hypotenuse 8 units longer than one leg. Write the formula for the length of the other leg in terms of the one leg.

A24. Write the formula for the area of a right triangle with one leg 6 inches and the hypotenuse x inches.

A25. A square is made from a string of length x. Write the formula for its area in terms of x.

A26. A circle is made from a string of length x. (The string becomes the circle itself, not its radius.) Write the formula for the area of the circle in terms of x.

Problems with Numerical Answers

A27. (A) Build the relevant formula for this problem: "Together Jose and Mike paid a total of $96 for the trip. Jose paid twice as much as Mike. How much did Mike pay?"
(B) Set up the equation and solve it.

A28. (A) Build the relevant formula for this problem: "Together Jose and Mike paid a total of $134 for the trip. Jose paid $10 more than Mike. How much did Mike pay?"
(B) Set up the equation and solve it.

B

B1.* The advice of this section is to "Build your own formula." How is "guess-and-check" supposed to help with indirect word problems? What is the point of doing the calculations with a guess that is probably not correct?

B2.* (A) ☺ If you don't know the relevant formula, the main advice of this section is to _____ _____ _____ _____ (fill in exactly four words).
(B) There are at least three other related pieces of advice. What are they?

Use Guess-and-Check to Help Build Your Own Formula.

B3. A daily car rental costs $50 with up to 100 free miles. After the first 100 miles, miles cost 35 cents each.
(A) What is the total cost for using 155 miles? [6900]
(B) Build a formula for the total cost in terms of the number of miles.

B4. The monthly cost of a phone is $29 includes 200 free minutes. After that, minutes are 7 cents each.
(A) What is the cost of using 250 minutes?
(B) What is the cost of using x minutes?

B5. Taxes are zero if income is less than $11,000. Taxes are 15 percent of the amount of income above $11,000.
(A) What is the tax on an income of $13,000? [300]
(B) Build a formula for taxes in terms of income.

B6. Photocopies cost 7 cents each for the first 20 of one original. After 20, the cost only 5 cents each.
(A) How much do 32 copies of one original cost? [200]
(B) Build a formula for the cost of photocopies in terms of the number of copies of one original.

B7. A right triangle has one leg twice as long as the other.
(A) How long is the hypotenuse if the shorter leg is 12? [27]
(B) Build a formula for the length of the hypotenuse in terms of the shorter leg.

B8. A point is on the line $y = 2x + 7$.
(A) How far is it from the origin if its x-value is 3? [13]
(B) Build a formula for how far it is from the origin (in terms of x alone).

B9. Regular mayonnaise has 100 calories per tablespoon. LiteMayo has only 40 calories per tablespoon (but it is not as good).
(A) If you mix 20 tablespoons of regular with 10 tablespoons of LiteMayo, how many calories per tablespoon will there be in the mixture? [80]
(B) If you mix x tablespoons of regular with 10 tablespoons of LiteMayo, how many calories per tablespoon will there be in the mixture?

B10. Animal pens are constructed from 100 feet of fence as in Example 12, except with three (instead of two) pens side by side.
(A) If the base is 20, what is the total area? [300]
(B) If the base is x, what is the total area?

B11. A length of string is made into a circle.
(A) If the length of the string is 12 (so the circumference is 12), what is the area of the circle?
(B) If the length of the string is x, what is the area of the circle?

B12. (A) An athlete runs 5 miles at 10 miles per hour and then walks 2 miles at 4 miles per hour. What is the athlete's average speed over the seven miles?
(B) An athlete runs 5 miles at 10 miles per hour and then walks x miles at 4 miles per hour. Build a formula for the athlete's average speed over the workout.

Build the Formula

B13. A circle is made from a string of length x. Write its area in terms of x, the circumference.

B14. One square has side twice the side, x, of a smaller square. Write the formula for how much more area the larger has.

B15. A circle is inscribed in a square (The square is just big enough to enclose the circle). Write the formula for the area of the region inside the square that is outside the circle (in terms of the side of the square).

B16. A circle circumscribes a square (The circle is just big enough to enclose the square.) Write the formula for the area inside the circle that is outside the square (in terms of the side of the square.)

B17. The hypotenuse is twice one leg, x, of a right triangle. Write the formula for its area.

B18. The hypotenuse is three units longer than one leg, x, of a right triangle. Write the formula for its area.

B19. A right triangle has hypotenuse 5 inches long. Write a formula for its area in terms of its base.

B20. An equilateral triangle has three equal sides, x. Write the formula for its height.

B21. An equilateral triangle has three equal sides, x. Write a formula for its area.

B22. A circle could be made from a length of string, or a square could be made from the same length of string. Write a formula for the difference in their areas.

B23. A string is 100 centimeters long. It is cut into two pieces and one piece is made into the circumference of a circle and the other piece is made into the perimeter of a square. Write a formula for the total area of the two figures. [Be clear about what "x" is.]

B24. An athlete runs 4 miles at 8 miles per hour and then walks x miles at 4 miles per hour. Write a formula for the athlete's average speed over the workout.

Area of Regions in the First Quadrant

B25. Sketch a line with negative slope, m, that crosses the y-axis at $y = 5$. The line and the axes form a triangle in the first quadrant.
  (A) Write the formula for the x-value where the line crosses the x-axis (in terms of m).
  (B) Write a formula for the area of the triangle.

B26. Sketch a line through $(2, 3)$ with negative slope. The line and the axes form a triangle in the first quadrant. Write the formula for the area of the triangle in terms of the slope.

B27. Sketch the curve $y = \sqrt{8 - x}$. It and the axes bound a region in the first quadrant. Sketch a rectangle within that region, where two sides are along the axes and one vertex is on the curve. Write a formula (with only one variable) for the area of the rectangle.

B28. Sketch the curve $y = 9 - x^2$. It and the axes bound a region in the first quadrant. Sketch a rectangle

within that region, where two sides are along the axes and one vertex is on the curve. Write a formula (with only one variable) for the area of the rectangle.

Density is Mass Divided by Volume

B29. Silver has density 10.5 grams per cc. Gold has density 19.3 grams per cc. Suppose you have 100 cc of gold.
  (A) If you alloy it with 20 cc of silver, what will be the density of the resulting alloy (mixture).
  (B) If you alloy it with x cc of silver, what will be the density of the resulting alloy?

B30. Silver has density 10.5 grams per cc. Gold has density 19.3 grams per cc. Suppose you have 500 grams of gold.
  (A) If you alloy it with 20 grams of silver, what will be the density of the resulting alloy (mixture).
  (B) If you alloy it with x grams of silver, what will be the density of the resulting alloy?

B31. Density is mass divided by volume. Silver has density 10.5 grams per cc. Copper has density 8.93 grams per cc. Suppose you have 100 cc of alloy (mixture).
  (A) If the alloy has 20 cc of copper, what will be the density of the resulting alloy.
  (B) If the alloy has x cc of copper, what will be the density of the resulting alloy?

B32. Density is mass divided by volume. Silver has density 10.5 grams per cc. Copper has density 8.93 grams per cc. Suppose you have 800 cc of alloy (mixture).
  (A) If the alloy has 100 cc of copper, what will be the density of the resulting alloy.
  (B) If the alloy has x grams of copper, what will be the density of the resulting alloy?

Distance

B33. Build a formula for the distance from $(2, 5)$ to points on the curve $y = x^2$, in terms of x alone.

B34. Build a formula for the distance from $(10, 1)$ to points on the line $y = 3x$, in terms of x alone.

B35. A point in the first quadrant is twice as far from the x-axis as it is from $(0, 3)$.
  (A) Build the formula for the distance of a point in the first quadrant from the x-axis.
  (B) Build the formula for the distance of a point in the first quadrant from $(0, 3)$.
  (C) Set up the equation implied by the original sentence [Do not bother to solve it.]

B36. A point P is somewhere in the plane. It is three times as far from $(-2, 0)$ as it is from $(2, 0)$.
- (A) Build the formula for the distance of any point from $(-2, 0)$
- (B) Build the formula for the distance of any point from $(-2, 0)$
- (C) Set up the equation for the location of the point P. [Do not bother to simplify or solve it.]

Problems that Use Formulas from Examples in this Section

B37. Suppose there are no taxes on the first $12,000 of income, but income in excess of $12,000 is taxed at a 15 percent rate. If taxes are 1/10 of income, what is the income?

B38. Density is mass divided by volume. Lead has density 11.3 grams per cubic centimeter (cc). Copper has density 8.93 grams per cc. If you alloy (mix) copper with 200 cc's of lead and the resulting alloy has density 9.6 grams per cc, how much copper did you mix in?

B39. A right triangle has one leg is twice the length of the other. The perimeter is 8. How long is the shorter leg?

B40. Phone calls cost 99 cents for all calls up to twenty minutes and 7 cents a minute for each minute after the first twenty. What is the length of a call that averages 5.5 cents per minute?

Various Problems with Numerical Answers

B41. An open-topped box is formed from a $8\frac{1}{2}$" by 11" sheet of cardboard by cutting squares out of the four corners and folding up the sides.
- (A) Obtain the formula for its volume.
- (B) Find the size of the cutout that maximizes the volume. [1.6]

B42. The cost of a rental car is $40 including up to 100 free miles. Miles after that are 27 cents each. If the average cost per mile is 32 cents, how many miles were driven?

B43. Suppose 100 linear feet of fence made three (3) animal pens as in Example 12 and Figure 6, except with three pens side by side instead of only two. What dimensions would maximize the area? [12]

B44. A rancher builds three pens using a long existing straight fence for one side. She uses 200 feet of

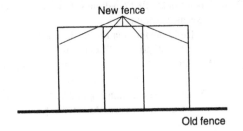

New fence

Old fence

new fencing and gates in an arrangement as pictured.
- (A) Build a formula for the area of the pens in terms of the side pictured vertically.
- (B) What are the dimensions that maximize the total area of the pens?

B45. A figure consists of a rectangle with a triangle on one end as in the figure. If the rectangle is 6 units high, the entire width of the bottom edge is 10 units, and the entire area is 40 square units, how wide is the rectangle? [3-3]

B46. A given length of string can be made into either the circumference of a circle or the perimeter of a square. If the area of the circle is 10 square inches more than the area of the square, how long is the string? [24]

B47. A rain gutter is formed from a sheet of metal 9 inches wide which is bent to form a trough. Three inches are mounted on the house, three inches form the bottom of the trough, and three inches form the lip (see the figure). You wonder if

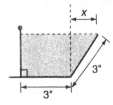

you bend the lip out a bit whether the cross-sectional area would be greater. The trough would be wider, but not so deep.
- (A) Build the formula for the cross-sectional (shaded) area in terms of how far out the lip is bent (x).
- (B) How far out should the lip be bent to maximize the area of the cross section?

B48. A picture (without its frame) is 24 inches by 16 inches. Around it is a frame of uniform width on all four sides.
- (A) Build a formula for the area of the frame in terms of its width.
- (B) If the surface area of the frame is half the surface area of the picture, how wide is the frame? [2.2]

B49. A field is in the shape of a square with a semicircular cap on one side (see the figure).

(A) Build a formula for the area of the field.

(B) Its area is 10,000 square feet. What is the length of a side of the square?   [85]

B50. A right triangle has hypotenuse 5 inches long. Its area is 4 square inches. How long are the other sides? [1.7 and . . .]

B51. A semicircular enclosure has perimeter (including a diameter) 20 centimeters. What is its area?   [24]

B52. Walking a hypotenuse saves distance compared to walking the legs of a right triangle. Suppose the only sidewalk from building *A* to building *B* has a right angle at point *C*, as pictured in Figure 5. Let *AC* be 400 yards and *CB* be 200 yards. Suppose you begin to walk from *A* toward *C*, but at point *D*, you leave the sidewalk and head straight for *B*. If this saves you 100 yards of walking, where is point *D*?

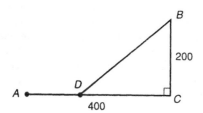

Mixtures

B53. A firm in Japan orders wheat with 15% protein content from your firm and they will take all you have. You have 100 tons of wheat on hand, but it has only 14.4% protein content, so you will buy some wheat with 16% protein content to mix in with your wheat to create wheat with exactly 15% protein content. How much wheat should you buy?   [60]

B54. You have 200 cubic yards of topsoil that contains 10% rocks. A customer will take all the topsoil you have if it contains at most 6% rocks. Therefore you decide to buy just the right amount of topsoil that is only 4% rocks so that mixture of the topsoil you have and the topsoil you buy is exactly 6% rocks. Set up the formula and then determine how much you should buy.

B55. Regular mayonnaise has 100 calories per tablespoon. LiteMayo has only 40 calories per tablespoon (but it is

not as good). How many tablespoons of LiteMayo should you add to 20 tablespoons of regular mayonnaise to make a mixture with 55 calories per tablespoon?

B56. Gold weighs 19.3 grams per cubic centimeter (cc), and silver weighs 10.5 grams per cc.

(A) If one cc of a mixture of gold and silver has .2 cc of gold (and the rest silver), how much does it weigh?

(B) Write a formula for the weight of one cc of a mixture of silver and gold in terms of the **volume** of gold in the cc.

(C) If one cc of mixture weighs 12.7 grams, what is the **weight** of gold in it?   [4.8 grams of gold]

B57. Janet walks and runs a total of 5 miles. She walks at 4 miles per hour and runs at 8 miles per hour. She wants to finish in exactly 50 minutes. How far should she run?

B58. Alvin runs 3 miles and walks 2 miles. He runs twice as fast as he walks and he finishes in exactly one hour. How fast does he walk?   [3.5]

Various

B59. One phone company offers long distance calls at 8 cents a minute and another offers them at 6 cents a minute, but with an additional $4.95 monthly fee. When is it worthwhile to use the company with the additional monthly fee?   [>250]

B60. Suppose there are no taxes on the first $14,000 of income and any income above $14,000 is taxed at a 15% rate. A tax-law change is proposed that there should be no taxes on the first $16,000 of income and the rate should be 17% on any income above $16,000. People with which incomes will pay less under the proposed changes?

B61. If the price of a WidgetCar is $12,000, then 100,000 will be sold. If the price is $13,000, then 70,000 will be sold. Assume the price-to-number sold relationship is a straight line.

(A) Write the price to number-sold relationship.

(B) Write the revenue if the price is "*x*" dollars.

(C) If the cost of selling each is $9000, find the profit if the selling price is *x*. (Profit is revenue minus cost.)

(D) Find the selling price that would maximize profit.   [12,000]

B62. A jet fighter shoots cannon shells at 4000 feet per second. An enemy plane flies a straight-line course perpendicular to the course of the fighter at 1000 feet per second. The enemy is 6000 feet directly in front of the fighter when the cannon is fired. Where should the cannon be aimed? (Express the answer in terms of how far along the projected straight-line path of the enemy plane the cannon should be aimed.) See the figure. [1500]

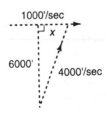

B63. The length plus girth of packages shipped by Universal Package Service must not exceed 150 inches. Your company wishes you to design the cylinder-shaped package with the largest possible volume that they can ship. The "length" is the height of the cylinder and the "girth" is the circumference of the circle. SET up all the relevant formulas and then Write an expression, with one variable, to maximize. Make it very clear which expression you intend to maximize. Last (and least) find the radius of the cylinder of maximum volume. [16]

B64. At this instant a jet fighter is flying north at 1000 miles per hour straight toward a site 200 miles away. Also at this instant an enemy plane 100 miles west of that site is flying east straight toward the site at 600 miles per hour. The jet's missile radar will lock on when they are 60 miles apart. SET UP (but do <u>not</u> bother to solve) the equation to solve the question "How long from now (in hours) until the jet's radar locks on?" [Just give a clear and correct equation.]

B65. Sue likes to walk in the desert. She walks at 2 miles per hour in the desert, and at 4 miles per hour on roads. She wants to get to the parking lot as quickly as possible when she is 1 mile from the road and the parking lot is 2 miles down the road, as pictured. Rather than walking straight toward her car, or straight toward the road, she will walk at an angle toward the road so she can spend more time walking faster on the road. Where should she aim to get to her car the quickest?

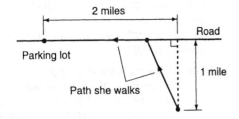

B66. Write the equation satisfied by all points (x, y) that are equidistant from the point $(0, 1)$ and the line $y = -1$.

B67. Write the equation satisfied by all points (x, y) such that the sum of the distances from $(1, 0)$ and $(-1, 0)$ to (x, y) is d, where $d > 2$.

B68. A warship is sailing due east at 25 miles per hour. When it is 3 miles north of a submarine, the submarine fires a torpedo which travels at 70 miles per hour. The torpedo is aimed so that it will meet the ship without changing course (if the ship does not change course). Where will it meet the ship? [1.1 miles east]

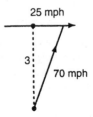

B69. The formula for the volume of a cone is $\left(\dfrac{1}{3}\right)Bh$, where "$B$" is the area of the base, which is the area of a circle, and h is the height. If you take a circular sheet of paper 10 inches in diameter and cut out a sector as in the picture, it can be shaped into a cone by taping together the two edges of the cutout. How much of the circumference should be cut out to make the cone of maximum volume?

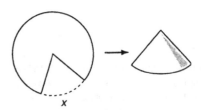

B70. A plane flies northeast from Bozeman. It leaves at noon and flies at a constant speed of 170 miles per hour. Assume Billings is 140 miles east of Bozeman. The plane will appear on the Billings radar when it is 110 miles from Billings. When will the plane be 110 miles from Billings?

Section 3-6 MORE ON WORD PROBLEMS

You can do problems you don't know how to do! The key is to

Start writing!

Formulas are the key to solving algebraic word problems. Build your own formula.

EXAMPLE 1   Computer widgets cost $8 each for the first 100. However, after the first 100 they cost only $5 each. You bought enough so that the average cost was $6.25. How many did you buy?

You bought x. (The value of x is to be determined later, but it is usually good to "**Name the answer**." This problem is not geometric and no picture is appropriate, so don't draw a picture.) Read the problem again. It mentions cost and average cost. **Build formulas** for the cost and for the average cost. If you buy more than 100,

$$C(x) = 8(100) + 5(x - 100) \text{ A key formula!}$$

This represents "$8 for each of the first 100 plus $5 for each after the first 100." Simplifying,

$$C(x) = 300 + 5x \text{ (if } x > 100).$$

The average cost per computer widget is the cost divided by the number of widgets:

$$\frac{C(x)}{x} = \frac{300 + 5x}{x} = \frac{300}{x} + 5 \text{ (if } x > 100).$$

Set this equal to the desired average cost ($6.25) and solve for x, which is the answer (Problem A12).   ▬

EXAMPLE 2   Tickets for adults cost $5 and other tickets cost $2. 120 tickets were sold for a total of $477. How many tickets were sold to adults?

Start writing!

Build your own formula!

Let the number of tickets sold to adults be x ("Name the answer"). The total cost would be the cost of the tickets for adults ($5x$) plus the cost of the other tickets. How many other tickets were there?

$120 - x$.

So total the cost formula is

$$C(x) = 5x + 2(120 - x).$$

Set $C(x) = 477$ and solve (Problem A2).

You can find that formula by using the operations from guess-and-check. Suppose you do not know how to do this problem. Take a guess at the answer (make it a simple guess so the calculations will be easy). Guess, say, 100 (a simple number) were sold to adults. Then the total cost would be

$$5(100) + 2(20) = 540.$$

The "guess" did not "check," but that is not the point. Use the calculation to help see the operations. Then **write** those operations in a formula:

$$C(x) = 5x + 2(120 - x).$$

Set that equal to 477 and solve. ▬

"Name the answer" and "Build your own formula" work on longer problems, too.

EXAMPLE 3 A cup of lentils has 646 calories of which 19 are from fat. Sausage is 50 calories per ounce of which 27 calories are from fat. A chef wants to add sausage to lentil soup, but does not want the calories from fat to exceed 10 percent of the calories. How many ounces of sausage can the chef add to a cup of lentils to have 10 percent of the total calories from fat?

Read it again, slowly. What will you **start writing**?

Name the answer. Let x be the number of ounces of sausage the chef can add.

How does "x" relate to the other quantities in the problem, "total calories" and "calories from fat"?

If the chef uses x ounces of sausage, the total number of calories will be

$646 + 50x$ [646 from the lentils and $50x$ from the sausage].

The number of calories from fat will be

$19 + 27x$ [19 from the lentils and $27x$ from the sausage].

The problem asserts that the number of calories from fat is 10 percent of the total calories.

$$19 + 27x = .10(646 + 50x).$$

Now the "word" part is over.

The hard part of this problem is not solving this equation, it is using algebra to express the operations in a formula. Once you have the formula, the problem is easy (Problem A3). ▬

■ Constraints

Relationships between variables which are normally independent are called constraints.

Many formulas have two independent variables. For example, the area of a rectangle is given by "$A = bh$" and the base and height are independent variables. Its perimeter is given by "$P = 2b + 2h$" with two independent variables. However, if the perimeter were given as, say, 40, then the base and height would no longer be independent because you could find the height from the base. The fact that the perimeter is 40 would be called a "constraint."

EXAMPLE 4 Express the area of a rectangle given that its perimeter is 40 inches.

Start writing!

Draw and label a picture (Figure 1). **Write down relevant formulas:**

$$A = bh, \text{ and}$$
$$P = 2b + 2h.$$

The given fact that $P = 40$ is a constraint:

$$2b + 2h = 40.$$

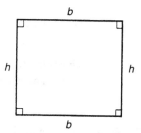

FIGURE 1 A rectangle with base b and height h

Therefore,

$2h = 40 - 2b$ and $h = 20 - b$.

Substituting for h, the area formula can be rewritten:

$$A = bh = b(20 - b).$$

This is the answer. Now there is only one independent variable, b. The constraint converts a two-variable formula for area into a one-variable formula.

The area of a rectangle with perimeter 40 inches is 60 square inches. What are the sides?

Use the formula you wrote. Solve $A = b(20 - b) = 60$, a quadratic equation (Problem A4)

MORE FROM EXAMPLE 4 Find the dimensions of the rectangle with maximum possible area, given its perimeter is 40 inches.

The area formula is $A = b(20 - b)$. Maximize this. With the formula, the "word" part of the word problem is over. ▬

EXAMPLE 5 Suppose the area of a rectangle is 100 square inches. Find its perimeter if one side is 8 inches.

Write down relevant formulas:

$A = bh$ and

$P = 2b + 2h$.

You are given $A = 100$ and $b = 8$. So $100 = bh = 8h$. Then $h = \dfrac{100}{8} = 12.5$.

Plug both b and h into the perimeter formula.

$$P = 2b + 2h = 2(8) + 2(12.5) = 41.$$

Suppose the area of a rectangle is 100 square inches. Find the dimensions which yield the minimum possible perimeter.

Again, the relevant formulas are

$A = bh$ and

$P = 2b + 2h$.

You want to minimize P given the constraint

$A = bh = 100$.

Using the constraint, you can solve for either b or h, say h. $h = \dfrac{100}{b}$. Then

$$P = 2b + 2(100/b).$$

This is the formula you want.

To minimize P, switch "b" to "x", graph it, and use guess-and-check (Problem A5).

————

EXAMPLE 6 Suppose the hypotenuse of a right triangle is 10 units. Find its base if the area is 20 square units.

Draw and label a picture (Figure 2). **Write down relevant formulas**. From the picture, the area formula and the Pythagorean Theorem leap to mind.

$$A = \left(\tfrac{1}{2}\right)ab = 20.$$

$$a^2 + b^2 = 10^2.$$

Now the words have been written in mathematical symbols. This is key to doing the problem.

These two equations with two unknowns can be solved several different ways. One way is to solve for a in the second:

$$a = \sqrt{10^2 - b^2}, \text{ and substitute this into the first:}$$

$$A = \left(\frac{1}{2}\right)b\sqrt{10^2 - b^2} = 20$$

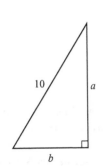

FIGURE 2 A right triangle with hypothenuse 10

This can be solved using guess and check (Problem A6). ————

There are many types of constraints. For example, general points can be anywhere in the plane, but a constraint may force points to be on a line.

EXAMPLE 7 Find a formula for the distance from the point (3, 1) in the plane to points on the line $y = 2x$.

Draw a picture (Figure 3).

The two relevant formulas are the distance formula for distances between any two points in the plane and the given formula ("$y = 2x$") for points on the line.

The distance between two points (x_1, y_1) and (x_2, y_2) is given in Theorem 3-3-2 by

$$d = \sqrt{(x_2 - x_1)^2 + (y_2 - y_1)^2}$$

Name the point on the line, say, (x, y). One point is (3, 1). What do we know about (x, y)?

The distance from (3, 1) to any point (x, y) is given by

$$d(x, y) = \sqrt{(x - 3)^2 + (y - 1)^2}.$$

But "y" is not just any y, it must be on the line $y = 2x$, which is the constraint. Substituting for y:

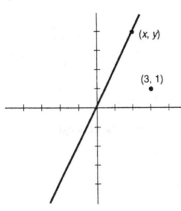

FIGURE 3 The point (3, 1) and the line $y = 2x$. $[-5, 5]$ by $[-5, 5]$

$$d(x) = \sqrt{(x - 3)^2 + (2x - 1)^2}.$$

This is the formula you want. It has only one variable.

Find the points on the line $y = 2x$ which are 4 units away from (3, 1).

The big step was to build the formula. Now, plug "4" in for "d" in the formula (Problem A7).

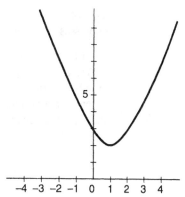

FIGURE 4 $d(x)$. The distance from (3, 1) to $(x, 2x)$ in terms of x. $[-5, 5]$ by $[0, 10]$

MORE FROM EXAMPLE 7

Find the point on the line $y = 2x$ closest to (3, 1).

The word "closest" tells us to minimize distance. To do this first build the distance formula. We did. To minimize it, you may graph it (Figure 4, Problem A13).   ▬▬

EXAMPLE 8

A wire 100 inches long will be cut into two lengths and each will be bent to form the perimeter of a square. Express the total area of the two squares in terms of where the wire is cut.

Draw and label a picture (Figure 5). The wire is cut at "x".

The relevant formulas are the formula for the area of a square, the formula for the perimeter of a square, and given fact that the perimeters of the two squares sum to 100 inches (which is the constraint).

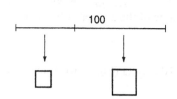

FIGURE 5 A wire 100 inches long cut and bent to form squares

$$A_{\text{square}} = s^2. \quad P_{\text{square}} = 4s.$$

If it is cut so one length is x, that is the perimeter of one square: $x = 4s$. The side of the one square would be $\dfrac{x}{4}$. The rest of the wire is $100 - x$, because the constraint is that the lengths sum to 100. The side of the other square would be $\dfrac{(100-x)}{4}$. Therefore, the total area of the two squares would be

$$T = \left(\frac{x}{4}\right)^2 + \left[\frac{(100-x)}{4}\right]^2.$$

You might wish to rewrite this as

$$T = \frac{x^2}{16} + \frac{x^2 - 200x + 10,000}{16} = \frac{2x^2 - 200x + 10,000}{16}.$$

This is the desired formula.

If you don't see where "$\dfrac{x}{4}$" came from, try a case with particular numbers (use guess-and-check). For example, if the wire were cut 40 inches from one end, there would be two pieces of lengths 40 inches and 60 inches. From them you could form two squares, of sides 10 inches $\left(10 = \dfrac{1}{4} \text{ of } 40 \right)$ and 15 inches $\left(15 = \dfrac{1}{4} \text{ of } 60 \right)$.

The total area would then be $10^2 + 15^2$ (square inches). The formula you want expresses those same operations with a cut at a general distance, x, instead of at 40.

A wire 100 inches long was cut into two lengths and each was bent to form the perimeter of a square. The total area of the two squares is 350 square inches. Where was the wire cut?

Now that you have the formula, the rest is relatively easy. The particular fact that the total area is 350 square inches can be used to set up an equation:

$$T = \frac{2x^2 - 200x + 10{,}000}{16} = 350.$$

MORE FROM EXAMPLE 8

This can be solved for x using the Quadratic Formula (Problem A8).

A wire 100 inches long will be cut into two lengths and each piece bent to form the perimeter of a square. Where should it be cut to minimize the total area of the two squares?

With the formula, this is easy. The minimum will occur at the vertex of the parabola, since the coefficient on "x^2" is positive (Problem A14). ▬

The next example illustrates that you may need to name and utilize an auxiliary variable to express the relationships. A variable is <u>auxiliary</u> when it assists in finding the answer even though is not the name of the answer or the name of another quantity mentioned in the problem.

EXAMPLE 9

The perimeter of a semicircular enclosure is 10 meters (Figure 6). What is its area?

The question mentions "perimeter" and "area," both of which are usually discussed in term of the radius or diameter. The radius is a useful "auxiliary" variable.

Use well-known formulas:

$$A_{circle} = \pi r^2. \quad C = 2\pi r.$$

$$\text{Therefore,} \quad A_{semicircle} = \frac{\pi r^2}{2}.$$

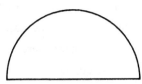

FIGURE 6 A semicircular enclosure

The perimeter is half a circumference of a circle plus a diameter:

$$P = \frac{(2\pi r)}{2} + 2r.$$

$$P = (\pi + 2)r.$$

Now use this formula and the fact the perimeter is 10 to solve for r and then plug r into the area formula (Problem A9). ▬

For the next example the formula for the relation between total cost, cost per unit area, and area is relevant.

cost = (cost per unit area) × (units of area).

EXAMPLE 10   Suppose a box with a square bottom is to contain 1000 cubic inches. The top costs 7 cents per square inch, the 4 sides cost 3 cents per square inch, and the bottom costs 5 cents per square inch. Assume there are no other costs. Find the dimensions of the box with minimum cost.

Read the problem again. Then start writing.

Draw a picture (Figure 7).

You need to build a formula for cost. The relevant formulas concern volume and area. The formula for the volume, V, of a box is

$$V = lwh.$$

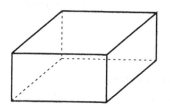

FIGURE 7 A box with a square bottom

(Using l for length, w for width, and h for height). In this problem, the box has a square base (which is a constraint), so "l" and "w" are equal. Let's use x for both l and w.

$$V = x^2h.$$

The problem did not mention the height, h, but you need to use this auxiliary variable. The volume is given as 1000, which is a constraint on x and h:

$$1000 = x^2h.$$

Now write the cost formula. There are 6 sides. The top and bottom each have area x^2. The 4 other sides each have area xh.

$$C = 7x^2 + 5x^2 + 3(4xh) = 12x^2 + 12xh.$$

You want a formula with only one variable. You can use the constraint to replace either x or h in the formula for cost. It is simpler to replace h. From the constraint,

$$h = \frac{1000}{x^2}, \text{ so}$$

$$C = 12x^2 + 12x\left(\frac{1000}{x^2}\right) = 12x^2 + \frac{12,000}{x}.$$

This is the formula you want. It has only one variable; "h" is gone. To minimize the image, graph it and use guess-and-check (Problem A10). ▬

EXAMPLE 11   The length plus girth of packages shipped by the Universal Package Service must not exceed 120 inches (Figure 8). The "length" is the length of the longest edge, and the "girth" is the perimeter around the side with the shorter two edges. Suppose the smaller side is a square. Find the box of maximum volume that can be shipped with the Universal Package Service.

Reread the problem. Relevant formulas are the volume of a box, and the given constraints. Let "x" be the side of the square. The volume of the box is

$$V = x^2h,$$

where "h" is the length.

The perimeter of the square side (the "girth") is given by $P = 4x$. The constraint is

$$4x + h = 120.$$

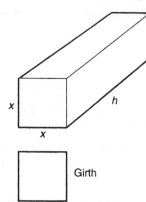

FIGURE 8 A box with a square side. The distance around the square is the girth

Therefore, $h = 120 - 4x$. Substituting for h in the volume formula,

$$V = x^2(120 - 4x).$$

This is the formula you want. It can be graphed and maximized using guess-and-check (Problem A11). Calculus techniques also work. ▬

CONCLUSION Formulas are the key to solving algebraic word problems. **Build your own formula**.

To do a word problem

1. Read and reread the problem
2. **Start writing (including drawing) what you know**
3. Name the answer (perhaps "x", and give letter names to other relevant quantities. Label your picture, if any)
4. **Keep writing—the answer will develop**
5. Write and use basic formulas to help build a formula for the problem

You can try to memorize how to do each type of problem, but that won't work because there are too many different types. Instead, even if you don't know how to do a problem, try writing about the operations. **Build your own formula**.

Terms: constraint, auxiliary variable.

Exercise 3-6

A

A1. Give four pieces of advice from the previous section about how to do word problems.

A2. Find the solution to Example 2.

A3. Find the solution to Example 3.

A4. Find the solution to "Example 4, continued."

A5. Find the solution to "Example 5, continued."

A6. Find the solution to Example 6.

A7. Find the solution to "Example 7, continued."

A8. Find the solution to "Example 8, continued."

A9. Find the solution to Example 9.

A10. Find the solution to Example 10.

A11. Find the solution to Example 11.

A12. Find the solution to Example 1.

A13. Find the solution to "More from Example 7."

A14. Find the solution to "More from Example 8."

A15. Together, A and B have $200. A has $40 more than B. How much does B have?

A16. A man is twice as old as his son and three times as old as his daughter.

(A) Build a formula for the sum of their ages in terms of the age of the man.
(B) Together, their ages sum to 66. How old is the man?

A17. A triangle has three sides. One is twice as long as the second, and the third is 24 centimeters long.
(A) Use "x" for the length of the shorter side and build a formula for the perimeter.
(B) If the perimeter is 54 centimeters, how long is the shorter side?

A18. A semicircular enclosure has perimeter 25. What is its area?

A19–A22. Set up the equation with one unknown. You are not asked to solve it.

A19. A point on the line $y = 2x + 1$ is 5 units from the point $(3, 4)$. Set up an equation with one unknown to solve "Where is the point on the line?"

A20. The perimeter of a rectangle is 28. Its diagonal is 10. Set up an equation with one unknown to solve "Find its sides."

A21. The area of a rectangle is 10. Its diagonal is 5. Set up an equation with one unknown to solve "Find its sides."

A22. A point on the parabola $y = x^2$ is 3 units from the point (0, 2). Set up an equation with one unknown to solve "Where is the point on the parabola?"

B

B1.* ☺ The main advice from last section was to "Build your own formula." What is the additional advice from this section?

B2. A sports club wishes to make an indoor 200 meter track in the shape of two straight parallel lines connected by semicircles on each end. The lines form two sides of a rectangle. Find the rectangle of largest possible area. [straight sides 50]

B3. A figure is in the shape of a square with two semicircular caps on opposite sides. Its perimeter is 100. Find its side. [19]

B4. Suppose the hypotenuse of a right triangle is twice the length of a leg and the area is 100. Find the leg. [11]

B5. Suppose the hypotenuse of a right triangle is twice the length of a leg and the perimeter is 40. Find the leg. [8.4]

B6. Suppose one leg of a right triangle is three times the other. The perimeter is 500. How long is the shorter leg? [70]

B7. An isosceles triangle has two equal sides of length 20 and area equal to 120 square units. How long is the other side? [13, . . .]

B8. A length of string could be either the circumference of a circle or the perimeter of a square. The circle would have 100 square centimeters more area than the square. How long is the string? [77]

B9. A circle and a square have the same area. The circumference of the circle is 20 units less than the perimeter of the square. What is the side of the square? [44]

B10. Suppose a semicircle is constructed inside a square using a side of the square as its diameter. The area in the square that is not in the semicircle is 45. Find the side of the square. [8.6]

B11. 100 centimeters of wire are cut into two lengths, one of which is bent to form a square and the other a circle. Express the total area of the two figures and find where the cut should be to minimize that total. [56 for the square]

B12. Find the point on the parabola $y = x^2$ which is closest to the point (5, 1).   [(1.5, . . .)]

B13. Find the point on the parabola $y = x(x - 1)$ which is closest to the point (0, 1). [(. . ., .75) Note: The .75 in this answer is the y-value, not the minimum distance, which is .56.]

B14. It costs $10 per foot to lay cable underwater, and $4 per foot to lay it over land along the shore. A lighthouse is 100 feet off shore (see the figure, L). To connect a terminal (T) 200 feet down the shore to the lighthouse with cable, it is cheaper to lay the cable from the lighthouse to a point, P, on shore somewhat toward the terminal, rather than lay it perpendicular to the shoreline. Where should the cable meet the shore to minimize the cost?   [44 feet toward the terminal]

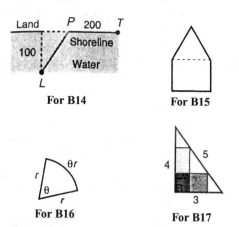

For B14   **For B15**

For B16   **For B17**

B15. The side of a large outdoor playhouse has the shape pictured in the figure. It is 13 feet high and 8 feet wide. The area of the side is 76 square feet. How tall are the vertical sides?   [6.0]

B16. (See the figure.) The area of a sector of a circle of radius r and central angle θ (in radians) is $A = \dfrac{\theta r^2}{2}$. The perimeter of the sector is $2r + \theta r$ (two radii plus an arc, see the figure). Suppose the perimeter is fixed at 100. Find the value of r which maximizes the area.

B17. Consider a 3-4-5 right triangle. Consider all possible rectangles inside the triangle that have two sides along the triangle's sides of length 3 and 4 (see the figure).
  (A) Use algebra to find the formula for the area of any such rectangle.
  (B) Which one has the largest area?

B18. 100 inches of wire are to be cut into two lengths. One is bent to form a rectangle with one side 10 inches long. The other forms a square. Find the cut which produces the minimum possible total area of the two figures. [60 for the rectangle]

B19. Let point P be somewhere on the x-axis. Consider the total of the distance from $(0, 5)$ to P and the distance from P to $(9, 3)$. Where should P be located to minimize that total?

B20. Consider the curve $y = 10 - \sqrt{x}$ in the first quadrant. Suppose a rectangle has one vertex at unknown point P on the curve, another vertex at $(0, 0)$, and two sides along the two axes. Where should P be located to maximize the area of the rectangle? $[P = (44, 3.3)]$

B21. Find the largest (in area) rectangle in the first quadrant with a vertex at $(0, 0)$, a vertex on the line that goes through $(5, 1)$ and $(1, 4)$, and two sides along the two axes. $[x = 3.2]$

B22. Jane walks and runs a total of 4 miles. She wants to finish in exactly 40 minutes. If she walks at 4 miles per hour and runs at 8 miles per hour, how many miles should she run?

B23. Find the set of all points (x, y) such that the distance from (x, y) to $(1, 0)$ is the same as the distance from (x, y) to the line $x = -1$. Simplify.

B24. Find the set of all points (x, y) such that the distance from (x, y) to $(1, 0)$ is the same as the distance from (x, y) to the line $y = -1$. Simplify.

B25. A jogger runs out at a constant speed and runs back 2 miles per hour faster. She averages 8 miles per hour over the 6 mile round trip. How fast does she run out? [7.1]

B26. The WidgetCan Company wishes to make the cheapest cylindrical metal can which can hold 2000 cubic centimeters. They reckon the cost of the metal as $1 per square meter (with no waste) and the cost of sealing the seams (around the top and bottom, and one up the side) as 1 cent per 10 centimeters. All other costs are considered the same, no matter what the shape of the can is. What are the dimensions of the cheapest can?

B27. Suppose you wish to mix 10% acid solution with 24% acid solution to make 10 liters of 14% acid solution.
(A) What is the constraint?
(B) How much of each should you use? [7.1 liters of 10% solution]

B28. Density is mass divided by volume. Gold weighs 19.3 grams per cubic centimeter, and silver weighs 10.5 grams per cubic centimeter. Suppose gold is worth $9.65 per gram and silver is worth $.17 per gram.

(A) Develop a formula for the value of a mixture of silver and gold in terms of its density. (Disregard the cost of refining it into its two components.)
(B) If a mixture weighs 12.7 grams per cubic centimeter, how much is a cubic centimeter worth?

B29. Density is mass divided by volume. Lead has density 11.3 grams per cubic centimeter (cc). Copper has density 8.93 grams per cc. If you alloy (mix) copper with 200 cc's of lead and the resulting alloy has density 9.5 grams per cc, how much copper did you mix in?

B30. A warship is sailing due west at 30 miles per hour. When it is 4 miles north of a submarine, the submarine fires a torpedo which travels at 90 miles per hour. The torpedo is aimed so that it will meet the ship without changing course (if the ship does not change course). Where will it meet the ship? [1.3 west]

B31. A fisher stands with her eye level six feet above the water, and a fish ten feet out is two feet below the water. Light travels 1.33 times as fast in air as in water. This causes refraction. That is, the light she sees will not come straight from the fish to her eye, but will appear to bend at the surface of the water. A principle of physics ("Fermat's Principle") says that the light she sees will have traveled the path of least time. In the figure, find x such that the time for light to travel from F to P to A is least.

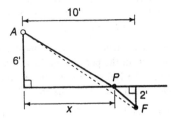

B32. Suppose the bottom and sides of an open-topped (no top) cubical box cost .3 cents per square inch (neglect other contributions to the cost). Find the volume of a box costing 75 cents. [350]

B33. Prove that the maximum area of a rectangle with a fixed perimeter occurs when the rectangle is a square. Use a general variable, P, to represent the fixed perimeter.

B34. In "Example 1, continued" can the "60 square inches" fact be regarded as a constraint?

B35. Chris has $50,000 of her money invested in Fund A, which has $\frac{9}{10}$ of its money in stocks and $\frac{1}{10}$ in cash. She has the remaining $30,000 of her money in Fund B, which has $\frac{1}{3}$ of its money in stocks and $\frac{2}{3}$ in cash.

   (A) What fraction of Chris's money is in stocks?   [.69]

   (B) She wishes that $\frac{3}{4}$ of her money were in stocks. How much of her invested money should she have in Fund A, with the rest in Fund B, so have $\frac{3}{4}$ of her money in stocks?   [59,000]

B36. Butter has 100 calories per tablespoon, all of which are from fat. Marie has prepared a dish that, without butter, has 600 calories of which 50 are from fat. How many tablespoons of butter can she add so that exactly $\frac{1}{5}$ the calories are from fat?

B37. ConcreteKing wants to make 20,000 pounds of concrete with 10 percent water by weight. They already have lots of pre-mix concrete that is 2 percent water by weight. How much water should be mixed in with how much of this pre-mix concrete to make 20,000 pounds of the 10 percent water mixture?

B38. Sketch a line with slope $m = -.4$ that crosses the y-axis at $y = b$. The line and the axes form a triangle in the first quadrant. Find b such that the area of the triangle is 10.   [2.8]

B39. A rental car costs $40 a day and comes with up to 100 free miles. However, if you use more than 100 miles, each mile over 100 costs 25 cents. When you take it back after only one day you find it cost an average of 30 cents per mile. How many miles did you drive?   [300]

B40. Write the equation of the circle with center (3, 2) that is tangent to the unit circle (The unit circle has equation $x^2 + y^2 = 1$).

B41.* (A) What is the purpose of algebraic notation?
   (B) What does this have to do with indirect word problems?

4

Powers

Section 4-1   POWERS AND POLYNOMIALS

x^2, x^3, and $x^{1/2}$ are "power" functions.

2^x, 10^x, and $\left(\frac{1}{2}\right)^x$ are "exponential" functions.

Power and exponential functions are closely related. This chapter concerns power functions. Exponential functions are the subject of the next chapter.

When a power is a positive integer, it is appropriate to think of *repeated multiplication*. The power gives the number of repeated factors in the product. For example, denote

$$2 \times 2 \times 2 \times 2 \text{ by } 2^4 \ (= 16), \text{ and}$$
$$b \times b \times b \text{ by } b^3.$$

Let $b^p = b \times b \times \ldots \times b$ [p factors]. For any p,

$$1^p = 1. \tag{4-1-0}$$

■ Products of Powers

The properties of integer powers follow from the repeated-multiplication interpretation. For example, a product of integer powers is easy to express in an alternative form.

$$b^2 b^3 = (b \times b) \times (b \times b \times b) = b^5.$$

Evidently, the power (5) of a product $(b^2 \times b^3 = b^5)$ is the sum of the powers $(2 + 3 = 5)$. This generalizes to:

$$(b^p)(b^r) = b^{p+r}. \tag{4-1-1}$$

The left side of this identity displays *powers and multiplication*, in that order. The right side displays *addition and a power*, in that order. Not only the order changes, but also the operations change.

EXAMPLE 1 $x^2 x^4 = x^6$ [not x^8].

$$(2x + 1)^3 (2x + 1)^4 = (2x + 1)^7 \quad [\text{not } (2x + 1)^{12}]. \qquad ■$$

■ Quotients of Powers

A quotient of powers is easy to simplify if the bases are the same. For example,

$$\frac{b^5}{b^2} = \frac{b \times b \times b \times b \times b}{b \times b} = b \times b \times b = b^3.$$

The power (3) of a quotient $\left(\dfrac{b^5}{b^2} = b^3\right)$ is the difference of the powers $(5 - 2 = 3)$. This generalizes to (if $b \neq 0$):

$$\frac{b^p}{b^r} = b^{p-r}. \tag{4-1-2A}$$

EXAMPLE 2 $\dfrac{x^6}{x^2} = x^4$ [not x^3]. ■

We can use this property to define what we mean by the 0 power and negative powers. First, look at the pattern for positive powers in this example of powers of 2.

power	power of 2	
3	$8 = 2^3$	
2	$4 = 2^2$	
1	$2 = 2^1$	
0	1	so $1 = 2^0$
-1	$\dfrac{1}{2}$	so $\dfrac{1}{2} = \dfrac{1}{2^1} = 2^{-1}$
-2	$\dfrac{1}{4}$	so $\dfrac{1}{4} = \dfrac{1}{2^2} = 2^{-2}$

For positive powers, when the power is increased by 1, the image is multiplied by 2, and when the power is decreased by 1, the image is divided by 2. Continuing that pattern determines the zero power and negative powers.

The power-function pattern illustrated here for base 2 holds for any base b. To *add* 1 to the power is to *multiply* the image by b, the base. To *subtract* 1 from the power is to *divide* the image by b.

Consider any base $b \neq 0$.

$$\frac{b}{b} = 1. \quad \text{But, by 4-1-2A,} \quad \frac{b}{b} = \frac{b^1}{b^1} = b^{1-1} = b^0. \quad \text{So } b^0 = 1.$$

Then,

$$\frac{1}{b^p} = \frac{b^0}{b^p} = b^{0-p} = b^{-p}.$$

$$b^0 = 1. \tag{4-1-2B}$$

$$b^{-p} = \frac{1}{b^p} = \left(\frac{1}{b}\right)^p. \tag{4-1-2C}$$

EXAMPLE 3 $x^{-1} = \dfrac{1}{x}. \quad x^{-2} = \dfrac{1}{x^2}$

$$b^{-2} = \frac{1}{b^2} = \frac{1}{b \times b} = \frac{1}{b} \times \frac{1}{b} = \left(\frac{1}{b}\right)^2.$$

$$x^3 x^{-4} = x^{-1} = \frac{1}{x}.$$

$$\frac{x^2}{x^{-1}} = x^{2-(-1)} = x^3.$$
■

EXAMPLE 4   Find $P(x)$ that satisfies the given equation.

$$x^4 P(x) = x^6. \qquad\qquad \text{Then } P(x) = x^2.$$
$$(x-1)^3 P(x) = (x-1)^7. \qquad \text{Then } P(x) = (x-1)^4.$$
$$x^4 P(x) = x^{-3}. \qquad\qquad \text{Then } P(x) = x^{-7}.$$
$$x^{-3} P(x) = x^{-2}. \qquad\qquad \text{Then } P(x) = x.$$
■

EXAMPLE 5   Solve $3x^2(1-x)^5 - 5x^3(1-x)^4 = 0$.

Use the Zero Product Rule. Factor the left side. Both terms have "x" to a power and "$1-x$" to a power. Factor out the largest powers the terms have in common, that is, the *lesser* power of each: x^2 and $(1-x)^4$. It equals

$$x^2(1-x)^4[3(1-x) - 5x] = 0$$

$$x^2(1-x)^4[3 - 8x] = 0.$$

$$x = 0 \quad \text{or} \quad x = 1 \quad \text{or} \quad x = \frac{3}{8}.$$
■

EXAMPLE 6 Solve $5x^{-2}(1-x)^4 + 2x^{-3}(1-x)^5 = 0$

The two terms have factors in common. Factor out the *lesser* power of each. It equals

$$x^{-3}(1-x)^4[5x + 2(1-x)] = 0$$

$$x^{-3}(1-x)^4[3x + 2] = 0$$

$$x = 1 \text{ or } x = \frac{-2}{3}.$$

Note that $x^{-3} = 0$ has no solution and does not contribute to the answer. ▬▬

■ Powers of Powers

Now consider powers of powers. If both powers are integers, the idea of repeated multiplication shows, for example,

$$(b^2)^3 = b^2 \times b^2 \times b^2 = (b \times b) \times (b \times b) \times (b \times b) = b^6.$$

In general,

$$(b^r)^p = b^{rp}. \tag{4-1-3}$$

EXAMPLE 7 $(2^3)^4 = 2^{12}.$

$(x^2)^5 = x^{10}.$

$\left(\left(\frac{1}{2}\right)^3\right)^2 = \left(\frac{1}{2}\right)^6 = 2^{-6}$, using 4-1-3 and then 4-1-2C. You may obtain

the same result another way: $\left(\left(\frac{1}{2}\right)^3\right)^2 = (2^{-3})^2 = 2^{-6}$, using 4-1-2C and

then 4-1-3. ▬▬

Powers distribute over multiplication. For example,

$$(2 \times 5)^3 = (2 \times 5)(2 \times 5)(2 \times 5)$$

$$= (2 \times 2 \times 2)(5 \times 5 \times 5)$$

$$= 2^3 5^3$$

In English the word "distribute" means to deal out or allot. The same meaning holds in Mathematics. The power in the expression $(ab)^2$ is distributed to each of the two factors, a and b, to form a^2b^2. There is a parallel with the Distributive Property of multiplication over addition: $a(b + c) = ab + ac$. The factor (a) which multiplies the sum $(b + c)$ is distributed to both terms to form $ab + ac$.

This generalizes to:

$$(ab)^p = a^p b^p. \tag{4-1-4A}$$

Similarly (Problem B57),

$$\left(\frac{a}{b}\right)^p = \frac{a^p}{b^p} \tag{4-1-4B}$$

EXAMPLE 8

$$(5x)^2 = 5^2x^2 = 25x^2.$$

$$\left(\frac{x}{3}\right)^2 = \frac{x^2}{3^2} = \frac{x^2}{9}.$$

$$\left(\frac{2}{x}\right)^3 = \frac{2^3}{x^3} = \frac{8}{x^3} = 8x^{-3}.$$

Negative numbers to *integer* powers are positive or negative depending upon the number of factors. For example,

$$(-7)(-7) = 7^2.$$
$$(-7)(-7)(-7) = -7^3.$$

Be careful with the order conventions. -7^2 is negative. It is $-(7^2)$. -7^2 is not $(-7)^2$. Powers are executed before multiplication, and the negative sign is treated like multiplication by -1. So $-7^4 = -(7)^4$, not $(-7)^4$.

These examples generalize to:

If n is even, $(-b)^n = b^n$.
If n is odd, $(-b)^n = -b^n = -(b^n)$.   (4-1-5)

Here are the properties repeated with base "x" replacing "b".

TABLE 4-1-6 Properties of Powers

For $x > 0$, $a > 0$, and any p and r,	
(4-1-1)	$(x^p)(x^r) = x^{p+r}$
(4-1-2A)	$\dfrac{x^p}{x^r} = x^{p-r}$
(4-1-2B)	$x^0 = 1$
(4-1-2C)	$x^{-p} = \dfrac{1}{x^p} = \left(\dfrac{1}{x}\right)^p$
(4-1-3)	$(x^r)^p = x^{rp}$
(4-1-4A)	$(ax)^p = a^p x^p$
(4-1-4B)	$\left(\dfrac{x}{a}\right)^p = \dfrac{x^p}{a^p}$
(4-1-5)	If n is even, $(-x)^n = x^n$. If n is odd, $(-x)^n = -x^n$.

Earlier in this section these properties were motivated with *integer* powers. Nevertheless, they hold for all *real-valued* powers if b or x is *positive*.

EXAMPLE 9   For positive values of x,

$$x^\pi x^2 = x^{\pi+2}, \qquad \text{by 4-1-1}$$

$$\frac{x^2}{x^{1.5}} = x^{.5} = \sqrt{x}, \quad \text{by 4-1-2A}$$

$$(x^{1.3})^2 = x^{2.6}, \qquad \text{by 4-1-3}$$

$$(3x)^{.7} = (3^{.7})x^{.7}, \qquad \text{by 4-1-4A}$$

$$\left(\frac{x}{5}\right)^{1/4} = \frac{x^{1/4}}{5^{1/4}}, \qquad \text{by 4-1-4B}$$

These fractional powers of x do not yield real numbers for negative values of x. If x is negative, the properties in the table still hold for *integer* powers. However, for fractional powers of negative numbers there are complications which will be discussed in Section 4-3.

Simplify $\qquad\qquad \dfrac{x^3 x^8}{x^4}.$

$$\frac{x^3 x^8}{x^4} = \frac{x^{11}}{x^4} = x^7.$$

■ Polynomials

Monomials are the simplest kinds of polynomials.

DEFINITION 4-1-7 A <u>monomial</u> is any expression of the form "cx^n", where n is a non-negative integer and $c \neq 0$. n is its <u>degree</u> and c is its <u>coefficient</u>.

> **Calculator Exercise 1:** Graph x^p for various positive integer values of p, including $p = 1, 2, 3, 4,$ and 5 (Problem A1). Which resemble each other?
> Which are one-to-one and therefore have unique solutions to "$x^p = c$"?

EXAMPLE 10 "$4.5x^3$" is a monomial of degree 3 with coefficient 4.5.

"$-x^2$" is a monomial of degree 2 with coefficient -1.

"7" is a monomial of degree zero, because it can be regarded as "$7x^0$," since $x^0 = 1$. The coefficient is 7. All numbers are monomials.

"$\dfrac{x^5}{3}$" is a monomial of degree 5 with coefficient $\dfrac{1}{3}$ (not 3, coefficients are *multiplicative* constants).

DEFINITION 4-1-8 A <u>polynomial</u> is either a monomial, or a sum or difference of two or more monomial terms. The <u>degree</u> of a polynomial is the highest power of any of its terms. The <u>leading coefficient</u> is the coefficient on the highest-power monomial term.

EXAMPLE 11 "$x^2 + 2x + 1$" is a polynomial of degree 2 with leading coefficient 1. All quadratics are polynomials of degree 2.

"$3x - 7$" is a polynomial of degree 1 with leading coefficient 3. All linear expressions of the form $mx + b$ where m is not zero are of degree 1.

"$7x^2 - 5x^3$" is a polynomial of degree 3 with leading coefficient -5. The terms "degree" and "leading coefficient" refer to the term with the highest power, which is not necessarily on the left.

"$3x^2(5x^4 + 2x + 12)$" is a sixth degree polynomial written in factored form, with leading coefficient 15.   ▬

■ Graphs of Polynomials

This section concentrates on cubic and higher-degree polynomials because we have already thoroughly discussed polynomials of degree 2 (quadratics) and degree 1 (lines).

> **Calculator Exercise 2:** What slope does the graph of x^p have at the origin for $p \geq 2$? Graph x^3, x^4 and x^5 on $[-2, 2]$ by $[-2, 2]$ and see. Which two points do all these graphs go through (Problem A2)?

EXAMPLE 12   Cubic polynomials ("cubics") are polynomials of degree 3. Unlike quadratics, they come in a variety of shapes. The basic cubic, x^3, is familiar (Figure 1).

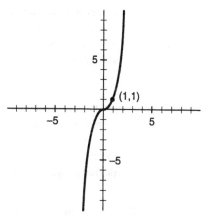

FIGURE 1 x^3. $[-10, 10]$ by $[-10, 10]$

Images in the first quadrant correspond to images in the third quadrant. When (x, y) is on the graph, so is $(-x, -y)$, because $(-x)^3 = -(x^3)$, by 4-1-5. The graph of x^3 is symmetric about the origin.   ▬

DEFINITION 4-1-9   Functions (such as f defined by $f(x) = x^3$) with the property that $f(-x) = -f(x)$ for all x are said to be <u>odd</u> functions. The term "odd" is used because all odd-power monomials have this property (4-1-5).

The graphs of odd functions are said to be <u>point-symmetric</u> about the origin. That is, if (a, b) is on the graph, so is $(-a, -b)$, which is equidistant from the origin and directly opposite (a, b) through the origin.

The graph of "$y = x$" ($x = x^1$, a first degree monomial) has this point symmetry. Graphs of odd functions are not limited to first and third quadrant points — they may exhibit point symmetry about the origin between second and fourth quadrant points (Figure 2).

The graph of $x^3 - 5x$ exhibits two <u>local extrema</u> (Figure 2. "Extremum" is singular. "Extrema" is plural).

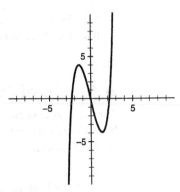

FIGURE 2 $x^3 - 5x$. $[-10, 10]$ by $[-10, 10]$

DEFINITION 4-1-10   A graph has a <u>local extremum</u> where it has either a "local maximum" or a "local minimum." A graph has a "<u>local maximum</u> at $x = x_0$" when there is an interval surrounding x_0 for which $f(x) \leq f(x_0)$ for all values of x in the interval. That is, the graph is locally highest there.

In Figure 2, the graph has a local maximum at $x = -1.29$, where the graph is higher than at nearby x values. This is not the overall highest point on the graph (which is why the maximum is only "local"). The key to a "local maximum at $x = x_0$" is that $f(x_0)$ is at least as great as $f(x)$ for all *nearby* x-values. Similarly, a graph has a <u>local minimum</u> at $x = x_0$ if $f(x_0) \leq f(x)$ for all values of x in some interval surrounding x_0. Figure 2 exhibits a local minimum at $x = 1.29$.

Quadratics have one local extremum. For example, the well-known graph of x^2 exhibits a local minimum at $x = 0$ (Figure 3-2-1), and the graph of $-x^2$ exhibits a local maximum at $x = 0$.

Cubics may have two local extrema, or no local extrema, but they cannot have exactly one local extremum or three or more local extrema. It takes a higher degree polynomial to have three or more local extrema.

EXAMPLE 13   The graph of x^4 (Figure 3) is somewhat similar to the graph of x^2. All even power monomials have some features in common. They are symmetric about the y-axis. If p is even, from 4-1-5,

$$(-x)^p = x^p,$$

which says that the images of x to the right of the y-axis are duplicated as images of $-x$ to the left of the y-axis

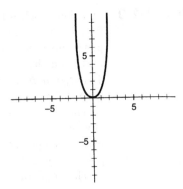

FIGURE 3 x^4. $[-10, 10]$ by $[-10, 10]$

> **A graph is symmetric . . .**
>
> about the y-axis iff $(-x, y)$ is on the graph whenever (x, y) is: $f(-x) = f(x)$.
>
> about the origin iff $(-x, -y)$ is on the graph whenever (x, y) is: $f(-x) = -f(x)$,
>
> about the x-axis iff $(x, -y)$ is on the graph whenever (x, y) is.
>
> Graphs of functions generally are not symmetric about the x-axis because that would require two y-values for a given x-value (unless $y = 0$), but functions have a unique y-value for each x-value.

DEFINITION 4-1-11   Functions (such as f defined by $f(x) = x^2$) with the property that $f(-x) = f(x)$ for all x are called <u>even</u> functions. The term "even" is used because all even-power monomials have this property (4-1-5). Graphs of even functions are symmetric about the y-axis.

The extreme case is the constant function, $f(x) = c$, which is graphed as a horizontal line which is clearly symmetric about the y-axis. Recall that $c = cx^0$, so it can be regarded as a constant times an even power of x (0 is an even number).

The graph of x^4 has only one local extremum (Figure 3). The graph of $x^4 - 5x^2$ on the other hand, has three (Figure 4).

THEOREM 4-1-12   A polynomial of degree $n > 0$ can have at most $n - 1$ local extrema. Odd degree polynomials have even numbers of local extrema, and even degree polynomials have odd numbers of local extrema.

If $P(x)$ is a polynomial of degree n, then the equation $P(x) = k$ can have at most n solutions. If $P(x)$ has m local extrema, then the equation $P(x) = k$ has at most $m + 1$ distinct solutions.

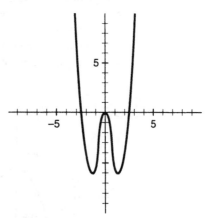

FIGURE 4 $x^4 - 5x^2$. $[-10, 10]$ by $[-10, 10]$

For example, lines of degree 1 have 0 local extrema. Quadratics (of degree 2) have 1 local extremum, and polynomials of degree 4 have either 1 (Figure 3) or 3 (Figure 4). Cubics may have 2 (Figure 2) or zero (Figure 1).

Quadratic equations (of degree 2) may have 2 solutions. Cubic equatons (of degree 3) may have 3 when they have 2 local extrema (Figure 2, with $k = 1$).

A polynomial equation of degree 4 could have 4 solutions when it has three local extrema (Figure 4, with $k = -2$), but at most only two solutions if it has only one local extremum (Figure 3).

EXAMPLE 14 Graph $x^3(x - 2)$.

This is a polynomial of degree 4. Instead of having 3 local extrema, it has 1 (Figure 5).

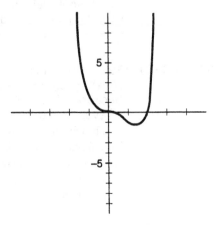

FIGURE 5 $x^3(x - 2)$. $[-5, 5]$ by $[-10, 10]$

Knowledge of the possible shapes of a graph can help you find a "representative" graph.

EXAMPLE 15 Graph $x^3 - 20x^2 + 150$. Then find the minimum for $x > 0$.

The picture on the standard scale is not very illuminating. Try it yourself. The picture consists of two nearly vertical slashes. The window is bad. A glance at the expression tells us that y-values can be as large as 150, so you must change the vertical scale. Try, say $-500 \leq y \leq 500$ (Figure 6).

If you didn't know what cubics look like, you might think Figure 6 is a representative graph. It's not.

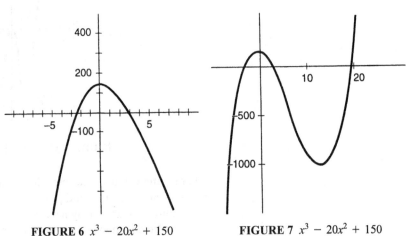

FIGURE 6 $x^3 - 20x^2 + 150$
$[-10, 10]$ by $[-500, 500]$

FIGURE 7 $x^3 - 20x^2 + 150$
$[-10, 30]$ by $[-1500, 500]$

Cubics can not have exactly one local extremum. If there is one, there must be a second. The window is still not right. Try again. In Figure 7 the scale is $[-10, 30]$ by $[1500, 500]$. The picture looks like a cubic. It must be representative. The minimum, for $x > 0$, is easy to find (Problem A27). ▬

■ End-Behavior

Some features of the graph of a polynomial are determined by its degree and its leading coefficient. If you "zoom out" far enough, the graph of any polynomial looks much like the graph of its highest-degree monomial. To express this clearly, we assign a name to the highest-degree monomial in a polynomial.

DEFINITION 4-1-11   The <u>end-behavior model</u> of a polynomial is its highest-degree monomial term.

EXAMPLE 16   The end-behavior model of "$x^2 + 2x + 1$" is "x^2".

The end-behavior model of "$7 - 5x^3$" is "$-5x^3$."

The end-behavior model of "$2x^{10} + 165x^8 + 942$" is "$2x^{10}$."

The end-behavior model of "$(x^2 - 4)(3x - 5)$" is "$3x^3$." ▬

The point of the end-behavior model is that **for large x, the polynomial behaves much like its leading term.** In Figure 6 the graph does not look like x^3 for large x's. Therefore, you know it is not a representative graph.

EXAMPLE 17   The end-behavior model of the polynomial $2x^3 - 7x - 3$ is $2x^3$. For large values of x, the graph looks much like the graph of $2x^3$. For small values of x the value of "$-7x + 3$" is substantial relative to the value of "$2x^3$". For small x's, the graphs of a polynomial and its end-behavior model need not be alike (Figure 8). But for large x's, their "end-behaviors" are alike (Figure 9).

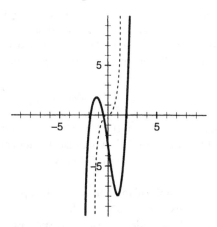

FIGURE 8 $2x^3 - 7x - 3$, and $2x^3$
(dotted). $[-10, 10]$ by $[-10, 10]$

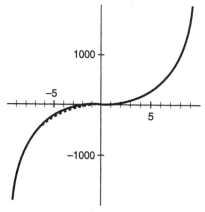

FIGURE 9 $2x^3 - 7x - 3$ and $2x^3$.
$[-10, 10]$ by $[-2000, 2000]$

▬

EXAMPLE 18 A fourth-degree polynomial may have three local extrema. For small x, the graph of $x^4 - 5x^2$ (Figure 4) is quite unlike the graph of x^4 (Figure 3), its end-behavior model. But for large x, they are very similar (Figure 10).

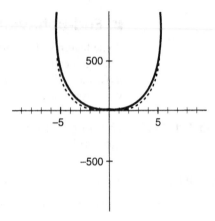

FIGURE 10 $x^4 - 5x^2$ (dotted) and x^4 (solid). $[-10, 10]$ by $[-1000, 1000]$ ▬

A crude indication of the behavior of the graph of a polynomial for large absolute values of x can be given with two arrows, one to indicate its behavior on the far left and one for the far right. For example, the end-behavior of the graph of x^2 might be described by "↑↑" and the graph of x^3 by "↓↑". The graph of $-x^2$ would be described by "↓↓" and the graph of $-x^3$ by "↑↓".

■ The Use of Polynomials

Polynomials can express important real-world functions. Here is an application of a polynomial in statistics.

EXAMPLE 19 Suppose voters are chosen at random from a huge pool of voters and asked whom they prefer, candidate A or candidate B. Probability theory gives the probability of any sequence of responses. For example, the probability of 5 responses being in favor of A, then B, then B, then A, and then A is approximately

$$x^3(1 - x)^2,$$

where x is the true fraction ($0 \leq x \leq 1$) of voters who would select candidate A (Figure 11). The power "3" is because 3 voters preferred A, and the power "2" is because 2 voters preferred B. This is a fifth-degree polynomial in factored form.

Here is an example of an important type of problem in statistics: Find x which maximizes that expression over the domain $0 \leq x \leq 1$.

Calculus will provide an exact answer. But you do not have to wait for calculus—you can obtain an answer graphically. As usual, the scale is important. If you use the standard scale the tiny region of interest with x between 0 and 1 will be hard to see. Figure 11 uses the window [0, 1] by [0, .1]. Note the small y-interval. The maximum occurs when $x = .6$ (Problem B39). Using calculus, it can be proven that this numerical result is exact.

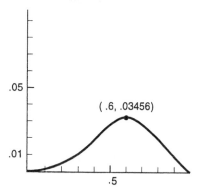

FIGURE 11 $x^3(1 - x)^2$. [0, 1] by [0, .1]

■ Approximation

Because integer powers come from repeated multiplication, polynomials can be evaluated using the four functions of arithmetic. Logarithmic, exponential, and trigonometric expressions cannot. We can exactly evaluate "2.3^4," but there is no way to exactly evaluate "log 2.3" or "$2^{2.3}$"or "sin 2.3" The sine function is very cumbersome to evaluate (except when you have a calculator). It used to be, a mere 40 years ago, that people looked up $\sin x$ in printed tables. Another approach was to use "approximating polynomials," which can be evaluated using only arithmetic operations that computers can do very rapidly.

When the sine curve is graphed on the domain in Figure 12, $-\pi \leq x \leq \pi$ (equivalent to $-180°$ to $180°$, but the angle is measured in radians), it almost looks like a polynomial—perhaps a cubic with two local extrema.

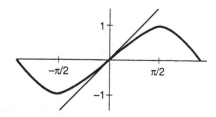

FIGURE 12 $y = \sin x$ and $y = x$. $-\pi \leq x \leq x$

EXAMPLE 20   Consider $\sin x$, where x is in radians. Near the origin the graph of $\sin x$ is almost a straight line, $y = x$ (Figure 12). For small values of x, $\sin x$ is nearly x.

$$\sin .1 = .0998, \text{ only .0002 off.}$$
$$\sin .2 = .1987, \text{ only .0013 off.}$$

The linear polynomial "x" can approximate $\sin x$ for x near 0.

Figure 12 shows that the linear approximation is not good if x is far from 0. For example, sin 1 $= .84$, not very close to 1. However, a cubic approximation can do better:

For small x,

$$\sin x \text{ is approximately } x - \frac{x^3}{6}.$$

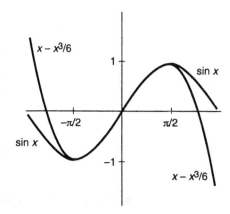

FIGURE 13 sin x and $\dfrac{x - x^3}{6}$. $[-\pi, \pi]$ by $[-2, 2]$

x	.1	.2	.5	1.0
sin x	.099833417	.198669331	.4794255	.84147
$x - \sin x$	.000166583	.001330669	.0205745	.15853
$x - \dfrac{x^3}{6}$	.099833333	.198666667	.4791667	.83333
$\sin x - (x - \dfrac{x^3}{6})$	.000000084	.000002664	.0002588	.00814

Higher degree polynomials can give still better approximations (Problem B41). ▬

CONCLUSION Integer powers can be interpreted as repeated multiplication. If the base is positive, all the properties of powers can be understood by considering repeated multiplication (even if the powers are not integers). If the base is negative, the usual properties of integer powers hold, but there are complications for fractional and irrational powers.

Polynomials are important because they can be evaluated using only arithmetic operations. Monomials serve as end-behavior models for all polynomials, and the graphs of monomials are fundamental. The end-behavior model helps determine if a graph is representative.

Terms: base, power, exponent, monomial, polynomial, odd function, point-symmetric about the origin, local extremum, local maximum, local minimum, even function, symmetric about the y-axis, end-behavior model.

Exercise 4-1

A

A1.* ☺ Do "Calculator Exercise 1."
  (A) Sketch the graphs of x^3, x^4, and x^5
  (B) Which two look alike?
  (C) Which of the three in part (A) have unique solutions to "$x^p = c$" for any c?

A2.* ☺ Do "Calculator Exercise 2."
  (A) All graphs of x^p for positive values of p go through the two points _____ and _____.
  (B) For all $p \geq 2$, the slope of the graph of x^p at the origin appears to be _____.

A3.* ☺ (A) Do all quadratics have similar shapes?
  (B) Do all cubics have similar shapes?

A4.* ☺ True or false?
  (A) x^2 and x^4 and x^6 have graphs with somewhat similar shapes.
  (B) x^3 and x^5 and x^7 have graphs with somewhat similar shapes.

☺ *Simplify to an equivalent expression with only one appearance of "x", assuming x is not zero.*

A5. (A) $\dfrac{x^5}{x^2}$ (B) $x^6 x^4$ (C) $(x^3)^7$

A6. (A) $\dfrac{2x^5}{8x}$ (B) $\dfrac{x^{-3}}{x}$ (C) $(x^{-4})x^5$

A7. (A) $\dfrac{(4x)^3}{(2x)^2}$ (B) $x^2 x^5$ (C) $\dfrac{x^2}{x^7}$

A8. (A) $\dfrac{x^4 x^6}{x^2}$ (B) $\dfrac{4x^4}{2x^6}$ (C) $(1-x)^2(1-x)^5$

☺ *Find P(x) given*

A9. $x^5 P(x) = x^{15}$

A10. $x^4 P(x) = x^7$

A11. $(x-2)^2 P(x) = (x-2)^3$

A12. $(\log x)^2 P(x) = (\log x)^5$

Factor

A13. $4x^3(1-x)^2 - x^4(1-x)$

A14. $5x^3(1-x)^3 - 3x^5(1-x)^2$

A15. $4(2x+1)^3(x-2)^2 + 2(x-2)(2x+1)^4$

A16. $2(x+3)(x-2)^3 + 3(x+3)^2(x-2)^2$

☺ *Give the end-behavior model of the polynomial.*

A17. $2x^3 - 6x^2$

A18. $4x^5 - 20x^2 + 14$

A19. $5 - x^2$

A20. $5x + 13 - 3x^3$

A21. $(x-3)(5x^2+4)$

A22. $(4x-1)(x+2)$

A23. $(2x+1)(x-3)(x-5)$

A24. $(x^2+3)(5x^2+7)$

☺ *If possible, simplify these to only one appearance of "x". If not possible, say so.*

A25. (A) $x^7 - x^4$ (B) $x^3(1-x^2)$ (C) $5x^2(x^6)$

A26. (A) $(7x)(3x^2)$ (B) $(5x)(5x+1)$ (C) $x(x^2+2)$

A27. Find, for $x > 0$, the minimum of $x^3 - 20x^2 + 150$. [-1000]

A28. (A) Give the first five powers of 2.
  (B) Approximately what magnitude is 2^{10}? [Powers of 2 are numbers you will see in articles about computers.]
  (C) Approximately what magnitude is 2^{20}?

Decide if the given expression is "odd" (4-1-9), "even" (4-1-11), or neither.

A29. x^4 **A30.** x^5

A31. $x^6 + 2x^2$ **A32.** $x^2 + 7$

A33. $x^3 + 5$ **A34.** $x^3 + 6x$

A35. $x^3 + 3x + 5$ **A36.** $x^3 + 3x^2$

A37. $x^4 + 3x^2 + 3$ **A38.** $x^4 + 4x^3$

A39.* Distinguish between "monomial" and "polynomial."

B

B1.* Sketch, roughly, all the shapes that graphs of monomials, x^n, can have for $n \geq 2$.

B2.* Sketch an example of each different type of shape graphs of cubic polynomials can have. [For this problem regard turning a graph upside down as worth another sketch, even if you do not regard it as a new shape.]

B3.* ☺ How many local extrema can the graph of a polynomial of degree n have?

B4.* ☺ What is the largest number of solutions the equation $P(x) = 0$ can have if $P(x)$ is a polynomial of degree n?

B5–6. *The figure gives the graph of x^3 in a certain window. One scale is given.*

B5. Window $[-10, 10]$ by $[-c, c]$. Find c.

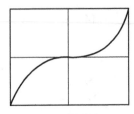

For B5–B6

B6. Window $[-c, c]$ by $[-10, 10]$. Find c.

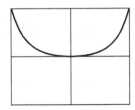

For B7–B8

B7–8. *The figure gives the graph of x^4 in a certain window. One scale is given.*

B7. Window $[-10, 10]$ by $[-c, c]$. Find c.

B8. Window $[-c, c]$ by $[-10, 10]$. Find c.

If possible, simplify to an equivalent expression with only one appearance of "x".

B9. $x^5(x^7)^3$

B10. $(1 - x)^3(1 - x)^5$

B11. $\dfrac{(2x)^4}{8x^6}$

B12. $\dfrac{x^2 x^{10}}{x^3}$

B13. $\dfrac{x^{-2}}{x^3}$

B14. $\dfrac{(x^4)^2}{x^{-2}}$

Here is the graph in a certain window. Find the window.

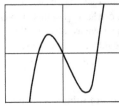

B15. $x(x - 6)(x + 4)$

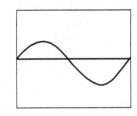

B16. $x(x - 50)(x - 100)$

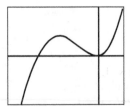

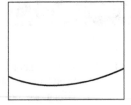

B17. $x^3 + 20x^2$

B18. $(x - 50)(x - 100)(x - 150)$
[Note: For B18, the graph is all in the bottom half of the window.]

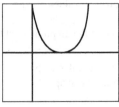

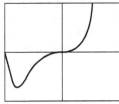

B19. $(10x - 1)^2$

B20. $x^4 + 20x^3 + 50$

☺ *Rewrite these using properties of powers if convenient, and, if not convenient, say so.*

B21. (A) $a^2 b^3$ (B) $a^3 b^3$ (C) $3^2 + x^2$

B22. (A) $5^3 6^4$ (B) $\dfrac{x^2}{x^3 + 1}$ (C) $(xy^2)^2$

B23. (A) $5^2 + x^2$ (B) $\left(\dfrac{a}{b}\right)^3$ (C) $x^2 + y^2$

B24. (A) $16x^4$ (B) $a^4 + b^4$ (C) $(x^2(x^3))^4$

Simplify.

B25. $\left(\dfrac{1}{9}\right)^n 3^{2n+1}$

B26. $\dfrac{5^{n+1}}{25^{n/2}}$

B27. $2^n\left(\dfrac{1}{2}\right)^{2n+1}$

B28. $\dfrac{4^n}{2^{3n}}$

☺ *Suppose a polynomial has the following characteristics (and no other extrema). What degree is it?*

B29. 2 local maxima and one local minimum.

B30. 1 local maximum and no local minima.

B31. 3 local minima and 2 local maxima.

B32. 2 local extrema.

Solve for p.

B33. $x^2(x^p) = x^6 x^3$, for all $x \geq 0$.

B34. $\dfrac{x^2}{x^p} = x^{-1}$, for all $x \geq 0$.

B35. $(x^2)^p = (x^3)(x^5)$, for all $x \geq 0$.

B36. $x^{2p} = x(x^4)$, for all $x \geq 0$.

B37. Consider a sheet of metal 12 inches by 20 inches. Cut squares out of each corner and fold up the sides to make an open-topped box. Find the size of the cutout squares that maximizes the volume of the box. [2.4]

B38. Consider a sheet of metal in the form of a right triangle with base 36 inches and height 20 inches. What are the dimensions of the rectangle of largest area that can be cut from that sheet?

B39. Find a window so that the expression $x^3(1 - x)^2$ in Example 19 exhibits the behavior you would expect of a fifth-degree polynomial (Figure 11 does not). Then sketch it and label the viewing rectangle.

B40. Near 0, $\cos x$ (in radians) can be approximated by $1 - \dfrac{x^2}{2} + \dfrac{x^4}{24}$. What is the maximum error of the approximation on the interval $[0, .5]$? [.000022]

B41. $x - \dfrac{x^3}{6} + \dfrac{x^5}{120}$ is approximately $\sin x$, for small x (in radians). Find the largest error used in approximating $\sin x$ by it when $0 \leq x \leq \dfrac{\pi}{2}$.

B42. A polynomial approximation of degree 2 to e^x near $x = 0$ is given by "$1 + x + \dfrac{x^2}{2}$" (Figure 15).

How accurate is it
(A) For $x = 0$?   (B) For $x = .1$?
(C) Find the largest value of x such that the error of the approximation is no more than 1% of the true value.

B43. Sketch the possible shapes of the graph of $ax^4 + bx^3 + cx^2 + dx + e$.

B44. Sketch the possible shapes of the graph of $ax^5 + bx^4 + cx^3 + dx^2 + ex + f$.

B45. See Example 19. The probability of a sequence of n randomly chosen voters responding to form a particular string of a choices for candidate A and the rest of the choices for B is given approximately by $P(x, a, n) = x^a(1 - x)^{n-a}$.
Find the x-value which maximizes it over $0 \leq x \leq 1$ for $a = 6$ and $n = 17$.

Reading and Writing Mathematics

B46. State the property of e^x that parallels
(A) 4-1-1   (B) 4-1-2A   (C) 4-1-2B   (D) 4-1-2C

B47.* (A) Define "end-behavior model."
(B) What is the flaw in this remark of a student? "The graph of a polynomial looks like the graph of its end-behavior model."

B48.* Explain, as if to a younger student who does not know, why any number (except possibly 0) to the 0 power is 1.

B49. What are the characteristics of the graph of an even function?

B50. What are the characteristics of the graph of an odd function?

B51. In Example 14, $x^3(x - 2)$ is a polynomial of degree 4, but it is not an "even" function. Why not? Look at the graphs of $x^2 + x$, $x^2 + x + 1$, and $x^2 + 1$. Which are even? Look at the graphs of $x^3 + x^2$, $x^3 + x$, and $x^3 + 1$. Which are odd? Generalize. That is, state a result that describes when a polynomial is even or odd.

B52. (A) Prove that a sum of two odd functions is an odd function.
(B) Prove that a product of two odd functions is an even function.

B53. (A) Prove that a sum of two even functions is an even function.
(B) Prove that a product of two even functions is an even function.
(C) True or false? The product of and odd and an even function is an even function.

B54. In calculus the idea of end-behavior is formalized after introducing the concept of a limit. In essence, the idea is that the values of the ratio of the polynomial to its end-behavior model approach 1 as x becomes large in absolute value. Let $P(x)$ be the polynomial and $E(x)$ be its end-behavior model.

As x goes to $\pm\infty$, $\dfrac{P(x)}{E(x)}$ goes to 1 .

Therefore, for large absolute values of x, the ratio of $P(x)$ to $E(x)$ is approximately 1. Use a general notation for a polynomial $P(x)$ and use algebra to rewrite $\dfrac{P(x)}{E(x)}$ to justify the previous sentence.

Explain why

B55.* $a^n a^m = a^{n+m}$ **B56.*** $(ab)^n = a^n b^n$

B57.* $\left(\dfrac{a}{b}\right)^n = \dfrac{a^n}{b^n}$

B58. The Binomial Theorem. Let n be a positive integer. Then $(a + b)^n$ can be written as a sum of $n + 1$ terms. In each term, the exponents add to n. The coefficient on the term $a^{n-k}b^k$ is "n choose k",

written $\begin{pmatrix} n \\ k \end{pmatrix} = \dfrac{n!}{k!(n-k)!}$, where $n!$ (pronounced

"n factorial") is the product of the first n integers. For example, $3! = 1(2)(3) = 6$ and $4! = 24$. For convenience, $0!$ is defined to be 1. Also, $1! = 1$, so the coefficient, n, on $a^{n-1}b$ can be regarded as $\dfrac{n}{1!}$. In summary,

$$(a + b)^n = a^n + na^{n-1}b + \left[\frac{n(n-1)}{2!}\right]a^{n-2}b^2$$

$$+ \left[\frac{n(n-1)(n-2)}{3!}\right]a^{n-3}b^3 + \ldots + b^n.$$

(A) Give $5!$. (B) Give the expansion of $(a + b)^3$.

(C) Give and simplify the coefficient on the a^2b^3 term of $(a + b)^5$.

(D) Explain how the coefficient "1" on the b^n term is "n choose k".

Section 4-2 POLYNOMIAL EQUATIONS

Let $P(x)$ denote a polynomial. The problem in this section is to solve the equation "$P(x) = c$," for constant c.

Polynomial expressions can be *evaluated* using only arithmetic operations. But polynomial equations cannot necessarily be *solved* using only arithmetic operations. Polynomial equations are often difficult to solve, and frequently we must be satisfied with approximate solutions. This section distinguishes the types that can be solved algebraically using roots or factoring. Of course, the guess-and-check method always works, so it is an important alternative even when algebraic methods would work, and it is necessary when algebraic methods do not work.

If the polynomial is linear, the equation can be solved using only arithmetic operations.

EXAMPLE 1 Solve $mx + b = c$ for x, for $m \neq 0$.

In this problem "m" and "b" are parameters of a line, and "x" is the unknown. By inverse-reverse thinking, $x = \dfrac{(c - b)}{m}$. The solution uses only arithmetic operations.

If $P(x)$ is not linear, the solution to "$P(x) = c$" is much more complex. Even such a simple quadratic as "$x^2 = 15$" cannot be solved using only arithmetic operations. You are so accustomed to using the square root symbol that you may take square roots for granted, but they are not trivial to compute. The first electronic calculator with square roots came out in 1965 and cost $1850 then!

In Mathematics we use the square root symbol to *name* the *non-negative* solution to "$x^2 = c$," for non-negative c. The idea of naming the solution to the equation "$x^n = c$" for integer values of n leads to the concept of an "nth root."

■ Solving Monomial Equations

Let $P(x)$ be an nth degree monomial $[ax^n]$, for $n \geq 2$. Then you can solve "$P(x) = c$" using the nth-root function. For example, the inverse of the third-power (cubing) function is commonly called the "cube root" function and sometimes given a special notation like the square root function, except with a tiny "3" in the crook of the radical symbol:

$$\sqrt[3]{x} = x^{1/3}.$$

The notation with a fractional power is preferable in calculus. We will use it. It is not easy to evaluate by hand, but it is on your calculator.

CALCULATOR EXERCISE 1   Check the keystrokes required by your calculator to obtain these results.

$$1000^{1/5} = 3.981.$$
$$(-300)^{1/3} = -6.694.$$

Evaluate $(-300)^{1/4}$. For this one some calculators return an error message. Even-degree roots of negative numbers are not real numbers.

CALCULATOR EXERCISE 2   Graph x^n for any odd n, say, $n = 3$ or 5 (Figure 1). Can you always solve "$x^n = c$"? Is there always exactly one solution, regardless of c? Then graph $x^{1/3}$ or $x^{1/5}$. What is its domain?

Now graph x^n for any even n, say $n = 4$ (Figure 2). Can you always solve "$x^n = c$"? Is it one-to-one? When does $x^4 = c$ have two solutions? No solutions? Graph $x^{1/4}$. What is its domain?

The nth-root function is the $\dfrac{1}{n}$ power function.

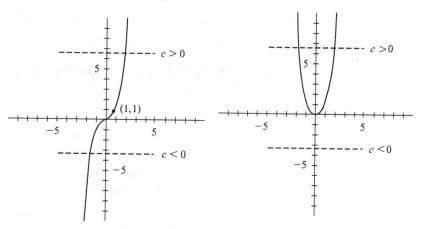

FIGURE 1 x^n, n odd, $n \geq 3$ FIGURE 2 x^n, n even, $n \geq 2$

THEOREM 4-2-1 (TO SOLVE "$x^n = c$")

Let n be a positive integer.

"$x^n = 0$" has one solution, $x = 0$.

For odd values of n, $x^n = c$ has exactly one solution, $x = c^{1/n}$, for any c.

For even values of n:
If $c > 0$, "$x^n = c$" has two solutions: $x = \pm c^{1/n}$.
If $c < 0$, "$x^n = c$" has no real-valued solutions.

This theorem has two distinct cases, depending on whether n is odd or even. If n is odd (Figure 1), x^n can be negative and the nth-root function can be defined for negative arguments. If n is even (Figure 2), x^n is non-negative (so $c \geq 0$), and the nth-root function is therefore not defined for negative arguments (however, it can be defined using complex numbers).

EXAMPLE 3 Solve $5x^3 = 12$.

By inverse-reverse thinking, first divide by 5. Then "uncube" with the one-third power function.

$$x^3 = \frac{12}{5} = 2.4 \text{ iff } x = 2.4^{1/3} = 1.339.$$ ▬

EXAMPLE 4 Solve $x^6 = 1200$.

Since 6 is even and 1200 is greater than 0, there will be two solutions, much like "$x^2 = 15$" has two solutions. $1200^{1/6} = 3.2598$. The two solutions are $x = \pm 3.2598$.

The reason the nth root is the $\dfrac{1}{n}$ power can be seen by inspecting the result about a power of powers (4-1-3): $(x^r)^p = x^{rp}$. Therefore, for $x > 0$,

$$(x^n)^{1/n} = x^{n(1/n)} = x^1 = x.$$

The squaring function has power 2, so this says its inverse has power $\dfrac{1}{2}$. The square root function is the $\dfrac{1}{2}$ power function. All the properties of power functions hold for square roots of positive numbers. ▬

■ Solving Polynomial Equations

Four types of polynomial equations are easy to solve algebraically—those with

1. lines, $mx + b = c$,
2. quadratics, $ax^2 + bx + c = 0$,
3. monomials, $ax^n = c$, and
4. polynomials already factored, or factorable, and set equal to zero, so that the Zero Product Rule applies.

All the other types of polynomial equations are substantially more difficult to solve algebraically.

There are formulas for solving polynomial equations of degree 3 (cubics) and degree 4 in terms of roots, but they are so long and messy that very few mathematics professors have bothered to learn them.

So, if the cubic and quartic formulas are too long to use, how does a mathematician solve a cubic equation? There are two ways, selected from the "Four Ways to Solve an Equation." One is to factor and use the Zero Product Rule. Some of this section discusses clever methods of factoring cubic and higher-degree polynomials. The other way is to use guess-and-check.

> <u>Guess-and-check</u> The easiest way to solve a polynomial equation of degree 3 or higher is to use guess-and-check.

EXAMPLE 5   Solve $x^3 - 4x - 1 = 0$.

Graph the cubic (Figure 3). Apparently, there are three solutions. You can discover each of the three solutions to any desired degree of accuracy. $x = -1.861$ and $x = -.254$ are two solutions. The third is left to the reader (Problem A11).

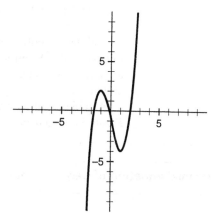

FIGURE 3 $x^3 - 4x - 1$ [$-10, 10$] by [$-10, 10$]

The only trick to guess-and-check is to remember the concept of a "representative" graph (which exhibits all relevant features of the function) and that a polynomial of degree n may have up to n zeros. Do not assume all the zeros will be exhibited on the standard scale.

EXAMPLE 6   Solve $\dfrac{x^3 + 20x^2 - 16x - 320}{100} = 0$.

From the graph (Figure 4, on the standard scale) it may appear that there are two solutions. But that graph is not a representative graph and changing the window gives a different picture (Figure 5). When you see a cubic polynomial you know from its end-behavior model that a representative graph of it cannot look like Figure 4. To the left of the window in Figure 4 the graph must come back down and cross the x-axis again.

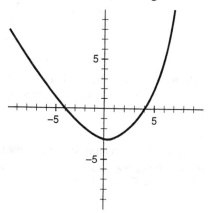 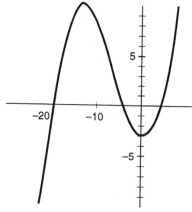

FIGURE 4 $\dfrac{x^3 + 20x^2 - 16x - 320}{100}$

[$-10, 10$] by [$-10, 10$]

FIGURE 5 $\dfrac{x^3 + 20x^2 - 16x - 320}{100}$

[$-30, 10$] by [$-10, 10$]

In Figure 5 you see there will be three solutions, the maximum number possible for a third degree polynomial (Problems A13 and 14). ▬

Cubic equations always have at least one real-valued solution. The range of every cubic expression is all real numbers, including zero. Their behavior is either ↓↑ or ↑↓, so their graphs must cross the horizontal line $y = c$ for every c, so there will be at least one solution to $P(x) = c$. This generalizes to all odd-degree polynomials: All odd-degree polynomial equations have at least one real-valued solution. But not even-degree polynomials. For example, you already know that quadratic equations do not necessarily have any real-valued solutions.

A cubic equation may have at most three real-valued solutions (Example 5, Figure 3). In general,

---

A polynomial equation of degree n may have at most n real-valued solutions.

Look at Figure 6. It shows that the number of possible solutions to a polynomial equation $P(x) = c$ is related to the number of local extrema of the graph of the expression $P(x)$. For example, a quartic can have at most three local extrema, and can yield 4 solutions in certain cases (for example, when $c = c_4$), two in others (when $c = c_2$), and none in others (when $c = c_0$). Also one or three solutions are possible as special cases when the line $y = c$ is tangent to the curve ($c = c_1$ or $c = c_3$).

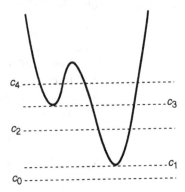

FIGURE 6 A quartic (polynomial of degree 4). $f(x) = c$ can have 4, 2, or no solutions. It can have 1 or 3 if the horizontal line $y = c$ is tangent to a bump

The Zero Product Rule
The first way to solve a high-degree polynomial is to use guess-and-check. The only simple remaining way to solve a polynomial of degree 3 or higher is to factor it and use the Zero Product Rule.

THE ZERO PRODUCT $ab = 0$ iff $a = 0$ or $b = 0$, for all a and b.
RULE 4-2-2 The Zero Product Rule has a natural extension to three or more factors.

EXAMPLE 7   Solve $(x - 2)(x + 3)(x - 7) = 0$.

The solution is obtained by setting each factor equal to zero. $x = 2$ or $x = -3$ or $x = 7$ (Figure 7).

The next example illustrates a type of problem that appears frequently in calculus. Sometimes a sum can be factored because the terms have factors in common.

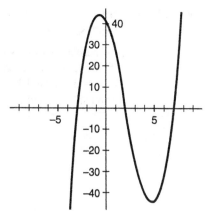

FIGURE 7 $(x - 2)(x + 3)(x - 7)$ $[-10, 10]$ by $[-50, 50]$ ▬

EXAMPLE 8   Solve $-2x^3(1 - x) + 3x^2(1 - x)^2 = 0$.

First factor out the common factors. Each has an "x^2" and each has a "$1 - x$".

$$-2x^3(1 - x) + 3x^2(1 - x)^2 = x^2(1 - x)[-2x + 3(1 - x)]$$
$$= x^2(1 - x)[3 - 5x].$$

Therefore, by the Zero Product Rule, the original equation is equivalent to

$$x^2 = 0 \quad \text{or} \quad 1 - x = 0 \quad \text{or} \quad 3 - 5x = 0.$$

The solution   is $x = 0$ or $x = 1$ or $x = \dfrac{3}{5}$.   ▬

■ Graphical Factoring Techniques

Occasionally, but only occasionally, polynomials of degree three or higher can be factored using integers. The Factor Theorem can be very helpful in these cases.

Equivalent Statements
$x = c$ is a solution to $P(x) = 0$.
c is a root of $P(x)$.
c is a zero of $P(x)$.
$P(c) = 0$.
c is an x-intercept of the graph of $P(x)$.
$x - c$ is a factor of $P(x)$.

THE FACTOR
THEOREM 4-2-3

"$x - c$" is a factor of the polynomial expression $P(x)$ iff **c** is a zero of $P(x)$.

Factoring cubic or higher-power polynomials is difficult unless one of the factors is "$x - c$" where c is an integer. The Factor Theorem makes these factors easy to find.

EXAMPLE 9

Part 1: Factor $x^3 - 3x^2 - 5x + 15$.

Part 2: Solve $x^3 - 3x^2 - 5x + 15 = 0$.

These are two very closely related problems. Graph the expression (Figure 8). There appears to be a zero at, or at least near, $x = 3$. By evaluating the expression for $x = 3$, we find 3 is a zero. That means that "$x - 3$" is a factor. The other factor must be a quadratic. We can find it by long division.

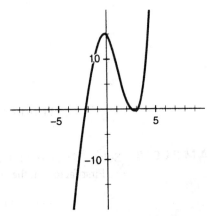

FIGURE 8 $x^3 - 3x^2 - 5x + 15$. [−10, 10] by [−20, 20]

Polynomial long division exactly parallels regular long division of integers. The layout is the same and the steps are the same. Also, the remainder will be zero if and only if the divisor is truly a factor.

$$
\begin{array}{r}
x^2 - 5 \\
x - 3 \overline{\smash{\big)}\, x^3 - 3x^2 - 5x + 15} \\
\underline{x^3 - 3x^2 } \\
0 - 5x + 15 \\
\underline{- 5x + 15} \\
0
\end{array}
$$

[For the first term, divide the lead terms. That is, divide x into x^3 to get x^2. Put the "x^2" above the line and multiply it by $x - 3$. Put the result below "$x^3 - 3x^2$." Subtract (this case yields 0). Bring down the next part, "$-5x + 15$". The lead term of "$x - 3$" goes into that -5 times. Put the -5 up above and multiply it times "$x - 3$" and subtract that from the previous line. This case yields 0. It does factor – there is no remainder.]

So, $\qquad x^3 - 3x^2 - 5x + 15 = (x - 3)(x^2 - 5)$.

The equation in Part 2 is equivalent to $x = 3$ or $x^2 - 5 = 0$. The solutions to "$x^2 - 5 = 0$" are $\sqrt{5}$ and $-\sqrt{5}$. $\sqrt{5}$ is not an integer, so the factor theorem tells us it cannot be factored further using only integers. If you are willing to use real numbers, it factors into $(x - 3)(x - \sqrt{5})(x + \sqrt{5})$. ▬

**EXAMPLE 9
ANOTHER WAY**

Factor $x^3 - 3x^2 - 5x + 15$ using integers.

Another method of factoring third-degree polynomials is called "grouping." This method requires examples that work (most do not), takes a lot of practice, and uses the Distributive Law.

By inspection, $x^3 - 3x^2 - 5x + 15 = x^2(x - 3) - 5(x - 3)$. Regrouping, this
$$= (x^2 - 5)(x - 3).$$
This "grouping" method rarely finds any factors that would not be found more easily by graphing and using the Factor Theorem. ▬

EXAMPLE 10

Solve $x^3 - 2x^2 - 5x - 12 = 0$ for all solutions, real and complex.

Without using the cubic formula, you have only two ways to go. Factor it, or use guess-and-check. You may try to factor this by guess-and-check, which is a fine method when it works. However, there is an intelligent way to guess.

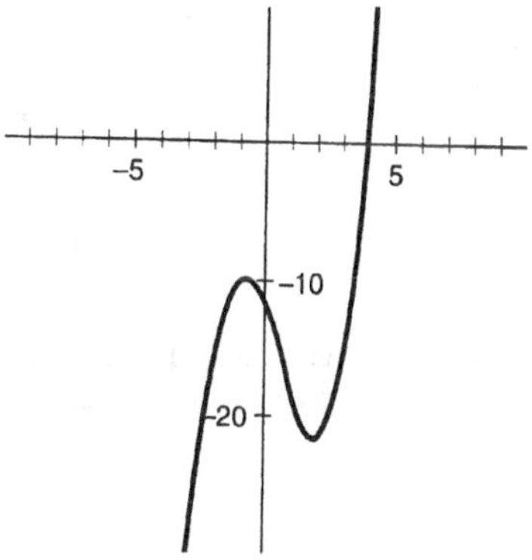

A graph can show you what to expect (Figure 9). Cubic equations have 1, 2, or 3 real-valued solutions, and this one appears to have 1. By the Factor Theorem, there will be a corresponding linear factor. Then the other factor must be a quadratic. Furthermore, that quadratic will not factor using real numbers.

Guess-and-check confirms that "$x = 4$" is a solution: $P(4) = 0$. Therefore "$x - 4$" is a factor. The other factor can be determined by polynomial long division:

FIGURE 9 $x^3 - 2x^2 - 5x + 12$.
$[-10, 10]$ by $[-30, 10]$

$$
\begin{array}{r}
x^2 + 2x + 3 \\
x - 4 \overline{\smash{\big)}\ x^3 - 2x^2 - 5x - 12} \\
\underline{x^3 - 4x^2} \\
2x^2 - 5x \\
\underline{2x^2 - 8x} \\
3x - 12 \\
\underline{3x - 12} \\
0
\end{array}
$$

So it factors into $(x - 4)(x^2 + 2x + 3)$. The solution is then the "$x = 4$" we already knew, together with the solutions to the quadratic equation $x^2 + 2x + 3 = 0$, which are easy to obtain:

$$x = 4 \text{ or } x = \frac{-2 \pm \sqrt{4 - 12}}{2}.$$

$$x = 4 \text{ or } x = -1 \pm i\sqrt{2}.$$

The last step uses $\sqrt{-8} = i\sqrt{8} = i\sqrt{4(2)} = 2i\sqrt{2}$. ▬

EXAMPLE 11 Find a cubic polynomial $P(x)$ that goes through $(2, 0)$, $(4, 0)$, $(-3, 0)$, and $(1, 4)$ (Figure 10).

The Factor Theorem and the zeros tell us the cubic is, in factored form,

$$k(x - 2)(x - 4)(x + 3),$$

where k remains to be determined. The zeros alone do not determine the constant factor. Use the point $(1, 4)$ to determine k. Substitute 1 for x.

$$4 = P(1) = k(1 - 2)(1 - 4)(1 + 3) = 12k.$$

Therefore $k = \dfrac{1}{3}$ and the cubic is

$$\left(\frac{1}{3}\right)(x - 2)(x - 4)(x + 3).$$ ▬

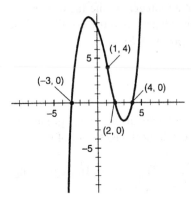

FIGURE 10 A cubic through 4 given points. $[-10, 10]$ by $[-10, 10]$

▪ Identities

The properties of powers given in 4-1-1 to 4-1-3 (see Table 4-1-6) are useful for reordering expressions.

EXAMPLE 12 Solve $x^2 x^3 = 1000$.

Rewrite it to exhibit only one appearance of "x" and use inverse-reverse.

$x^5 = 1000$, by Property 4-1-1 $\left[x^p x^r = x^{p+r}\right]$.

$x = 1000^{1/5} = 3.981.$ ▬

EXAMPLE 13 Solve $\dfrac{x^7}{x^4} = .04$.

By Property 4-1-2 $\left[\dfrac{x^p}{x^r} = x^{p-r}\right]$, $\dfrac{x^7}{x^4} = x^3 = .04$. $x = .04^{1/3} = .342.$ ▬

EXAMPLE 14 Solve $(x^2)^3 = 12$.

$(x^2)^3 = x^6 = 12$, by Property 4-1-3 $[(x^r)^p = x^{rp}]$.

$x = \pm 12^{1/6} = \pm 1.513$ Remember that even powers yield two solutions. ▬

EXAMPLE 15 Solve $x^3 + 2x^2 = 7$.

A sum of terms with different powers (here, 3 and 2) does not simplify into a single power function. Moving the 7 to the other side yields a cubic polynomial with no integer factors. There is no shortcut to solving this equation. Use guess-and-check (Problem A12). ▬

EXAMPLE 16 Solve $x^3 + 2x^2 = 0$.

This equation is much different from the previous equation, despite having the same polynomial on the left, because the right side is zero. Factor out x^2 and use the Zero Product Rule.

$x^3 + 2x^2 = 0$ iff $x^2(x + 2) = 0$ iff $x^2 = 0$ or $x + 2 = 0$.

The solution is $x = 0$ or $x = -2$. ▬

■ Factoring in Integers

The problem is still to solve "$P(x) = 0$" where $P(x)$ is a cubic or higher-degree polynomial. Remember that any polynomial equation can be solved for its real-valued solutions using guess-and-check, so think of that possibility first. This subsection concerns finding algebraic, exact, solutions and complex-valued solutions when the leading coefficient is an integer that is not 1. The techniques are based on the Factor Theorem.

EXAMPLE 17 $6x^2 - x - 35 = (3x + 7)(2x - 5)$.

Note that the "3" and the "2" in the factors divide the leading coefficient "6." Also, the "7" and the "-5" divide the constant, "-35." The next theorem says that this type of behavior always happens when you are dealing with *integer* coefficients. The leading coefficients of the factors must divide the leading coefficient of the original polynomial. The constant terms of the factors must divide the constant of the polynomial.

Solve $6x^2 - x - 35 = 0$.

From the previous part, $6x^2 - x - 35 = (3x + 7)(2x - 5)$. Therefore, by the Zero Product Rule, $x = \dfrac{-7}{3}$ or $x = \dfrac{5}{2}$.

Note that the solutions to $P(x) = 0$ have denominators that divide the leading coefficient, 6, and numerators that divide the constant, -35. This type of behavior always happens when you are dealing with a polynomial with *integer* coefficients. The upcoming "Rational Zeros" theorem states this idea in general. ▬

THEOREM ON FACTORING POLYNOMIALS WITH INTEGER COEFFICIENTS 4-2-4

Suppose $P(x)$ is a polynomial with integer coefficients, leading coefficient a, and constant term c.

If $P(x)$ factors into factors with integer coefficients and "$jx - k$" is one of the factors, then j divides the leading coefficient, a, and k divides the constant, c. ["j" and "k" denote integers in this subsection.]

This theorem describes all the potential factors with integer coefficients (Problem B46). If none are actually factors, then $P(x)$ has no linear factors with integer coefficients.

EXAMPLE 18   Factor "$2x^3 - x^2 - x - 3$" using integers.

According to the theorem, any linear factor "$jx - k$" must have j that divides 2 (that is, $j = \pm 1$ or ± 2) and k that divides -3 (that is, $k = \pm 1$ or ± 3). Therefore, any linear factor must be on this list:

$$x - 1, x + 1, x - 3, x + 3, 2x - 1, 2x + 1, 2x - 3, \text{ or } 2x + 3.$$

Now the problem can be solved using trial-and-error among these possibilities. [You do not need to try the negatives of all of these expressions because you can treat any negative sign as on the other factor, since $(-a)b = a(-b)$].

That looks like a lot of long division to try. Can't the Factor Theorem help? (We will continue this problem shortly.)

A theorem very much like the Factor Theorem can help. By the Zero Product Rule, linear factors "$jx - k$" and rational zeros, $x = \dfrac{k}{j}$ go together. ▬

THEOREM 4-2-5 (THE RATIONAL ZEROS THEOREM)

Suppose $P(x)$ is a polynomial with integer coefficients, leading coefficient a, and constant term c.

A) Suppose $x = \dfrac{k}{j}$ is a rational solution in lowest terms to the equation "$P(x) = 0$." Then the denominator j divides the leading coefficient a and the numerator k divides the constant term c.

B) Every rational solution corresponds to a linear factor with integer coefficients, and vice versa.

Therefore, you can find the factors by looking for zeros at the rational possibilities. Instead of trying the whole list of possible linear factors, all you need to try is the linear factors which would yield rational zeros where they appear to be on the graph.

EXAMPLE 18 CONTINUED

You want to factor $2x^3 - x^2 - x - 3$ using integers, but there are more possible factors than you are happy to try. Use the Rational Zeros Theorem to narrow the search.

Graph the expression (Figure 11) and look for rational zeros with denominator 1 or 2, (because they are the only possibilities that divide the leading coefficient, 2). The numerator of the zero must be ± 1 or ± 3. Figure 11 shows the only zero is between 1 and 2. Therefore, the only viable candidate for a zero is $\dfrac{3}{2}$ and the only viable factor is the one with a zero at $\dfrac{3}{2}$: $2x - 3$. If that expression is not a factor, it does not have a linear factor with integer coefficients.

Long division shows it is a factor. You obtain

$$2x^3 - x^2 - x - 3 = (2x - 3)(x^2 + x + 1).$$

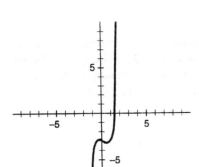

FIGURE 11 $2x^3 - x^2 - x - 3$. $[-10, 10]$ by $[-10, 10]$

Now you can solve for all zeros, real and complex. $x = \dfrac{3}{2}$ is one zero. From the quadratic factor (using the Quadratic Formula), the other two zeros are:

$$\frac{-1 \pm \sqrt{1^2 - 4(1)(1)}}{2} = -\frac{1}{2} \pm \frac{i\sqrt{3}}{2}.$$

EXAMPLE 19

Solve $2x^5 - 5x + 5 = 0$.

The first idea is to graph it (Figure 12). The graph tells us to expect only one real-valued solution. You can use guess-and-check to find it to any desired degree of accuracy. However, if you want an algebraic solution, you can hope this factors in integers. If it does not factor in integers, you will be stuck (Not every problem has an algebraic solution).

To factor it in integers, look for a rational solution with denominator ± 1 or ± 2 (the only divisors of the leading coefficient, 2). Also, its numerator must be ± 1 or ± 5 (the only divisors of the constant term, 5). The candidates are

± 1, ± 5, $\pm\dfrac{1}{2}$, and $\pm\dfrac{5}{2}$. You can check all of these candidates using guess-and-check, or, cleverly, you can narrow the search by checking only the candidates near the solution you can see on the graph. The solution is between -2 and -1. There are no candidates there—so you conclude it does not factor in integers. You are stuck. Be happy with the one solution you can find using guess-and-check (Problem B45).

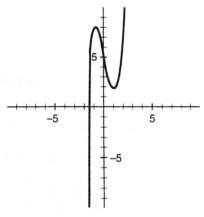

FIGURE 12 $2x^5 - 5x + 5$. $[-10, 10]$ by $[-10, 10]$

COROLLARY 4-2-6 TO THEOREM 4-2-5A If $P(x)$ is a polynomial with integer-valued coefficients and its leading coefficient is 1, then the only *rational* solutions to "$P(x) = 0$" are integer-valued solutions; there are no rational solutions that are not integers (that is, there are no solutions such as $\dfrac{1}{2}$ or $-\dfrac{5}{3}$, but there might be solutions such as $\sqrt{2}$).

EXAMPLE 20 Factor "$x^3 + 3x^2 - 4x - 8$" in integers.

If it factors in integers, there will be an integer zero. Graph it and see (Figure 13).

You might want to zoom in to assure yourself that the three zeros are definitely not integers. Since they are not, the cubic does not factor in integers. The problem cannot be done.

There is a theorem, the "Fundamental Theorem of Algebra," which says that any polynomial of degree n with real-valued coefficients is equivalent to a product of a constant and n linear factors of the form "$x - c$". Unfortunately, the c's may be complex numbers and the theorem does not say how to find them. We do not study complex numbers here because they are best studied after trigonometry, which, perhaps surprisingly, is important for understanding complex numbers.

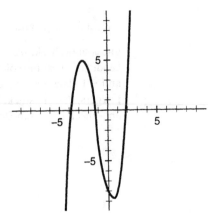

FIGURE 13 $x^3 + 3x^2 - 4x - 8$ $[-10, 10]$ by $[-10, 10]$ ▬

CONCLUSION Polynomial equations that are linear or quadratic are easy to solve algebraically. The only other easy cases occur when the polynomial is a monomial, or is factored or factorable. Graphs can help find integer or rational solutions, and they correspond to factors with integer coefficients.

Any polynomial equation can be solved for all its real-valued solutions using guess-and-check. The Factor Theorem helps solve cubic and higher-degree polynomials algebraically whenever one solution is easily found. The Rational Zeros Theorem can help find algebraic solutions.

Terms: nth-root function, factor (verb and noun), zero (of an expression), Factor Theorem, Rational Zeros Theorem.

Exercise 4-2

A

A1.* State the Zero Product Rule.

A2.* Which equation-solving method always works to solve polynomial equations?

A3. ☺ True or false? The zeros of "$(x - 3)(x + 2)$" and "$5(x - 3)(x + 2)$" are the same.

A4. ☺ True or false? The zeros of "$(x + 5)(x - 1)$" and "$-2(x + 5)(x - 1)$" are the same.

A5. True or false? The Factor Theorem determines constant factors.

A6.* ☺ True or false? The cubic formula is easy to memorize.

Solve graphically.

A7. $x^3 + 2x = 8$ [1.7]

A8. $3x^3 + x = 1$ [.54]

A9. $x^4 - 2x^3 = 5$ [2.4, . . .]

A10. $x^4 + 5x^3 = 6$ [1.0, . . .]

A11. Find the third solution in Example 5.

A12. $x^3 + 2x^2 = 7$ (Example 15).

A13. Find the leftmost solution to the equation in Example 6 (Figure 5).

A14. Find the rightmost solution to the equation in Example 6 (Figure 5).

Use the Quadratic Formula or a graph to factor the expression using integers.

A15. $6x^2 - 11x - 10$

A16. $8x^2 + 30x - 27$

A17.* The Factor Theorem is closely related to a rule we have used a lot. Which one?

A18. There is another place in Figure 6 where the right value of "*c*" would yield three solutions. Where?

A19. Solve $mx + b = c$ for x.

A20. Let $f(x) = k(x - 5)(x^2 + 3)$ and $f(2) = 7$. Find k.

A21. Let $f(x) = k(x - 3)(x + 4)(x + 5)$ and $f(-1) = 11$. Find k.

B

B1.* Four types of polynomial equations are relatively easy to solve algebraically, compared to all the rest. Which ones?

B2.* ☺ Let n be an integer greater than or equal to 2 and $c \neq 0$.

  (A) $x^n = c$ has two real-valued solutions if
  _____.
  (B) It has one solution if _____.
  (C) It has no real-valued solutions if
  _____.

B3.* ☺ (A) How many local extrema can the graph of a polynomial of degree n have?

  (B) The number of local extrema affects the number of possible solutions to a polynomial equation. If the number of local extrema is m, how many solutions could there be?

B4.* Use the square root symbol to rewrite properties 4-1-3, 4-1-4A, and 4-1-4B (when $p = \dfrac{1}{2}$, as in Table 4-1-6).

B5. (A) Suppose $P(x)$ has a factor of $5x + 17$. Where will it have a zero?

  (B) Suppose $P(x)$ has integer coefficients and $\dfrac{5}{2}$ is a zero. Give a factor with integer coefficients.

B6. (A) Suppose $P(x)$ has a factor of $3x - 4$. Where will it have a zero?

  (B) Suppose $P(x)$ has integer coefficients and $\dfrac{13}{4}$ is a zero. Give a factor with integer coefficients.

Here is the graph in a certain window. Find the window.

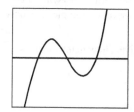
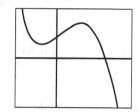

B7. $(x - 20)(x - 30)(x - 40)$ **B8.** $10x^2 - 4x^3 + 50x + 200$

Factor into two factors.

B9. $x^3 - 5x^2 + 8x - 16$

B10. $x^3 - x - 6$

B11. $x^3 - x^2 - 3x + 6$

B12. $x^3 - 5x^2 - 4x - 12$

Solve algebraically.

B13. $3x^2(1 - x)^4 - 4x^3(1 - x)^3 = 0$

B14. $5x^4(1 - x)^3 - 3x^5(1 - x)^2 = 0$

B15. $4x^3(2x + 1)^3 + 6x^4(2x + 1)^2 = 0$

B16. $2(x + 1)(x - 2)^3 + 3(x + 1)^2 (x - 2)^2 = 0$

B17. True or false: Constant factors are determined by the Factor Theorem.

B18. Explain why, if $P(1.5) = 0$, then $2x - 3$ is a factor of the polynomial $P(x)$.

Algebraically solve for all solutions.

B19. $x^3 - x^2 - 11x - 10 = 0 \; [4.2, \ldots, \ldots]$

B20. $x^3 + 5x^2 - x - 20 = 0 \; [1.8, \ldots, \ldots]$

B21. $x^3 - 5x^2 + x + 12 = 0 \; [2.3, \ldots, \ldots]$

B22. $x^3 - 21x^2 + 17x + 60 = 0$

B23. $x^4 + x^2 - 3 = 0 \; [1.1, \ldots]$

B24. $x^4 + 2x^2 - 5 = 0$

B25. $x + \sqrt{x} = 5 \; [3.2]$

B26. $x - 2\sqrt{x} = 7 \; [15]$

Use the Quadratic Formula and the Factor Theorem to factor these into two factors.

B27. $x^2 + x - 7$

B28. $2x^2 - 4x - 3$

Find a polynomial P with the given properties.

B29. P is quadratic, with zeros at 4 and -2 such that $P(1) = 3$.

B30. P is quadratic, with zeros at -1 and 2 such that $P(0) = 4$.

B31. P is cubic, with zeros at 2, 3, and -1 such that $P(0) = 5$.

B32. P is cubic, with zeros at -3, 1, and 2 such that $P(-1) = 6$.

B33. P is cubic with integer coefficients, with zeros at $3, 4, \dfrac{1}{2}$.

B34. P is cubic with integer coefficients, with zeros at 1, 5, and $\dfrac{3}{4}$.

Here are graphs of quadratics. Find the quadratic by noting its zeros and one other value.

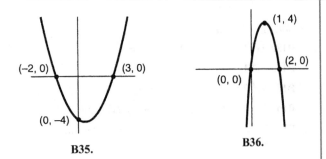

B35. **B36.**

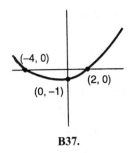

B37.

B38. Algebraically solve for all solutions:
$x^4 + x^3 + 2x + 2 = 0$.

Reading and Writing Mathematics

B39.* Use fractional notation for the nth root (as the $\dfrac{1}{n}$ power in place of p) and rewrite the properties 4-1-3, 4-1-4A, and 4-1-4B (from Table 4-1-6).

B40.* Why is factoring so important in solving equations?

B41.* "A quotient is zero if and only if the top is zero and the bottom is not."
State that result abstractly using dummy variables.

B42. Why is there an essential difference between the methods of solving
"$x^5 + 5x^3 = 1$" and "$x^5 + 5x^3 = 0$"?

B43. Someone might say "$x^2 - 3x - 17$ does not factor." Is he right? Discuss this technically.

B44. (A) Solve $nx^{n-1}(1 - x)^m - mx^n(1 - x)^{m-1} = 0$.

(B) This equation can be used to find the extreme of the expression $x^n(1 - x)^m$, where m and n are integers. Which solution corresponds to the maximum of the expression for x in the interval $[0,1]$?

B45. (A hard problem) In Example 19 there were no integer-valued solutions, so we could not factor the expression using the methods of this section. That means there will be some complex-valued solutions we cannot find using the methods of this section. But there must be a way to find the complex-valued solutions. Think of one. [You need not actually use it; just explain how it would work, in theory.]

B46. Prove Theorem 4-2-4 by assuming
$P(x) = (jx - k)Q(x)$, for some polynomial $Q(x)$ with integer coefficients and multiplying the right hand side out.

B47. The Remainder Theorem states, "When a polynomial $P(x)$ is divided by $x - c$, the remainder is $P(c)$." The remainder, $R(x)$, satisfies

$$P(x) = (x - c)Q(x) + R(x),$$

for some polynomial $Q(x)$.

(A) Verify it for $x^3 - 2x^2 - 14$ divided by $x - 3$.

(B) Prove the Remainder Theorem.

B48. Descartes's Rule of Signs.
Let $f(x) = a_n x^n + a_{n-1}x^{n-1} + \ldots + a_2 x^2 + a_1 x + a_0$, where $a_n \neq 0$. Then the number of positive real zeros of f is either equal to the number of changes of signs of the coefficients, or less than that by an even integer. (For example, if a_3 and a_2 are of different signs, count that as one change of sign of the coefficients. The quadratic "$x^2 - 3x + 5$" has two changes of sign.) Then create $f(-x)$, The number of negative zeros of f is either equal to the number of changes of sign of the coefficients of $f(-x)$, or less than that by an even integer.

Let all $a, b, c, \ldots$ all be greater than zero. According to Descartes's Law, what is the maximum number of *positive* zeros possible for these?

(A) $ax^3 - bx^2 - cx - d$

(B) $ax^3 + bx^2 + cx + d$

(C) $ax^3 - bx^2 + cx - d$

(D) Find an example of the form in (c) with three positive zeros, and a second example with only one positive zero.

B49. (See B48) If $f(x)$ is as given, how many changes of sign does $f(-x)$ have?

(A) $ax^3 - bx^2 - cx - d$

(B) $ax^3 + bx^2 + cx + d$

(C) $ax^3 - bx^2 + cx - d$

B50. (A) Solve the inequality $x^3 < 50$.

   (B) Write, symbolically, the theorem which expresses how to solve all similar problems. (Your theorem should include other similar powers, as well as numbers other than 50.)

B51. (A) Solve the inequality $x^2 < 50$.

   (B) Write, symbolically, the theorem which expresses how to solve all similar problems. (Your theorem should include other similar powers, as well as numbers other than 50.)

B52. Complete the theorem: Let $P(x)$ be a polynomial. If $P(b) = 0$, then _____ is a factor of ____.

Section 4-3 FRACTIONAL POWERS

The square root is by far the most important fractional power. A common technique for solving equations with square roots is to "square both sides."

EXAMPLE 1 Solve $x - 1 = \sqrt{x + 11}$.

Squaring both sides, $x^2 - 2x + 1 = x + 11$

$$x^2 - 3x - 10 = 0 \quad \text{[use the Quadratic Formula or factor]}$$

$$(x - 5)(x + 2) = 0$$

$$x = 5 \text{ or } x = -2.$$

But this is not the right answer. The complete solution is $x = 5$. The other value, $x = -2$, does not solve the original equation.

The steps were right, so what went wrong?

The steps created a sequence of equations, but not a sequence of *equivalent* equations, because the squaring function is not one-to-one. Whenever you employ the procedure "Square both sides," you will find all solutions, but you may unintentionally add a solution–an <u>extraneous</u> solution. ▬

DEFINITION 4-3-1 Suppose an equation is solved with work that creates a sequence of equations. An <u>extraneous</u> solution is one that satisfies the last equation that does not actually solve the original equation. Extraneous solutions can be eliminated by checking all the solutions to the terminal equation back in the original equation and discarding those that do not really work.

   In Example 1, 5 works $(5 - 1 = \sqrt{5 + 11})$, but -2 does not $(-2 - 1 \neq \sqrt{-2 + 11})$. "$-2$" is extraneous. The solution is $x = 5$.

After squaring both sides, eliminate extraneous solutions by checking.

A simpler example shows why this can happen. Consider the equation "$x = 3$." Square both sides: "$x^2 = 9$." The new equation has two solutions (including $x = -3$, which does not solve the original equation). Squaring introduced an extraneous solution.

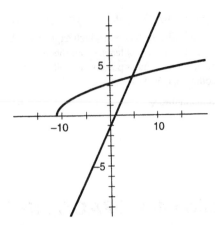

FIGURE 1 $x - 1$ and $\sqrt{x + 11}$. $[-20, 20]$ by $[-10, 10]$

The graph of the two expressions in Example 1 shows what happened (Figure 1). The graphs intersect only once, although they would intersect again if the parabola were complete, instead of just the part with non-negative y-values. The extraneous solution is the one that would have been there if the negative part of the parabola had been included, but it is not included because square roots are never negative.

Squaring both sides *may* introduce an extraneous solution, however it does not necessarily introduce an extraneous solution.

EXAMPLE 2 Solve $4\sqrt{x - 2} = x$.

Square both sides to obtain: $16(x - 2) = x^2$, and then $x^2 - 16x + 32 = 0$. According to the Quadratic Formula,

$$x = \frac{16 \pm \sqrt{16^2 - 4(1)(32)}}{2} = 2.34 \text{ or } 13.66.$$

Neither solution is extraneous. Figure 2 shows the two original expressions on the standard scale. Only one solution is evident, but the graph is not representative. On the scale $[0, 20]$ by $[0, 20]$ (Figure 3) the two intersections can be seen.

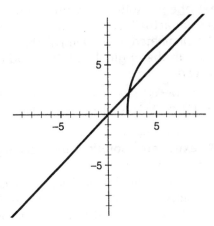

FIGURE 2 $4\sqrt{x - 2}$ and x $[-10, 10]$ by $[-10, 10]$

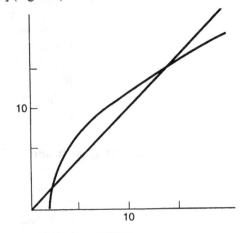

FIGURE 3 $4\sqrt{x - 2}$ and x $[0, 20]$ by $[0, 20]$ ■

THEOREM 4-3-2   If $x = b$, then $x^2 = b^2$.
ON SQUARING   If $\sqrt{x} = b$, then $x = b^2$.

■ Reading Mathematics

This theorem is about the operation "Square both sides." It is not stated with "iff." It does **not** say the new equation is equivalent to the original equation. On the other hand, the two equations might be equivalent, but the process does not guarantee it. You know by checking.

When a theorem describes a process for changing an equation (such as squaring both sides) there are two possibilities.

1. If the process always produces an equivalent equation, the theorem is stated with "if and only if", or
2. **If the process may introduce extraneous solutions, the theorem is stated with "if . . . , then . . .". Extraneous solutions are eliminated by checking.**

EXAMPLE 3   Here is a theorem. What process is it about?

THEOREM 4-3-3   (A) For $c \neq 0$, $a = b$ iff $ca = cb$ iff $\dfrac{a}{c} = \dfrac{b}{c}$.
(B) If $a = b$, then $ca = cb$.

This theorem is about the process of multiplying or dividing both sides of an equation ("$a = b$") by an expression ("c").

Now, use the theorem to solve $\dfrac{x^2 + x - 6}{x - 2} = 2x - 1$.

You may "multiply through by $x - 2$," (using Part B, with "c" as "$x - 2$").

$$x^2 + x - 6 = (2x - 1)(x - 2)$$
$$= 2x^2 - 5x + 2.$$
$$0 = x^2 - 6x + 8 = (x - 2)(x - 4).$$

$x = 2$ or $x = 4$, by the Zero Product Rule.

The solution may appear to be "$x = 2$ or $x = 4$," but it is not. In the initial equation, the expression on the left is undefined at $x = 2$ because of division by zero. Thus $x = 2$ cannot be a solution; it is extraneous. The solution is "$x = 4$."

Part B of the theorem warns us about the possibility of extraneous solutions by using the connective "if . . . , then . . .". Part A, which asserts equivalence, does *not* apply because it requires the multiplier to be non-zero and "$x - 2$" might be zero.

Multiplying through by an expression such as "$x - 2$" will preserve solutions in most examples, but not in this one. When $x = 2$, multiplying by "$x - 2$" is multiplying by 0, and multiplying both sides by 0 will convert any equation, even a false one, into the true equation "$0 = 0$."

■

EXAMPLE 4 Solve $x(x - 5) = 2x$.

You might be tempted to cancel "x" to obtain "$x - 5 = 2$" and then "$x = 7$," but that would be wrong. Canceling a factor on both sides is legal *if* it is non-zero, by 4-3-3, Part A. But "x" can be zero, and when it is, canceling "x" is dividing by zero, which is not legal. If $x = 0$, both sides are zero and the equation is true, so $x = 0$ is a solution. The complete solution is "$x = 0$ or $x = 7$." The relevant theorem is closely related to the Zero Product Rule (Problem B70). ▬

THEOREM 4-3-4 THE RULE ON CANCELING $ca = cb$ iff $a = b$ or $c = 0$.

EXAMPLE 4 REVISITED Solve $x(x - 5) = 2x$.

By the theorem, with "x" in the place of "c", $x - 5 = 2$ or $x = 0$. The solution is then $x = 7$ or $x = 0$. ▬

■ Square Roots

Sometimes, to eliminate a square root you may wish to reorganize the equation first.

EXAMPLE 5 Solve $2 + \sqrt{x} = x$.

Squaring immediately does not remove the square root.

Watch! $(2 + \sqrt{x})^2 = x^2$, $4 + 4\sqrt{x} + x = x^2$. The square root did not go away!

Instead, isolate the square root first:

$$2 + \sqrt{x} = x \text{ iff } \sqrt{x} = x - 2.$$

Now square:

$$x = x^2 - 4x + 4. \text{ [This way, the square root is gone.]}$$
$$0 = x^2 - 5x + 4.$$
$$0 = (x - 4)(x - 1).$$
$$x = 4 \text{ or } x = 1.$$

If you quit here you will have made a mistake. Whenever you square, you must check for extraneous solutions. Check $x = 4$. It works. Check $x = 1$. It does not. The solution is $x = 4$.

Why did the extraneous solution $x = 1$ appear? The equation we squared was "$\sqrt{x} = x - 2$." Checking $x = 1$ in that one, it says "$\sqrt{1} = 1 - 2$," that is, "$1 = -1$," which is false. It is false, but its square is true. Squaring loses track of signs, which is how extraneous solutions can appear.

EXAMPLE 5 ANOTHER WAY Solve "$2 + \sqrt{x} = x$."

Give $\sqrt{x}$ a new name, say, $\sqrt{x} = w$. Then $x = w^2$ and the equation can be rewritten as "$2 + w = w^2$." Solve this for w. Then solve $\sqrt{x} = w$.

$$2 + w = w^2. \quad w^2 - w - 2 = 0.$$
$$(w - 2)(w + 1) = 0. \quad w = 2 \text{ or } w = -1.$$

Substituting for w: $\sqrt{x} = 2$ or $\sqrt{x} = -1$. The equation "$\sqrt{x} = -1$" has no real-valued solutions, so the solution is $x = 4$. ▬

The distance formula employs the square root function.

EXAMPLE 6   Where are the points on the line $y = x$ that are 4 units from the point $(5, 2)$? See Figure 4.

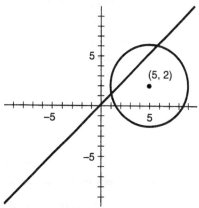

FIGURE 4 $y = x$, and a circle of radius 4 centered at $(5, 2)$. $[-10, 10]$ by $[-10, 10]$

According to the distance formula (3-1-12), the distance from (x, y) to $(5, 2)$ is given by

$$\sqrt{(x - 5)^2 + (y - 2)^2}.$$

The constraint is that the point must be on the line $y = x$, so, replacing y with x, the distance from $(5, 2)$ to the point, in terms of x, is

$$\sqrt{(x - 5)^2 + (x - 2)^2} = \sqrt{x^2 - 10x + 25 + x^2 - 4x + 4}$$
$$= \sqrt{2x^2 - 14x + 29}.$$

Set this equal to 4 and solve. Squaring both sides,

$$2x^2 - 14x + 29 = 16.$$

This can be solved using the Quadratic Formula (Problem A2), which will give two solutions. By Figure 4 and geometry, you can see that there should be two solutions, so they are right.   ▬

EXAMPLE 7   Find all points P equidistant from $(0, 1)$ and the line $y = -1$ (Figure 5).

The problem uses the term "equidistant" to assert two distances are equal. You first need two formulas for distance. Let P be denoted by (x, y).

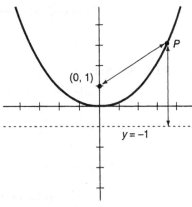

FIGURE 5   The point $(0, 1)$, the line $y = -1$, and P, a point equidistant from both. $[-5, 5]$ by $[-5, 5]$

The distance from (x, y) to the line $y = -1$ is simply $y + 1$ for y above the line. The distance from (x, y) to the point $(0, 1)$ is $\sqrt{x^2 + (y - 1)^2}$. So

$$y + 1 = \sqrt{x^2 + (y - 1)^2}.$$

Squaring,

$$y^2 + 2y + 1 = x^2 + (y - 1)^2$$
$$= x^2 + y^2 - 2y + 1.$$
$$4y = x^2. \quad y = \frac{x^2}{4}.$$

This is a parabola. The set of points equidistant from any line (the "directrix") and any point (the "focus") not on the line is a parabola. ▬

Sometimes equations have x's inside two square root symbols. For these you may need to square twice (Problems B11 and B12).

EXAMPLE 8 Solve $1 + \sqrt{2x} = \sqrt{x + 7}$.

Square both sides: $1 + 2\sqrt{2x} + 2x = x + 7$.

Isolate the square root: $2\sqrt{2x} = 6 - x$.

Now square again. $8x = 36 - 12x + x^2$.

Solve that quadratic: $x = 2$ or $x = 18$.

Checking the original equation for extraneous solutions, you see $x = 2$ works and $x = 18$ is extraneous. Therefore, $x = 2$ is the solution. ▬

■ Fractional Powers

The $\frac{1}{n}$ power is sometimes called the nth root. You have already seen properties of the nth root function. Theorem 4-3-5 extends Theorem 4-2-1 to include negative integers.

THEOREM 4-3-5 For odd integer values of n, $x^n = c$ iff $x = c^{1/n}$.

For even integer values of $n \geq 2$, $x^n = c$ iff $c \geq 0$ and $x = \pm c^{1/n}$.

For even integer values of $n \leq -2$, $x^n = c$ iff $c > 0$ and $x = \pm c^{1/n}$.

EXAMPLE 9 Solve $x^5 = 100$.

Take the fifth root. $x = 100^{1/5} = 2.51$.

Solve $x^{-4} = 4.2$. Take the $\frac{1}{(-4)}$ power and remember the plus or minus.

$x = \pm 4.2^{(1/-4)} = \pm .6985$. ▬

The same idea works for non-integer and negative powers if the unknown is positive.

THEOREM 4-3-6   For $x > 0$ and $p \neq 0$, "$x^p = c$" is equivalent to "$x = c^{1/p}$."

The domain is given as $x > 0$ because complex numbers are needed to define negative numbers to general powers.

Powers do not have to be integers.

Calculator Exercise 1: Graph x^p for various values of p, including fractions and irrational numbers. Try $p = \frac{1}{2}, \frac{1}{3}, \frac{2}{3}, \frac{3}{4}, \frac{-1}{5}$, and .41. Which yield no points to the left of the y-axis (Problem B71)?

EXAMPLE 10   Solve $x^{4.7} = 1000$.

Take the $\dfrac{1}{4.7}$ power.

$$x = 1000^{1/4.7} = 4.348.$$

Powers do not have to be positive.

EXAMPLE 11   Solve $x^{-1.2} = 5$.

Take the $\dfrac{1}{(-1.2)}$ power. Theorem 4-3-6 permits negative powers.

$$x = 5^{1/(-1.2)} = .2615.$$

EXAMPLE 12   Solve $\sqrt[3]{x} = 7.8$.

The cube root is the $\dfrac{1}{3}$ power. The equation is $x^{1/3} = 7.8$. The reciprocal of $\dfrac{1}{3}$ is 3. Take the third power:

$$x^{1/3} = 7.8 \text{ iff } x = 7.8^3 = 474.55.$$

■ Graphs

The graphs of the most important root functions, the square root (Figure 6) and the cube root (Figure 7), resemble the graphs of x^2 and x^3, except the x- and

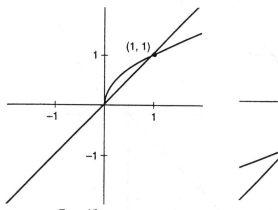

FIGURE 6 $\sqrt{x} = x^{1/2}$ and x. $[-2, 2]$ by $[-2, 2]$ **FIGURE 7** $x^{1/3}$ and x. $[-2, 2]$ by $[-2, 2]$

y-axes are switched and the square root is only half a parabola, since the square root is defined to be non-negative. The graph of "$y = x$" is included for comparison in each case.

The graphs of all even root functions resemble the graph of $x^{1/2}$, and the graphs of all odd root functions resemble the graph of $x^{1/3}$ (Figures 8 and 9).

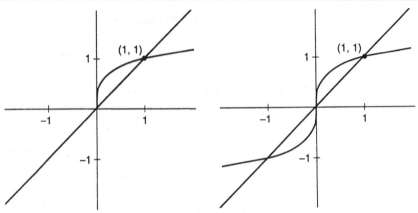

FIGURE 8 $x^{1/4}$ and x. $[-2, 2]$ by $[-2, 2]$ **FIGURE 9** $x^{1/5}$ and x. $[-2, 2]$ by $[-2, 2]$

There is a big difference between even root functions and odd root functions: the domain of even root functions does not include negative numbers (unless complex numbers are permitted as images). [This is because even powers of x are non-negative, so even roots are defined only for non-negative numbers.] Therefore, when discussing real-valued powers it is natural to exclude negative numbers so that all the powers will exist. For positive numbers, the shape of the power curve is easy to anticipate from the approximate size of the power. Powers between 0 and 1 are <u>concave down</u> like the square root curve (for example, the .41 power in Figure 10). Powers greater than 1 are <u>concave up</u> like the x^2 curve (Figure 11). Negative powers have a vertical asymptote at $x = 0$ and are concave up like the reciprocal function, $\dfrac{1}{x}$ (Figure 12). By the way, there is nothing special about the powers .41, 1.41, and $-.41$, they are just arbitrary examples.

Rational powers arise naturally when different integer powers are combined in the same equation.

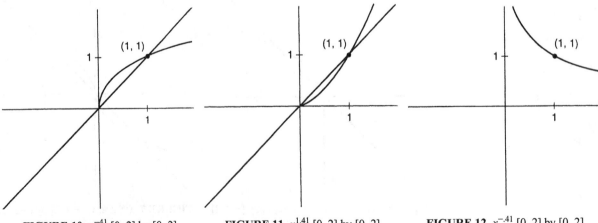

FIGURE 10 $x^{.41}$ $[0, 2]$ by $[0, 2]$ **FIGURE 11** $x^{1.41}$ $[0, 2]$ by $[0, 2]$ **FIGURE 12** $x^{-.41}$ $[0, 2]$ by $[0, 2]$

EXAMPLE 13 Kepler's Third Law of planetary motion says: $T^2 = ca^3$, where T is the time required for a planet to orbit the sun in an elliptical orbit and "a" is the "a" of the equation of the ellipse in 3-3-5 where $a > b$. If the orbit is a circle, the value of a is just the radius of the orbit. The letter c is a constant which takes into account the force of gravity and the units employed to express time and distance. Solve for T in terms of a.

All variables in the equation are positive, so you do not need to worry about negative solutions or square roots of negative numbers.

$$T = \sqrt{ca^3} = (ca^3)^{1/2} = c^{1/2}a^{3/2}.$$

The final steps used properties of powers which also apply to nth roots when $p = \dfrac{1}{n}$. Reformulated for roots:

THEOREM 4-3-7 For $a > 0, b > 0,$ and any $p \neq 0$,

$$(ab)^{1/p} = a^{1/p}b^{1/p} \qquad \text{[see 4-1-4A] and}$$

$$\left(\frac{a}{b}\right)^{1/p} = \frac{a^{1/p}}{b^{1/p}} \qquad \text{[see 4-1-4B]}$$

$$(b^r)^{1/p} = b^{r/p} = (b^{1/p})^r \qquad \text{[see 4-1-3].}$$

Theorem 4-3-7 is not new. It restates Theorems 4-1-4 and 4-1-3 using letters differently to emphasize different applications.

**EXAMPLE 13
CONTINUED** Solve for a in terms of T: $T^2 = ca^3$.

The expression "ca^3" is a product. So, first divide.

$$\frac{T^2}{c} = a^3, \text{ and } a = \left(\frac{T^2}{c}\right)^{1/3} = \frac{(T^2)^{1/3}}{c^{1/3}} = \frac{T^{2/3}}{c^{1/3}}.$$

EXAMPLE 14 Give the surface area, A, of a cubical box in terms of its volume, V.

We usually express area and volume in terms of the length of an edge, s. Because a cube has 6 sides, $A = 6s^2$. For the volume, $V = s^3$. Use this equation to isolate s: $s = V^{1/3}$. Substituting this in for s in the area formula,

$$A = 6(V^{1/3})^2, \text{ which is } A = 6V^{2/3}, \text{ (Figure 13).}$$

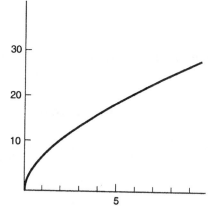

FIGURE 13 $A = 6V^{2/3}$. [0, 10] by [0, 40]

Complications

There are complications when the base is negative. The next example shows what can go wrong, even in an expression as simple as "x".

EXAMPLE 15 $x = \left(\sqrt{x}\right)^2$ is not always true. If x is negative, the left side is defined and the right side is not, because the square root of x is not real if x is negative, so the operations cannot be executed in that order for negative x. Do not misapply Theorem 4-3-7 by using it on negative numbers. Anytime a fractional power has an even denominator, like $\frac{1}{2}$ has denominator 2, this type of difficulty will occur. ▬

■ Identities

Identities can be used to simplify expressions.

EXAMPLE 16 Solve $\dfrac{x^4}{\sqrt{x}} = 30$.

Simplify the left side first: $\dfrac{x^4}{\sqrt{x}} = \dfrac{x^4}{x^{1/2}} = x^{3.5} = 30$. $x = 30^{1/3.5} = 2.64$. ▬

EXAMPLE 17 Solve $\dfrac{\sqrt{x^3}}{(2x)^2} = 4.3$.

$$\frac{\sqrt{x^3}}{(2x)^2} = \frac{\left(x^3\right)^{1/2}}{4x^2} = \frac{x^{3/2}}{4x^2} = \frac{x^{-1/2}}{4} = \frac{1}{4\sqrt{x}} = 4.3.$$

$$\frac{1}{4} = 4.3\sqrt{x}. \quad .05814 = \sqrt{x}. \quad x = .00338.$$ ▬

EXAMPLE 18 Solve for p. $x^{1/3} x^p = x^{4/3}$.

The left side is $x^{1/3+p}$. $\dfrac{1}{3} + p = \dfrac{4}{3}$. $p = 1$. $x^{1/3} x^1 = x^{4/3}$.

Solve for p: $x^{-1/3} x^p = x^{5/3}$.

The left side is $x^{-1/3+p}$. $\dfrac{-1}{3} + p = \dfrac{5}{3}$. $p = 2$. $x^{-1/3}x^2 = x^{5/3}$. ▬

EXAMPLE 19 Solve $x^{1/3}(x-3)^3 + x^{4/3}(x-3)^2 = 0$.

Use the Zero Product Rule. Factor it. There are powers of x and powers of $x - 3$.

Factor out the **lesser** power of each.

$$x^{1/3}(x-3)^2\left[(x-3) + x\right] = 0.$$

So, $x = 0$ or $x = 3$ or $2x - 3 = 0$.

$$x = 0 \text{ or } x = 3 \text{ or } x = \frac{3}{2}$$ ▬

EXAMPLE 20 Solve $\left(\dfrac{3}{2}\right)\sqrt{x}\,(1-x)^3 - 3x^{3/2}(1-x)^2 = 0$.

Both terms have a factor of $(1-x)^2$. Also, $\sqrt{x} = x^{1/2}$, so both have a factor of $x^{1/2}$, because $x^{3/2} = x^{1/2}x^1$, by 4-1-1.

$$x^{1/2}(1-x)^2\left[\left(\frac{3}{2}\right)(1-x) - 3x\right] = 0$$

$$x^{1/2}(1-x)^2\left[\frac{3}{2} - \frac{9x}{2}\right] = 0.$$

$$x = 0 \text{ or } x = 1 \text{ or } x = \frac{1}{3}.$$ ▄

Negative powers can be treated the same way.

EXAMPLE 21 Solve $\left(\dfrac{-1}{3}\right)x^{1/2}(1-x)^{-2/3} + \left(\dfrac{1}{2}\right)x^{-1/2}(1-x)^{1/3} = 0$.

Pull out a factor with the **lesser** of the powers on the similar factors. For "$x^{1/2}$" and "$x^{-1/2}$," the lesser power is $\dfrac{-1}{2}$. Factor out $x^{-1/2}$.

Then the "$x^{1/2}$" term is $x^{-1/2}x$, which leaves behind an "x" on the first term. For "$(1-x)^{-2/3}$" and "$(1-x)^{1/3}$," the lesser power is $\dfrac{-2}{3}$. Factor out $(1-x)^{2/3}$. Because $(1-x)^{1/3} = (1-x)^{-2/3}(1-x)$, it leaves behind a "$(1-x)$" on the second term.

$$x^{-1/2}(1-x)^{-2/3}\left[\left(\frac{-1}{3}\right)x + \left(\frac{1}{2}\right)(1-x)\right] = 0.$$

$$x^{-1/2}(1-x)^{-2/3}\left[\frac{1}{2} - \frac{5x}{6}\right] = 0.$$

Only the last factor yields a solution, because variables to negative powers are never zero. Think of negative powers as being a type of division (because $x^{-p} = \dfrac{1}{x^p}$, which is never zero).

$$\frac{1}{2} - \frac{5x}{6} = 0 \text{ iff } \frac{5x}{6} = \frac{1}{2} \text{ iff } x = \frac{6}{10} = \frac{3}{5}. \text{ The solution is } x = \frac{3}{5}.$$ ▄

EXAMPLE 22 Solve algebraically: $\dfrac{\left(\dfrac{4}{3}\right)x^{1/3}(1+x)^2 - 2x^{4/3}(1+x)}{(1+x)^4} = 0$.

Set the top equal to zero and solve, remembering that "$x = -1$" can not be a solution because the bottom cannot be zero. So, factor the top.

$$\left(\frac{4}{3}\right)x^{1/3}(1 + x)^2 - 2x^{4/3}(1 + x) = (1 + x)\left[\left(\frac{4}{3}\right)x^{1/3}(1 + x) - 2x^{1/3}\,x^{3/3}\right]$$

$$= (1 + x)x^{1/3}\left[\left(\frac{4}{3}\right)(1 + x) - 2x\right]$$

$$= (1 + x)x^{1/3}\left[\left(\frac{4}{3}\right) - \left(\frac{2}{3}\right)x\right].$$

Now the Zero Product Rule applies: $x^{1/3} = 0$ or $\left(\frac{4}{3}\right) - \left(\frac{2}{3}\right)x = 0$ or $1 + x = 0$. The solution is $x = 0$ or $x = 2$. (x cannot be -1.) ▬

CONCLUSION Squaring may introduce extraneous solutions. Also, multiplying by an expression with a variable may introduce extraneous solutions. When the relevant theorem is stated with "if . . . , then . . ." (as opposed to "iff"), extraneous solutions may arise. They can be eliminated by checking.

For positive bases power functions with real-valued powers have properties which parallel the properties of power functions with integer powers. Not all the parallels hold for negative bases, but we rarely need to use negative bases.

Terms: extraneous solution, iff, if . . . then . . ., nth root, concave up, concave down.

Exercise 4-3

A

A1.* To solve an equation is to find the values of the unknown that make the equation <u>true</u>.
  (A) An extraneous solution is a value of the unknown that makes the terminal equation _____ but makes the original equation _____.
  (B) True or false: Extraneous solutions never arise unless you make a mistake.

A2. Solve the equation in Example 6:
$$2x^2 - 14x + 29 = 16.$$

Evaluate these expressions and report <u>at least three</u> significant digits. To help you check your keystrokes, the value is given with two significant digits in brackets.

A3. (A) $4^{2.5}$ [32]　　　　(B) $4^{-2.5}$ [.031]

A4. (A) $2^{3.4}$ [11]　　　　(B) $2^{-3.4}$ [.095]

A5. (A) $.6^3$ [.22]　　　　(B) $.6^{-3}$ [4.6]

A6. (A) $.3^{2.1}$ [.080]　　　　(B) $.3^{-2.1}$ [13]

A7. (A) $100^{1/5}$ [2.5]　　　　(B) $100^{-1/5}$ [.40]

A8. (A) $1000^{1/7}$ [2.7]　　　　(B) $1000^{-1/7}$ [.37]

A9. (A) $.02^5$ [3.2×10^{-9}]　　(B) $.02^{-5}$ [310,000,000]

A10. (A) $.15^3$ [.0034]　　　　(B) $.15^{-3}$ [300]

Solve algebraically for a positive solution and evaluate the solution as a decimal.

A11. $x^{3.4} = 16$ [2.3]　　　**A12.** $x^{4.5} = 300$ [3.6]

A13. $x^{-2.3} = 20$ [.27]　　　**A14.** $x^{-3.2} = 50$ [.29]

A15. $x^{1/5} = 1.43$ [6.0]　　　**A16.** $x^{1/4} = 4.2$ [310]

A17. $x^{1/3} = .4$ [.064]　　　**A18.** $x^{1/5} = .2$ [.00032]

A19. $x^{2/3} = 10$ [32]

A20. $x^{3/4} = 100$ [460]

A21. $(1 + r)^{12} = 1.23$ [.017]

A22. $(1 + r)^{25} = 8$ [.087]

A23. $\left(1 + \dfrac{r}{2}\right)^4 = 1.18$ [.084]

A24. $\left(1 + \dfrac{r}{12}\right)^{36} = 1.6$ [.16]

☺ *Simplify to only one appearance of "x" and one power, if possible. Assume x > 0.*

A25. $(x^{1/2})x$ **A26.** $(x^{2/3})x$ **A27.** $\dfrac{x^{1/2}}{x}$

A28. $\dfrac{x}{x^{1/2}}$ **A29.** $(x^{1/3})^2$ **A30.** $(x^{3/2})^{1/2}$

A31. $\left(x^6\right)^{1/2}$ **A32.** $(x^2)^{1/3}$ **A33.** $(x^{1/3})^{1/2}$

A34. $\left(x^{1/2}\right)^{1/4}$ **A35.** $\left(\sqrt{x}\right)^6$ **A36.** $\dfrac{x}{\sqrt{x}}$

A37. $\dfrac{x^2}{\sqrt{x}}$ **A38.** $\dfrac{\sqrt{x}}{x}$ **A39.** $(a^2 + b^2)^{1/2}$

A40. $(x^2 + 4)^{1/2}$ **A41.** $x^{1/3} + y^{1/3}$ **A42.** $(16 - x^2)^{1/2}$

A43. Find the equation of the circle with center $(1, 3)$ through the point $(2, 6)$.

A44. Find the equation of the circle with center $(-2, 5)$ through the point $(0, 0)$.

A45. ☺ Which graph, $\sqrt{x}$, x^2. or $\dfrac{1}{x}$, do the following resemble the most for $x > 0$?

(A) $x^{.68}$ (B) $x^{1.2}$ (C) x^2 (D) $x^{-.5}$

A46. ☺ Which graph, $\sqrt{x}$, x^2, or $\dfrac{1}{x}$, do the following resemble the most for $x > 0$?

(A) $x^{4/5}$ (B) $x^{1.6}$ (C) $x^{-.5}$ (D) x^{-2}

☺ *Over the interval 1 < x, which is larger?*

A47. x or x^{-1} **A48.** x^2 or x^{-2} **A49.** $x^{1/2}$ or x

A50. $x^{1/3}$ or x **A51.** x^2 or x **A52.** x^2 or x^3

A53. $x^{1/2}$ or $x^{1/4}$ **A54.** $x^{1/3}$ or $x^{2/3}$

A55. If $p > 0$, x^p or x^{-p}

A56. Generalize from Exercises A39–A47 to determine a simple way to decide which is larger over the interval $1 < x$: x^p or x^r.

☺ *Over the interval 0 < x < 1, which is larger?*

A57. x or x^{-1} **A58.** x^2 or x^{-2} **A59.** $x^{1/2}$ or x

A60. $x^{1/3}$ or x **A61.** x^2 or x **A62.** x^2 or x^3

A63. $x^{1/2}$ or $x^{1/4}$ **A64.** $x^{1/3}$ or $x^{2/3}$

A65. If $p > 0$, x^p or x^{-p}

A66. Generalize from Exercises A49–A58 to determine a simple way to decide which is larger over the interval $0 < x < 1$: x^p or x^r.

Reading and Writing Mathematics

A67. In the equation just after Theorem 4-3-3, identify the expressions corresponding to the letters "a", "b", and "c" of the theorem.

A68. Expand the square: $(a + b + c)^2$ [You should obtain 6 terms.]

A69. Expand the square: $\left(x - \sqrt{a}\right)^2$.

B

B1.* (A) Define "extraneous" solution.
  (B) What operations might cause one to occur?
  (C) Do you have to make a mistake for an extraneous solution to occur?
  (D) How can you eliminate extraneous solutions?

B2. Suppose a process is applied to both sides of an equation to create a new equation.
  (A) Which term (connective) is used to state the corresponding theorem if the process never yields extraneous solutions?
  (B) Which term (connective) is used to state the corresponding theorem if the process may yield extraneous solutions?

Solve algebraically.

B3. $x = \sqrt{x + 20}$

B4. $x - 4 = \sqrt{x - 2}$

B5. $2\sqrt{x} + 3 = x$

B6. $2\sqrt{x} + 1 = x$ [5.8]

B7. $\sqrt{x + 5} = 7 - x$

B8. $\sqrt{10 - x} = x + 10$

B9. $\sqrt{20 - x} = 14 - x$ [11]

B10. $\sqrt{6x - 5} = 16 - x$

B11. $\sqrt{x} + \sqrt{x + 5} = 10$ [23]

B12. $\sqrt{x} + \sqrt{x + 10} = 9$ [16]

Solve algebraically.

B13. $x^2(x^{3.2}) = 1000$ [3.8]

B14. $(x^2)^{2.3} = 600$ [4.0]

B15. $x^3 = 3.7x^{1.7}$ [2.7, . . .]

B16. $x^2(x^5) = 50x^{3.1}$

Factor.

B17. $x^{1/2}(2 + x)^{5/2} + 2x^{3/2}(2 + x)^{3/2}$

B18. $4x^{11/6}(3 + x)^{3/4} + x^{5/6}(3 + x)^{7/4}$

B19. $x^{-1/2}(1 - x)^{7/2} + 3x^{-3/2}(1 - x)^{9/2}$

B20. $5x^{-1/3} + 2x^{2/3}$

Solve algebraically.

B21. $-6\sqrt{x}\,(1 - 2x)^2 + \left(\dfrac{1}{2}\right)(1 - 2x)^3 x^{-1/2} = 0$

    [two solutions]

B22. $\left(\dfrac{5}{2}\right)(1 - x)^2\, x^{3/2} - 2x^{5/2}\,(1 - x) = 0$ $[.56, \ldots]$

B23. $\left(\dfrac{1}{3}\right)x^{1/4}(x + 3)^{-2/3} - \left(\dfrac{1}{4}\right)x^{-3/4}(x + 3)^{1/3} = 0$

    [one solution]

B24. $\left(\dfrac{5}{3}\right)x^{2/3}\,(1 - x)^{3/4} - \left(\dfrac{3}{4}\right)x^{5/3}\,(1 - x)^{-1/4} = 0$

    $[\ldots, .69]$

B25. ☺ Which of these processes, when applied to both sides of an equation, necessarily yield an equivalent equation?

  (A) Add 5   (B) Add x
  (C) Multiply by 2   (D) Multiply by x
  (E) Square   (F) Divide by 4
  (G) Subtract $x - 2$ (H) Divide by x

B26. ☺ Which of these processes, when applied to both sides of an equation, necessarily yield an equivalent equation?
  (A) Multiply by $x + 1$
  (B) Add $x + 1$
  (C) Divide by $3x$
  (D) Subtract $2x$
  (E) Cancel a common factor of $12x$
  (F) Add 2 and then square

B27. ☺ Which of these processes, when applied to both sides of an equation, necessarily yield an equivalent equation?
  (A) Multiply by $x^2 + 1$
  (B) Subtract $x^2 + 1$
  (C) Square
  (D) Add 2 and then multiply by 3
  (E) Multiply by $x + 3$

B28. ☺ Which of these processes, when applied to both sides of an equation, necessarily yield an equivalent equation?
  (A) Multiply by $3x$
  (B) Add $3x$

  (C) Cancel a common factor of $x - 3$
  (D) Square
  (E) Cube

☺ *Are the two equations necessarily equivalent? ("a" and "b" hold places for expressions that may have "x" in them.)*

B29. "$a = b$" and "$a + c = b + c$"

B30. "$a = b$" and "$ca = cb$"

B31. "$\sqrt{x} = c$" and "$x = c^2$"

B32. "$a = b$" and "$a^2 = b^2$"

B33. "$a = b$" and "$3a = 3b$"

B34. "$a = b$" and "$\dfrac{a}{x} = \dfrac{b}{x}$"

B35. "$a = b$" and "$\dfrac{a}{5} = \dfrac{b}{5}$"

B36. "$2a = b$" and "$4a^2 = b^2$"

B37. "$a = b$" and "$a - x = b - x$"

B38. "$x^2 = 9$" and "$x = 3$"

B39.* ☺ (A) Name two processes for solving equations that may lead to extraneous solutions.
  (B) Do they always create extraneous solutions?
  (C) How can you eliminate extraneous solutions?

B40.* To solve an equation, multiplying both sides by an expression may introduce extraneous solutions.
  (A) Why?
  (B) Create a very simple illuminating example.
  (C) State the relevant theorem from this section.

Give the domain.

B41. (A) $x^{1.5}$ (B) $x^{1/3}$ (C) x^{-2} (D) $x^{-1/2}$

B42. (A) x^{-4} (B) $x^{2.3}$ (C) $x^{1/4}$ (D) $x^{1/5}$

B43. Use your calculator to graph $x^{1/3}$. Now use it to graph $x^{2/6}$ (entering the operations in the given order). Are they different graphs? Use the following idea to explain why they may be different. $x^{2/6} = \left(x^2\right)^{1/6}$ for all x, but $\left(x^2\right)^{1/6}$ is not $(x^{1/6})^2$ for all x.

In the next two, your final formula should not mention the side.

B44. Express the surface area of a cube as a function of its volume.

B45. Express the volume of a cube as a function of its surface area.

The surface area of a sphere is $4\pi r^2$. The volume of a sphere is $\left(\dfrac{4}{3}\right)\pi r^3$.

B46. (A) Express the surface area in terms of the volume.

(B) Express your answer to part (a) as a decimal number (with at least 3 significant digits) times some power of V.

B47. (A) Express the volume in terms of the surface area.

(B) Express your answer to part (a) as a decimal number (with at least 3 significant digits) times some power of S.

B48. The moon and satellites obey Kepler's Third Law (Example 14). If the moon orbits the Earth in 27.3 days and is 58 times as far away from the center of the earth as a satellite, how long does it take the satellite to orbit the earth?

B49. Express the distance from the point (a, b) to the line $x = c$.

B50. Express the distance from the point (a, b) to the line $y = c$.

B51. Set up the equation. Let A (left) and B (right) be fixed points on a line. Let M be on the line and vary between A and B. Let P be on the perpendicular to the line at M such that $(MP)^2 = (AM)(MB)$. Find the equation for all such points P.

B52. Set up the equation. Let A (left) and B (right) be fixed points on a line. Let M vary to the right of B. Let P be on the perpendicular to the line at M such that $MP^2 = AM(MB)$. Find the equation for all such points P.

B53. Let point A be the point $(-3, 0)$ and point B be $(3, 0)$. Find all points P such that the sum of the distances from P to A and B is 10. [Generalized in B54.]

B54. [Generalizes B53.] Let A be $(-c, 0)$ and B be $(c, 0)$ Set up the equation for all points P such that the sum of the distances AP and BP is $2a$, where $2a > 2c > 0$. [This yields an ellipse which can be written in the form 3-2-10. Continued in B55.]

B55. Simplify the equation obtained in B54. Let $b^2 = a^2 - c^2$ and obtain the form of an ellipse in 3-2-10: $\dfrac{x^2}{a^2} + \dfrac{y^2}{b^2} = 1$.

B56. Let A be $(-c, 0)$ and B be $(c, 0)$ Set up the equation for all points P such that the difference of the distances AP and BP is $\pm 2a$, where $0 < 2a < 2c$. [This yields a hyperbola. Continued in B57.]

B57. Simplify the equation obtained in B56. Let $b^2 = a^2 - c^2$ and obtain the form: $\dfrac{x^2}{a^2} - \dfrac{y^2}{b^2} = 1$.

B58. Find an equation for the set of all points that are twice as far from the y-axis as from $(3, 0)$. Simplify.

☺ *The two equations are not necessarily equivalent. Find numerical values of the variables that make one of the equations true and the other false.*

B59. "$\sqrt{x} = c$" and "$x = c^2$"

B60. "$x^2 = 16$" and "$x = 4$"

B61. "$ca = cb$" and "$a = b$"

B62. "$ab = c$" and "$a = \dfrac{c}{b}$"

B63. (A) Sketch, on the standard scale, the graphs of the two expressions in Example 5: "$2 + \sqrt{x}$" and "x".

(B) Mark (perhaps with dashes) the rest of the parabola of which only a part is visible.

(C) Where, on your sketch, is the point corresponding to the extraneous solution?

B64. Consider the equation $\sqrt{x} = mx + b$. Sketch the two expressions (for various m and b) to illustrate that there can be one solution, two solutions, or no solutions.

B65. Give an example such that "$x^p = c$" and "$x = c^{1/p}$" are not equivalent.

B66. Prove, for non-negative a and b: $\dfrac{a + b}{2} \geq \sqrt{ab}$.

B67. Example 13 shows how to compute $x^{3.7}$ by treating 3.7 as $3 + \left(\dfrac{1}{2}\right) + \left(\dfrac{1}{5}\right)$. Show how to do it another equivalent way by using Property 4-1-3.

B68. x^c can be defined for irrational c by taking the "limit" of the values of x^r where r varies over rational values approaching c. Give a sequence of rational r's approaching π.

B69. Which power functions have negative numbers in their domain (without using complex numbers)?

B70. Prove Theorem 4-3-4, the Rule on Canceling, from the Zero Product Rule.

B71. Do Calculator Exercise 1: Use your graphing calculator to look at the graph of x^p for various values of p, including fractions and irrational numbers. Try $p = \dfrac{1}{2}, \dfrac{1}{3}, \dfrac{2}{3}, \dfrac{3}{4}, \dfrac{-1}{5}$, and .41. Which yield no points to the left of the y-axis? Generalize.

Section 4-4 PERCENTS, MONEY, AND COMPOUNDING

The term "percent" occurs in the context of multiplication. When you see or hear the word "percent," think *multiplication*. For example, the phrase "70 percent of the students" refers to multiplying the number of students by .70. When an amount goes "up 10 percent," the final amount is obtained by *multiplying* by 1.10. To "take 20 percent off of the price" yields a price obtained by *multiplying* the price by .80.

Repeated multiplication by the same factor produces powers. For example, when money grows at 8% per year for 5 years there will be 5 factors of 1.08 and the total growth factor will be 1.08^5. The idea of such "compound" interest makes the connection between percents and powers, which are the subjects of this chapter.

■ Review of Percents

"Percent" means, in its Latin derivation, "by the hundred." It is a way of comparing two quantities where the standard of comparison is "100 percent."

To compare b to the standard a, the percent p is given by

$$\frac{b}{a} = \frac{p}{100} \quad \text{Therefore, } p = 100\left(\frac{b}{a}\right). \tag{4-4-1}$$

The standard is in the denominator.

EXAMPLE 1   Compare 12 to 15.

$$\frac{12}{15} = \frac{p}{100}. \quad 100\left(\frac{12}{15}\right) = p. \quad p = 80 \text{ (percent)}.$$

"12 is 80 percent of 15."

You may simply compute the ratio $\frac{12}{15} = .80$ and move the decimal point over two places (which is what the factor of 100 does).   ▬

Amounts less than the standard are less than 100 percent, and amounts greater than the standard are greater than 100 percent.

EXAMPLE 2   Compare 15 to 12.

$$\frac{15}{12} = \frac{p}{100}. \quad p = 100\left(\frac{15}{12}\right) = 125 \text{ (percent)}.$$

"15 is 125 percent of 12." $\frac{15}{12} = 1.25$, which is 125 percent.   ▬

" b is p percent of a" means $b = \left(\frac{p}{100}\right) a.$

"$b = ca$" means "b is $100c$ percent of a. $\tag{4-4-2}$

EXAMPLE 3   "b is 65 percent of a" means $b = \left(\dfrac{65}{100}\right)a$. That is, $b = .65a$.

If $k = .078x$, then k is 7.8 percent of x.   ▬

Think of percents as expressing a functional relationship using multiplication. The standard plays the role of the argument, a.

Mathematicians usually convert fractions like "$\dfrac{p}{100}$" to decimals.

EXAMPLE 4   The function signaled by the phrase "25 percent of" is "Multiply by .25," because $\dfrac{25}{100} = .25$.

25 percent of 600 is $.25(600) = 150$.
The function signaled by "72 percent of" is "Multiply by .72."
The function signaled by "220 percent of" is "Multiply by 2.20."   ▬

EXAMPLE 5   Compare the area of a circle inscribed in a square to the area of the square (Figure 1).

The area of the circle is given by $A = \pi r^2$. The side of the surrounding square is $2r$, so the area of the square is $(2r)^2 = 4r^2$. That is the standard.

$$\frac{\pi r^2}{4r^2} = \frac{\pi}{4} = .785$$

The circle has 78.5% of the area of the square.

You can also do this with any particular radius. Pick any r, say $r = 10$. Then the area of the circle would be $\pi(10)^2 = 314.16$ and the area of the square would be $20^2 = 400$. Compare these two, with the area of the square as the standard in the denominator: $\dfrac{314.16}{400} = .785$. The circle has 78.5% of the area of the square.   ▬

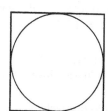

FIGURE 1 A circle inscribed in a square

■ Incorporating Change

Often percents are used in the context of change. "Sales went *up* 12 percent." "During the sale prices are knocked *down* 35 percent." "The index of leading indicators *gained* .3 percent last month." "The Dow Jones stock market index *lost* .4 percent last week."

EXAMPLE 6   A 9 percent increase corresponds to a final amount that is "109 percent of" the original. The standard is automatically 100%, so the increase of 9% brings the total up to 109%. The final amount compares to the original amount by the function, "Multiply by 1.09."   ▬

EXAMPLE 7   The phrase "80 percent of" corresponds to a 20 percent decrease, which is equivalent to "20 percent off." The words "off" and "of" can be mistaken for each other. Their spelling differs by only one letter, but they have radically different interpretations. "y is 20 percent of x" means $y = .2x$. "y is 20 percent off x," means $y = .8x$. The word "off" refers to the percent being a decrease. That second "f" in "off" makes a big difference.   ▬

EXAMPLE 8 Express as a rule the function which relates the new amount to the standard.

phrase	rule
"gain 20%"	multiply by 1.2
"down 10%"	multiply by .9
"70% off"	multiply by .3
"up 6%"	multiply by 1.06
"20% discount"	multiply by .8

To "increase," "gain," or "go up," p percent means the final amount is the standard multiplied by a factor of $1 + \dfrac{p}{100}$.   (4-4-3A)

To "decrease," "lose," or "go down," p percent means the final amount is the standard multiplied by a factor of $1 - \dfrac{p}{100}$.   (4-4-3B)

EXAMPLE 9 Express as a formula: "B is 30% more than A."
   Decide what is the standard. B is being compared to A, so A is the standard. A is therefore regarded as 100%, so B is 130% of A: $B = 1.3A$.

EXAMPLE 10 Skirts originally priced $80 are now $60. How much off are they?
   There are two kinds of answers to this problem. Thinking additively, since $80 is $20 more than $60, they are $20 off.
   But it is common and often important to think multiplicatively. Think of $60 as some multiple of $80.

$$60 = 80x. \quad x = \frac{60}{80} = .75.$$

Therefore the skirts are "25% off."
   Another, equivalent, approach is to think of "$20 off" compared to a standard of $80. The factor is obtained from the equation: $20 = 80x$. So $x = .25$, which is a second way to obtain the answer, "25 percent off."

EXAMPLE 11 At 20 percent off a jacket costs $100. What did it cost originally?
   Many people would erroneously reply, "$120." They have reversed the image and the argument.
   The phrase "20 percent off" refers to the original price as argument (the standard). The cost is .80 times the original cost. $100 = .8c$ yields $c = 125$ (dollars).

Sometimes it is good not to have to have a particular standard.

EXAMPLE 12 Suppose sidewalks are laid out perpendicular to each other to form a square grid (Figure 2). If you walk across the grass on the diagonal of a square rather than on the sidewalks, how much distance do you save?
   Obviously, the distance you save depends upon the size of the square. But, since all squares are geometrically similar, you can express the distance saved as a fraction of the standard distance. Percents are good for this.

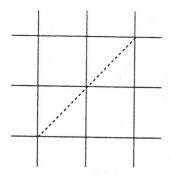

FIGURE 2 Sidewalks laid on a square grid and a diagonal shortcut

You could do this by naming the side "*s*" and doing the calculation in general. However, you will get the same result if you pick any particular side, say *s* = 1 or 100.

If the sidewalks are 1 unit apart, then the diagonal is $\sqrt{2}$ units, by the Pythagorean Theorem. Instead of walking 2 units (the standard) you could walk $\sqrt{2}$ units. The standard goes in the denominator: $\dfrac{\sqrt{2}}{2} = .707 = 70.7$ percent. The savings is 29.3 percent.

How much further is it to walk on the sidewalks rather than cut across on the grass?

Now the standard has changed. The standard is the length of the diagonal, $\sqrt{2}$. The standard goes in the denominator: $\dfrac{2}{\sqrt{2}} = 1.414$. It is 41.4 percent further.

An increase of 41.4 percent corresponds to a decrease of 29.3 percent. The numbers are not the same. In general, an increase of *p* percent does not correspond to a decrease of *p* percent. Percent is not an additive concept! ▬

■ Composition of Functions

One trick to percent problems is that the standard can change within a single problem.

EXAMPLE 13   Sales of widgets went up 30 percent in 2000 and down 20 percent in 2001. How did they change over the two years?

Do not add or subtract. The context for percents is *multiplication*, not addition or subtraction. It did **not** go "up 10 percent," even though 30 − 20 = 10.

The factor over two years is: (1.30)(.80) = 1.04. Sales went up 4 percent. ▬

EXAMPLE 14   Which was a better investment, stocks which went up 11.5 percent and were sold for a 1.5 percent commission or antiques which went up 30 percent and were sold for a 20 percent commission? Note that the difference is 10 percentage points (not 10 percent) in each case.

You are comparing two growth factors. Build your own formula for each.

The factor for stocks is 1.115(.985) = 1.098275, for a 9.8275 percent return. The factor for antiques is 1.30(.80) = 1.04, for a 4 percent return. Stocks were the better investment. ▬

EXAMPLE 15   *B* is 30% more than *A*. How much less than *B* is *A*?

The phrase "30% more than" is used to compare *B* to *A*. The question asks us to compare *A* to *B*.

The first sentence is expressed by:

$$B = 1.3A.$$

Therefore, solving for *A*,

$$A = \frac{B}{1.3}.$$

To express the answer you must convert this division to multiplication:

$$A = .769B.$$

So A is 76.9% of B, so A is 23.1% less than B (because $100 - 76.9 = 23.1$).

Another way is to treat A as 100 (100 percent). Then B is 30% more, or $1.3(100) = 130$. Now compare A to B, with B as the standard in the denominator:

$$\frac{100}{130} = .769, \text{ or } 76.9\%. \text{ Therefore } A \text{ is } 23.1\% \text{ less than } B. \quad \blacksquare$$

The cumulative effect of successive percentage changes is *not* additive. The cumulative effect of successive changes is obtained by multiplication.

EXAMPLE 16 An increase of 20% was followed by a decrease of 20%. What was the cumulative effect?

Do not add to get 0. Percentage changes are multiplicative, not additive. The factor is given by

$$1.2(.8) = .96,$$

which corresponds to a decrease of 4%.

You may also do this on a scale of 100. An increase of 20% would increase 100 to 120. A decrease of 20% would yield $(.8)120 = 96$. Down 4 on a scale of 100 is down 4%. $\quad \blacksquare$

For many word problems it is a good idea to

Build your own formula.

EXAMPLE 17 You are about to invest $100,000 in two funds, Fund A which is 85% in stocks, and Fund B which 45% in stocks. You want your investments to be 70% in stocks. How much should you invest in Fund B?

Let the amount you invest in Fund B be called x. Each of the three percentages mentioned has a different standard. The standard for Fund B is x. The stocks in Fund B are worth $.45x$. The standard for Fund A is $100,000 - x$. The stocks in Fund A are worth $.85(100,000 - x)$. You want the total to be 70 percent in stocks: $.70(100,000)$. Therefore

$$.45x + .85(100,000 - x) = .70(100,000)$$

$$-.40x + 85,000 = 70,000$$

$$-.40x = -15,000. \quad x = 37,500 \text{ (dollars in Fund B)}.$$

Invest $37,500 in Fund B and the rest ($62,500) in Fund A. $\quad \blacksquare$

EXAMPLE 18 The diameter of Pizza A is 30 percent greater than the diameter of Pizza B. How much greater is the area of Pizza A?

You can do this with conveniently chosen numbers. Let the smaller pizza be, say, a 10-inch pizza. Then the larger pizza will be a 13-inch pizza. Their areas are $\pi\left(\dfrac{10}{2}\right)^2 = 25\pi$ and $\pi\left(\dfrac{13}{2}\right)^2 = 42.25\pi$.   The area of the 10-inch pizza is the standard:

$$\frac{42.25\pi}{25\pi} = 1.69. \quad \text{The larger pizza has 69 percent more area.} \quad \blacksquare$$

■ Money and Compounding

Percents are often used in the context of money and investments. The "rate of growth" or "rate of return" on invested money is expressed in percent *per year*. For example, if the rate of return on $100 is 10 percent per year, it would grow to $110 in one year. However, the growth of money over time periods of greater than or less than one year is not so simple.

To "compound" money means to compute percentage increases periodically based on the current amount of money rather than on the original amount. The original amount of money invested is called the principal. As the money grows in value, investors want percentage increases calculated based on the larger current amount and not on the smaller principal. This is the idea of compounding—periodic readjusting of the standard on which the percentages are calculated.

EXAMPLE 19 Suppose $1000 is invested at 8% rate of interest, compounded yearly. How much will there be after 3 years?

The term "compounded yearly" means that each year the amount is multiplied by factor of 1.08. Three years yield three such factors.

1000	[original]
1000(1.08)	[after 1 year]
1000(1.08)(1.08)	[after 2 years]
1000(1.08)(1.08)(1.08)	[after 3 years]

Of course, repeated multiplication can be expressed with powers. The final amount is $1000(1.08)^3$. This comes to $1259.71.

Note that the money grew by a factor of $1.08^3 = 1.25971$. It grew 25.971 percent, which is more than 3 times 8 percent = 24 percent. Over a short time period the difference between compounding and not compounding is not very great. However, over a longer time period the effect of compounding can be phenomenal. ▬

EXAMPLE 20 In 2007 a collectible 1801 half dollar in splendid condition was worth $25,000. This sounds like it would have made a super investment to save that coin in 1801. Would it?

Suppose someone in 1801 had been able to invest 50 cents at 8% per year. What would it be worth in 2007?

The time, $t = 2007 - 1801 = 206$ (years). There would be 206 factors of 1.08. The value of 50 cents would be $.50(1.08)^{206} = \$3,839,402.42!$

Maybe an increase from 50 cents to $25,000 is not so super after all (Problem A52).

This makes the point that compounding over a long period of time can produce dramatic results, although, of course, it would be rare for an investment to be committed for 50 years, much less 206 years. ▬

Let an amount P (for principal) be invested at an interest rate of i per time period and compounded for n time periods. At the end of n time periods its amount, A, satisfies

$$A = P(1 + i)^n. \qquad (4\text{-}4\text{-}4)$$

Money is usually compounded more frequently than once a year. If so, the cited percentage rate *per year* (the so-called " annual percentage rate" or " APR") must be converted to a percentage rate *per time period*. The rate per time period, i, is defined to be the rate per year, r, divided by the number of time periods per year.

EXAMPLE 21   Suppose an investment yields 12% per year compounded monthly. How does it grow? What actually happens in one year? Does it grow 12%?

Because compounding is "monthly," time is reckoned in months and the rate "per year" must be converted to a rate "per month." That is, "12% per year compounded monthly" is actually computed as "1% per month compounded monthly."

Each month the investment will grow by $\dfrac{12\%}{12} = 1\%$, that is, it will change by a factor of 1.01. Let the initial amount be P and the number of months be m. The formula for the amount after m months is $A = P(1.01)^m$.

After 1 year the amount is 1.01^{12} times the original amount. $1.01^{12} = 1.1268$, which corresponds to an annual growth of 12.68%, not 12%. The real growth rate is higher than the nominal rate.   ▬

DEFINITION 4-4-5   The name of a growth rate for investments is the "nominal" rate called the <u>Annual Percentage Rate</u>, abbreviated APR. The actual growth rate for one year is called the <u>effective rate</u> or the <u>Annual Percentage Yield</u> (APY).

Given the APR, to compute the effective rate simply determine the change over one year.

So, the same interest rate is described two ways. The APR alone (here, 12%) does not describe the growth; you also need the number of times per year it is compounded. The real growth in one year is given by the effective rate (Annual Percentage Yield).

Compound Interest Formula
When money is compounded, it grows according to the "compound interest formula" given by

$$A = P\left(1 + \frac{r}{k}\right)^{kt} \tag{4-4-6}$$

where   P is the principal (the amount invested)
r is the Annual Percentage Rate
k is the number of times per year the money is compounded
t is the number of years $\left(\text{where } t \text{ is a multiple of } \dfrac{1}{k}\right)$, and
A is the amount to which the money grows.

There are three parameters, P, r, and k. t is the argument and A the image.

EXAMPLE 22   Find the future value of $10,000 invested at 8% compounded quarterly for 10 years.

The problem says: $P = 10,000$, $r = .08$, and $t = 10$. To compound money "quarterly" means to compound it four times per year, so $k = 4$. The rate per time period is one-fourth the annual rate.

$$A = \$10,000\left(1 + \frac{.08}{4}\right)^{4(10)} = \$22,080.40$$

What is the effective rate (Annual Percentage Yield) corresponding to 8% APR compounded quarterly?

The phrase "8% compounded quarterly" means $\frac{8\%}{4} = 2\%$ per quarter, for four quarters a year. The money grows by a factor of $1.02^4 = 1.08243$.

The Annual Percentage Yield is $1.0824 - 1 = .0824 = 8.24\%$. This is somewhat higher than the APR of 8%. ▬

The factor giving the actual increase in one year gives the <u>effective</u> <u>rate</u> or <u>Annual Percentage Yield</u> (APY).

$$\text{APY} = r_{\text{effective}} = \left(1 + \frac{r}{k}\right)^k - 1. \qquad (4\text{-}4\text{-}7)$$

EXAMPLE 23 Some credit cards charge interest at 18% per year compounded monthly. What is their effective rate?

$$\left(1 + \frac{.18}{12}\right)^{12} = 1.1956. \quad \text{Their effective rate is 19.56\% (per year).} \quad ▬$$

EXAMPLE 24 Suppose the yearly growth rates of an investment over the past three years were 10%, 43%, and 15%. a) Find the cumulative growth. b) What was the average (compound) growth rate?

The actual growth factor was $(1 + .10)(1 + .43)(1 + .15) = 1.80895$, for a total growth of 80.9%.

In this context the word "average" does not mean to add three rates and divide by three. The intention is to find a single annual growth rate that, over three years, would have produced the same result. Growth over three years at a constant annual rate r would produce a growth factor

$$(1 + r)(1 + r)(1 + r) = (1 + r)^3.$$

Set this equal to the actual growth factor and solve for r.

$$(1 + r)^3 = 1.80895,$$

$$1 + r = (1.80895)^{1/3} = 1.2185,$$

so $r = 21.85\%$. Money invested at the fixed annual rate of 21.85% would have produced the same three-year return. ▬

The <u>average</u> (compound) <u>rate</u> of n successive rates, $r_1, r_2, \ldots, r_n$, is the single rate, r, which when repeated n times, yields the same cumulative result. That is, the average is the solution for r to

$$(1 + r)^n = (1 + r_1)(1 + r_2)\ldots(1 + r_n) \text{ or }$$

$$1 + r = [(1 + r_1)(1 + r_2)\ldots(1 + r_n)]^{1/n}. \qquad (4\text{-}4\text{-}8)$$

EXAMPLE 25 Suppose an investment goes up 10% one year, up 45% the next, down 35% the third, and up 12% the fourth. a) What is the cumulative change over those four years? b) What is the average yearly growth rate?

$$(1 + .10)(1 + .45)(1 + -.35)(1 + .12) = 1.10(1.45)(.65)(1.12) = 1.16116.$$

The investment went up 16.1%.

$$(1 + r)^4 = 1.16116.$$

$$1 + r = 1.16116^{1/4} = 1.0381.$$

$$r = 3.81\%.$$

The average yearly change was 3.81%. ▬

EXAMPLE 26 Money invested at a fixed rate per year tripled in 11 years. What was the annual rate?

The growth factor is $(1 + r)^{11}$. The word "tripled" tells us this was 3. So solve

$$(1 + r)^{11} = 3,$$

$$1 + r = 3^{1/11} = 1.10503, \text{ and } r = 10.503\%.$$

The annual compound rate was 10.503 percent. ▬

In Formula 4-4-6, $A = P\left(1 + \dfrac{r}{k}\right)^{kt}$, if time begins now ($t = 0$ now) you may think of the principal "P" as the " present value" of the investment and "A" as its "future value" in t years. $1000 in the future is worth less than $1000 now. Everyone would rather win $1000 cash now than $1000 to be paid in 10 years. If an amount is to be paid in the future, we can determine its present value by solving for P in Formula 4-4-6,

$$A = P\left(1 + \frac{r}{k}\right)^{kt}.$$

Let P be the present value and A be the future value after t years of an amount of money invested at an annual rate, r, compounded k times per year.

$$P = A\left(1 + \frac{r}{k}\right)^{-kt}. \tag{4-4-9}$$

EXAMPLE 27 Suppose a tax-free retirement fund has $200,000 invested at 10% per year and inflation is projected to average 6% per year. Find the amount after 20 years and then find the present value of that amount, discounted for inflation.

The dollar amount will grow substantially over 20 years. The growth rate is 10%.

$$A = \$200,000(1.10)^{20} = \$1,345,500.$$

This number is pretty impressive. But a dollar will not buy as much in 20 years as it does now. What is the present value of that amount? The inflation rate is 6%.

$$P = \$1,345,500(1.06)^{-20} = \$419,533.26.$$

Still a nice amount, but not so impressive. ▬

■ Annuities

An <u>annuity</u> is of a sequence of payments or deposits.

EXAMPLE 28 John wins a $1,000,000 lottery and finds out he does not really get $1,000,000 now. He gets $50,000 now and payments of $50,000 each year for 19 more years. John actually wins an annuity.

Suppose payments, each of amount R, are made at equal time intervals beginning one time period from now and continuing for n payments. The <u>future value</u>, S, of the annuity (the value just as the last payment has been made at time n, at rate i per time period) is given by

$$S = \frac{R\big((1 + i)^n - 1\big)}{i}. \tag{4-4-10}$$

The <u>present value</u>, A, of the annuity (one time period before the first payment) is given by

$$A = \frac{R\big(1 - (1 + i)^{-n}\big)}{i}. \tag{4-4-11}$$

John, who won the "million dollar" lottery, wants all his money now. So he agrees to sell his annuity for a lump sum payment. How much is the lump sum?

The value of his "million" dollars depends upon the going interest rate. If the interest rate is high, payments promised in the distant future are worth less than money now. Suppose the rate is 9.5% per year. He gets $50,000 now and 19 more payments in an annuity. According to 4-4-11, the total is

$$\$50,000 + \frac{\$50,000[1 - (1 + .095)^{-19}]}{.095} = \$482,477.92.$$

That amount of money invested now at 9.5% could yield the same sequence of payments. I expect that John will be disappointed that his "million" is worth less than half of that (and, this does not even count taxes!). ▬

CONCLUSION Percents are a convenient way to express the functional relationship "Multiply by a constant." Because the context for percents is multiplication and not addition, successive percentage changes are not totaled by adding and not averaged by dividing.

Compounding of money invested at a fixed rate produces amounts given by power functions that are effectively repeated multiplication.

Terms: percent, principal, annual percentage rate, compound interest, annual percentage yield, effective rate, present value, average growth rate, annuity.

Exercise 4-4

A

A1.* What does "percent" mean, in its Latin derivation?

A2.* ☺ The context for "percent" is (pick one)
(A) addition   (B) multiplication   (C) subtraction
(D) powers

A3.* ☺ Which percents refer to less than the standard?

A4.* ☺ Write the numbers as percents (assuming the context is multiplication).
(A) 1   (B) $\frac{1}{2}$ (C) 2   (D) $\frac{1}{4}$

☺ *Express, as a rule (for example, "Multiply by 1.3") the function which relates the original amount to the final amount, as expressed by the given phrase:*

A5. "87 percent of"   **A6.** "up 15 percent"

A7. "down 5%"   **A8.** "40 percent off"

A9. "gained 33 percent"   **A10.** "12% of"

A11. "25% discount"   **A12.** "14% more than"

A13. "70 percent less than"   **A14.** "70 percent off"

☺ *y is what percent of x? Be careful with A21 through A24 where the relationship is not written in the natural order.*

A15. $y = .82x$ **A16.** $y = 2.1x$

A17. $y = .07x$ **A18.** $y = .005x$

A19. $y = 1.3x$ **A20.** $y = .3x$

A21. $x = 4y$ **A22.** $x = 8y$

A23. $x = .5y$ **A24.** $x = .4y$

An amount changes from x to y. What is the percent change?

A25. $x = 12$, $y = 13$.   **A26.** $x = 500$, $y = 510$

A27. $x = 1000$, $y = 1100$ **A28.** $x = 2500$, $y = 2000$

A29. $x = 9000$, $y = 9500$ **A30.** $x = .05$, $y = .07$

A31. $x = .075$, $y = .067$ **A32.** $x = 14.2$, $y = 15.1$

Give the functional relationship expressed by using appropriate letters for the variables.

A33. "Sales of video games in 1993 were 1300 percent of what they were in 1980."

A34. "55 percent of the students in Freshman math classes are female."

A35. "54 percent of eligible voters voted."

A36. "Liquidation sale: All hand tools priced at 20 percent of manufacturer's suggested retail price."

A37. ☺ (A) "Sales this year are 250% of last year's." How much are sales up?
(B) "Sales this year are 96% of last year's." How much are sales down?

A38. ☺ (A) "The market is 125% of what it was last year." How much is the market up?
(B) "Revenue this year is only 80% of what it was last year." How much is revenue down?

A39. The market index went from 415.7 to 401.8. What was the percent change?

A40. Gina's salary just went up from $21,125 per year to $22,456 per year. What percentage raise did she get? [6.3%]

A41. Employees just got an across-the-board $970 per year raise. For a friend, that works out to a 4.2% raise. How much did the friend earn before the raise? [$23,000]

A42. Suppose there was an across-the-board tax rebate of 10% of taxes. A friend's taxes were, after the rebate, $4590. What were they before the rebate?

A43. What was the original price of a jacket that sold for $114 at "40% off"?

A44. Including a 6% sales tax a car cost $13,668.70. What was the pre-tax price?

A45. Suppose a country has a population of 15,000,000 and the population grows at a rate of 3% per year. How many people will it have in 20 years? [27,000,000]

A46. Suppose $10,000 is invested at 10% compounded monthly. What will be the amount after ten years? [$27,000]

A47. Money invested at a fixed rate per year doubled in 9 years. What was the annual rate?

A48. Solve $5000\left(1 + \dfrac{r}{2}\right)^{24} = 5600$, for $r > 0$. [.0095]

What is the effective rate of

A49. 6% compounded monthly?

A50. 10% compounded quarterly?

A51. Which is larger after a year? Money compounded at 10.2% monthly or 10.3% semiannually?

A52. What is the average annual growth rate of the half dollar in Example 20?

A53. (A) How many parameters does the compound-interest formula (4.4.6) have?
(B) List them.

A54.* You like to give 15% tips when you eat in a restaurant. Explain, in English, how you can approximate, in your head, 15% of amounts such as $8 or $20.

A55. After the down payment, Sadie still owes $19,000 on her car. She will pay it with 60 months of equal payments, at an annual percentage rate of 9.6%. How big are her monthly payments? [400]

B

B1.* ☺ (A) What operation is usually signaled by "percent"?
(B) If there are two successive percents in a problem, what operation is required to find their cumulative effect?
(C) What operation is sometimes erroneously used to find the cumulative effect of two operations expressed in percents?

B2. ☺ (A) What percent of the side is the perimeter of a square?

 (B) What percent of the perimeter is the side of a square?

B3. (A) What percent of the diameter is the circumference of a circle?

 (B) What percent of the circumference is the diameter of a circle? [32%]

B4. A friend asks you which of her employer's payment plans she should choose. On Plan A she would get $40 a day plus 2 percent of all her sales. On Plan B she would get $20 a day plus 5 percent of all her sales. When is Plan B the better option for her?

What is the cumulative change of

B5. an increase of 10% followed by an increase of 15%?

B6. a loss of 70% followed by an increase of 100%?

B7. a loss of 10% followed by an increase of 10%?

B8. a gain of 70% followed by a loss of 50%?

B9. A course grade is based on 3 unit exams of 100 points each and a final of 200 points. If a student averages 86% on the first three exams, give the formula for how her 200-point final will affect her average.

B10. A length of wire can be made into the circumference of a circle or the perimeter of a square. How much more area does the circle have than the square?

B11. A path is in the shape of a quarter circle from point A to point B. How much shorter is it straight from A to B than around the path? [10%]

B12. John pays no taxes on the first $12,000 of his income. His taxes are 15% of the amount of his income above $12,000. If his taxes are 6% of his income, what is his income? SET UP the relevant formula and equation, and then solve.

B13. Income and taxable income are not the same. Suppose the first $12,500 of income is not taxable, and income in excess of that amount is taxable at 15%. At what income level would taxes amount to 10% of income?

B14. Suppose that under the current tax system the first $15,000 of income is not taxed, and any income over that is taxed at 15%. A possible revision has the first $16,000 not taxed, but any income over that taxed at 18%. People with which incomes will pay less under the proposed revision?

B15. Your friend has $100,000 to invest and is comparing tax-free municipal bonds at 6% to other bonds at 8%. Which has a higher net return? Assume that taxes take 28% of the interest on the other bonds.

B16. Suppose the first $12,000 of taxable income is taxed at 15%, and that any taxable income above that is taxed at a 28% rate. At what taxable income level would taxes be 20% of taxable income?

B17. Suppose a tax-free retirement fund has $200,000 invested at 10% per year and inflation is projected to average 6% per year. Find the amount after 20 years and then find the present value of that amount, discounted for inflation. [$1,300,000, $420,000]

B18. You have some round logs that are four times as long as their diameter. To make them burn better, you decide to split them lengthwise in quarters. How much does that increase their surface area? [110%]

B19. What percent of the volume of a cube is filled by the largest sphere $\left(V = \left(\dfrac{4}{3}\right)\pi r^3\right)$ that fits in it?

Compounding and Average Rates

B20. The cost of that computer has dropped 20% twice since it was introduced at $1000. What does it cost now?

B21. Over four years an investment went up 8%, down 12%, up 22%, and up 15%.

 (A) What was the cumulative change?

 (B) What was the average (compound) growth rate? [up 7.5%]

B22. Three annual increases were 10%, 15%, and 50%.

 (A) What was the total change?

 (B) What was the average annual change? [24%]

B23. Which yields the greater cumulative increase:

 (A) an increase of 10% followed by an increase of 20%, or

 (B) an increase of 20% followed by an increase of 10%?

B24. (A) An investment went up 12% one year and 42% the next. How much did it go up over the two years?

 (B) What was its average annual growth rate? [59, 26]

B25. (A) The cost of computer memory went down 30% one year and 60% the next. How much did it go down over the two years?

 (B) What was its average annual rate of change? [72, 47% down]

B26. (A) An investment went up 34% one year and down 12% the next. How much did it change over the two years?

 (B) What was its average annual rate of change? [18% up, 8.6%]

B27. Box *A* is 40% heavier than Box *B*. How much lighter is Box *B* than Box *A*? [29%]

B28. One year the second quarter earnings of IBM fell 92%. How much would they have to go up to reach the previous level?

B29. Profits are down 50% this year. What increase will bring profits back up to last year's level?

B30. Over the last two years the Widget Mutual Fund went up 60%. Last year alone it went up 35%. How much did it go up the year before last? [19%]

B31. Worldwide Widget stock went down 10% in the last 5 years, and in the first of those 5 years it went up 20%. How much has it gone down in the last 4 years? [25%]

B32. On September 15, 1998, the Brazilian stock market index went up 18%, but it was still down 45% for the year. How much had it been down that year prior to that day's jump?

B33. Wheat cost 200 times as much in 300 A.D. as it did in 150 A.D. What was the average annual rate of inflation of the cost? [3.6%]

B34. "Profits were down 40 percent last year, but they are up 55 percent this year." How are company profits doing compared with two years ago?

B35. Clyde has investments worth $100,000 at the beginning of the year. After 8 months he adds $20,000 to his investments. At the end of the year they are worth $140,000. What annual percentage yield would produce that growth? Assume monthly compounding. [Hint: Use an unknown APR and convert to APY after you find the APR. Use guess-and-check to solve the equation.] [.19 APY.]

B36. Suppose your investments make 10% and taxes take 28% of the increase in value of your investments.
(A) How much do your investments increase, after taxes?
(B) Suppose inflation is 4%. Adjusted for inflation, how much do your investments increase? [See also the next two problems.] [up 3.1%]

B37. Suppose your investments make 20% and taxes take 28% of the increase in value of your investments.
(A) After taxes, how much do your investments increase?
(B) Suppose inflation is 14%. Adjusted for inflation, how much do your investments increase? [See also the previous problem and the next problem.] [up .35%]

B38. Compare the results of the previous two problems. If your investments make a constant *p*% over the inflation rate, are you better off with low or high inflation?

B39. On Sept. 1, 1993 you could buy 104 Brazilian cruzeiros for a dollar. On Nov. 23, you could buy 217. The time interval is 83 days.
(A) What was the daily inflation rate?
(B) If that rate of inflation kept up for 365 days, what would be the annual effective inflation rate? (Assume a constant dollar.)

Mixtures

B40. Rose invested in several mutual funds. 15% of her investment was in a fund that went down 65%. But, overall, her investments went up 40%. How did her other funds (the other 85%) do? [up 59%]

B41. Kristi invested in two mutual funds. She put some money in one and it went up 8% and, at the same time, she put twice as much money in another and it went up 12.6%. How much did her money go up, overall? [11%]

B42. Add an amount of 5% salt solution to 10 ounces of 8% salt solution. How much does it take to make a 6% salt solution? [20]

B43. Jane's house has 120° hot water and 45° cold water. How should they be mixed to make 95° water? [67% hot]

Annuities See Example 28.

B44. If the interest rate is 12%, what is the present value of John's "million" dollars?

B45. If the interest rate is 5%, what is the present value of John's "million" dollars?

B46. Erin invests, beginning next month, $50 a month for the next 30 years (30 × 12 payments).
(A) What is the sum of the payments?
(B) If the yield is a constant 8% per year, what will be the value of the annuity at the end of 30 years? [$75,000]
(C) If the inflation rate is 5%, what would be the present purchasing power of that amount? [$17,000] [See also the next problem.]

B47. [Compare with the previous problem.] Erin invests, beginning one year from now, $600 a year for the next 30 years.
(A) What is the sum of his payments?
(B) If the yield is a constant 8% per year and all increases are taxed each year at a 28% federal rate and a 10% state rate, what will be the value at the end of 30 years?
(C) If the inflation rate is 5%, what would be the current purchasing power of that amount?

B48. Here is a problem which helps explain why bond prices move in the opposite direction of interest rates. Suppose a $1000, 8% bond really gives you interest payments, calculated at 8% per year, every year for 30 years beginning one year from when you buy it, and, in addition, at the end of the 30 years you get $1000. If the current interest rate drops to 7% after two years (and there are still 28 interest payments left, plus the $1000), what is the present value of the bond then? [$1100]

B49. (Why bond prices move in the opposite direction of interest rates.) Suppose a $1000, 10% bond really gives you interest payments, calculated at 10% per year, every year for 20 years beginning one year from when you buy it, and, in addition, at the end of the 20 years you get $1000. If the current interest rate goes up to 12% after three years (and there are still 17 interest payments left, plus the $1000), what is the present value of the bond then? [$860]

Money

B50. Darren carries credit card debt of $2000 for one year because each month he pays off all but exactly $2000. If his card charges 1.5% per month, how much interest on that $2000 does he pay in one year?

B51. Melinda gets a $4000 credit card bill. It charges 15.9% APR. If she pays the minimum of $50 month and charges nothing at all that month, how much will she owe next month?

B52. (Leverage) Suppose you can control an investment without paying for it all. For example, suppose you buy stock by immediately paying 20 percent and agreeing to pay interest on the remaining amount at 1 percent per month. Of course, when you sell it you will pay off the balance. Suppose in one month the stock is up 5 percent, and you sell it for a 1.5 percent commission. How did you do?

Taxes

B53. [The advantage of tax-deferred investing.]
   (A) An investment grows at 10% per year (every year), but at the end of each year 40% of that year's increase it taken away to pay taxes, leaving the rest to continue to grow the next year. Write a mathematical model for the amount (after taxes) after any number of years.
   (B) Write a model for an alternative investment that grows at the same rate but the increase is taxed only once at the end when the money is withdrawn.
   (C) Compare the two final after-tax amounts after 30 years if the initial investment was $10,000.

B54. Perhaps you have heard that some paintings have sold for phenomenal values (up to $80 million!) to Japanese collectors. A major part of the reason is the Japanese tax system which encourages the purchase of art. Here is why.

   Suppose you are old and rich and wish to pass your estate down to your heirs. But the death-tax rate is over 50 percent, so, without some scheme, your heirs will get less than half. However, the value of art (as opposed to other assets such as land) is generally determined by the owner, not the government, and the government seems to accept art valued at only 10% of its actual cost. Thus, by borrowing against other assets and buying art with the money, the apparent value of the estate can be decreased and thus death taxes decreased. Even after the heirs pay a 20 percent commission to have the art sold, they should come out well ahead.

   Find out how well this scheme works by making the following assumptions and determining the percentage of the value of the estate that the heirs can retain. Suppose the death-tax rate is 55 percent and 70 percent of the value of the estate is art.

Reading and Writing Mathematics

B55.* Explain the difference between "200 percent more than" and "200 percent of".

B56. Box A is $p\%$ heavier than Box B. How much lighter is Box B than Box A?

B57. (A) Let an amount change p percent and then change q percent. What is the total change? The actual percentage change is $p + q + \left(\dfrac{pq}{100}\right)$. Prove it.
   (B) Find the cumulative effect of a 5% change followed by a 10% change two ways: using part (a), and by directly computing the cumulative multiplicative factor.

B58. This year's rainfall is 125% of last year's. Last year's was 40% of normal. How does this year's compare to normal?

Money

B59. A "Zero Coupon" Bond pays $100,000 after 20 years and nothing in the meantime.
   (A) If the going interest rate is 8% (per year), what is it worth now?
   (B) Suppose that 2 years after you buy the bond (with 18 years left before getting the $100,000), the interest rate has gone up to 10% (per year). What is the bond worth then?
   (C) The interest rate went up. Did the value of the bond go up?

B60. [Difficult, but realistic and with an important point] You are going to buy a $18,000 car. The dealer offers you $2000 cash back or nothing down and 0% financing for 60 months. If you take the financing, starting in one month you will pay one-sixtieth of $18,000 each month for 60 months. In some sense that is not really "0% financing" because you could have bought the car for, effectively, $16,000 and you will be making $18,000 in payments. Use the "present value" formula to determine the actual finance rate [Hint: Do not expect to solve it algebraically.]

Section 4-5 RATIONAL FUNCTIONS

A "rational" function can be evaluated using only the four arithmetic operations.

DEFINITION 4-5-1 An expression is <u>rational</u> if and only if it is equivalent to a quotient of polynomials.

EXAMPLE 1 $\dfrac{x^2 - 1}{2x - 3}$ is a rational expression.

$\dfrac{1}{x}$ and $\dfrac{1}{x^2}$ are rational expressions.

$\dfrac{\sqrt{x}}{x + 1}$ is not a rational expression—the numerator is not a polynomial. ▬

■ Solving Rational Equations

A primary concern with division is to avoid division by zero. Numbers for which the denominator is zero are not in the domain of a rational function.

The most important features of the rational function $\dfrac{P(x)}{Q(x)}$ occur when $P(x) = 0$ or $Q(x) = 0$.

When the denominator is zero the expression is undefined. Near those x-values the behavior of the graph is remarkable and will be discussed shortly. Of course, the zeros of a function are always important.

THEOREM 4-5-2 (ZERO QUOTIENT RULE) $\dfrac{P(x)}{Q(x)} = 0$ iff $P(x) = 0$ and $Q(x) \neq 0$.

EXAMPLE 2 Solve $\dfrac{x - 2}{x - 1} = 0$.

The solution is simply $x = 2$, from the top. The "$Q(x) \neq 0$" part of the theorem tells us to check $x = 2$ to make sure the bottom is not zero there. It is not. ▬

In many problems forgetting to check when the denominator is zero does not cause mistakes, because the solution ruled out was not found anyway. But, sometimes, you do need to worry about the denominator.

Check to make sure the denominator is not zero.

EXAMPLE 3   Solve $\dfrac{x^2 - 4}{x - 2} = 0$.

The top is zero when $x = 2$ or $x = -2$. But $x = 2$ is not in the domain, because it would make the bottom zero. Therefore, the complete solution is $x = -2$. "$x = 2$" is ruled out by the part of Theorem 4-5-2 that says that "$Q(x) \neq 0$."

You might factor "$x^2 - 4$" into "$(x - 2)(x + 2)$" and cancel the "$x - 2$" factors in the top and the bottom, but this approach requires caution. Canceling may change the domain and make solutions that should be ruled out look legal.   ▬

EXAMPLE 4   Solve $\dfrac{(x - 2)^2(x - 3)}{(x - 2)} = 0$.

Canceling an "$x - 2$" from the both the numerator and denominator is illegal. In the original expression it is clear that $x = 2$ is illegal. But, after canceling, the equation would read "$(x - 2)(x - 3) = 0$," in which it is not clear that $x = 2$ is illegal. This new equation has solution $x = 2$ or $x = 3$, which is *not* equivalent to the original equation. The only solution is $x = 3$.   ▬

EXAMPLE 5   In calculus, problems like the following occur: Solve

$$\frac{3(x - 1)^2(x - 2)^2 - 2(x - 2)^3(x - 1)}{(x - 1)^4} = 0.$$

The first step is to set the numerator equal to zero and solve, remembering to exclude any resulting solutions for which the denominator would also be zero.

$$3(x - 1)^2(x - 2)^2 - 2(x - 2)^3(x - 1) = 0$$

iff $(x - 1)(x - 2)^2[3(x - 1) - 2(x - 2)] = 0$

iff $(x - 1)(x - 2)^2[3x - 3 - 2x + 4] = 0$

iff $(x - 1)(x - 2)^2[x + 1] = 0$

iff $x = 1$ or $x = 2$ or $x = -1$.

The solution $x = 1$ is ruled out by the denominator, so the solution to the original equation is $x = 2$ or $x = -1$.   ▬

■ Difference Quotients

Examples in which the top and the bottom are zero for the same values of x occur in the definition of the derivative in calculus. Derivatives of functions are defined in terms of "difference quotients."

DEFINITION 4-5-3 The <u>difference</u> <u>quotient</u> of the function f at x and a is

$$\frac{f(x) - f(a)}{x - a}$$

If $f(x)$ is a polynomial or a rational function, this difference quotient is a rational function.

EXAMPLE 6 Let $f(x) = x^2$. Find and simplify the difference quotient.

$$\frac{f(x) - f(a)}{x - a} = \frac{x^2 - a^2}{x - a}$$

$$= \frac{(x - a)(x + a)}{x - a}$$

$$= x + a, \text{ if } x \neq a.$$

It is critical that x cannot equal a. Otherwise the original difference quotient would be zero over zero, which is undefined. Everywhere except at a the difference quotient has the relatively nice form "$x + a$." In calculus, we are interested in the difference quotient "as x goes to a," in which case "$x + a$" goes to $2a$. This limit calculation yields the derivative of x^2 at a (The derivative is $2a$). ▬

EXAMPLE 7 Let $f(x) = \dfrac{1}{x}$. Find and simplify the difference quotient.

$$\frac{f(x) - f(a)}{x - a} = \frac{\dfrac{1}{x} - \dfrac{1}{a}}{x - a} \quad \text{[Now, find a common denominator for the top.]}$$

$$= \frac{\left(\dfrac{a - x}{xa}\right)}{x - a} = \frac{\dfrac{-(x - a)}{xa}}{x - a} \quad \text{["xa" is a common denominator]}$$

$$= \frac{-1}{xa}, \text{ if } x \neq a.$$

Again, x cannot equal a in any difference quotient, but x could equal a in the last expression, if we did not remember to rule it out. ▬

When the Quotient is not Zero
To solve a rational equation it is common to "Multiply through by the denominator." Again, you must be careful that the denominator is not zero.

EXAMPLE 8 Solve $\dfrac{3x - 5}{x + 1} = 2$.

Multiply through by the denominator:

$$3x - 5 = 2(x + 1)$$
$$= 2x + 2.$$
$$x = 7.$$

Technically, you must remember to check that this potential solution does not make the original denominator zero. It doesn't, so it is right. The relevant result is next. ▬

THEOREM 4-5-4 $\dfrac{a}{b} = x$ iff $a = bx$ and $b \neq 0$.

The "$b \neq 0$" part tells us that any solution to the "$a = bx$" part that makes $b = 0$ must be ruled out.

This is virtually the definition of division! What is $\dfrac{6}{2}$? The theorem says "$\dfrac{6}{2} = x$ iff $6 = 2x$." Knowing multiplication, you know $6 = 2(3)$, so $\dfrac{6}{2} = 3$. Division is defined to be the inverse of multiplication.

EXAMPLE 9   Solve $\dfrac{x^2 - x - 2}{x - 2} = 5$.

Multiplying through by "$x - 2$," we obtain

$$x^2 - x - 2 = 5(x - 2),$$

but this is not guaranteed to be equivalent to the original equation. Proceed as usual and rule out "$x = 2$" if it occurs.

$$
\begin{array}{ll}
x^2 - x - 2 = 5x - 10 & \\
x^2 - 6x + 8 = 0 & \text{[consolidating like terms]} \\
(x - 2)(x - 4) = 0 & \text{[factoring]} \\
x = 2 \text{ or } x = 4 & \text{[by the Zero Product Rule]}.
\end{array}
$$

But "$x = 2$" is extraneous and must be omitted, according to Theorem 4-5-4. The complete solution is "$x = 4$." ▬

If two rational functions are combined in one equation, you may treat them like fractions and multiply through by a common denominator.

EXAMPLE 10   Solve $\dfrac{1}{x - 2} + \dfrac{2}{x - 1} = 3$.

The product of the denominators is a common denominator. Multiplying through by "$(x - 2)(x - 1)$",

$$1(x - 1) + 2(x - 2) = 3(x - 1)(x - 2).$$

Now consolidate like terms, solve, and check the solutions to make sure they do not include "$x = 1$" or "$x = 2$" (Problem A34). ▬

▪ Graphs

Rational functions can have complicated graphs. That is part of the reason you study them—they can represent a wide variety of relationships that cannot be represented by polynomials. For example, the graphs of rational functions can display vertical asymptotes which cannot be a feature of the graphs of polynomials.

DEFINITION 4-5-5 A vertical line $x = a$ is said to be a <u>vertical asymptote</u> of the graph of $f(x)$ if, as x approaches a (from either the left or right), $f(x)$ is positive and becomes arbitrarily large or $f(x)$ is negative and becomes arbitrarily large in absolute value.

The reciprocal function exhibits the classic case of a vertical asymptote.

EXAMPLE 11 $\frac{1}{x}$ has a vertical asymptote at $x = 0$ (Figure 1). As x nears 0 from the right, x gets very small, so $\frac{1}{x}$ gets very large. For example, when $x = .01 = \frac{1}{100}$, then $\frac{1}{x} = 100$, which is large. Similarly, as x nears 0 from the left, $\frac{1}{x}$ is negative and becomes arbitrarily large in absolute value. For example, when $x = \frac{-1}{100} < 0, \frac{1}{x} = -100$ which is a large negative number. $\frac{1}{x}$ is undefined at $x = 0$; 0 is not in the domain.

The reciprocal function is an odd function (the power -1 is an odd number):

$$f(-x) = \frac{1}{(-x)} = \frac{-1}{x} = -f(x).$$

Its graph exhibits point symmetry about the origin (Definition 4-1-9).

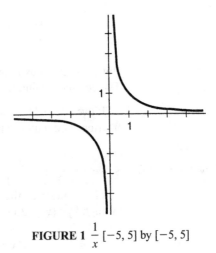

FIGURE 1 $\frac{1}{x}$ $[-5, 5]$ by $[-5, 5]$

EXAMPLE 12 $\frac{1}{x^2}$ has a vertical asymptote at $x = 0$, which is a zero of the denominator (Figure 2).

$\frac{1}{x^2}$ is undefined at $x = 0$. As x nears 0 from either side, x^2 is positive and very small, so $\frac{1}{x^2}$ is positive and very large.

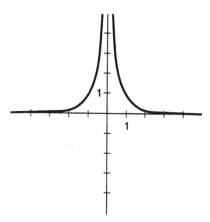

FIGURE 2 $\dfrac{1}{x^2}$ $[-5, 5]$ by $[-5, 5]$

$f(x) = \dfrac{1}{x^2}$ is an even function (-2 is an even number): $f(-x) = f(x)$. Its graph exhibits symmetry about the y-axis (Definition 4-1-11). ▬

In rational functions, vertical asymptotes usually occur where the denominator is zero.

EXAMPLE 13   Graph $\dfrac{2}{x + 3}$.

Expect a vertical asymptote at $x = -3$. There are no zeros.

This is a shift and a scale change applied to the reciprocal function. Let $f(x) = \dfrac{1}{x}$. This is $2f(x + 3)$. Therefore its graph is much like the graph of $\dfrac{1}{x}$, but three units to the left and expanded vertically by a factor of 2 (Figure 3).

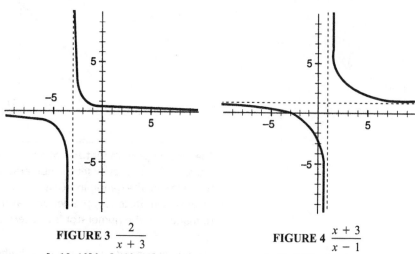

FIGURE 3 $\dfrac{2}{x + 3}$

$[-10, 10]$ by $[-10, 10]$

FIGURE 4 $\dfrac{x + 3}{x - 1}$

$[-10, 10]$ by $[-10, 10]$

EXAMPLE 14 Graph $\dfrac{x + 3}{x - 1}$

Expect a vertical asymptote at $x = 1$. The zero is $x = -3$ (Figure 4). ▬

The only cases where there is not necessarily a vertical asymptote when the denominator equals 0 is when the numerator also equals 0 at the same value.

> **Expect a vertical asymptote when the denominator is zero.**
> (There are some exceptions when the numerator is zero for the same value of x.)

EXAMPLE 15 Graph $\dfrac{(x - 2)(x - 5)}{x - 5}$.

At first, you expect a vertical asymptote at $x = 5$. But $x = 5$ is also a zero of the numerator, so the top and the bottom have a common factor (which is obvious here, but which would not be so obvious if the top were multiplied out).

$$\frac{(x - 2)(x - 5)}{x - 5} = x - 2, \text{ if } x \neq 5.$$

Therefore, the rational expression is nearly identical to "$x - 2$," a line. The difference is only that there is a tiny hole in the line at $x = 5$ where the rational function is not defined. Holes that are only one point wide are too small to see, and graphing calculators usually do not notice them, but, on a man-made graph, you can put a tiny open circle to indicate the existence of a hole (Figure 5).

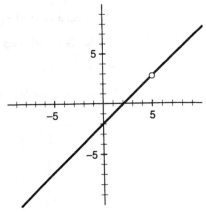

FIGURE 5 $\dfrac{(x - 2)(x - 5)}{x - 5}$ $[-10, 10]$ by $[-10, 10]$ ▬

> **The most important features of a rational function are its zeros and asymptotes.** To find the zeros, set the numerator equal to zero and solve (and check to make sure the denominator is not zero for the same x-values). To find the vertical asymptotes, set the denominator equal to zero and solve (and check to make sure the numerator is not zero at the same x-values).

If the numerator and denominator are higher-degree polynomials, the graph can be quite complex. We separate out for special consideration "end behavior" and behavior near asymptotes.

■ End-Behavior

The behavior of a rational function for large absolute values of x is easy to determine by looking at the quotient of the leading terms in the numerator and denominator.

DEFINITION 4-5-6   Let a rational function, $R(x)$, be the quotient of polynomials $P(x)$ and $Q(x)$. Suppose $P(x)$ has leading term $p(x) = ax^n$ of degree n and $Q(x)$ has leading term $q(x) = bx^m$ of degree m. Then the rational function

$$\frac{P(x)}{Q(x)} \text{ has } \underline{\text{end-behavior model}} \ \frac{ax^n}{bx^m} = \left(\frac{a}{b}\right)x^{n-m}.$$

EXAMPLE 16   Determine the end-behavior models of these functions.

$$\frac{3x^2 + 2x + 5}{x - 4} \text{ has end-behavior model } \frac{3x^2}{x} = 3x.$$

$$\frac{x^5 - 2x + 1}{4x^2 + 5} \text{ has end-behavior model } \frac{x^5}{4x^2} = \frac{x^3}{4}.$$

$$\frac{1 - x}{x^2 + 5} \text{ has end behavior model } \frac{-x}{x^2} = \frac{-1}{x}.$$

$$\frac{(2x + 3)(x - 7)}{x^2 - 2x - 5} \text{ has end-behavior model } \frac{2x^2}{x^2} = 2.$$

For large absolute values of x, the graph of a rational function resembles the graph of its end-behavior model.

EXAMPLE 17   Find the end-behavior model of $\dfrac{(x - 2)(x + 3)}{x - 1}$ and compare its graph to the graph of its end-behavior model.

The end-behavior model is simply $\dfrac{x^2}{x} = x$, a diagonal line. The rational function has zeros at 2 and -3, and a vertical asymptote at $x = 1$. For small x we do not expect the graph of the rational function to be like the graph of its end-behavior model, but for large x the similarity is evident (Figure 6).

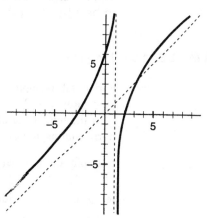

FIGURE 6 x and $\dfrac{(x - 2)(x + 3)}{x - 1}$ $[-10, 10]$ by $[-10, 10]$

End-behavior models are helpful for large absolute values of x, but are not intended to help for small values of x. However, they can sometimes indicate when a graph is not "representative" because the window is too small.

EXAMPLE 18   The end-behavior model of $\dfrac{2x^3 + 12x - 20}{6 - x}$ is $\dfrac{2x^3}{-x} = -2x^2$. On a moderate-scale graph its end-behavior is not clear (Figure 7). It almost looks like a cubic in that window. But that cannot be a representative graph, because the end-behavior is not indicated correctly. If you zoom out far enough, you should be able to see the end-behavior. The end-behavior refers only to large absolute values of x, and what is "large" depends on the particular rational function. For large x, but not for small x, the graph resembles the graph of $-2x^2$ (Figure 8).

Sometimes the graph of a rational function approaches a horizontal line.

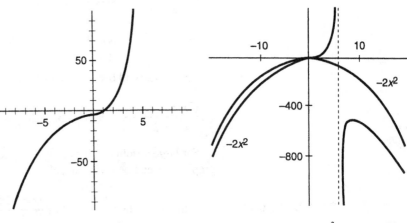

FIGURE 7 $f(x) = \dfrac{2x^3 + 12x - 20}{6 - x}$
$[-10, 10]$ by $[-100, 100]$

FIGURE 8 $f(x) = \dfrac{2x^3 + 12x - 20}{6 - x}$
and $-2x^2$. $[-20, 20]$ by $[-1200, 400]$

DEFINITION 4-5-7   If, as x becomes large, or as x becomes large and negative, the graph of a function approaches a horizontal line, that horizontal line is said to be a <u>horizontal asymptote</u> of the graph.

EXAMPLE 19   The graphs of $\dfrac{1}{x}$ (Figure 1), $\dfrac{1}{x^2}$ (Figure 2) and $\dfrac{2}{x + 3}$ (Figure 3) have the horizontal asymptote $y = 0$. If the denominator has higher degree than the numerator, $y = 0$ will be a horizontal asymptote.

If the numerator and denominator have the same degree, $y = c$ will be a horizontal asymptote where c is the quotient of the leading coefficients. For example, the rational function "$\dfrac{x + 3}{x - 1}$" (Example 10, Figure 4) is a quotient terms of degree 1 with leading coefficients 1 and 1, so their quotient is $\dfrac{1}{1} = 1$, and the horizontal asymptote is $y = 1$.

EXAMPLE 20   Determine the major features of the graph of $\dfrac{2x^2 + x - 7}{x^2 - 2x - 5}$.

To find the zeros, solve $2x^2 + x - 7 = 0$. It does not factor easily, so use the Quadratic Formula to obtain $x = \dfrac{\left(-1 \pm \sqrt{57}\right)}{4} = 1.63$ or -2.14 (Figure 9).

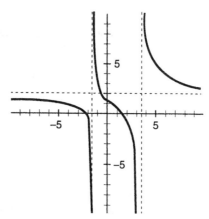

FIGURE 9 $\dfrac{2x^2 + x - 7}{x^2 - 2x - 5}$ $[-10, 10]$ by $[-10, 10]$

To find the vertical asymptotes, solve $x^2 - 2x - 5 = 0$. It does not factor easily, so use the Quadratic Formula. $x = 1 \pm \sqrt{6}$. The two equations have no solutions in common—the rational function cannot be reduced. It has vertical asymptotes at $x = 1 \pm \sqrt{6}$.

Now consider the end-behavior model: $\dfrac{2x^2}{x^2} = 2$. It has a horizontal asymptote: $y = 2$. These ideas isolate only the major features of the graph. Evaluation of particular values provides the details. ▄▄▄

Horizontal Asymptotes

For a rational function, there will be a horizontal asymptote if and only if the degree of the numerator is less than or equal to the degree of the denominator.

If the end-behavior model is a constant, c, then $y = c$ is a horizontal asymptote (Figures 4 and 9).

If the end-behavior model is $\dfrac{c}{x^n}$ for $n \geq 1$, then $y = 0$ is a horizontal asymptote (Figures 1, 2, and 3).

The "major" features of a graph of a rational function include zeros and asymptotes.

EXAMPLE 21   Graph $\dfrac{(x - 2)(x + 5)}{2x + 3}$ and identify its major features.

The zeros occur when $x = 2$ or $x = -5$ (neither makes the denominator 0).

The denominator is zero when $2x + 3 = 0$, that is, when $x = \dfrac{-3}{2}$. Because the

top is not zero for the same x, $x = \dfrac{-3}{2}$ is a vertical asymptote and the zeros of

the numerator are the zeros of the rational function (Figure 10). The end-behavior

model is $\dfrac{x^2}{2x} = \dfrac{x}{2}$, a line (Figure 10, dashes).

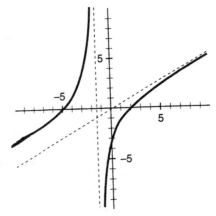

FIGURE 10 $\dfrac{(x - 2)(x + 5)}{2x + 3}$ $[-10, 10]$ by $[-10, 10]$

CALCULATOR EXERCISE 1   Graphing calculators may draw graphs with artifacts. When graphing Figure 10, many produce a vertical line at $x = \dfrac{-3}{2}$ that should not be there. Does yours? (Problem A33.)

Such artifacts are produced because the calculator fills in the graph by connecting dot to dot. A dot (pixel) high to the left of the asymptote is connected to a dot low to the right. For most functions, connecting dot to dot is the right thing to do, but not for this rational function.

■ Behavior Near Vertical Asymptotes

Once a vertical asymptote has been discovered, the next question is whether the graph goes up or down along it on either side. Of course, graphic calculators have taken most of the work out of graphing functions, but you should try to understand *why* rational functions behave near asymptotes as they do.

EXAMPLE 22   Consider the simple rational function $\dfrac{1}{x}$. There is a vertical asymptote at $x = 0$ (Figure 1). As x nears 0 from the left, x is negative and small and the numerator is positive, so the quotient will be negative and large. Therefore the graph goes

down along the asymptote from the left. As x nears 0 from the right, x is positive and small, so the quotient is positive and large and the graph goes **up** along the asymptote.

The graph of $\dfrac{1}{x^2}$ was different (Figure 2). It went up along both sides of the asymptote because the denominator x^2 is positive and the quotient is positive on both sides of the asymptote. ▬

Even without a graphing calculator, you can distinguish the ups from the downs by the sign of the quotient as x approaches the asymptotic x-value.

EXAMPLE 23 Graph $\dfrac{(x-3)(x+1)}{x^2-4}$. Explain the behavior of the graph near the vertical asymptotes.

The two zeros of the numerator are $x = 3$ and $x = -1$. The denominator is zero at different places, $x = 2$ and at $x = -2$, so the graph has vertical asymptotes at $x = 2$ and $x = -2$. The end-behavior model is $\dfrac{x^2}{x^2} = 1$, so it has a horizontal asymptote at $y = 1$ (Figure 11).

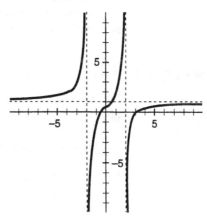

FIGURE 11 $\dfrac{(x-3)(x+1)}{x^2-4}$ $[-10, 10]$ by $[-10, 10]$

How does the graph behave near the vertical asymptotes?

Near $x = 2$, the numerator is near its value *at* $x = 2$: $(2 - 3)(2 + 1) = -3$. The numerator is negative near $x = 2$. Just to the right, the denominator is small and positive, so the quotient (which is negative over positive) is large and negative; the graph goes down along the asymptote (Figure 11).

Near $x = 2$, but just to its left, the denominator is small and negative, so the quotient is large and positive. Therefore, the graph goes up along the asymptote. The behavior at the other asymptote is determined in a similar manner. ▬

To determine whether a graph goes up or down along an asymptote $x = a$, determine the sign of the quotient near $x = a$. Near a the numerator will have the same sign it has at a, and the sign of the denominator can be determined by considering the signs of its factors in the two cases, $x < a$ and $x > a$.

■ Uses of Rational Functions

Rational functions are very useful in higher mathematics. They express sums, help evaluate special functions by approximating them, and are used to define special functions in calculus.

EXAMPLE 24

$$1 + x + x^2 + \ldots + x^n = \frac{1 - x^{n+1}}{1 - x}, \text{ for } |x| \neq 1. \qquad (4\text{-}5\text{-}8)$$

The left side is a "geometric series" (A series is a sum of terms; it is "geometric" when each term is a constant times the previous term). The right side gives the sum as a rational function (Problem B40).

The unending infinite geometric series has a famous sum.

$$1 + x + x^2 + \ldots + x^n + \ldots = \frac{1}{1 - x}, \text{ for } |x| < 1. \qquad (4\text{-}5\text{-}9)$$

For example,

$$\text{when } x = \frac{1}{2}, 1 + \frac{1}{2} + \frac{1}{4} + \frac{1}{8} + \frac{1}{16} + \ldots = \frac{1}{\left(1 - \frac{1}{2}\right)} = 2. \qquad \blacksquare$$

Polynomials can closely approximate many functions, but not functions that have a vertical asymptote. Rational functions can be used to approximate functions with vertical asymptotes.

EXAMPLE 25

Figure 12 displays the graph of $\tan x$ on $0 \leq x \leq \pi$ (angle x in radians). $x = \frac{\pi}{2}$ is a vertical asymptote. Because no polynomial has a vertical asymptote, it is not possible to closely approximate $\tan x$ near $\frac{\pi}{2}$ with a polynomial. However,

advanced mathematics tells us that the rational function $\dfrac{1}{\left(\dfrac{\pi}{2} - x\right)}$ is close to

$\tan x$ for x close to $\frac{\pi}{2}$.

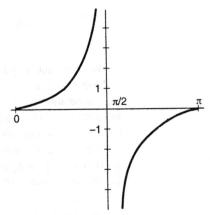

FIGURE 12 $\tan x$. $[0, \pi]$ by $[-5, 5]$

x	$\tan x$	$\dfrac{1}{\left(\dfrac{\pi}{2} - x\right)}$
$\dfrac{\pi}{2} - .1$	9.96664	10
$\dfrac{\pi}{2} - .01$	99.9666	100

Although $\tan x$ is hard to evaluate, the rational approximation near $\dfrac{\pi}{2}$ is easy to evaluate. ▬

EXAMPLE 26

Sue runs four miles every day, two miles out and two miles back. She likes to speed up on the return trip, so she runs back three miles per hour faster than she runs out. If her run takes 30 minutes, how fast does she run out?

The problem gives the distance and elapsed time, and asks for the rate. So distance, time, and rate are related. How are they related?

The basic formula is "distance equals rate times time" ($d = rt$). The question asks "How fast does she run out?" **Name the answer**. Let her rate out (in miles per hour) be the unknown, r. Then her rate back is $r + 3$. The time is given, so you need the basic formula rewritten to give time: $t = \dfrac{d}{r}$. The total time it takes is her time out $\left(\dfrac{2}{r}\right)$ plus her time back $\left(\dfrac{2}{r+3}\right)$. Then the formula for the total time is:

$$T = \frac{2}{r} + \frac{2}{r + 3}.$$

Writing this rational formula is the hard part. Now the word problem tells us the total time is also $\dfrac{1}{2}$ hour (do not use "30" if you use miles per *hour*). Set these two expressions equal and solve for r (Problem B27). ▬

CONCLUSION

Rational functions are the most complex type of function that can be evaluated using only arithmetic operations. Since their behavior can be so complex, before solving rational equations it is a good idea to use your calculator to graph them and see their behavior. Be sure to choose a window that yields a representative graph. If you use algebraic methods to solve a rational equation you can check your answer with the graph. The most important features of the rational function $\dfrac{P(x)}{Q(x)}$ occur when $P(x) = 0$ or $Q(x) = 0$.

Terms: rational expression, vertical asymptote, end-behavior model, horizontal asymptote.

Exercise 4-5

A

A1.* (A) How do you solve an equation of the form

"$\dfrac{a}{b} = 0$"?

(B) What part can cause trouble?

(C) State the relevant theorem (using variables)

A2.* The major features of the graph of a rational function include its _____ and its _____.

Identify all the zeros and the vertical and horizontal asymptotes.

A3. ☺ $\dfrac{x - 5}{x + 1}$

A4. ☺ $\dfrac{x + 2}{2x - 3}$

A5. ☺ $\dfrac{3x + 1}{x - 3}$

A6. $3 + \dfrac{x}{x - 4}$

A7. ☺ $\dfrac{(x - 3)(x - 10)}{x + 2}$

A8. ☺ $\dfrac{x - 9}{(x + 1)(x + 2)}$

A9. $\dfrac{x^2 + 5x + 6}{x}$

A10. $\dfrac{x + 1}{x^2 + x}$

A11. $\dfrac{x^2 - 3x + 1}{2x^2 - 5}$

A12. $6\left(\dfrac{x^2 + 3}{x^2 - x - 6}\right)$

A13. $\dfrac{2}{x} + 3(x - 4)$

A14. $x + \dfrac{1}{x}$

A15. $\dfrac{3}{2 + \dfrac{1}{x}}$

A16. $\dfrac{x - 1}{x + \dfrac{1}{x}}$

A17. Solve $\dfrac{1}{x} - \dfrac{1}{x - 2} = 3$. $[1.6, \ldots]$

A18. Solve $\dfrac{3 - x}{x^2 - 2} = 4$. $[1.5, \ldots]$

☺ *Give the end-behavior models.*

A19. $\dfrac{x^3}{x - 2}$

A20. $\dfrac{5 - x}{x^2}$

A21. $\dfrac{x}{x + 3}$

A22. $\dfrac{x - 2}{x}$

A23. $\dfrac{2 - x - 3x^2}{(x - 1)(x + 5)}$

A24. $\dfrac{(2x - 3)(x + 1)}{5x(x + 12)}$

Here are the given graphs in certain windows. Find the windows.

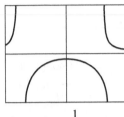 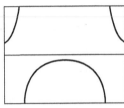

A25. $\dfrac{1}{(x - 20)(x + 30)}$

A26. $\dfrac{1}{(x - 10)(x - 50)}$

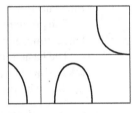 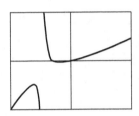

A27. $\dfrac{1}{x(x - 5)(x - 30)}$

A28. $\dfrac{x^4 + 50x}{x + 5}$

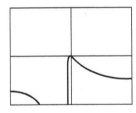 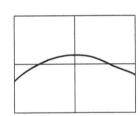

A29. $\dfrac{100 - 25x^2}{x^2 + 50x + 600}$

A30. $\dfrac{100 - 25x^2}{x^2 + 50x + 600}$

A31. Find any c such that the window $[-c, c]$ by $[-c, c]$ makes $\dfrac{x^2 - 20}{x - 20}$ look much like x on a square scale, where it exits the window at the lower left and upper right.

A32. Find c such that the window $[-20, 20]$ by $[-c, c]$ makes $\dfrac{x^4 - 16}{x - 3}$ look much like x^3 in $[-1, 1]$ by $[-1, 1]$ where it exits the window at the lower left and upper right.

A33. (A) What model graphing calculator do you use?

(B) When it graphs Figure 10, does it produce an artifact by making a vertical line at $x = \dfrac{-3}{2}$?

A34. Solve "$1(x - 1) + 2(x - 2) = 3(x - 1)(x - 2)$" from Example 10.

A35.* Define "rational" expression.

B

B1.* ☺ Where are the most interesting places on the graph of a rational function?

B2.* (A) ☺ Looking at the expression for a rational function, how can you tell where the vertical asymptotes are likely to be?
(B) Will there always be a vertical asymptote where you expected (from your answer to part (a))?
(C) Looking at the expression for a rational function, how can you tell where to expect zeros?
(D) Will there always be zeros where you expected them (from your answer to part (c))?

B3.* (A) ☺ Looking at the expression for a rational function, how can you tell whether there are horizontal asymptotes and where they will be?
(B) Will there always be a horizontal asymptote where you expected (from your answer to part (a))?

B4.* ☺ The text says to "expect" a vertical asymptote of a rational function where there is a zero of the denominator. Why doesn't it say "There will be a vertical asymptote where there is a zero of the denominator"?

Give the simplest rational function with

B5. A zero at 2, a vertical asymptote at −3, and a horizontal asymptote at 4.

B6. A zero at 0, a vertical asymptote at 1, and a horizontal asymptote at −1.

B7. ☺ Zeros at 2 and 5 and a vertical asymptote at 1.

B8. ☺ A zero at −3 and vertical asymptotes at 0 and 4.

B9. Zeros at 2 and 0, vertical asymptotes at 1 and 3, and a horizontal asymptote at 5.

B10. [A hard problem] Vertical asymptotes at 0 and 2 and it goes up along both sides of $x = 0$ and down along both sides of $x = 2$.

B11. A vertical asymptote at 3, zeros at 0 and 5, and a horizontal asymptote at 2. It has no other zeros or vertical asymptotes.

B12. A vertical asymptotes at 3 and −2, a zero at 5, and a horizontal asymptote at −1. It has no other zeros or vertical asymptotes.

The given graph is of P(x). Grid lines are one unit apart. Copy this original graph and sketch over it the corresponding graph of its reciprocal, $\frac{1}{P(x)}$.

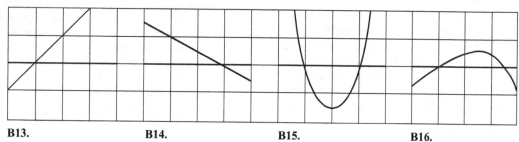

B13.   **B14.**   **B15.**   **B16.**

Solve algebraically

B17. $\dfrac{1}{x - 1} = \dfrac{3}{x}$ **B18.** $\dfrac{2x - 5}{4 - x} = 12$ [3.8]

B19. $\dfrac{x^2 - 1}{x + 2} = 1$ [2.3, . . .]   **B20.** $\dfrac{10}{x} = \dfrac{x - 7}{x - 1}$ [16, . . .]

B21. (A) Do the long division of $x - 2$ into $x + 1$.
(B) Sketch the graph of $\dfrac{x + 1}{x - 2}$.
(C) Use your result from part (a) to explain the appearance of the graph of $\dfrac{x + 1}{x - 2}$.

B22. (A) Do the long division of $x - 1$ into $2x - 5$.
(B) Sketch the graph of $\dfrac{2x - 5}{x - 1}$.
(C) Use your result from part (a) to explain the appearance of the graph of $\dfrac{2x - 5}{x - 1}$.

B23. (A) Do the long division of x into $x^2 + 1$.
(B) Sketch the graph of $\dfrac{x^2 + 1}{x}$.
(C) Explain, from your result in part (a), the appearance of the graph of $\dfrac{x^2 + 1}{x}$.

B24. (A) Do the long division of $2x - 1$ into $6x^2 + x - 4$.

   (B) Sketch the graph of $\dfrac{6x^2 + x - 4}{2x - 1}$.

   (C) Explain, from your result in part (a), the appearance of the graph of $\dfrac{6x^2 + x - 4}{2x - 1}$.

Difference Quotient

B25. Find and simplify the difference quotient (at x and a) of $f(x) = x^3$.

B26. Find and simplify the difference quotient (at x and a) of $f(x) = x^2 + x$.

B27. Solve the rational equation in Example 26.

B28. Solve $\dfrac{x - 3}{x + 2} + \dfrac{x - 5}{x + 1} = 7.$ $[-3.8, \ldots]$

B29. Find the slope of the line through the points on the graph of $\dfrac{1}{x}$ where $x = 1$ and $x = 1 + h$. Simplify.

B30. Find the slope of the line through the points on the graph of $\dfrac{1}{x}$ where $x = a$ and $x = b$. Simplify.

B31. A car travels 100 miles at an average speed of 40 miles per hour and then another 100 miles at an average speed of 60 miles per hour. What is the average speed over the 200 miles? [Hint: It is not 50 miles per hour.]

B32. The formula $d = rt$ holds for a *constant* rate over a distance, or for an *average* rate over a distance. The average rate over a two-part trip may not be the numerical average of the rates of the two parts. Suppose a round trip consists of traveling a distance d out at an average rate of r and then the same distance back at an average rate of s. What is the average rate for the whole trip $\left[\text{Hint: It is not } \dfrac{r + s}{2}\right]$.

B33. Teresa runs 8 miles round trip each day. She runs out at a constant rate and runs back 2 miles per hour faster. If she averages 7 miles per hour, how fast does she run out?

   Before doing the work, make a reasonable guess. Then do the detailed work and calculations. How close was your guess?

B34. [See 4-5-9] Graph $1 + x + x^2 + x^3 + x^4$ and $\dfrac{1}{1 - x}$ on $[0, 1]$ by $[0, 10]$.

   (A) Where are they within .1 of each other?
   (B) What changes if you add $x^5 + x^6$ to the sum?

Reading and Writing Mathematics

B35. Theorem 4-5-4, "$\dfrac{a}{b} = c$ iff $a = bc$ and $b \neq 0$," is actually the definition of division. Prove by counterexample that "$\dfrac{a}{b} = c$" is *not* equivalent to "$a = bc$."

B36.* (A) State (with variables) the theorem about solving a rational equation by multiplying through by the denominator.
   (B) What part can cause trouble?

B37. (A) Show that the graph of a quotient of linear polynomials usually resembles the graph of $\dfrac{1}{x}$ (shifted or rescaled) (Use a general quotient of linear polynomials).
   (B) When wouldn't it?

B38. When the end-behavior model is a diagonal line (such as $y = x$ in Example 17), the graph will parallel the line, but not necessarily come closer and closer to it. There will be a <u>slant asymptote</u> (also called an "oblique asymptote") that the graph does come closer and closer to, but that line (asymptote) is not quite the same as the end-behavior model. Find the line that the graph in Example 17 does come closer and closer to as x gets large.

B39. If we know e^x, for all x, then for $b > 0$, all powers (including irrational powers) of b can be defined using $b^p = (e^{\ln b})^p = e^{p \ln b}$ (using Property 4-1-3). Thus b^p can be evaluated as $e^{p \ln b}$, using the exponential function without needing the "repeated multiplication" idea which motivated the definition of powers but which really worked only for integer powers. Suppose we know e^x for all x.

   (A) How would $4.3^{2.8}$ be evaluated with this approach?
   (B) Give $x^{2.3}$ with this approach.

B40. Prove:
$$1 + x + x^2 + \ldots + x^n = \frac{1 - x^{n+1}}{1 - x}, \text{ for } x \neq 1.$$

B41. Use 4-5-9 to rewrite $\dfrac{1}{1 - x^2}$ as an infinite series for $x^2 < 1$.

B42. Use 4-5-9 to rewrite $\dfrac{1}{x}$ as an infinite series for $0 < x < 2$.

B43. Use 4-5-9 to rewrite $\dfrac{1}{1 + x^2}$ as an infinite series for $|x| < 1$.

Solve for a and b [This way to rewrite a fraction is called "partial fractions." See Problem B46.] [Hint: Find a common denominator and match coefficients in the numerators.]

B44. "$\dfrac{5x - 7}{x^2 - 3x + 2} = \dfrac{a}{x - 1} + \dfrac{b}{x - 2}$, for all x."

B45. "$\dfrac{x + 1}{x^2 - x - 6} = \dfrac{a}{x + 2} + \dfrac{b}{x - 3}$, for all x."

B46. **Partial Fractions**. To add fractions we use a common denominator. For example,

$$\frac{2}{x - 1} + \frac{3}{x - 2} = \frac{2(x - 2) + 3(x - 1)}{(x - 1)(x - 2)}$$

$$= \frac{5x - 7}{(x - 1)(x - 2)}.$$

In this example a sum of terms with constant numerators and linear denominators is turned into a quotient with a linear numerator and a quadratic denominator. Sometimes we wish to reverse the process. (The reverse process is called "partial fractions.")

(A) Find A and B such that

$$\frac{x - 4}{(x - 1)(x - 2)} = \frac{A}{x - 1} + \frac{B}{x - 2}.$$

(B) Solve for A and B in terms of a, b, m, and k in

$$\frac{mx + k}{(x - a)(x - b)} = \frac{A}{x - a} + \frac{B}{x - b}.$$

(C) Is there always a solution?

Section 4-6 INEQUALITIES

This section discusses the various ways to solve an inequality. Some operations that work for solving equations do not necessarily work for solving inequalities.

1. Multiplying or dividing both sides of an inequality by a **negative** number does not preserve the direction of an inequality–it reverses it!

2. Multiplying or dividing both sides of an inequality by an **expression with a variable** *often does not* preserve the solutions to an inequality.

■ Graphical Solutions

The solution to an inequality can be approximated graphically. To solve "$f(x) > 0$" graph "$y = f(x)$" and find the x-values of the points that are above the x-axis ("above" is "greater than" for y-values).

EXAMPLE 1 Figure 1 is a representative graph of f. Solve "$f(x) > 0$."

Report the x-values of all the points that are above the x-axis. Points on the far left and far right are above the x-axis. They have x-values less than (to the left of) -3 or greater than (to the right of) 2. The solution is: $x < -3$ or $x > 2$.

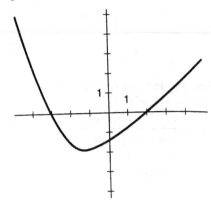

FIGURE 1 $f(x)$, without an algebraic representation. $[-5, 5]$ by $[-5, 5]$ ▬

Guess-and-Check
A graphing calculator makes solving inequalities easy.

EXAMPLE 2 Solve "$x^2 < 2x + 3$."

One way (not the only way) is to graph both expressions "x^2" and "$2x + 3$" (Figure 2). Look for points on the graph of "x^2" which are below the corresponding points on the graph of "$2x + 3$" (because "below" corresponds to "less than" for images). Then, read off the x-values of those points.

The solution is $-1 < x < 3$.

Be very clear about one thing: The *points* themselves are not solutions, it is *the x-values of the points* that are solutions.

Rather than compare two y-values to each other, it is easier to compare their difference to zero, which is justified by the next theorem.

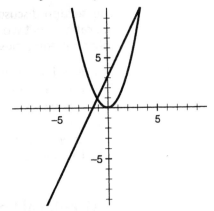

FIGURE 2 x^2 and $2x + 3$. $[-10, 10]$ by $[-10, 10]$ ▬

THEOREM 4-6-1 For all a, b, and c,

A) $a < b$ iff $a + c < b + c$.

B) $a < b$ iff $a - b < 0$.

This tells us that, for solving inequalities, adding and subtracting work just like they do for equations. But, beware, multiplying and dividing do not (Examples 4 through 8, later).

**EXAMPLE 2
REVISITED**

Solve "$x^2 - 2x < 3$."

This is equivalent to

$$\text{"}x^2 - 2x - 3 < 0\text{."}$$

Now graph "$y = x^2 - 2x - 3$" (Figure 3). The solution corresponds to the x-values of the points below the x-axis: $-1 < x < 3$.

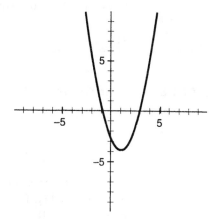

FIGURE 3 $x^2 - 2x - 3$. $[-10, 10]$ by $[-10, 10]$

There are still more ways to solve this inequality that we will discuss after Theorem 4-6-2.

The "guess-and-check" method will work for any inequality.

EXAMPLE 3

Solve "$\tan x < 2.1x$" for x in $\left[0, \dfrac{\pi}{2}\right)$ (in radians).

There is no algebraic method of solving this, but guess-and-check works.

First put everything on one side to obtain "$\tan x - 2.1x < 0$." Then graph "$y = \tan x - 2.1x$" and compare the y-values to 0 (Figure 4). The solution is $0 < x < 1.190$.

FIGURE 4 $y = \tan x - 2.1x$ $\left[0, \dfrac{\pi}{2}\right]$ by $[-1, 1]$

■ Solving Linear Inequalities

The processes which can be used to solve a linear inequality are *not* perfectly parallel to the processes that you use to solve equations. Inequalities are more complex than equalities.

Multiplying or dividing or squaring both sides of an inequality do *not* work just like they do for equations.

EXAMPLE 4 $2x < 8$ is equivalent to   [dividing by 2]

$x < 4$ [because $2 > 0$, the direction is preserved].

However $-2x < 6$ is equivalent to   [dividing by -2]

$x > -3$ [because $-2 < 0$, the direction is reversed].

The direction of the inequality is *reversed* after dividing by a negative number.

EXAMPLE 5   Solve $8 - 4x > 10$.

$8 - 4x > 10$ iff $-4x > 2$ [subtracting 8, Theorem 4-6-1]

$$\text{iff } x < \frac{2}{-4} = \frac{-1}{2}.$$ [changing the direction]

If a number is *negative*, division or multiplication by it *reverses* the direction of the inequality. For example, $2 < 3$, but $-2 > -3$ (because -2 is to the *right* of -3). If $x > 5$, then $-x < -5$. Figure 5 illustrates the fact that multiplying both sides of an inequality by -1 reverses the direction of the inequality.

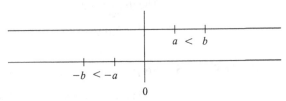

FIGURE 5 On the upper line, $a < b$, which means a is to the left of b. On the lower line, $-a > -b$ ($-a$ is to the *right* of $-b$). Equivalently, $-b < -a$

THEOREM 4-6-2   on multiplying both sides of an inequality by a number:

A) Let $c > 0$. Then $a < b$ iff $ca < cb$, and $a < b$ iff $\dfrac{a}{c} < \dfrac{b}{c}$.

B) Let $c < 0$. Then $a < b$ iff $ca > cb$, and $a < b$ iff $\dfrac{a}{c} > \dfrac{b}{c}$.

The trick to this theorem is part B. If you multiply or divide by a variable which could be negative, things usually go wrong.

EXAMPLE 6   Solve $\dfrac{2}{x} < 1$.

Your first thought to simplify this might be to "multiply through by x" to obtain "$2 < x$." This new inequality would be simpler, but it wouldn't be *equivalent*.

Multiplying through by x would be a big blunder. You may not multiply both sides of an *in*equality by "x", *unless you already know whether "x" is positive or negative.*

To do this problem requires a new approach. One approach that works is to separately consider two cases as in Theorem 4-6-2 above.

<u>Case I</u> $(x > 0)$: If $x > 0$, you may "Multiply through by x" to convert "$\frac{2}{x} < 1$" to "$2 < x$." (The direction of the inequality is preserved because x is positive.) The numbers that satisfy both conditions ($x > 0$ and $2 < x$) solve the inequality, so "$2 < x$" is part of the solution.

<u>Case II</u> $(x < 0)$: If $x < 0$, you may "Multiply through by x," but you must remember to change the direction of the inequality (as in Part B of the theorem) to obtain "$2 > x$." The numbers that satisfy both conditions ($x < 0$ and $2 > x$) solve the inequality, so "$x < 0$" is part of the solution.

The solution is "$x < 0$ or $2 < x$."   ▬

EXAMPLE 6
ANOTHER WAY

Solve $\frac{2}{x} < 1$.

$$\frac{2}{x} - 1 < 0. \; \frac{2 - x}{x} < 0.$$

The interesting places are at $x = 0$ and $x = 2$. The graph (Figure 6, for solving "$\frac{2}{x} - 1 < 0$") is below the x-axis in two distinct regions. The upcoming "Theorem on Zeros and Signs" says the solution will be intervals with those endpoints.

Reading the graph, the solution is (as before) "$x < 0$ or $2 < x$."

But, if you illegally "multiply through by x" to convert "$\frac{2}{x} < 1$" into "$2 < x$" the answer is wrong because you lose track of all the negative solutions (because multiplying through by a negative number should have changed the direction of the inequality by Theorem 4-6-2B).

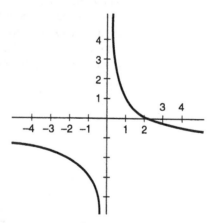

FIGURE 6 $\frac{2}{x} - 1$ $[-5, 5]$ by $[-5, 5]$ ▬

Dividing both sides of an inequality by an expression with a variable is also dangerous.

EXAMPLE 7   Solve "$x^2 < 2x$."

Your first thought might be to simplify this by canceling the factor of x on both sides. The resulting inequality ($x < 2$) would then be simpler, but it wouldn't be *equivalent*. Canceling "x" would be a big blunder. You may not divide both sides of an inequality by an expression, *unless you already know whether the expression is positive or negative.*

Here is a correct way to solve the problem.

$$x^2 < 2x \text{ iff } x^2 - 2x < 0.$$
$$\text{iff } x(x - 2) < 0.$$

The expression on the left is a polynomial (Figure 7). Now the problem is easy because you can simply look at the graph and read off the answer.

The zeros of $x(x - 2)$ are $x = 0$ and $x = 2$. Only the middle interval yields negative images. The solution is $0 < x < 2$.

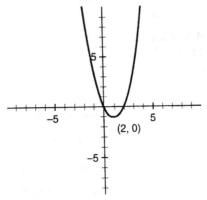

FIGURE 7 $y = x^2 - 2x$ [−10, 10] by [−10, 10]

■ Zeros and Inequalities

The zeros of a polynomial divide the x-axis into intervals. Inside each of these intervals the polynomial can not change sign, because it would have to cross the axis at a zero to do so. For example, in Figure 8, the three zeros divide the line into four intervals. Inside any one of those intervals the polynomial is either positive everywhere or negative everywhere.

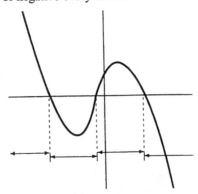

FIGURE 8 The zeros of a polynomial divide the line into intervals. In the interior of each interval, it is either positive for every x-value or negative for every x-value

THEOREM ON ZEROS AND SIGNS 4-6-3

A) The zeros of a polynomial $P(x)$ divide the number line into intervals. The sign of $P(x)$ at any point in the interior of an interval is the sign of $P(x)$ for all points in the interior of that interval.

B) Let $R(x)$ be a rational expression, $\dfrac{P(x)}{Q(x)}$. The zeros of P together with the zeros of Q divide the number line into intervals. The sign of $R(x)$ at any point in the interior of an interval is the sign of $R(x)$ for all points in the interior of that interval.

Use this theorem with a graph. This is my favorite way to solve polynomial or rational function inequalities.

- Put everything on one side (so an expression can be compared to zero).
- Graph it and find the zeros and vertical asymptotes algebraically, which are endpoints of intervals.
- Note from the graph which of those intervals yield solutions to the inequality.
- Check the endpoints separately.

EXAMPLE 8   Solve $\dfrac{x+3}{x-1} \leq 0$.

Graph it (Figure 9). The most important features of the graph are the zero at $x = -3$ and the vertical asymptote at $x = 1$. The three intervals of the theorem are "$x < -3$," "$-3 < x < 1$," and "$1 < x$."

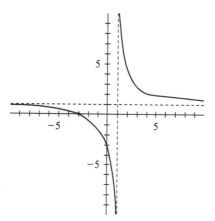

FIGURE 9 $\dfrac{x+3}{x-1}$ $[-10, 10]$ by $[-10, 10]$

The graph shows the negative images occur in the interval $-3 < x < 1$. According to the Theorem on Zeros and Signs, the solution includes the interior of that interval. The endpoints must be checked separately. Try -3. It is a zero, and, because the inequality is "$\leq$", the zero is a solution. Try 1. It is not in the domain and does not solve the inequality. So the complete solution is: $-3 \leq x < 1$.

Wouldn't it be easier to just "Multiply through by $x - 1$" and eliminate the quotient? It would be *easier*, but it would be *wrong*. The expression "$x - 1$" can be either positive or negative, and when it is negative it would be wrong to leave the direction of the inequality unchanged. By the way, you would erroneously get "$x + 3 < 0$" and then "$x < -3$," which is completely wrong. ▬

The idea of "Zeros and Signs" works only for comparing expressions to *zero*.

EXAMPLE 9   Solve $(x - 1)(x + 3) \geq 1$.

In many examples you are pleased to see factored form, but in this one the factoring is useless. Factored form is useful if and only if the expression is compared to *zero*.

$$(x - 1)(x + 3) \geq 1 \text{ iff } x^2 + 2x - 3 \geq 1 \quad \text{[multiplying out]}$$

$$\text{iff } x^2 + 2x - 4 \geq 0 \quad \text{[subtracting 1].}$$

Solve this inequality instead.

Now the zeros of $x^2 + 2x - 4$ are relevant. It factors, but not using integers. Use the quadratic formula to find the zeros: $x = 1 \pm \sqrt{5}$. In decimal form, the zeros are approximately -3.236 and 1.236.

Graph it (Figure 10). From the picture, the middle interval gives the negative values, and the two outer intervals give the positive values. Because the inequality has this expression greater than or equal to zero, you want the positive values. The solution is

$$x \leq -1 - \sqrt{5} \text{ or } x \geq -1 + \sqrt{5},$$

where the endpoints are included because the inequality uses "≥ 0," which includes the zeros.

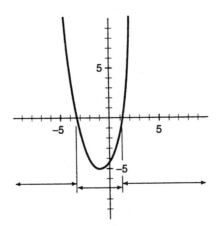

FIGURE 10 $x^2 + 2x - 4$ $[-10, 10]$ by $[-10, 10]$ ▬

EXAMPLE 10   Solve $\dfrac{x^2 + 3x - 5}{(x + 1)(x - 3)} \geq 0$.

Graph it (Figure 11). You see the zeros of the top as zeros, and the zeros of the bottom as vertical asymptotes.

The zeros of the denominator are $x = -1$ and $x = 3$. The zeros of the numerator can be determined by the Quadratic Formula.

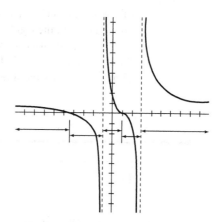

FIGURE 11 $\dfrac{x^2 + 3x - 5}{(x + 1)(x - 3)}$ [−10, 10] by [−10, 10]

$$x = \frac{-3 \pm \sqrt{29}}{2} = 1.19 \text{ or } -4.19.$$

The four zeros divide the line into 5 intervals. The graph shows where the expression is positive. Checking the endpoints separately, you see that the zeros of the numerator are included and the zeros of the denominator are excluded. The solution is

$$x \le \frac{-3 - \sqrt{29}}{2} \quad \text{or} \quad -1 < x \le \frac{-3 + \sqrt{29}}{2} \quad \text{or} \quad x > 3.$$

This is an algebraic solution because we found the endpoints of the intervals algebraically.

Interval Notation

To <u>solve</u> an inequality means to find the values of the variable (often "*x*") that make the inequality *true*. Solutions may be expressed in two ways. One way is to give an equivalent inequality which exhibits the *x*-values. The other is to consolidate those *x*-values into a single set.

EXAMPLE 11 The solution to "$2x < 10$" can be expressed as the equivalent inequality, "$x < 5$," which exhibits the solutions. The solution can also be expressed as a set which, in "interval notation," can be written "$(-\infty, 5)$."

Here is "interval notation" for sets. The rules are

1. if the endpoint is <u>included</u>, use a square bracket: "[" or "]"

2. if the endpoint is <u>excluded</u>, use a parenthesis: "(" or ")".

interval	set	interval	set
$a < x < b$	(a, b)	$x < b$	$(-\infty, b)$
$a \le x \le b$	$[a, b]$	$x \le b$	$(-\infty, b]$
$a < x \le b$	$(a, b]$	$x > a$	(a, ∞)
$a \le x < b$	$[a, b)$	$x \ge a$	$[a, \infty)$

The symbols "∞" (called "infinity") and "$-\infty$" are not real numbers and therefore can not be included. The symbols "∞" and "$-\infty$" are used to take the place of the other (non-existent) endpoint when the interval has only one real-number endpoint (as in the right column).

Sentences such as "$a < x < b$" express two inequalities connected by "and": "$a < x$" and "$x < b$." The word "and" means both inequalities apply to x. Figure 12 shades the three regions.

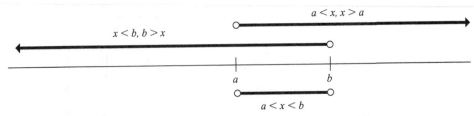

FIGURE 12 "$a < x$", "$x < b$", and "$a < x < b$."

If x belongs to either of two intervals the proper connective to use is "or", not "and". For example, Figure 13 illustrates the region of the number line in which "$x < -2$ or $x > -2$." In interval notation, that region is the set $(-\infty, -2) \cup (2, \infty)$ ["union" ($\cup$) connects sets the way "or" connects sentences.]

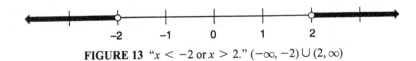

FIGURE 13 "$x < -2$ or $x > 2$." $(-\infty, -2) \cup (2, \infty)$

The region consists of two intervals. A particular "x" cannot be in both. It can be in one *or* in the other. The way to express this is "$x < -2$ or $2 < x$," using "or". This region is *not* described by "$x < -2$ *and* $x > 2$" – there are no such x's.

A string of inequalities is always interpreted as inequalities connected by "and," never "or." Therefore it is not permitted to state "$2 < x < -2$." There are no such x's.   ▬

■ Absolute Values

On the number line, the inequality "$a < b$" means a is to the *left of b*. It does not mean a is "smaller than" b (unless both are positive). $-1,000,000$ is less than 5, but we would not call it small (unless you have been thinking about the national debt). The size of a number is determined by its absolute value.

DEFINITION 4-6-4   The <u>absolute value</u> of a real number, x, is denoted by $|x|$, which is read "the absolute value of x." It is defined by $|x| = x$ if $x \geq 0$ and $|x| = -x$ if $x < 0$.

EXAMPLE 12 $|7.6| = 7.6$. $|-7.6| = 7.6$. $|0| = 0$. $|7 - 5| = 2$. $|5 - 7| = 2$.   ▬

In the definition, when $x < 0$, "$-x$" is positive, so the absolute value of x is always non-negative.

The absolute value of x is the distance of x from the origin, 0. On the number line, real numbers express *directed* distance, not just distance. For example, -5 is 5 units to the *left* of the origin.

The absolute value of "$x - a$" is the <u>distance</u> from x to a on the number line. So you may read "$|x - a|$" aloud as "the distance from x to a," which is the same as "$|a - x|$", "the distance from a to x." In calculus, absolute values are used to express the idea of numbers being "close to" one another.

EXAMPLE 13   Which numbers are "close" to 5?

The distance between x and 5 is $|x - 5|$. The inequality "$|x - 5| < d$" describes the numbers that are within d units of 5. The smaller the d, the closer x is forced to be.

THEOREM ON ABSOLUTE VALUES 4-6-5

<u>A</u>) For all x and c, $|x| < c$ iff $-c < x < c$ [Figure 14A].
<u>B</u>) For all x and c, $|x| > c$ iff $x < -c$ or $x > c$ [Figure 14B].

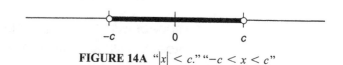

FIGURE 14A "$|x| < c$." "$-c < x < c$"

FIGURE 14B "$|x| > c$." "$x < -c$ or $x > c$"

Read "$-c < x < c$" as "negative c is less than x is less than c," or "negative c is less than x *and* x is less than c." The connective "and" is not written in "$-c < x < c$," but it is really there.

Solve $|x - 5| < 1$. Shade the interval on a number line.

$|x - 5| < 1$ iff $-1 < x - 5 < 1$,   [by the Theorem on Absolute Values, 4-6-5]

     iff $4 < x < 6$ [by Theorem 4-6-1A twice, adding 5 to all sides of both inequalities.]

The solution is the interval $4 < x < 6$, all numbers within 1 unit of 5 (Figure 15).

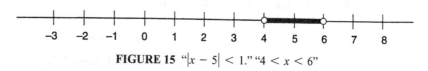

FIGURE 15 "$|x - 5| < 1$." "$4 < x < 6$"

Solve $|x - 5| < d$.

$|x - 5| < d$ iff $-d < x - 5 < d$ [by the Theorem on Absolute Values]

     iff $5 - d < x < 5 + d$ [by Theorem 4-6-1, twice.]

The inequality describes an interval on the number line of width $2d$ about a central point. In this case, the center of the interval is 5.   ▬

A major purpose of absolute values in calculus is to describe intervals centered about a point.

THEOREM 4-6-6   A) $|x - c| < d$ iff $c - d < x < c + d$.

B) "$a < x < b$" can be written in the form "$|x - c| < d$"

where $c = \dfrac{a + b}{2}$ and $d = \dfrac{b - a}{2}$.

This theorem gives alternatives for expressing "All the numbers within d units of c." The interval is of width $2d$, so d is its "half-width." Its center is c (Figure 16).

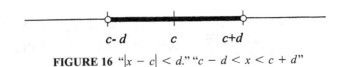

$$c\text{-}d \qquad c \qquad c\text{+}d$$

FIGURE 16 "$|x - c| < d$." "$c - d < x < c + d$"

PROOF OF PART A: $|x - c| < d$ iff $-d < x - c < d$ iff $c - d < x < c + d$.

PROOF OF PART B: From Part A, $a = c - d$ and $b = c + d$. Adding those two,

$$a + b = 2c, \text{ so } c = \frac{a + b}{2}.$$ Subtracting the first from the second,

$$b - a = 2d, \text{ so } d = \frac{b - a}{2}.$$

EXAMPLE 14   Use absolute value notation to express all numbers within 2 units of 3.4. Then find the interval in interval notation.

All numbers within 2 units of 3.4 can be written "$|x - 3.4| < 2$."
By Theorem 4-6-5A, "$|x - 3.4| < 2$." is equivalent to $3.4 - 2 < x < 3.4 + 2$. The interval is $(1.4, 5.4)$. ▬

EXAMPLE 15   Express the interval $(2.1, 2.2)$ in the form "$|x - c| < d$."

The center is $c = \dfrac{2.1 + 2.2}{2} = 2.15$ and the half-width is

$d = \dfrac{2.2 - 2.1}{2} = .05$. The interval can be written "$|x - 2.15| < .05$." ▬

EXAMPLE 16   Solve $|x - 3| > 2$.

$$|x - 3| > 2 \text{ iff } x - 3 < -2 \text{ or } x - 3 > 2 \quad [\text{by Theorem 4-6-5B}]$$
$$\text{iff } x < 1 \text{ or } x > 5 \text{ (Figure 17)}.$$

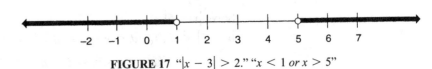

FIGURE 17 "$|x - 3| > 2$." "$x < 1 \text{ or } x > 5$"

This expresses all the points on the number line that are greater than 2 units away from the location of "3". There are two separate intervals, which is expressed with "or". In interval notation, the set of solutions is $(-\infty, 1) \cup (5, \infty)$. ▬

EXAMPLE 17   Solve $|3 - x| < 1$.

$$|3 - x| < 1 \text{ iff } -1 < 3 - x < 1 \quad \text{[the Theorem on Absolute Values]}$$
$$\text{iff } -4 < -x < -2 \quad \text{[subtracting 3 from all sides]}$$
$$\text{iff } 4 > x > 2 \quad \text{[dividing by } -1, \text{ a negative number]}.$$

Do not forget to change the direction of the inequality for the last line. If you wish to rewrite the final interval as "$2 < x < 4$," you may.

Using the interpretation of absolute values of a difference as the distance between the two numbers on the number line, you can see that $|3 - x|$ expresses the distance between 3 and x, which will be less than 1 if x is between 2 and 4.   ▬

EXAMPLE 18   Sue's employer offered to pay her either of two ways. On Plan A she would get $25 a day plus 4 percent of her sales. On Plan B she would get $10 a day plus 7 percent of her sales. When is Plan B the better option for her?

First, build a formula for her pay under Plan A, and a separate formula for her pay under Plan B. Let "x" denote the dollar value of her daily sales. Under Plan A, her daily pay is:

$$25 + .04x \text{ (dollars)}.$$

Under plan B her daily pay is:

$$10 + .07x \text{ (dollars)}.$$

Plan B is better for her when her pay under Plan B is greater:

$$10 + .07x > 25 + .4x.$$

To solve this, subtract 10 from both sides and subtract $.04x$ from both sides

$$.03x > 15.$$

Now divide by .03, which is a positive number.

$$x > \frac{15}{.03} = 500 \text{ (dollars)}.$$

Plan B is better if she sells more than $500 per day.

It is critical that the division was by a positive number (.03) so that the direction was preserved.   ▬

CONCLUSION   Some of the processes that work for solving equations do not necessarily work for solving inequalities.

1. Multiplying by a negative number **reverses** the direction of the inequality.
2. Multiplying by an expression is not correct unless you know in advance whether the expression is positive or negative. Otherwise you cannot tell the direction of the resulting inequality.
3. Squaring both sides is not justified.

To solve an inequality, guess-and-check will always work. However, it is not an algebraic method. The advice, "Put everything on one side," is usually appropriate. The Theorem on Zeros and Signs is helpful when the zeros and asymptotes can be found. If they are found algebraically, the theorem yields an algebraic solution, even if a graph is used to help see where the expression is positive or negative.

Terms: inequality, solution, interval notation, and, or, union, $\cup$, absolute values.

Exercise 4-6

A

A1.* To solve an inequality, you may multiply both sides by a _____ number and preserve the direction of the inequality. However, if the number you multiply by is _____, you must remember to switch the _____.

A2. ☺ Suppose you apply the following operations to both sides of an inequality and leave the direction of the inequality unchanged. Which operations always preserve the solutions?

(A) Add 2. (B) Subtract 2.
(C) Add −2. (D) Subtract −2.
(E) Multiply by 2. (F) Divide by 2.
(G) Multiply by −2. (H) Divide by −2.
(I) Square. (J) Add x.
(K) Multiply by x. (L) Divide by x.

A3. ☺ Suppose you apply the following operations to both sides of an inequality and leave the direction of the inequality unchanged. Which operations always preserve the solutions?

(A) Add $x + 1$ (B) Multiply by $x + 1$
(C) Square (D) Divide by $2x$.
(E) Cancel a common factor of x.
(F) Eliminate a common denominator of 5.

A4. ☺ Suppose you apply the following operations to both sides of an inequality and leave the direction of the inequality unchanged. Which operations always preserve the solutions?

(A) Divide by −3x. (B) Subtract −3x.
(C) Add 1 and then square.
(D) Cancel a common factor of $x + 2$
(E) Eliminate a common denominator of x.

Solve algebraically.

A5. $3x - 5 < 13$. **A6.** $-4x < 20$.

A7. $10 - 2x < 40$. **A8.** $5x < 8 - 2x$.

Solve algebraically.

A9. $|x - 9| < 1$. **A10.** $|x - 4| \le 3$.

A11. $|x + 2| > 1$. **A12.** $|2x - 1| < 7$.

A13. $|5 - x| < 1$. **A14.** $|20 - 4x| < 1$.

A15. $|x - 20| > .01$. **A16.** $.2 < |x - 2|$.

☺ *The given graph of f(x) is representative. Read the graph to solve the inequality.*

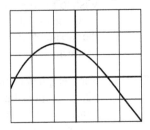 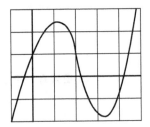

A17. $f(x) < 1$ **A18.** $f(x) > 2$

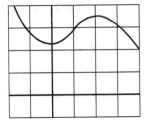

A19. $f(x) \le 3$

Find the natural domain of f if f(x) is as given.

A20. $\sqrt{5 - 2x}$ **A21.** $\sqrt{1 - \dfrac{3}{x}}$

A22. $\sqrt{16 - x^2}$ **A23.** $\sqrt{x^2 - 25}$

A24. $\sqrt{x(x + 5)}$ **A25.** $\sqrt{\dfrac{x - 1}{x + 1}}$

Graph the two expressions in the equation and shade the solutions.

A26. $|x - 5| < 1$ **A27.** $|x - 2| < .5$

Reading and Writing Mathematics

A28.* ☺ The sentence "$a < x < b$" abbreviates two inequalities connected by the word _____.

☺ *Describe in English the solution to*

A29.* "$|x - a| < d$." **A30.*** "$|x - a| > d$."

A31.* The solution to an equation is likely to be one or more points (numbers). The solution to an inequality is likely to be one or more _____.

A32.* Give a single inequality describing the points within d units of a.

A33.* Give a single inequality describing the points further than d units from a.

A34. Here is a "**sign pattern**" theorem: $ab > 0$ iff $[(a > 0 \text{ and } b > 0) \text{ or } (a < 0 \text{ and } b < 0)]$. Rephrase it in English beginning, "A product is positive if and only if . . ."

A35. (A) Which is easier to solve algebraically?
"$(x - 5)(x + 7) > 0$" or
"$(x - 5)(x + 7) > 2$"?
(B) Why?

☺ Decide if the two inequalities are equivalent (Answer "Yes" or "No").

A36. $x < 3$. $x^2 < 9$. **A37.** $x < 3$. $2x < 6$

A38. $x < 3$. $x^2 < 3x$. **A39.** $\dfrac{x}{3} < 2$. $x < 6$.

A40. $\dfrac{3}{x} < 2$. $3 < 2x$

A41. $2x < x + 3$. $x < 3$.

A42. $(x - 2)(x + 3) < 2x(x - 2)$. $x + 3 < 2x$.

A43. $3x + 2(x - 2) < 7 + 2(x - 2)$. $3x < 7$.

B

B1.* (A) To solve an inequality, why is it dangerous to multiply or divide both sides of an inequality by an expression with a variable?
(B) What can you do to avoid the danger?

B2. ☺ What is the difference in meaning between the sentences "$x < 2$ and $x > 3$" and "$x < 2$ or $x > 3$"?

Write in the form "$|x - c| < d$":

B3. $2.4 < x < 2.8$ **B4.** $6.8 < x < 7.0$

B5. $40 < x < 42$ **B6.** $-5 < x < -4$

Find the relevant endpoints algebraically and use the Theorem on Zeros and Signs to solve.

B7. $\dfrac{x + 5}{x - 4} > 0$. **B8.** $\dfrac{x - 6}{3 - x} \leq 0$.

B9. $\dfrac{x - 3}{x + 4} \leq 0$.

B10. $(x - 1)(x + 3)(x - 5) > 0$.

B11. $\dfrac{(x - 1)(x - 2)}{(x - 3)} \geq 0$.

B12. $\dfrac{x + 2}{(x - 1)(x - 5)} \leq 0$.

Find the relevant endpoints algebraically and use the Theorem on Zeros and Signs to solve.

B13. $\dfrac{3}{x} > 4$. **B14.** $\dfrac{5}{x} < 15$.

B15. $\dfrac{12}{x - 3} \leq 2$. **B16.** $\dfrac{1}{x - 1} < .2$.

Solve algebraically:

B17. $|1 - 3x| < 5$. **B18.** $|x^2 - 4| < 1$.

B19. $|9 - x^2| < 1$. **B20.** $|4 - 2x| < 1$.

Find the relevant endpoints algebraically and use the Theorem on Zeros and Signs to solve.

B21. $\dfrac{2}{x - 1} < 6$. **B22.** $\dfrac{10}{x + 3} < 2$.

B23. $\dfrac{x - 1}{x - 4} < 2$. **B24.** $\dfrac{2x + 1}{x - 1} < 1$.

Find the relevant endpoints algebraically and use the Theorem on Zeros and Signs to solve.

B25. $(x - 2.3)(x + 4)(2x + 3) \geq 0$.

B26. $\dfrac{(x^2 + 3)(x - 20)}{(x + 100)} \leq 0$. **B27.** $\dfrac{x^2}{x - 5} > -3$.

B28. $\dfrac{3}{x} > 2 - x$. **B29.** $\dfrac{x^2 - x - 7}{(x - 3)(x + 6)} > 0$.

B30. $\dfrac{x^2 + 2x - 13}{(x + 5)(x - 1)} > 0$. **B31.** $\dfrac{x}{x - 5} < 2$.

B32. $\dfrac{x^2}{x + 3} < 1$. **B33.** $\dfrac{1}{x - 1} \leq x$.

B34. $\dfrac{1}{x} + \dfrac{1}{x - 2} \leq 3$. **B35.** Solve $2 + x \geq \dfrac{2}{2 - x}$

B36. $1 + \dfrac{1}{x} \leq 5 + x$

B37.* True or False:
(A) "$x < a$" is equivalent to "$x^2 < a^2$."
(B) "$xf(x) < xg(x)$" is equivalent to "$f(x) < g(x)$."
(C) "$\dfrac{f(x)}{x} < \dfrac{g(x)}{x}$" is equivalent to "$f(x) < g(x)$."

(D) "$a < b$" is equivalent to "$ax < bx$."

(E) "$x > 0$" iff "$\frac{1}{x} > 0$."

(F) "$\frac{1}{x} > 2$" iff "$x < \frac{1}{2}$."

(G) "$\frac{1}{f(x)} > \frac{1}{g(x)}$" iff "$f(x) < g(x)$."

B38.* True or False?

(A) $|x - a| = |a - x|$ (B) $|x| = -|x|$

(C) $|x| = |-x|$ (D) $|x - a| = |x| - a$.

B39. True or false?

(A) $|b| = |-b|$ (B) $|kx| = k|x|$

(C) $|x + a| = |x| + |a|$.

B40. An employer gives you a choice of pay plans. Under Plan A you get $30 a day plus 4% of your sales. Under Plan B you gets $16 a day plus 6% of your sales. When is Plan B better for you?

B41. You are asked to advise your firm whether they should rent a photocopy machine on Plan A or Plan B. On Plan A they would pay $200 per month plus 3 cents per copy. On Plan B they would pay $300 per month but only 2.2 cents per copy. When is Plan B better for your firm?

B42. Each month, cell phone service with Company A costs $20, with 60 free minutes per month and 10 cents per minute after the first 60. Each month, cell phone service with Company B costs $30, with 100 free minutes per month and 8 cents per minute after the first 100. You are thinking of signing up for cell phone service with the cheaper company. How much use would make the service with Company A cheaper?

B43. A student "solved" the inequality "$\frac{6}{x} > 5$" by multiplying through by x to obtain "$6 > 5x$" and then "$\frac{6}{5} > x$." What went wrong and why?

Reading and Writing Mathematics

B44.* Equations and inequalities are different. What are some important differences in their solution processes?

B45. The solution to "$x^2 = 4$" is not "$x = 2$ and $x = -2$".

(A) Why not?

(B) The solution is "$x = 2$ or $x = -2$."

(C) What is the difference between these two similar sentences?

B46.* What is the difference between the uses of "or" and "and" to connect inequalities such as "$x < a$" and "$x > b$"?

B47.* To solve the inequality "$\frac{f(x)}{g(x)} > h(x)$" a student might multiply both sides by $g(x)$ to obtain "$f(x) > h(x)g(x)$." What can go wrong?

B48.* To solve the inequality "$f(x)g(x) > f(x)h(x)$" a student might divide both sides by $f(x)$ to obtain "$g(x) > h(x)$." What can go wrong?

B49. (A) State a "sign pattern" theorem for solving "$\frac{a}{b} > 0$" that resembles the theorem in Problem A34.

(B) Use it to solve "$\frac{x}{x - 2} > 0$."

B50. Suppose a student says the solution to "$x^2 > 9$" is "$x > 3$ and $x < -3$." The use of "and" is wrong.

(A) Why?

(B) What word should be used in place of "and"?

B51. (A) Solve the inequality $x^2 < 50$.

(B) Write, symbolically, the theorem which expresses how to solve all similar problems. (Your theorem should include other similar powers, as well as numbers other than 50.)

B52. (A) Solve the inequality $x^2 < 50$.

(B) Write, symbolically, the theorem which expresses how to solve all similar problems. (You theorem should include other similar powers, as well as numbers other than 50.)

B53. Inequalities can be solved in the form "$f(x) < g(x)$" by guess-and-check. Why bother, even when using guess-and-check, to "put everything on one side"?

Here is a theorem: $|a| < |b|$ iff $a^2 < b^2$. Use it to solve these algebraically.

B54. $|x| < |x - 3|$.   **B55.** $|x - 2| < |2x|$.

B56. Prove "$a < b$" is not equivalent to $a^2 < b^2$."

(A) What does this tell you about solving an inequality by squaring both sides?

(B) Give an illuminating example.

B57. Use symbolic notation and inequalities with absolute values to restate: "If x is within d units of a, then $f(x)$ is within c units of $f(a)$."

B58. A student "solved" the inequality
"$x(x - 4) < 3(x - 4)$" by dividing through by
"$x - 4$" to obtain "$x - 3$." What went wrong
and why?

B59. State a theorem which allows us to do the first step
in solving $(2x - 5)^2 \leq 17$. [No proof or solution
required.]

B60. State a theorem which allows us to do the first step
in solving $(2x - 5)^2 \leq 17$. [No proof or solution
required.]

B61. (A) Solve algebraically: $x^5 > 20$.
　　　(B) State a theorem which justifies your step
　　　　　[We have stated no theorems about powers and
　　　　　inequalities, but you can guess one].

B62. Suppose you want to solve "$x^2 > 4$". What is wrong
with the answer "$2 < x < -2$"?

B63. (A) Solve $x - 1 < \sqrt{x + 11}$ by graphing both
　　　　　sides and reading the graph.
　　　(B) Take that same inequality, square both sides,
　　　　　simplify, and solve the resulting quadratic.
　　　(C) The two methods do not yield the same result.
　　　　　Which one yields the right result? What went
　　　　　wrong with the other one? How could you have
　　　　　noticed when the wrong one went wrong?

5 Exponential and Logarithmic Functions

Section 5-1 EXPONENTS AND LOGARITHMS

This section introduces exponential functions and their inverses, logarithmic functions. Exponential functions are closely related to power functions because both have the form "b^p".

x^2 and x^3 are "power" functions.

10^x and e^x and $\left(\frac{1}{2}\right)^x$ are "exponential" functions.

EXAMPLE 1 Here are some basic facts about the exponential function with base 10.

$10^1 = 10$

$10^2 = 100$

$10^3 = 1,000$ (one thousand). 10^3 has three 0's to the left of the decimal point.

$10^6 = 1,000,000$ (one million). Read "10^6" as "ten to the sixth," which is short for "ten to the sixth power." 10^6 has six 0's to the left of the decimal point.

$10^9 = 1,000,000,000$ (one billion), and

$10^{12} = 1,000,000,000,000$ (one trillion).

$10^0 = 1$

$$10^{-1} = \frac{1}{10} = .1$$

$$10^{-2} = \frac{1}{100} = .01, \text{ and}$$

$10^{-3} = \frac{1}{1000} = .001.$ Read "10^{-3}" as "ten to the negative 3." Note that 10^{-3} has only *two* 0's (not three) to the right of the decimal point.

10^9 is huge. 10^{-3} is tiny. We cannot draw a pleasing graph of 10^x because no uniform vertical scale can handle the rapid changes in y-values. In a standard window two inches high (Figure 1), the image of $x = 10$ at the right edge would be 10^{10}, which would be 16,000 miles above the top of the window! Expanding the vertical interval helps, but not much. Figure 2, with window $[-4, 4]$ by $[0, 1000]$ is legible only for x-values between 1 and 3.

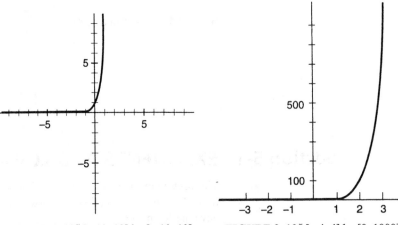

FIGURE 1 10^x $[-10, 10]$ by $[-10, 10]$ **FIGURE 2** 10^x $[-4, 4]$ by $[0, 1000]$

The images are never negative. The range of the exponential function is all positive real numbers. For example, 10^{-3} is not negative; it is positive and small.

We can rewrite all the power function facts of Section 4-1 to emphasize exponential functions.

EXAMPLE 2 $10^4 = 10,000.$

$x^4 = 10,000$ when $x = 10$ (A power function fact).

$10^x = 10,000$ when $x = 4$ (An exponential function fact).

$10^3 10^2 = 10^5$

$x^3 x^2 = x^5$ (A power function fact).

$10^x 10^y = 10^{x+y}$ (An exponential function fact).

■ Properties of Exponential Functions

All the important properties of exponential functions hold regardless of the base.

CALCULATOR EXERCISE 1   Look at your calculator's graphs of 2^x, e^x, and 10^x. Except for the horizontal scale, they are exactly the same. Even the graph of 1.1^x, which is much wider, differs only in its horizontal scale (Problem B44).

Here are the power function facts rewritten as exponential function facts. Only the position of "x" has changed.

	Facts about exponential functions Table of 5-1-1 through 5-1-3			
	"x" Emphasizing Power Functions	"x" Emphasizing Exponential Functions		
	For $x > 0$	base 10	base b, for $b > 0$	
(4-1-1)	$x^p x^r = x^{p+r}$	$10^x 10^y = 10^{x+y}$	$b^x b^y = b^{x+y}$	(5-1-1)
(4-1-2)	$\dfrac{x^p}{x^r} = x^{p-r}$	$\dfrac{10^x}{10^y} = 10^{x-y}$	$\dfrac{b^x}{b^y} = b^{x-y}$	(5-1-2)
(4-1-2B)	$x^0 = 1$	$10^0 = 1$	$b^0 = 1$	(5-1-2B)
(4-1-2C)	$x^{-p} = \dfrac{1}{x^p} = \left(\dfrac{1}{x}\right)^p$	$10^{-x} = \dfrac{1}{10^x} = \left(\dfrac{1}{10}\right)^x$	$b^{-x} = \dfrac{1}{b^x} = \left(\dfrac{1}{b}\right)^x$	(5-1-2C)
(4-1-3)	$(x^r)^p = x^{rp}$	$(10^x)^y = 10^{xy}$	$(b^x)^y = b^{xy}$	(5-1-3)

> The base, b, of an exponential function is always a positive number. Regardless of the choice of $b > 0$,
> The domain of b^x is all real numbers
> The range of b^x is all positive numbers.
> If $b > 1$, the graph is increasing and resembles the graphs for bases 10, e, and 2 (Figures 1, 2, 3, and 4).
> If $b < 1$, the graph is decreasing and resembles the graph for $\left(\dfrac{1}{2}\right)^x$ (Figure 5).

EXAMPLE 3 $10^x (10^3) = 10^{x+3}$ [5-1-1]

$\dfrac{10^x}{10^2} = 10^{x-2}$ [5-1-2]

$\dfrac{1}{2^x} = \left(\dfrac{1}{2}\right)^x$ [5-1-2C]

$(10^2)^x = 10^{2x}$ [5-1-3]

■ Logarithms

Logarithmic functions are defined as inverses of exponential functions. By definition,

Logs are inverses of exponentials.

Therefore, exponential and logarithmic function facts come in pairs. See this in the next example where the first fact is in "exponential form" and the second is the same fact in "logarithmic form."

E X A M P L E   4   Inspect these facts closely to see that **"logs are exponents."**

"exponential form"	"logarithmic form"	
$10^1 = 10$	$\log 10 = 1$	
$10^2 = 100$	$\log 100 = 2$	
$10^3 = 1000$	$\log 1000 = 3$	log of one thousand equals three.
$10^6 = 1,000,000$	$\log 1,000,000 = 6$	log of one million equals six.
$10^{-1} = \dfrac{1}{10}$	$\log\left(\dfrac{1}{10}\right) = -1$	log of one tenth equals negative one.
$\sqrt{10} = 10^{1/2} = 3.162$	$\log 3.162 = \dfrac{1}{2}$	
$10^0 = 1$	$\log 1 = 0$	log of one is zero.
$10^x = c$ iff	$\log c = x$	(This is the definition of logs, next.)

> Logarithms with base 10 are sometimes called "common" logarithms. "Natural" logarithms are base $e = 2.718. \ldots$. Read "e^x" as "e to the x." Read "ln y" as "the natural log of y" or just "log y." In higher math, "ln" is usually pronounced the same way as "log". If you are reading them, the difference is clear. If you are hearing them, it is not.

DEFINITION 5-1-4

In general,

$10^x = y$ iff $\log y = x$.
$e^x = y$ iff $\ln y = x$.
$b^x = y$ iff $\log_b y = x$.

Logs are exponents.

DEFINITION 5-1-4B (INVERSE FUNCTION VERSION)

For all x, $\log(10^x) = x$.

For all $x > 0$, $10^{\log x} = x$.

For all x, $\ln(e^x) = x$.

For all $x > 0$, $e^{\ln x} = x$.

E X A M P L E   5   Solve $10^x = 1000$.

$10^x = 1000$ iff $x = \log 1000$ [This is immediate from 5-1-4.]
$= 3$.

Another way is to "Take the log of both sides."

$$10^x = 1000.$$

$$\log(10^x) = \log(1000).$$
$$x = \log(1000) \qquad \text{[by 5-1-4B, } \log(10^x) = x.\text{]}$$
$$x = 3.$$

Solve $10^x = 50$.

$$x = \log 50 \qquad \text{[from the definition, or taking log of both sides.]}$$
$$= 1.699 \qquad \text{[from a calculator.]}$$

The ".699" part of the answer is not obvious, but it is easy to see that log 50 is somewhere between 1 and 2 because 50 is between $10 = 10^1$ and $100 = 10^2$. ▬

CALCULATOR EXERCISE 2   Use your calculator to illuminate the inverse relationship between exponentials and logs. Try these pairs.

$10^{3.4} = 2511.886$. Take the log of that: $\log 2511.886 = 3.4$ [returns the original].

$\log 55 = 1.74036$. Take 10 to that power: $10^{1.74036} = 55$ [back to the original].

Next is a version of Definition 5-1-4 with the placeholders "x" and "y" switched to emphasize equations in which *log* is applied to the unknown.

DEFINITION 5-1-4C $\log x = y$ iff $x = 10^y$.

EXAMPLE 6   Solve $\log x = 5$.   [Now use 5-1-4C or "Take 10 to both sides."]

$$x = 10^5 \; [10^{\log x} = x.]$$
$$= 100,000.$$

Solve $\log x = 3.7$.   [Now use 5-1-4C, or take 10 to both sides.]

$$x = 10^{3.7} = 5011.8.$$

Obtaining this exact value requires a calculator, but the order of magnitude of the solution is clear from properties of powers. Because 3.7 is between 3 and 4, the answer is between $10^3 = 1000$ and $10^4 = 10,000$.

Solve $\log x = -2.3$.

$$x = 10^{-2.3} = .00501.$$

Because -2.3 is between -2 and -3, the answer is between $10^{-2} = .01$ and $10^{-3} = .001$. ▬

Graphs

Any inverse can be graphed by using the fact that (b, a) is on a graph of the inverse if and only if (a, b) is on the graph of the function (Section 2-3). The graph of log is the reflection of the graph of 10^x through the line $y = x$ (Figure 3). Both are increasing functions. 10^x grows extremely rapidly; $\log x$ grows extremely slowly.

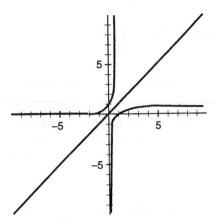

FIGURE 3 10^x, $\log x$ and $y = x$.
$[-10, 10]$ by $[-10, 10]$

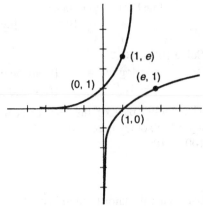

FIGURE 4 e^x and $\ln x$ $[-5, 5]$ by $[-5, 5]$

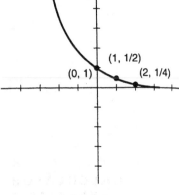

FIGURE 5 $\left(\dfrac{1}{2}\right)^x$ $[-5, 5]$ by $[-5, 5]$

CALCULATOR EXERCISE 3 Try to evaluate $\log 0$ and $\log(-2)$. Your calculator will not give a real number. Then use your calculator to graph "$\log x$" and "$\ln x$". Are the graphs similar? Are there points to the left of the y-axis? Are there points below the x-axis? This tells you about the domain and range of "log" (Problem A36).

■ Solving Equations

Logarithms are useful for solving for unknown exponents. The key properties are in the next theorems (All assume $c > 0$ and $d > 0$).

THEOREM 5-1-0 log is one-to-one and 10^x is one-to-one, so we may take log of both sides or exponentiate both sides of an equation and preserve its solutions.

THEOREM 5-1-1L (LOG VERSION) $\log(cd) = \log c + \log d$.

"The log of a product is the sum of the logs."

EXAMPLE 7 $\log(7x) = \log 7 + \log x$.
$\log(x(1 - x)) = \log x + \log(1 - x)$.

THEOREM 5-1-2L (LOG VERSION) $\log\left(\dfrac{c}{d}\right) = \log c - \log d$

"The log of a quotient is the difference of the logs."

EXAMPLE 8 $\log\left(\dfrac{x}{3}\right) = \log x - \log 3$.

$\log\left(\dfrac{1}{x}\right) = \log 1 - \log x = -\log x$.

$\log\left(\dfrac{x}{1 - x}\right) = \log x - \log(1 - x)$.

THEOREM 5-1-3L $\log(c^x) = x \log c.$ $\log(x^p) = p \log x.$
(LOG VERSION)

"The log of an exponential is the exponent times the log of the base."

EXAMPLE 9 $\log 10^3 = 3 \log 10$ $[= 3(1) = 3]$

$\log(7^{2.4}) = 2.4 \log 7.$

$\log(1.08^x) = x \log 1.08.$

$\log(x^3) = 3 \log x.$

$\log(cb^x) = \log c + x \log b$ [The log of a product is the sum of the logs]   ▬

EXAMPLE 10   Solve $2^x = 20.$

Take logs. $\log(2^x) = \log 20$ [log is one-to-one]

$x \log 2 = \log 20$ [5-1-3L, log of an exponential]

$x = \dfrac{\log 20}{\log 2}$

It does not matter which base you use. Try common (base 10) logarithms:

$$\frac{\log 20}{\log 2} = \frac{1.301}{.301} = 4.322.$$

Try natural (base e) logarithms:

$$\frac{\ln 20}{\ln 2} = \frac{2.996}{.693} = 4.322. \text{ The result is the same either way.}$$

Theoretically, we could do this problem by taking logs base 2 ($x = \log_2 20$), but logs base 2 are not available on most calculators. Work with base 10 or base e.   ▬

EXAMPLE 11   Solve $1000(1.006)^x = 1200.$

By inverse-reverse thinking, divide by 1000 first.

$$1.006^x = \frac{1200}{1000} = 1.2$$

Taking logs of both sides, $\log(1.006^x) = \log 1.2$

$x \log 1.006 = \log 1.2$ [log of an exponential, 5-1-3L]

$$x = \frac{\log 1.2}{\log 1.006} = 30.48.$$

Taking logs right away: $\log 1000 + x \log 1.006 = \log 1200.$ Solving for x (using inverse-reverse) yields the same result.   ▬

EXAMPLE 12   Solve for r: $2 = (1 + r)^{4.5}.$

The unknown is not in an exponent. Don't take logs. Use inverse-reverse and undo the 4.5 power:

$2^{1/4.5} = 1 + r.$ Therefore, $r = 2^{1/4.5} - 1.$ In decimal form, $r = .1665.$

Logs are useful for solving for unknown exponents. In this example, the exponent is not unknown. Use roots. ▬

Logs are useful for solving for unknown exponents.

EXAMPLE 13 Solve for d: $7\left(2^{3/d}\right) = 35$.

The unknown is in the exponent. You want to take logs, but may divide first.

$$2^{3/d} = \frac{35}{7} = 5.$$

$$\log\left(2^{3/d}\right) = \log 5$$

$$\left(\frac{3}{d}\right)\log 2 = \log 5 \quad \text{[log of an exponential, 5-1-3L]}$$

$$\frac{3 \log 2}{\log 5} = d \quad \text{[isolating } d\text{]}$$

$$d = 1.29.$$

EXAMPLE 13 ANOTHER WAY Taking logs right away: $\log 7 + \left(\dfrac{3}{d}\right)\log 2 = \log 35$.

Then solving for d with inverse-reverse yields the same result. ▬

EXAMPLE 14 Solve $1.06^x = 0.7(1.09)^x$.

Because the unknown is an exponent, take logs.

$\log(1.06^x) = \log(0.7(1.09)^x)$.　　　[The log of the product on the right is a sum.]

$x \log 1.06 = \log 0.7 + \log(1.09)^x$ [using log of an exponential on the left side and log of a product is a sum, on the right]

$x \log 1.06 = \log 0.7 + x \log 1.09$ [using log of an exponential]

$x(\log 1.06 - \log 1.09) = \log 0.7$ [putting all the terms with x on one side]

$$x = \frac{\log 0.7}{\log 1.06 - \log 1.09} = 12.78$$

EXAMPLE 14 ANOTHER WAY Solve $1.06^x = 0.7(1.09)^x$.

Divide both sides by 1.09^x,

$$\frac{1.06^x}{1.09^x} = 0.7 \text{ and then } \left(\frac{1.06}{1.09}\right)^x = 0.7, \text{ using 4-1-4B.}$$

Taking logs,

$$x \log\left(\frac{1.06}{1.09}\right) = \log 0.7 \quad x = \frac{\log 0.7}{\log\left(\dfrac{1.06}{1.09}\right)} = 12.78. \quad ▬$$

EXAMPLE 15   Solve $30 \left(2^{x/5}\right) = 2^{x/4}$.

$$\log 30 + \left(\frac{x}{5}\right)\log 2 = \left(\frac{x}{4}\right)\log 2.$$ [Using both log of a product and log of an exponential on the left, and log of an exponential on the right]

Now, consolidate like terms. Put the terms with "x" together.

$$\log 30 = \left(\frac{x}{4}\right)\log 2 - \left(\frac{x}{5}\right)\log 2$$

$$\log 30 = x\left[\frac{\log 2}{4} - \frac{\log 2}{5}\right]$$

$$x = \frac{\log 30}{\left[\dfrac{\log 2}{4} - \dfrac{\log 2}{5}\right]} = 98.138.$$

EXAMPLE 16   How many years does it take for $10,000 invested at 11% compounded quarterly to become $30,000?

The formula is for compounding money is Formula 4-4-6: $A = P\left(1 + \dfrac{r}{k}\right)^{kt}$, where, in this problem, $A = 30,000, P = 10,000, r = .11, k = 4$, and t is unknown. So the problem is to solve for t in:

$$30,000 = 10,000\left(1 + \frac{.11}{4}\right)^{4t}$$

First divide by 10,000 to obtain

$$3 = \left(1 + \frac{.11}{4}\right)^{4t}$$

Now take logs:

$$\log 3 = 4t \log\left(1 + \frac{.11}{4}\right)$$

$$t = 10.12 \text{ (years)}$$

■ Deriving the Properties of Logarithms

Exponential facts and logarithmic facts come in pairs. We can reformulate exponential facts as log facts using the definition.

(5-1-4, again) $10^x = y$ iff $x = \log y$ **Logs are exponents.**

(Here is 5-1-1, again) $(10^p)(10^r) = 10^{p+r}$.

You can see the exponents, so you can see the logs. Treat the left side as a product, cd. Its log is visible as the exponent $p + r$ on the right. But this is simply the sum of the logs of the factors c and d. Therefore,

(5-1-1L, log version) $\log(cd) = \log c + \log d$.

Here are the details of that proof. Let $c = 10^p$ and $d = 10^r$. Then

$$(10^p)(10^r) = 10^{p+r} \qquad \text{[5-1-1, again]}$$

$$= c \quad d \qquad \text{[our new notation].}$$

Taking logs,

$$\log(c\,d) = \log(10^{p+r}) = p + r = \log c + \log d.$$

Similarly, the log of a quotient can be rewritten. Recall

(5-1-2, again) $\qquad \dfrac{10^p}{10^r} = 10^{p-r}.$

The left side is a quotient $\left(\dfrac{c}{d}, \text{ where } c = 10^p \text{ and } d = 10^r\right)$, and its log is visible on the right side:

$$\log(10^{p-r}) = p - r = \log c - \log d.$$

This is the log of the numerator, c, minus the log of the denominator, d.

(5-1-2L, log version) $\qquad \log\left(\dfrac{c}{d}\right) = \log c - \log d.$

Property 5-1-2B says $10^0 = 1$. In logarithmic form, this is

(5-1-2BL, log version) $\qquad \log 1 = 0.$

(5-1-2CL, log version) $\qquad \log\left(\dfrac{1}{c}\right) = -\log c.$

"The log of a reciprocal is the negative of the log."

EXAMPLE 17 $\quad \log\left(\dfrac{1}{2}\right) = -\log 2 = -.301.$ ▬

(Here is 5-1-3, again) $\quad (b^p)^x = b^{px}$ We use it to rewrite logs of powers.

$$(10^r)^p = 10^{rp}. \qquad \text{[5-1-3 with different letters]}$$

$$\log\left((10^r)^p\right) = \log(10^{rp})$$

$$= rp \qquad \text{[logs are exponents]}$$

$$= p \log(10^r) \qquad [r = \log(10^r)]$$

Now let $c = 10^r$.

$$\log(c^p) = p \log c.$$

(5-1-3L, log version) $\qquad \log(c^p) = p \log c \quad \text{or} \quad \log(c^x) = x \log c.$

EXAMPLE 18 Use properties of logarithms to rewrite: $\log[20x^5(1-x)^7]$.
[Without logs of powers, logs of products, or logs of quotients.]

The order of operations is critical. As usual, work with the last operation first. Inside the log expression, multiplication is last, so use the log of a product from 5-1-1L.

$$\log[20x^5(1 - x)^7] = \log 20 + \log(x^5) + \log[(1 - x)^7]$$
$$= \log 20 + 5 \log x + 7 \log(1 - x), \text{ by 5-1-3L.}$$

There is no simplification of the logarithm of a difference, so this is the final expression.

EXAMPLE 19   Solve $\log x + \log(5x) = 3$.

$\log x + \log 5 + \log x = 3$ [log of a product is the sum of the logs]

$2 \log x = 3 - \log 5$. $\log x = \dfrac{(3 - \log 5)}{2} = 1.1505$. $x = 10^{1.1505} = 14.1$.

EXAMPLE 19
ANOTHER WAY

Solve $\log x + \log(5x) = 3$.

Use 5-1-1L to get: $\log x + \log(5x) = \log(5x^2) = 3$.

Now take 10 to both sides: $5x^2 = 10^3 = 1000$.

$x^2 = \pm\sqrt{200} = \pm14.1$. The negative solution is extraneous. $x = 14.1$.

Pairs of exponential and logarithmic facts Table 5-1-6 (of 5-1-1 through 5-1-4)			
	Exponential form (base 10)	**Logarithmic form (assume x, c, and d are positive)**	
(5-1-4)	$10^x = y$	$\log y = x$	(5-1-4)
(5-1-4inv)	$10^{\log x} = x$	$\log(10^x) = x$	(5-1-4inv)
(5-1-1)	$10^x 10^y = 10^{x+y}$	$\log(cd) = \log c + \log d$	(5-1-1L)
(5-1-2)	$\dfrac{10^x}{10^y} = 10^{x-y}$	$\log\left(\dfrac{c}{d}\right) = \log c - \log d$	(5-1-2L)
(5-1-2B)	$10^0 = 1$	$\log 1 = 0$	(5-1-2BL)
(5-1-2C)	$10^{-x} = \left(\dfrac{1}{10^x}\right)$	$\log\left(\dfrac{1}{c}\right) = -\log c$	(5-1-2CL)
(5-1-3)	$(10^p)^x = 10^{px}$	$\log(c^x) = x \log c$	(5-1-3L)
10^x has domain all real numbers		log has range all real numbers	
10^x has range all positive numbers		log has domain all positive numbers	

The properties in this table also hold for base e and any other positive base.

■ Scientific Notation

Base 10 is used for "scientific" notation for numbers—especially very large or very small numbers where the number of zeros to the left or right of the decimal point is unwieldy. The idea is to rewrite a number as a number between 1 and 10, times 10 to the appropriate power.

EXAMPLE 20

$$123 = 1.23 \times 10^2.$$

$$1,200,000 = 1.2 \times 10^6.$$

$$.00000005 = 5 \times 10^{-8}.$$

In chemistry, Avogadro's number (the number of atoms in a gram-molecule) $= 6.023 \times 10^{23}$. No one wants to write, or read, a 24-digit number.

In astronomy, a "light year" is a unit of length equivalent to approximately 5.88×10^{12} miles, which is approximately 9.47×10^{12} kilometers. ▬

CONCLUSION Power functions and exponential functions are closely related. Properties of power functions can be rewritten as properties of exponential functions just by changing letters. Logarithmic functions are inverses of exponential functions. **Logs are exponents**. Definition 5-1-4 gives the correspondence of exponential and logarithmic forms, and properties of these functions come in pairs (Table 5-1-6). For solving equations, the most important property of logs is 5-1-3L: $\log(c^x) = x \log c$. It is useful for solving for unknown exponents.

Terms: base, power, exponent, scientific notation, exponential function, logarithmic function, exponential form, logarithmic form.

Exercise 5-1

A

A1.* ☺ In a complete sentence with exactly three words, what are logs?

A2.* ☺ Properties of exponential functions are just properties of _____ functions with the letters switched.

A3.* ☺ Write this in "logarithmic form": $b^a = c$.

A4.* ☺ There are only two commonly-used types of logarithms. Which two?

A5.* ☺ The log of a product is the _____ of the logs.

A6.* ☺ The log of a quotient is the _____ of the logs.

A7.* ☺ Let n be a positive integer.
  (A) 10^n has _____ zeros to the left of the decimal point.
  (B) 10^{-n} has _____ zeros to the right of the decimal point.

A8. ☺ Say if each is a property of logs. If not, fix the right side, if possible. If there is no other expression equivalent to the left side, say so. (Assume all arguments are positive.)

(A) $\log\left(\dfrac{x}{4}\right) = \dfrac{\log x}{4}$.

(B) $\log\left(\dfrac{x}{4}\right) = \dfrac{\log x}{\log 4}$.

(C) $\log(x - 2) = \log x - \log 2$.

(D) $\log(\sqrt{x}) = \sqrt{(\log x)}$.

(E) $\log(x^2) = (\log x)^2$.

A9. ☺ Say if each is a property of logs. If not, fix the right side, if possible. If there is no other expression equivalent to the left side, say so. (Assume all arguments are positive.)

(A) $\log(x - 5) = \log x - \log 5$.

(B) $\log\left(\dfrac{x}{7}\right) = \log x - \log 7$.

(C) $\log(x + 2) = \log x + \log 2$.

(D) $\log(5x^2) = 2\log(5x)$.

(E) $\log\left(\dfrac{3}{x}\right) = \dfrac{\log 3}{\log x}$.

(F) $\log(2x) = 2\log x$.

A10.* ☺ True or false? There are properties of logarithms for expressing logs of
(A) products   (B) sums   (C) quotients
(D) differences   (E) powers.

A11.* ☺ True or false? There are properties of logarithms for rewriting
(A) differences of logs   (B) products of logs
(C) sums of logs   (D) quotients of logs.

A12. ☺ Write in logarithmic form:
(A) $10^4 = 10{,}000$.   (B) $10^{-2} = .01$.

A13. ☺ Write in logarithmic form:
(A) $10^{2.5} = 316.2$.   (B) $10^{.3} = 1.995$.

A14. ☺ Write in exponential form:
(A) $\log 100 = 2$.   (B) $\log(.1) = -1$.

A15. ☺ Write in exponential form:
(A) $\log 50 = 1.699$.   (B) $6 = \log 1{,}000{,}000$.

A16. ☺ Write in exponential form:
(A) $\log 150 = 2.176$.   (B) $\log(.015) = -1.824$.

Solve

A17. $\left(1 + \dfrac{r}{2}\right)^8 = 1.35$, for $r > 0$ [.076]

A18. $(1 + r)^{20} = 5$, for $r > 0$ [.084]

A19. $(1.005)^t = 2$ [140]

A20. $(1.03)^t = 1.5$ [14]

A21. $1000(1.005)^{12t} = 1500$ [6.8]

A22. $500(1.03)^{2t} = 1000$ [12]

A23. $\log(2x) = 1.3$

A24. $\log(x + 2) = .43$

A25. $2 \log(3x) = 5$

A26. $\log\left(\dfrac{x}{5}\right) = 2.5$ [1600]

A27. $10^{x+1} = 1000$ [2.0]

A28. $10^{x/3} = 30$ [4.4]

Use properties of logs to rewrite the expression as an equivalent expression without logs of products, logs of quotients, or logs of powers. (Assume all arguments are positive.)

A29. ☺ $\log(3x)$

A30. ☺ $\log(1.05^x)$

A31. ☺ $\log(x^5)$

A32. $\log[x(2x + 1)^2]$

A33. $\log\left[\dfrac{x}{(5x + 1)}\right]$

A34. $\log\left(\dfrac{7}{x^2}\right)$

A35. ☺ Answer these without using a calculator.
(A) The common name for 10^6 is _____.
(B) $\log 1000 = $ _____.
(C) $\log 10 = $ _____.
(D) $\log 0.001 = $ _____.
(E) $10^3 10^6 = $ _____.
(F) If $10^x = 1$, $x = $ _____.

A36. Do Calculator Exercise 3. That is, use your calculator to graph "$\log x$" and "$\ln x$".
(A) In what way are the graphs similar?
(B) Are there points to the left of the y-axis?
(C) Are there points below the x-axis?
(D) Try to evaluate $\log 0$ and $\log(-2)$. What does all this tell you about the domain and range of "log"?

☺ *Write out the pronunciation of these.*

A37. 10^5 **A38.** e^x **A39.** $\log x$

A40. $10^x 10^y = 10^{x+y}$ **A41.** $\log_2 x$ **A42.** $\ln x$

A43. $\log_2(x^5)$ **A44.** $\log(x^7)$ **A45.** $(\log x)^7$

A46. $\log\left(\dfrac{x}{2}\right)$ **A47.** $\dfrac{\log x}{2}$ **A48.** $\log_2 x$

☺ *Write these numbers in scientific notation.*

A49. 176   **A50.** 8764

A51. .00345   **A52.** .00754

A53. 2,000,000   **A54.** 23,000,000

A55. .0000000568   **A56.** .000000323

B

B1.* ☺ Here are three power-function facts. Give the associated exponential-function facts.
(A) $x^p x^r = x^{p+r}$ (B) $\dfrac{x^p}{x^r} = x^{p-r}$ (C) $(x^r)^p = x^{rp}$

B2.* ☺ Here are three exponential-function facts. Give the associated logarithmic-function facts.
(A) $10^x 10^y = 10^{x+y}$ (B) $\dfrac{10^x}{10^y} = 10^{x-y}$
(C) $(10^p)^x = 10^{px}$

Solve algebraically for a positive solution:

B3. $1600 = 1000\left(2^{t/12}\right)$ [8.1]

B4. $700 = 1000\left(\dfrac{1}{2}\right)^{40/d}$ [78]

B5. $(1 + r)^{20} = 14$ [.14]

B6. $\left(1 + \dfrac{r}{2}\right)^{8} = 1.56$ [.11]

B7. $1.04^{t} = 1.3(1.03^{t})$ [27]

B8. $1000\left(2^{t/5}\right) = 2^{t/4}$ [200]

B9. $45\left(2^{t/30}\right) = 12\left(2^{t/20}\right)$ [110]

B10. $5000\left(1 + \dfrac{.08}{4}\right)^{4t} = 4500\left(1 + \dfrac{.10}{12}\right)^{12t}$ [5.2]

B11. $\left(\dfrac{1}{2}\right)^{5/h} = .3$ [2.9]

B12. $(1.03)^{t} = .01(1.07)^{t}$ [120]

B13. $.8(1.10)^{t} = (1.085)^{t}$ [16]

B14. $2^{t/30} = 10\left(2^{t/40}\right)$ [400]

B15. $\dfrac{1.06^{x}}{1.04^{x}} = 5$ [84]

B16. $\dfrac{2^{x}}{3^{x}} = .5$ [1.7]

B17. $\dfrac{2^{x}}{3^{2x}} = .01$ [3.1]

B18. $10^{x} = 2^{x+1}$ [.43]

B19. $\log(x^{2} - 5) = 2$ [10]

B20. $\log\left(\dfrac{x}{5}\right) = 2.5$ [1600]

B21. $\log\left[\dfrac{x}{(1 - x)}\right] = -1$ [.091]

B22. $\log(3^{x}) = 2$ [4.2]

Use properties of logs to rewrite the expression as an equivalent expression without logs of products, logs of quotients, or logs of powers. (Assume all arguments are positive.)

B23. $\log[x^{3}(1 - x)^{6}]$

B24. $\log\left[\dfrac{x^{2}}{(1 - x)^{3}}\right]$

B25. $\log[20x^{-5}(10 - x^{2})]$

B26. $\log\left[\dfrac{(1 + x)}{(2 + x)^{3}}\right]$

Solve algebraically:

B27. $\log(x^{5}) + \log(x^{2}) = 3$ [2.7]

B28. $\log x + \log(3x) = 2$ [5.8]

B29. Find the approximate doubling time of money invested at 7% compounded monthly. [9.9 years]

B30. Find the approximate doubling time of a population that grows 2% per year. [35]

B31. Investment Plan A returns 10% compounded monthly. Investment Plan B returns 12% compounded monthly, but it costs 5% in broker's fees (so your principal is only 95% as much). After how many years will Plan B be better than Plan A? [2.6]

B32. Investment Plan A returns 8% compounded quarterly. Investment Plan B returns 9.2% compounded monthly, but there is a 10% penalty for taking the money out before you retire. Suppose you will want your money before you retire. How long do you need to wait before the amount under Plan B is better, even after the 10% penalty? [8.5]

B33. Population A is growing at 3% per year (compounded yearly). Population B starts out 40% larger, but is only growing at 1% per year (compounded yearly). If these rates remain the same, in how many years will the Population A equal Population B? [17]

Reading and Writing Mathematics

B34. Determine and state a theorem for solving: $Ac^{x} = B$ (state it using logs base 10, and assume A, B, and c are greater than 0).

B35. State, symbolically, a pattern which relates $10^{.4}$ to $10^{1.4}$, and $10^{4.67}$ to $10^{5.67}$, etc.

B36. State, symbolically, a pattern which relates log 2 to log 20, log 200, etc.

B37.* (A) Write in exponential form: $\log a = b$.

(B) Write in logarithmic form: $10^{c} = d$.

B38. State the identity which expresses how to evaluate b^{p} in terms of the natural exponential function.

B39. Example 11 fits the pattern: $b^{x} = c$. State a theorem that gives the solution to this.

B40. Look at a calculator's graph of b^{-x} for some $b > 1$.

How does the graph of b^{-x} compare to the graph of b^{x}?

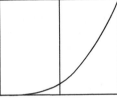

The figure to the right is the graph of the given exponential function in the window $[-c, c]$ by $[0, 10]$. It exits the window at the upper right corner. Find c.

B41. 10^{x} **B42.** 2^{x} **B43.** e^{x} **B44.** 1.1^{x}

Section 5-2   BASE 2 AND BASE e

All exponential functions have much in common. For example, both base 2 and base e work for modeling exponential growth.

■ Base 2

The exponential function with base 2 is given by $f(x) = 2^x$. It describes "exponential growth" in which the amount doubles every unit of time (where "x" is time, Figure 1).

Whenever x increases by 1, $f(x)$ doubles:

$$x = 0\ 1\ 2\ 3\ 4\ 5\ 6\ 7\ldots$$
$$f(x) = 1\ 2\ 4\ 8\ 16\ 32\ 64\ 128\ldots$$
$$f(10) = 2^{10} = 1024.\ f(20) = 2^{20} = 1{,}048{,}576.$$

The growth is rapid, although not as rapid as the growth of 10^x.

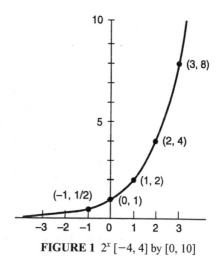

FIGURE 1 2^x $[-4, 4]$ by $[0, 10]$

The term <u>mathematical model</u> means a mathematical formula intended to quantify some real-world phenomenon.

Doubling-time Model
Let a population, $P(t)$, grow exponentially such that it doubles every d units of time. Then

$$P(t) = P(0)\ 2^{t/d}. \tag{5-2-1}$$

t is the time (the argument).
$P(t)$ is the population at time t (the image).
$P(0)$ is the population at time 0 (a parameter, the initial population).
d is the doubling time (a parameter, the time in which the population doubles).

The factor $2^{t/d}$ is the "growth factor." When $t = d$, the factor will be $2^{d/d} = 2^1 = 2$. That is why d is the "doubling time"(Figure 2).

Time:	0	d	$2d$	$3d$	$4d$
Factor:	1	2	4	8	16

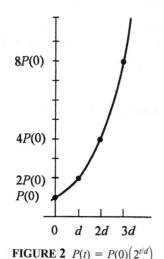

FIGURE 2 $P(t) = P(0)\left(2^{t/d}\right)$

EXAMPLE 1 Suppose a bacterial population is initially 10,000 and doubles every thirty minutes. Give the growth model.

Use the doubling-time model, $P(t) = P(0)\, 2^{t/d}$, with $P(0) = 10,000$ and $d = 30$ (minutes): $P(t) = 10,000(2^{t/30})$.

Dividing the argument of 2^t by 30 *before* applying the function makes the graph 30 times as wide. Effectively, this just relabels the horizontal scale (Figure 3).

Here is why. In the expression $P(0)(2^t)$, $t = 0$ yields $P(0)$, the initial population. When the time becomes 1, $2^1 = 2$, and the population will have doubled. To describe doubling in 30 units of time, we want the exponent to be 1 when $t = 30$, so use $\dfrac{t}{30}$ for the exponent. Time $t = 60$ yields exponent 2 and the population quadruples in 60 units of time.

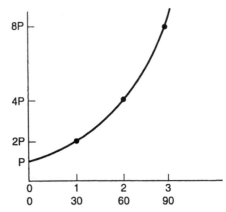

FIGURE 3 $P(t) = P(2^t)$ (upper scale) and $P(t) = P(2^{t/30})$ (lower scale)

Determine the population after 80 minutes.
Plug in $t = 80$.

$$P(80) = 10,000(2^{80/30}) = 63,496 \text{ (about 63,500)}.$$

How long will it take for this population to triple?
The question asks when $P(t) = P(0)(3)$. When will the growth factor will be 3?

$$P(t) = 10,000(2^{t/30}) = (10,000)(3)$$
$$2^{t/30} = 3.$$

Taking logs (any base, say, base 10) of both sides and using 5-1-3L:

$$\log(2^{t/30}) = \log 3$$

$$\left(\frac{t}{30}\right)\log 2 = \log 3$$

$$t = \frac{30 \log 3}{\log 2} = 47.55 \text{ (minutes)}.$$

The population will triple every 47.55 minutes.

EXAMPLE 2   A bacterial population doubles every 35 minutes and is 2,000 at 3:00 PM. How many bacteria will there be at 4:10 PM?

4:10 PM is 70 minutes after 3:00—exactly 2 doubling times later. Therefore the population will have doubled twice (grown by a factor of 4). It will be 8,000.

How many will there be at 5:00 PM?

Convert clock time to elapsed time. In exponential problems you can choose time 0 to be any convenient time. Here, let 3:00 PM be time 0. Then $P(0) = 2000$ and $d = 35$ are given. The model is: $P(t) = 2000(2^{t/35})$. What is t at 5:00 PM? 5:00 PM, which is two hours after 3:00 PM, corresponds to $t = 120$ (minutes). $P(120) = 2{,}000(2^{120/35}) = 21{,}534$ (about 21,500). ▬

EXAMPLE 3   A population undergoing exponential growth grows 10 percent in three hours. What is its doubling time?

Use the doubling-time model: $P(t) = P(0)2^{t/d}$. What do you know?

You know that when $t = 3$ the growth factor is 1.1 ("grows 10 percent in three hours"). That is, $1.1 = 2^{3/d}$. Take logs to solve for the unknown exponent.

$$\log 1.1 = \left(\frac{3}{d}\right)\log 2. \quad d = \frac{3 \log 2}{\log 1.1} = 21.8 \text{ (hours)}.$$

EXAMPLE 3 ANOTHER WAY   We can build a custom model. The growth factor $(1.1)^{t/3}$ fits the given facts. When $t = 3$, $(1.1)^{t/3}$ becomes $(1.1)^{3/3} = 1.1$, which is the factor for growth of 10 percent.

When does the population double? It doubles when $(1.1)^{t/3} = 2$. Take logs.

$$\left(\frac{t}{3}\right)\log 1.1 = \log 2. \quad t = \frac{3 \log 2}{\log 1.1} = 21.8 \text{ (hours), as before.} \quad ▬$$

■ Base $\dfrac{1}{2}$

Base $\dfrac{1}{2}$ is convenient for "half-life" models, where, instead of growth and increase, there is decay and decrease.

Half-Life Model
Let an amount, $A(t)$, decrease exponentially.

$$A(t) = A(0)\left(\frac{1}{2}\right)^{t/h}. \tag{5-2-2}$$

t is the time (the argument).
$A(t)$ is the amount at time t (the image).
$A(0)$ is the amount at time 0 (a parameter, the initial amount).
h is the half-life (a parameter, the time after which half remains).

The factor "$\left(\dfrac{1}{2}\right)^{t/h}$" is the decay factor.

Time:	0	h	$2h$	$3h$	$4h$
Factor:	1	$\dfrac{1}{2}$	$\dfrac{1}{4}$	$\dfrac{1}{8}$	$\dfrac{1}{16}$

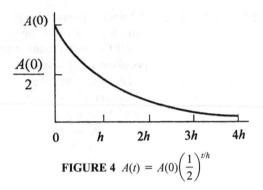

FIGURE 4 $A(t) = A(0)\left(\dfrac{1}{2}\right)^{t/h}$

EXAMPLE 4 The half-life of Carbon 14, used to date archaeological materials, is 5730 years. Give the model.

$$A(t) = A(0)\left(\frac{1}{2}\right)^{t/5730}.$$

This model is just a scale change of $A(0)\left(\dfrac{1}{2}\right)^{t}$. The factor $\left(\dfrac{1}{2}\right)^{t}$ is $\dfrac{1}{2}$ when $t = 1$. If we want a factor of $\dfrac{1}{2}$ when t is some other number, say 5730, replace t by $\dfrac{t}{5730}$. Then when $t = 5730$, the exponent will be $\dfrac{5730}{5730} = 1$, yielding a factor of $\left(\dfrac{1}{2}\right)^{1} = \dfrac{1}{2}$.

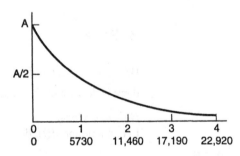

FIGURE 5 $A\left(\dfrac{1}{2}\right)^{t}$ (upper scale) and $A\left(\dfrac{1}{2}\right)^{t/5370}$ (lower scale)

If the initial amount is .012 grams, how much will remain after $2 \times 5730 = 11,460$ years?

This is exactly two half-lives. Therefore, one half of one half, that is, one quarter, of the original amount will remain: .003 grams.

How much will remain after 4000 years?

Use the formula: $A(t) = A(0)\left(\dfrac{1}{2}\right)^{t/h}$, where $A(0)$, t, and h are given.

$$A(4000) = .012\left(\frac{1}{2}\right)^{4000/5730} = .0074 \text{ (grams)}.$$

Suppose the half life of a substance is 5730 years and .00036 grams remain after 2000 years. How many grams existed originally?

Given the formula, word problems are easy. Just plug in for the given quantities.

General model:
$$A(t) = A(0)\left(\frac{1}{2}\right)^{t/h}$$

With given facts:
$$.00036 = A(2000) = A(0)\left(\frac{1}{2}\right)^{2000/5730}$$

We want to know the original amount, $A(0)$.

$$A(0) = \frac{.00036}{\left(\dfrac{1}{2}\right)^{2000/5730}} = .00046.$$

Originally, there existed .00046 grams. ▬

EXAMPLE 5 Suppose the half life of a substance is 90 years. When will only $\dfrac{1}{1000}$ of the original amount remain?

The problem asks when $A(t) = A(0)\left(\dfrac{1}{1000}\right)$.

The decay factor, $\left(\dfrac{1}{2}\right)^{t/h}$, is to be $\dfrac{1}{1000}$. h is given as 90. So solve

$$\left(\frac{1}{2}\right)^{t/90} = \frac{1}{1000}.$$

Take logarithms.
$$\left(\frac{t}{90}\right)\log\left(\frac{1}{2}\right) = \log\left(\frac{1}{1000}\right).$$

$$t = \frac{90\log\left(\dfrac{1}{1000}\right)}{\log\left(\dfrac{1}{2}\right)} = 897 \text{ years.}$$

One one-thousandth of the original amount will remain after 897 years. ▬

EXAMPLE 6 Suppose a substance undergoes rapid radioactive decay. Only $\dfrac{1}{10}$ the original amount remains after 20 seconds. What is its half life?

Using the half-life model, $A(t) = A(0)\left(\dfrac{1}{2}\right)^{t/h}$.

$$A(20) = A(0)\left(\frac{1}{2}\right)^{20/h} = A(0)\left(\frac{1}{10}\right).$$

$$\left(\frac{1}{2}\right)^{20/h} = \frac{1}{10}.$$

$$\left(\frac{20}{h}\right)\log\left(\frac{1}{2}\right) = \log\left(\frac{1}{10}\right).$$

$$h = \frac{20\log\left(\dfrac{1}{2}\right)}{\log\left(\dfrac{1}{10}\right)} = 6.02 \text{ (seconds).}$$

EXAMPLE 6
ANOTHER WAY

You may build a custom model for "Only one-tenth the original amount remains after 20 seconds."

By a scale change, the decay factor is $\left(\dfrac{1}{10}\right)^{t/20}$. That factor is $\dfrac{1}{10}$ when $t = 20$.

What is the half life? Solve for the time when the decay factor is one half.

$$\left(\frac{1}{10}\right)^{t/20} = \frac{1}{2}.$$

$$\left(\frac{t}{20}\right)\log\left(\frac{1}{10}\right) = \log\left(\frac{1}{2}\right).$$

$$t = \frac{20\log\left(\dfrac{1}{2}\right)}{\log\left(\dfrac{1}{10}\right)} = 6.02 \text{ (seconds), as before.} \qquad \blacksquare$$

■ Compound Interest

Money problems serve to illustrate the close relationship between exponential functions and power functions.

The formula for compound interest was given in Section 4-4 on Percents. Here it is again.

$$A = P\left(1 + \frac{r}{k}\right)^{kt} \tag{5-2-3}$$

where A is the amount at time t,

  t is time in years,

  P is the principal (the original amount of money),

  r is the annual percentage rate (APR), and

  k is the number of times per year the money is compounded.

To solve for t in the compound interest model, take logarithms of both sides using any convenient base (base 10 or base e).

EXAMPLE 7

How long does it take $1000 to triple when invested at 12% compounded monthly?

Plug into Formula 5-2-3 and solve

$3000 = 1000(1.01)^{12t}$ for t (in years).

$3 = 1.01^{12t}$.

$\log 3 = \log(1.01^{12t}) = 12t \log 1.01$.

$t = \dfrac{\log 3}{12 \log 1.01} = 9.2 \text{ years.}$

Actually, the formula is only valid when t expresses an integer number of months, and 9.2 years is not precisely an integer number of months. But it is close enough to answer the spirit of the question. $\qquad \blacksquare$

EXAMPLE 8 How long does it take money to double when invested at 8 percent compounded yearly?

This problem asks us to solve: $2 = 1.08^t$.

Taking logs, $\log 2 = t \log 1.08$, and $t = \dfrac{\log 2}{\log 1.08} = 9.01$ years.   ▬

EXAMPLE 9 Money compounded monthly grew from 10,000 to 14,000 in 3 years. What was the annual percentage rate?

This problem asks us to solve for r in: $14{,}000 = 10{,}000\left(1 + \dfrac{r}{12}\right)^{36}$.

The first step is to divide by 10,000.

$$1.4 = \left(1 + \frac{r}{12}\right)^{36}.$$

There is no need to take logs. Logs are great for solving for unknowns that are in the exponent. This unknown is not. Use inverse-reverse and take roots.

$$1.4^{1/36} = 1 + \frac{r}{12}.$$

$$1.00939 = 1 + \frac{r}{12}. \quad .00939 = \frac{r}{12}. \quad r = 12(.00939) = .1127.$$

The rate was 11.27%.   ▬

■ Base e

In calculus, base e ($= 2.71828\ldots$) is the preferred base for exponential and logarithmic functions. This may seem like an awkward number, but it really does have some wonderful properties. Logarithms with base e are said to be "natural logarithms" and base e really is "natural" in calculus. Some reasons will be discussed in the subsection "Why e?" of Section 5-4.

Everything we did with base 2 or base $\dfrac{1}{2}$ can be done with base e.

All the general facts about exponential functions and logarithms hold for base e. The facts are repeated here, but in the notation used for base e.

Pairs of exponential and logarithmic facts Table 5-2-4 (of 5-1-1L through 5-1-4invL)			
Exponential form (base e)		**Logarithmic form (assume c and d are positive)**	
(5-1-4)	$e^x = y$	$\ln y = x$	(5-1-4)
(5-1-4inv)	$e^{\ln x} = x$	$\ln(e^x) = x$	(5-1-4inv)
(5-1-1)	$e^x e^y = e^{x+y}$	$\ln(cd) = \ln c + \ln d$	(5-1-1L)
(5-1-2)	$\dfrac{e^x}{e^y} = e^{x-y}$	$\ln\left(\dfrac{c}{d}\right) = \ln c - \ln d$	(5-1-2L)
(5-1-2B)	$e^0 = 1$	$\ln 1 = 0$	(5-1-2BL)
(5-1-2C)	$e^{-x} = \left(\dfrac{1}{e^x}\right)$	$\ln\left(\dfrac{1}{c}\right) = -\ln c$	(5-1-2CL)
(5-1-3)	$(e^p)^x = e^{px}$	$\ln(c^x) = x \ln c$	(5-1-3L)
e^x has domain all real numbers		$\ln$ has range all real numbers	
e^x has range all positive numbers		$\ln$ has domain all positive numbers	

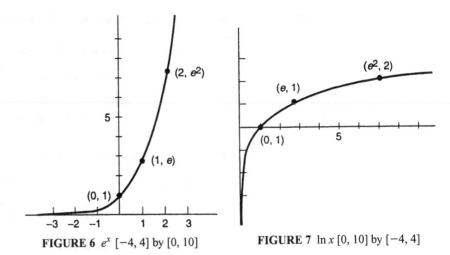

FIGURE 6 e^x $[-4, 4]$ by $[0, 10]$ **FIGURE 7** $\ln x$ $[0, 10]$ by $[-4, 4]$

EXAMPLE 10   Solve $18 = e^{3x}$.

Take (natural) logarithms: $\ln 18 = 3x$. $x = \dfrac{\ln 18}{3} = .963$.   ▬

■ Exponential Model

The <u>exponential model</u> (base e) for an amount, $P(t)$, at time t is:

$$P(t) = P(0)e^{kt}. \qquad (5\text{-}2\text{-}5)$$

It has two parameters, $P(0)$, the initial amount, and k, the exponential growth rate parameter. k is the slope of the graph of $y = e^{kt}$ at $(0,1)$.

The "exponential model" is a general model which can do everything we did with base 2 or base $\dfrac{1}{2}$.

It includes both the doubling-time model (base 2) and the half-life model $\left(\text{base } \dfrac{1}{2}\right)$. It is the preferred model in calculus.

EXAMPLE 11   A bacterial population doubles every 30 minutes and there are initially 10,000 bacteria. Give a formula for the population growth. When will there be 60,000 bacteria?

We used the doubling-time model for this in Example 1. But here, to illustrate that base e is versatile, we do the problem using base e. The model is: $P(t) = P(0)e^{kt}$.

The problem states that $P(0) = 10,000$. Furthermore, the doubling time tells us that when $t = 30$ minutes, $e^{kt} = e^{30k} = 2$. Solve for k by taking natural logarithms,

$$30k = \ln 2 \text{ and } k = \frac{\ln 2}{30} = .0231.$$

> For the same problem in Example 1 we got the formula
> $P(t) = 10,000(2^{t/30})$.
>   It may look different, but it is equivalent to the one here:
> $P(t) = 10,000e^{.023lt}$.

This determines the formula. It is
$$P(t) = 10,000e^{.023lt}.$$
To answer, "When will there be 60,000 bacteria?" Solve for t in
$$60,000 = 10,000e^{.023lt}.$$
$$6 = e^{.023lt}; \ln 6 = .0231t.$$
$$t = \frac{\ln 6}{.0231} = 77.6 \text{ (minutes)}.$$

The exponential model includes the doubling-time model. The growth factors in the two models are e^{kt} and $2^{t/d}$. They may look different, but they are exactly the same if k is chosen properly.

$$2^{t/d} = e^{kt} \text{ for all } t \text{ iff } k = \frac{\ln 2}{d}. \tag{5-2-6}$$

Proof

$2^{t/d} = e^{kt}$ for all t iff $\ln\left(2^{t/d}\right) = \ln\left(e^{kt}\right)$ for all t iff $\left(\frac{t}{d}\right)\ln 2 = kt$ for all t iff

$\left(\frac{1}{d}\right)\ln 2 = k.$

> In the exponential model the growth factor is e^{kt}, which equals $(e^k)^t$ [by 5-1-3].
>   If $k > 0$, there will be growth because e^k is greater than 1 when k is positive.
>   If $k < 0$, there will be decay because e^k is less than 1 when k is negative.

We can use the exponential model to do half-life problems

EXAMPLE 12 A substance undergoing radioactive decay has a half-life of 90 years. Give the formula for the amount at any given time.

This problem is easy using base $\frac{1}{2}$, but, here we show it can be done with base e.

The model, in terms of base e, is: $P(t) = P(0)e^{kt}$. The only unknown is k, since we are letting the initial amount be general, $P(0)$. The half-life tells us $P(90) = \left(\frac{1}{2}\right)P(0)$, that is, the decay factor e^{kt} equals $\frac{1}{2}$ at time 90. This can be used to solve for k.

$$\frac{1}{2} = e^{90k}, \; \ln\left(\frac{1}{2}\right) = 90k, \; k = \ln\frac{\left(\frac{1}{2}\right)}{90} = \frac{-.693}{90} = -.00770.$$

k is negative because the amount is declining. The model is

$$P(t) = P(0)e^{-.00770t}. \qquad \blacksquare$$

EXAMPLE 13 Suppose the population of a country is growing at a fixed rate of 3% per year. Model its population using the exponential model.

    The most straightforward model is: $P(t) = P(0)(1.03)^t$, where t is the number of years. This can be interpreted as an exponential model, 5-2-5 where $e^k = 1.03$.

$$P(0)e^{kt} = P(0)(e^k)^t = P(0)1.03^t$$

Simply adjust k so that $e^k = 1.03$.
The required value of k is $\ln 1.03 = .0296$. So, $P(t) = P(0)e^{.0296t}$. $\blacksquare$

Continuous-Compounding Model
The growth of money depends upon the annual rate and how frequently it is compounded. Sometimes it is compounded "continuously"! (What this really means is discussed in 5-4-7.) The continuous compounding model for the amount of money, A, at time t is

$$A = Pe^{rt}, \qquad (5\text{-}2\text{-}7)$$

with two parameters, P, the principle (the initial amount), and r, the (exponential) growth rate (This is the APR, annual percentage rate, but not the "effective" rate).

EXAMPLE 14 $5,000 is deposited in an account. It earns an annual percentage rate of 7.65% compounded continuously. How much will it be in 4 years?

$$\text{Plug in. } A = 5,000 \; e^{.0765(4)} = 6,789.91 \text{ (dollars).} \qquad \blacksquare$$

EXAMPLE 15 How long will it take money to double if it is invested at an annual rate of 10% compounded continuously?

    This problem asks you to solve $2 = e^{.1t}$ for t.

    Take natural logarithms [5-1-4], $\ln 2 \doteq .1t$, and $t = \dfrac{\ln 2}{.1} = 6.93$ years.

    What is the effective rate of money compounded continuously at a 10 percent annual percentage rate?

    In one year, the multiplicative factor will be $e^{.1} = 1.10517$, for an effective rate of 10.517 percent.

    As illustrated by this example, effective rates are generally slightly greater than annual percentage rates. $\blacksquare$

■ Changing Bases

Any exponential function can be rewritten with base *e*. For example, the doubling-time model (5-2-1) is equivalent to the exponential model (5-2-5) because the growth factor $2^{t/d}$ can be written as e^{kt} if k is chosen properly.

EXAMPLE 16   Rewrite $2^{t/8}$ as e^{kt}. (Find k.)

We want $e^{kt} = 2^{t/8}$.

Take logs. $kt = \left(\dfrac{t}{8}\right)\ln 2$

$$k = \frac{\ln 2}{8} = .0866$$

So, $2^{t/8} = e^{.0866t}$, for all t.   ▬

The graph of 2^x is just like the graph of 10^x and the graph of e^x, only with a horizontal scale change—the graph of 2^x is wider (Figure 8).

Recall $e = 10^{\log e}$. Therefore,

$$e^x = (10^{\log e})^x = 10^{(\log e)x} = 10^{.434x} = 10^{x/2.3}.$$

$$e^x = 10^{x/2.3}. \tag{5-2-8}$$

Therefore, by the scale-change ideas of Section 2-2, the graph of e^x is approximately 2.3 times as wide as the graph of 10^x (Figure 8). (Section 5-4 explains some of the "natural" properties of base e.)

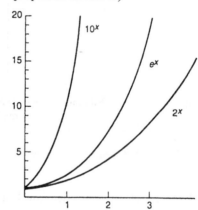

FIGURE 8 10^x (left), e^x and 2^x (right). [0, 4] by [0, 20]

Logarithms

Your calculator will evaluate 2 to any power using the general power key. However, most calculators do not have a key for logarithms base 2. To work with logs to unusual bases, convert to exponential form using Definition 5-1-4.

EXAMPLE 17   Solve $\log_2 x = 2.7$.

The given problem is in logarithmic form. Rewrite it in exponential form:

$$x = 2^{2.7} = 6.498. \quad ▬$$

EXAMPLE 18   Find $\log_2 x$ in terms of logs base 10. Logs are inverses of exponentials, so $2^{\log_2 x} = x$. Take log of both sides (base 10):

$$(\log_2 x)\log 2 = \log x.$$

$$\log_2 x = \frac{\log x}{\log 2}$$

$$= \frac{\log x}{.301} = 3.32 \log x.$$

The graph of $\log_2 x$ is about 3.32 times as high as the graph of $\log x$ (Figure 9). This argument can be repeated using logarithms with base *e* ("ln") or any other base (Problem B65 gives the result).

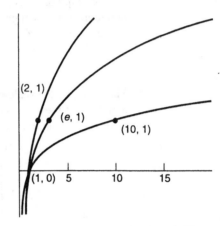

FIGURE 9 $\log_2 x$ (upper), ln *x* (middle), log *x* (lower). They are constant (vertical) multiples of each other. [0, 20] by [−1, 3]

CONCLUSION Exponential growth and decay are common real-world phenomena that are modeled using exponential functions. Although base 10 has substantial advantages, there are often good reasons to use base 2 and base *e*. Any exponential model with any base can be rewritten in terms of base *e*, which is the most important base in calculus. Logarithms undo exponentiation, so they are used for solving equations in which the unknown is in an exponent.

All exponential functions with base greater than 1 are much alike—they differ only in their horizontal scale. Similarly, all logarithmic functions with base greater than 1 are much alike—they differ only in vertical scale.

Terms: exponential growth, mathematical model, doubling time, half-life, compound interest, *e*, continuous compounding, exponential models.

Exercise 5-2

A

A1.* ☺ What type of problem can logs solve that arithmetic operations and taking powers or roots cannot?

A2. (A) Give the first 5 integer powers of 2.
  (B) Give round numbers close to 2^{10} and 2^{20}.

Solve algebraically.

A3. $2^x = 5$ [2.3]

A4. $1700 = 1000(2^{t/5})$ [3.8]

A5. $3 = 2^{40/d}$ [25]

A6. $\left(\dfrac{1}{2}\right)^{50/h} = .01$ [7.5]

A7. $20 = 100\left(\dfrac{1}{2}\right)^{t/30}$ [70]

A8. $1.3 = 2^{6/d}$ [16]

A9. $2 = e^{5k}$ [.14]

A10. $3600 = 2000(e^{t/20})$ [12]

A11. $\dfrac{1}{2} = e^{-k}$ [.69]

A12. $2^{x/3} = 5$ [7.0]

☺ *Write in exponential form.*

A13. $\ln x = 2$

A14. $\ln 5 = 1.61$

A15. $\log_2 x = 3.4$

A16. $\log_2 8 = 3$

A17. $\log_2\left(\dfrac{1}{4}\right) = -2$

A18. $\log_2 1024 = 10$

A19. $\log_2\left(\dfrac{1}{64}\right) = -6$

A20. $\ln c = x.$

A21. $\log_{16}(x) = c$

A22. $\log_{16}(1000) = 2.4914$

A23. ☺ True or false (for positive a and b)?

  (A) $\log\left(\dfrac{a}{b}\right) = \dfrac{\log a}{\log b}$
  (B) $\sqrt{(ab)} = \left(\sqrt{a}\right)\left(\sqrt{b}\right)$
  (C) $e^{a+b} = e^a + e^b$

A24. ☺ True or false (for positive x and y)?
  (A) $(xy)^2 = x^2 y^2$

  (B) $\log(x - y) = \log x - \log y$
  (C) $\dfrac{1}{e^x} = e^{-x}$

A25. ☺ True or false (for positive x and y)?
  (A) $\log(2x) = 2\log x$
  (B) $\log(x^2) = 2\log x$
  (C) $\log\left(\dfrac{x}{y}\right) = \log x - \log y$

A26. ☺ True or false (for positive a and b)?
  (A) $(e^x)^2 = e^{2x}$
  (B) $e^{xy} = (e^x)(e^y)$
  (C) $\log(ab^x) = x\log(ab)$

A27. ☺ (A) How many parameters does the doubling-time model (5-2-1) have?
  (B) List them.

A28. ☺ (A) How many parameters does the half-life model (5-2-2) have?
  (B) List them.

A29. ☺ (A) How many parameters does the compound-interest formula (5-2-3) have?
  (B) List them.

A30. ☺ (A) How many parameters does the continuous-compounding model (5-2-7) have?
  (B) List them.

A31. ☺ (A) How many parameters does the exponential model (5-2-5) have?
  (B) List them.

Solve algebraically.

A32. $\dfrac{50}{1 + e^{-x}} = 30.$

A33. $\dfrac{24}{4 + 6e^{-x}} = 5$

A34. $\dfrac{20}{2 + 8e^{-x}} = 4$

B

B1.* Sketch the graph of the doubling-time model. Label the parameters on it, and label the three most interesting points on the graph.

B2.* Sketch the graph of the half-life model. Label the parameters on it, and label the three most interesting points on the graph.

B3.* ☺ In the exponential growth model, if the doubling time is d, what is the quadrupling time (time until there are four times as many)?

B4.* ☺ In the exponential decay model, if the half-life is h, what is the time until there are one quarter as many?

Solve algebraically.

B5. $1.20^t = 10(1.15)^t$ [54]

B6. $3^x = 100(2^x)$ [11]

B7. $3(2^{t/20}) = 2^{t/19}$ [600]

B8. $.95(1.08)^t = 1.06^t$ [2.7]

B9. $1000(2^{t/30}) = 2^{t/29}$ [8700]

B10. $1.2(1.02)^{4t} = 1.008^{12t}$ [11]

B11. $.8\,e^{.08t} = e^{.07t}$ [22]

B12. $1.4e^{-2t} = e^{-1.6t}$ [.84]

B13. $1000\left(\dfrac{1}{2}\right)^{t/50} = 40\left(\dfrac{1}{2}\right)^{t/70}$ [810]

B14. $500\left(1 + \dfrac{.08}{12}\right)^{12t} = 485\,e^{.08t}$ [110]

Solve algebraically.

B15. $\log_2 x = 5.1$ [34]

B16. $\log_2 x = -3.4$ [.095]

B17. $\log x + \log(3x) = 6.2$ [730]

B18. $\ln x + \ln(x^2) = 4$ [3.8]

B19. Using the doubling-time model, when will there be 10^6 bacteria if there are now 1000 bacteria and their doubling time is 35 minutes?   [5.8 hours]

B20. Using the doubling-time model, if there were 10^4 bacteria four hours ago and there are now 5×10^6 bacteria, what is the doubling time?   [.45 hours]

B21. Use the half-life model. If $\dfrac{1}{15}$ of the original amount remains after 12 seconds, what is the half life?   [3.1]

B22. Use the half-life model. If the half-life is 5730 years and .762 of the original amount remains, how much time has elapsed?   [2200]

B23. The half-life of Potassium 40 is 1.31 billion years. If 92.7 percent of the original Potassium 40 remains in a rock, how old is it?   [.14 billion years]

B24. The half-life of Potassium 40 is 1.31 billion years. If 35 percent of the original Potassium 40 remains in a rock, how old is it?   [2.0 billion years]

B25. Use the doubling-time model or the exponential growth model. If the initial population is 10,000 and it grows 60 percent every 40 minutes, give a model for the population at any time.

B26. Use the half-life model or the exponential model. If the amount of a substance decreases 10 percent every two hours, give a model for the amount at any time.

B27. In the middle of the second century AD a modius of wheat (about $18\frac{1}{2}$ quarts) cost $\frac{1}{2}$ denarius. In 301 it cost 100 denarii. What was the average annual effective inflation rate over those 150 years?   [3.6%]

B28. During the 30 years of the reign of the Roman emperor Constantine the Great (AD 307–337), the intrinsic value of the copper coin denomination dropped to $\frac{1}{6}$ its original value. What was the average yearly rate of inflation over that period?   [6.2%]

B29. In AD 50 a Roman widget cost 12 denarii. In AD 210 it cost 45 denarii. What was the average annual rate of change in its cost?

B30. (Hyperinflation) In September 1995 a dollar bought 50,000 karbovantsiv (the Ukranian currency unit). One month later it bought 135,000. (Small bank note denominations like 100 karbovantsiv were being recycled into toilet paper.) Express the equivalent effective annual inflation rate if that monthly rate of inflation kept up for a whole year.

Write as an exponential function base e, that is, find k such that e^{kt} is equivalent to this.

B31. $2^{t/d}$.

B32. $\left(\dfrac{1}{2}\right)^{t/h}$.

B33. $(.90)^t$

B34. $(1.02)^t$

Two populations

B35. Population A is 40% larger than Population B, but Population B is growing at a 3% rate (compounded continuously), whereas Population A is growing at only a 1% rate (compounded continuously). When will the two populations be equal?   [17]

B36. There is now ten times as much of substance C as of substance D, but substance C is decaying with a half-life of 120 minutes and substance D is decaying with a half-life of 200 minutes. When will the two amounts be equal?   [1000 minutes]

B37. Bacteria of type A double every 30 minutes. Bacteria of type B double every 29 minutes (that is more frequently), but they begin with only 1 for every 1000 of type A. How long before they are 1000 times as common as type A? [Express your answer in the most comprehensible units: seconds, minutes, hours, days, or years.] [12]

B38. Suppose you are considering two alternative investments. One is to invest in a "cash equivalent" fund offering 8% compounded continuously. The other is to invest in collectable art, which you are willing to guess will appreciate at 12% per year compounded continuously. But there is a 20% commission to sell the art you buy. That is, you realize only 80% of its "value." Since 12% is greater than 8%, you figure art will be the better investment in the long run. How long do you have to hold on to the art before it begins to outperform the cash equivalent investment? [5.6]

B39. Economy A is 70% the size of Economy B. If A's economy grew at an exponential rate of 4% and B's economy grew at 2%, in how many years would A's economy overtake B's? [18]

B40. Clyde invested $40,000 at the beginning of the year in a mutual fund. After 8 months he added $10,000 to his investment. At the end of the year, the total was $62,000. What was the average annual effective growth rate (Assume continuous compounding and find e^r). [Do not expect to solve this algebraically.] [.28]

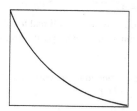

The graph above is $\left(\frac{1}{2}\right)^x$ *pictured in the window [0, 4] by [0, 1]. Graph these and find the window that makes them look just like that graph. (Make the left and bottom edges x = 0 and y = 0, and find a and c such that the window is [0, a] by [0, c]. (You may use guess and check.)*

B41. $20\left(\frac{1}{2}\right)^x$

B42. $\left(\frac{1}{2}\right)^{x/10}$

B43. $\left(\frac{1}{2}\right)^{4x}$

B44. $.1\left(\frac{1}{2}\right)^{x/50}$

B45. e^{-x}

B46. $5e^{-4x}$

B47. You were thinking of investing your money at p% per year compounded continuously, but you calculated that if you could get a 1% higher rate, your money would double exactly $\frac{1}{2}$ year sooner. What was p?

Reading and Writing Mathematics

B48. State the logarithmic fact corresponding to the given exponential fact.

(A) $e^{a+b} = e^a e^b$

(B) $e^{a-b} = \dfrac{e^a}{e^b}$

(C) $(e^a)^b = e^{ab}$

(D) $e^{-b} = \dfrac{1}{e^b}$

B49.* State the exponential fact corresponding to the given logarithmic fact.

(A) $\log(a^x) = x \log a$

(B) $\log a + \log b = \log(ab)$

(C) $\log a - \log b = \log\left(\dfrac{a}{b}\right)$

(D) $\log\left(\dfrac{1}{x}\right) = -\log x$

B50. Complete the identity: $\ln\left(\dfrac{1}{x}\right) = $ _____.

B51. Read Example 16 to answer this. The graph of e^x is approximately _____ times as _____ as the graph of 10^x.

B52. The graph of $\ln x$ is approximately _____ times as _____ as the graph of $\log x$.

B53. Graph $\left(1 + \dfrac{1}{x}\right)^x$ and tell what happens as x goes to infinity.

B54. Example 11 is a particular example of how the exponential model includes the half-life model 5-2-2. Prove that, if h is the half-life, then $k = \dfrac{-\ln 2}{h}$.

B55. Solve the following equation: $d = cb^a$ for the unknown variable, assuming the other values are known. If logarithms are required, use only logarithms base e or base 10.
(A) a is unknown
(B) b is unknown
(C) c is unknown.

B56. In the exponential growth model, if the doubling time is d, what is the time until there are k times as many?

B57. In the exponential growth model, what is the relationship between doubling time and tripling time?

B58. In the half-life model, if the half life is h, how long until there are $\dfrac{1}{k}$ times as many?

The graph of e^{kx} is exactly the graph of $\left(\dfrac{1}{2}\right)^x$ if k is chosen correctly.

B59. Find k algebraically. $[-.69]$

B60. Find k by graphing both graphs in the window $[0, 4]$ by $[0, 1]$ and using guess-and-check to find the k that matches the two graphs. [Great accuracy is not required. k will be negative.]

B61. Interpret this in logarithmic form:
$2^a = (10^{\log 2})^a = 10^{a \log 2}$. Which theorem results?

B62. Example 15 showed that $2^{t/8} = e^{kt}$ if k is chosen properly. Find the same k another way by using $2 = e^{\ln 2}$ and expanding and simplifying $2^{t/8}$.

B63. Let $k = e^x = (10^{\log e})^x = 10^{(\log e)x} = k$. Logs are exponents. Read this to identify $\ln k$ and $\log k$. Simplify to find the relationship between $\ln k$ and $\log k$.

B64. Let $c^{\log_c x} = x$. Take logs to derive the general change-of base result for logarithms:

$$\log_c x = \frac{\log_b x}{\log_b c}.$$

B65. (A) Prove: $\log_b c = \dfrac{1}{\log_c b}$.

(B) Confirm that $\log e = \dfrac{1}{\ln 10}$.

B66. $\dfrac{\log a}{\log b} = \dfrac{\ln a}{\ln b}$, for all positive a and b. Prove it.

B67. (A) Prove the **change-of-base result for exponentials**: $b^x = c^{(\log_c b)x}$

(B) Use it to explain how the graph of b^x compares to the graph of 10^x.

B68.* $f(kx)$ has a graph which is $\dfrac{1}{k}$ times as wide as the graph of $f(x)$. Explain why (as if you were teaching a student about Section 2-2 on composition).

B69. Solve for b: $b^{x+1} = 5\, b^x$.

B70. Solve for c: $\log(cx) - \log x = 1$.

B71. Let $y = x^2$, $Y = \ln y$ and $X = \ln x$. Now plot Y against X. What does the graph look like?

B72. Let $y = 2^x$ and $Y = \ln y$. Now plot Y against x. What does the graph look like?

B73. Let $y = cb^x$. Take logs of both sides to obtain a linear relationship between $Y = \ln y$ and x. If $Y = mx + b$, identify m and b.

B74. Let $y = cx^p$. Take logs of both sides to obtain a linear relationship between $Y = \ln y$ and $X = \ln x$. If $Y = mX + b$, identify m and b.

B75. How many digits does 950^{950} have? [Hint: Think about logs if your calculator won't do it.]

B76. In banking there is a "Rule of 72" that approximates the doubling time of money invested at various rates. It says that if the interest rate is p percent per year, the doubling time is about $\dfrac{72}{p}$ years. For example, if the interest rate is 8%, the doubling time is about 9 years $\left(\dfrac{72}{8} = 9\right)$.

(A) Compute the doubling time in terms of p.

(B) Compare the result of the "Rule of 72" to the actual doubling time for money compounded yearly at 4%.

(C) Use part (a) to explain why the Rule of 72 works. [Use the idea that $\ln(1 + x)$ is very near x, for small x. Note that "72" is used because it is evenly divisible by so many likely interest rates: 1,2,3,4,6,8,9,12,15,18, and 24.]

B77.* State the logarithmic fact corresponding to the given exponential fact.

(A) $e^0 = 1$.

(B) the domain of e^x is all real numbers.

(C) the range of e^x is all positive numbers.

(D) $e^a = c$.

B78.* State the exponential fact corresponding to the given logarithmic fact.

(A) $\ln 1 = 0$.

(B) The range of $\ln x$ is all real numbers.

(C) the domain of $\ln x$ is all positive numbers.

(D) $\ln c = a$.

B79.* State the logarithmic fact corresponding to the given exponential fact.

(A) $10^{a+b} = 10^a\, 10^b$

(B) $10^{a-b} = \dfrac{10^a}{10^b}$

(C) $(10^a)^b = 10^{ab}$

(D) $10^{-b} = \dfrac{1}{(10^b)}$

B80.* State the exponential fact corresponding to the given logarithmic fact.

(A) $\ln(a^x) = x \ln a$

(B) $\ln a + \ln b = \ln(ab)$

(C) $\ln a - \ln b = \ln\left(\dfrac{a}{b}\right)$

(D) $\ln\left(\dfrac{1}{x}\right) = -\ln x$

Frequent compounding and continuous compounding. For each problem evaluate two things. Evaluate the growth factor for one year in the "compound interest" model 5-2-3 and evaluate the growth factor for one year in the "continuous compounding" model 5-2-7. Say if they are similar.

B81. $r = .1$ and $k = 365$.

B82. $r = .05$ and $k = 12$.

B83. $r = .2$ and $k = 365$.

B84. $r = 1$ and $k = 52$.

Section 5-3   APPLICATIONS

In many applications it is appropriate to use the logarithm of a quantity, rather than the original quantity itself. Using "$\log x$" instead of x is a " change-of-variable."

EXAMPLE 1   How can you report the power of an earthquake? A weak earthquake might have waves with amplitude one-tenth millimeter or less. A very strong earthquake may have waves with amplitude one million times as large. Rather than report the amplitude itself, the Richter scale reports the logarithm of the amplitude.

Let a be the amplitude of the earthquake, measured in microns, where 1000 microns equal one millimeter and 1,000,000 microns equals one meter. Define

$$R(a) = \log a. \tag{5-3-1}$$

Then report the Richter scale value, $R(a)$, rather than a.

For example, a strong earthquake with waves with amplitude 1 meter would be reported as a 6, because 1 meter $= 1,000,000$ microns and $\log(1,000,000) = 6$.

A moderately weak earthquake with waves of amplitude 10 millimeters would be reported as a 4 on the Richter scale because 10 millimeters is 10,000 microns and $\log(10,000) = 4$. An earthquake with waves 10 times as large—amplitude 100 millimeters—would be reported as a 5. A very strong earthquake with waves of amplitude 10 meters ($= 10$ meters $\times$ 1,000,000 microns per meter $= 10,000,000$ microns) would be reported as a 7 because $\log(10,000,000) = 7$.

An increase in amplitude by a factor of 10 corresponds to an additive increase of 1 on the Richter scale. An additive increase of 2 on the Richter scale corresponds to an increase in amplitude by two factors of 10, that is, by a factor of 100.

An earthquake has waves with amplitude 1.7 meters. What is its Richter-scale value?

1.7 meters $= 1.7$ meters $\times$ (1,000,000 microns per meter) $= 1,700,000$ microns. $\log(1,700,000) = 6.23$. Call it a 6.2 on the Richter scale.

If an earthquake is a 3.8 on the Richter scale, how big are the waves?

$\log(a) = 3.8$. $a = 10^{3.8} = 6310$ microns $= 6310$ microns $\times$ (1 mm/1000 microns) $= 6.31$ millimeters, which is also .00631 meter.

**EXAMPLE 1
CONTINUED FURTHER**

One earthquake is a 4.6 on the Richter scale and a second has waves with 50 times the amplitude of the first. What is the second on the Richter scale?

The amplitude of a 4.6 earthquake is $10^{4.6} = 39,800$ microns. The amplitude of the second is 50 times that, $39,800(50) = 1,990,000$ microns. $\log(1,990,000) = 6.3$. The second is a 6.3 on the Richter scale. ■

We can use properties of logs to do this type of problem in general.

If one earthquake has waves with amplitude a and another has waves k times as large, how do they compare on the Richter scale?

The second has waves with amplitude ka. Its Richter scale value is $R(ka) = \log(ka) = \log k + \log a = \log k + R(a)$. Therefore, $R(ka) - R(a) = \log k$. Their Richter scale values differ by $\log k$.

Conversely, if their Richter scale values differ by d, $d = \log k$ and $k = 10^d$. The one has waves 10^d times as large as the other. This table gives the relationships.

THEOREM 5-3-2

If the waves of one earthquake have amplitude k times those of another, their Richter scale values differ by $\log k$.

A difference of d on the Richter scale corresponds to one having waves with 10^d times the amplitude of the waves of the other.

Amplitude	a	10^R	multiply by k	multiply by 10^d
Richter scale value	$\log a$	R	add $\log k$	add d

**MORE ABOUT
EXAMPLE 1**

If a second earthquake has waves 50 times as large as a first, how do their Richter scale values compare?

By the first part of the theorem, the larger has a Richter scale value that is $\log 50 = 1.7$ higher.

For example, if the first was 3.2, the second was 4.9. If the first was 4.6, the second was 6.3.

**STILL MORE ABOUT
EXAMPLE 1**

How much larger are the waves of an earthquake that is .5 higher on the Richter scale?

By the second part of the theorem, the larger has waves $10^{.5} = 3.1$ times as large.

For example, the waves of a 6.7 earthquake are about 3.1 times the size of the waves of a 6.2 earthquake.

**EXAMPLE 1
GRAPHS**

Suppose you were writing an article on earthquakes and you wished to illustrate, with a graph, the magnitudes of several famous earthquakes. We change variables to the Richter scale because plotting their amplitudes would not work well.

Some famous earthquakes have waves only $\dfrac{1}{100}$ the size of others. Plotting the amplitude would make the smaller ones look puny. Plot R instead. ■

EXAMPLE 2

The loudness of sound varies widely. In a quiet forest we can hear leaves rustling. At the other extreme, the loud sounds at a rock concert may be 10,000,000,000 (ten billion) times as loud as leaves rustling. To report the volume of sound we do not want a scale with 10 billion different levels of volume! The decibel scale was devised to take this wide range of values and convert it into a manageable range. The idea is to avoid reporting the actual volume in terms of the energy of sound, but to report a logarithmic function of the volume. Let v be the volume (energy of the sound) and $D(v)$ be the corresponding number of decibels.

$$D(v) = 10 \log\left(\frac{v}{v_0}\right) \tag{5-3-3}$$

where v_0 is chosen to be the volume of the faintest sound that can be heard by healthy young people.

Table of Sound Volumes		
Cause	Multiple of the faintest volume	Decibel level
faintest sound	1	0
leaves rustling	30	15
whisper	1000	30
normal conversation	1,000,000	60
vacuum cleaner	10,000,000	70
jackhammer	10,000,000,000	100
painful sound	$\geq 1,000,000,000,000$	≥ 120

EXAMPLE 2 CONTINUED   A sound is 5000 times as loud as the faintest sound healthy young people can hear. What is its decibel value?

$10 \log(5000) = 37$. It is 37 decibels.

A sound is 95 decibels. How does it compare to the faintest sound healthy young people can hear?

$$10\log\left(\frac{v}{v_0}\right) = 95. \quad \log\left(\frac{v}{v_0}\right) = 9.5. \quad \frac{v}{v_0} = 10^{9.5} = 3{,}160{,}000{,}000.$$

The 95 decibel sound is over 3 billion times as loud.

One sound is 65 decibels and a second is 90 decibels. The volume of the second is how many times the volume of the first?

For the first, $10\log\left(\dfrac{v_1}{v_0}\right) = 65$, so $\dfrac{v_1}{v_0} = 10^{6.5}$.

For the second, $10\log\left(\dfrac{v_2}{v_0}\right) = 90$, so $\dfrac{v_2}{v_0} = 10^{9.0}$.

$\dfrac{10^{9.0}}{10^{6.5}} = 10^{2.5} = 316$.   The louder is about 316 times as loud.

If one sound has volume v and a second sound has k times the volume, how do they compare on the decibel scale?

Let v be the volume of the one sound. The second has volume kv.

$$D(kv) = 10\log\left(\frac{kv}{v_0}\right) \qquad \text{[by Definition 5-3-3]}$$

$$= 10\left[\log(k) + \log\left(\frac{v}{v_0}\right)\right] \qquad \text{[by 5-1-1L, log of a product]}$$

$$= 10\log k + 10\log\left(\frac{v}{v_0}\right)$$

$$= 10\log k + D(v).$$

Therefore, the difference on the decibel scale is 10 times the log of the factor relating their volumes. This proves the first half of the next theorem. The second half is Problem B22.

THEOREM 5-3-4 If a louder sound is k times as loud as a quieter sound, they differ on the decibel scale by 10 log k.
If two sounds differ on the decibel scale by d, the louder is $10^{d/10}$ times as loud as the quieter.

Volume	v	multiply by k	multiply by $10^{d/10}$
decibel scale value	$10\log\left(\frac{v}{v_0}\right)$	add 10 log k	add d

MORE ABOUT EXAMPLE 2

When the volume dial is adjusted, a car radio displays the volume in decibels. The smallest increase it records is 2 decibels. What change in volume does that correspond to?

According to the second part of the theorem, the change is by a factor of $10^{2/10} = 1.58$. The smallest increase it records is by a factor of 1.58, a 58% increase. This type of increase in volume is noticeable, but rather minimal and certainly not dramatic.

STILL MORE ABOUT EXAMPLE 2

If one sound is 500 times as loud as a second sound, how do their decibel levels compare?

By the first part of the theorem, their decibel levels differ by $10\log 500 = 27$ decibels. ▬

In chemistry, the measure of acidity called "pH" is on a similar logarithmic scale (Problems B11–13.)

When Changes are Multiplicative
To analyze investments it is appropriate to record changes by giving the **factor** that produces the change, rather than by giving the difference of the values.

EXAMPLE 3 Suppose Stock A went up $2 last year and Stock B went up $1. Which was the better investment?

What is the standard of comparison for "better"? If Stock A started the year at $100 and went up $2, it went up 2 percent. If Stock B started the year at $5 and went up $1, it went up 20 percent. If you had invested $1000 in each, Stock B would have earned $200 and Stock A only $20. Stock B was by far the better investment.

Investment performance is properly described, not by the amount earned, but by the **percent** earned. ▬

Reporting Stock Market Gains
Invested money tends to grow over time. How can a graph exhibit useful information about the growth rate of money?

We want a graph to show the **factor** by which the money grows, not the amount by which it grows. In the context of investments, a change from $10,000 to $20,000 is equivalent to a change from $100,000 to $200,000, because in both cases the money doubled. The usual type of graph would be misleading by showing the second change 10 times as large. A "change-of-scale" to a logarithmic scale can fix that.

EXAMPLE 4   To graph the value of an investment of $10,000, label the first horizontal line "$10,000," the second "$20,000," and the third "$40,000" (not "$30,000"). Label each line double the preceding line (Figure 1). The fourth is $80,000, the fifth $160,000, etc. Then a change of one vertical unit means the money doubled, regardless of the amount.

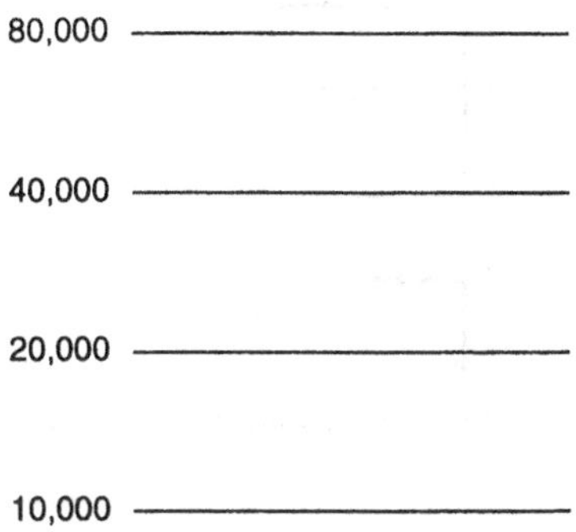

The points are computed using logs. Let the value at time t be $v(t)$. Compare that to the initial $10,000: $\frac{v(t)}{10,000}$. Take the log of that.

Here is a table of the value of the Templeton Fund:

80,000 ————————————

40,000 ————————————

20,000 ————————————

10,000 ————————————

FIGURE 1 A logarithmic vertical scale

t	Value, $v(t)$	$\log_2\left[\frac{v(t)}{10,000}\right] = L(t)$
1955	10,000	0
1960	18,000	.85
1965	25,000	1.32
1970	46,000	2.20
1975	102,000	3.35
1980	345,000	5.1
1985	660,000	6.05
1990	1,531,000	7.26

The first horizontal line corresponds to "0" in the $L(t)$ column ($\log 1 = 0$). Now plot L but label it "v" anyway (Figure 2).

Figure 2 does not use a "uniform" vertical scale, it uses a "logarithmic" scale. It is a "semi-log" graph. "Semi" means "half." The horizontal scale is uniform, but the vertical half is logarithmic.

Money graphs tend to emphasize growth by a factor of 2. In many other applications growth by a factor of 10 is emphasized. You can buy such graph paper, called "semi-log" graph paper, lined and ready to go (Figure 3).

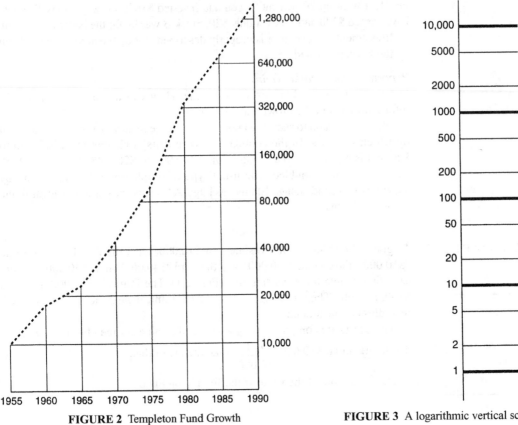

FIGURE 2 Templeton Fund Growth

FIGURE 3 A logarithmic vertical scale, base 10 ▬▬

■ Exponential Model

Money compounded continuously at a constant rate of growth fits the Exponential Model of growth. Of course, the growth rate of money does not remain constant over 35 years; rates go up and down. But, for comparison, the Exponential Model is appropriate. Let's see how a constant growth rate would look on a logarithmic scale such as in Figures 1, 2, and 3.

The exponential model can be converted to a linear model by a change of variable. Here is the exponential model 5-2-7 again with "P" for "$P(0)$":

$$P(t) = P\,e^{kt}. \qquad (5\text{-}3\text{-}5)$$

The parameters are k and P. The dependent variable is $P(t)$. Watch what happens if we, instead, use the natural log of the variable.

$$\begin{aligned}
\ln[P(t)] &= \ln(P\,e^{kt}) \\
&= \ln P + \ln(e^{kt}) \qquad &\text{[log of a product, 5-1-1L]} \\
&= \ln P + kt \qquad &\text{[inverses, 5-1-4L]}.
\end{aligned}$$

Now call $\ln[P(t)] = y(t)$ and $\ln P = y_0$. Then

$$y(t) = y_0 + kt. \qquad (5\text{-}3\text{-}6)$$

This is the Slope-Intercept form of a line with variable t and slope k (3-1-7).

This tells us that graphing $\ln[P(t)]$ instead of $P(t)$ itself yields a line (which has *constant* slope instead of the ever-increasing slope of an exponential growth curve).

If the Templeton Fund growth rate had been constant, the graph in Figure 2 would be a straight line. It appears to tend upward like a straight line, but there are many irregularities over the short term. That's the way it is with money.

EXAMPLE 5   Figure 4 illustrates the growth of a bacterial colony on a logarithmic scale. Time is given in hours. The horizontal lines are one logarithmic unit (base 2) apart, so they represent a doubling in numbers of bacteria. What is the doubling time?

The graph is a straight line, which means the exponential growth is at a constant rate. The doubling time is visible on the graph because it is the time between crossing one horizontal line and the next. It appears to be about $\frac{1}{2}$ hour. The semi-log graph makes it easy.

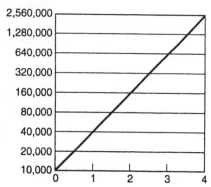

FIGURE 4 Bacterial growth on a semi-log scale

■ Putting an Exponential Curve Through Two Points

Linear models ($y = mx + b$) are so important that you study how to put a line through two points. The slope is a critical concept. Many mathematical models use the exponential function. The exponential growth rate is a critical concept.

The exponential model 5-2-7 and 5-3-5 is repeated here:

The Exponential Model

$$P(t) = P(0)\, e^{kt}.$$

This is a family of models with two parameters, $P(0)$ (the initial population), and k (the growth rate). Usually with two facts we can determine two parameters.

EXAMPLE 6   After 2 hours a population undergoing exponential growth is 100,000 and after 3.7 hours it is 980,000. Find the growth rate and the initial number.

The question asks you to find the specific model from the family of exponential models. The problem is to find $P(0)$ and k to identify the exact formula.

Plug the given information into the general model:

$$P(2) = 100,000 = P(0)\, e^{2k}.$$

$$P(3.7) = 980,000 = P(0)\, e^{3.7k}.$$

This is a system of two equations with two unknowns. Since both expressions on the right have $P(0)$ as a factor, eliminate $P(0)$ by dividing.

$$\frac{980{,}000}{100{,}000} = \frac{P(0)\, e^{3.7k}}{P(0)\, e^{2k}} = \frac{e^{3.7k}}{e^{2k}}$$
$$= e^{3.7k - 2k} = e^{1.7k}.$$

So, $9.8 = e^{1.7k}$. Take the natural logarithm of both sides,

$$\ln 9.8 = 1.7k. \quad k = \frac{\ln 9.8}{1.7} = 1.343.$$

Now, plug this back into either original equation to find $P(0)$.

$$100{,}000 = P(0)\, e^{2(1.343)}. \quad P(0) = \frac{100{,}000}{e^{2(1.343)}} = 6821.$$

So the specific model is

$$P(t) = 6821\, e^{1.343t}.$$

This formula can be used to answer any question about what the population will be at any time (as long as exponential growth continues at the same rate).

The equation for the growth rate k (obtained by dividing the original two equations) ended up using the *difference* in times ($3.7 - 2 = 1.7$) and the *factor* by which the population grew $\left(\dfrac{980{,}000}{100{,}000} = 9.8\right)$. The initial population canceled out. This always happens. Time zero plays no special role in these problems. We can pick time zero to be whenever we want without changing anything essential.

Example 6 mentioned a population of 100,000 "after 2 hours." Naturally we called that time 2, that is $t = 2$. But, in exponential model problems, we can choose time zero wherever we like. We could let $t = 0$ refer to the time when there were 100,000, in which case two hours earlier would have been time -2.

**EXAMPLE 6
ANOTHER WAY**

Reread the problem. Choose the time scale such that time 0 is when there was 100,000. According to the example, there were 980,000 "after 3.7 hours," which would then be at $t = 1.7$. Then the question asks about the number two hours before there were 100,000. That would be at time $t = -2$. Here is the model:

$$P(t) = P(0)\, e^{kt}.$$

We know $P(0)$ because we chose time 0 to correspond to 100,000. So the model is

$$P(t) = 100{,}000\, e^{kt},$$

where only k remains to be discovered. Use the fact that $P(1.7) = 980{,}000$.

$$980{,}000 = 100{,}000\, e^{1.7k}.$$

$$9.8 = e^{1.7k}. \quad \ln 9.8 = 1.7k. \quad k = \frac{(\ln 9.8)}{1.7} = 1.343,$$

as before. So, $P(t) = 100{,}000\, e^{1.343t}$,

Now, to find the "initial" number, plug in $t = -2$:

$$P(-2) = 100{,}000\, e^{1.343(-2)} = 6821,$$

as before.

Putting an exponential curve through two points is particularly simple when we locate one of the points at $t = 0$, because then $P(0)$ is given. Then the other point is used to create one equation for the growth rate, k. The only trick is to remember where the horizontal scale begins so that times are put in the proper position relative to time $t = 0$ (here, "3.7 hours later" became $t = 1.7$).

If you do not want to adjust time zero, you can always use the method of Example 6. Here it is in abstract form.

EXAMPLE 7 Fit the exponential model through the two points (s, u) and (t, v).
The two equations are

$$u = P(0)\, e^{ks} \text{ and } v = P(0)\, e^{kt}.$$

Divide to obtain

$$\frac{v}{u} = \frac{e^{kt}}{e^{ks}} = e^{k(t-s)}.$$

Solve for k by taking natural logs

$$\ln\!\left(\frac{v}{u}\right) = k(t - s). \quad k = \frac{\ln\!\left(\frac{v}{u}\right)}{(t - s)}.$$

Now use either of the original two equations to solve for P_0.

CONCLUSION Logarithms convert multiplication to addition. When comparisons use multiplication rather than addition, we may wish to "change variables" or "change scales" using logarithms. The Richter and decibel scales are well-known examples.

Suppose x and y are related. When adding 1 to any x always **adds** m to y, the result is a line with slope m. However, when adding 1 to any x always **multiplies** y by k the result is an exponential. The exponential model (5-3-5) can be converted into a linear model (5-3-6) by taking logarithms. This converts graphs with curves into straight-line graphs that are easy to interpret because the growth rate is simply the slope of the line. The growth of invested money should be graphed using this change-of-scale.

There is a straightforward method of putting an exponential curve through two points (Example 7).

Terms: micron, Richter scale, decibel, exponential model, change of variable, change of scale.

Exercise 5-3

A

A1.* ☺ On the Richter scale, a difference of 1 corresponds to what in terms of the size (amplitude) of the waves of the earthquakes?

A2. ☺ If the waves of an earthquake are 100 times the size (amplitude) of those of 4.3 earthquake, what does it register on the Richter scale?

A3. If the waves of an earthquake are twice the size (amplitude) of those of 4.3 earthquake, what does it register on the Richter scale?   [4.6]

A4. How do the waves of a 5.2 earthquake compare to those of a 4.6 earthquake?

A5. How do the waves of a 7.8 earthquake compare to those of a 5.2 earthquake?   [400 times]

A6. ☺ If one noise is 10 times as loud as another, how do they differ on the decibel scale?

A7. ☺ An 80 decibel noise is how many times as loud as a 60 decibel noise?

A8. A 90 decibel noise is how many times as loud as a 55 decibel noise?   [3200 times]

A9. If a noise is twice as loud as a 50 decibel noise, how many decibels is it?   [53]

A10. If a noise is 1000 times as loud as a 50 decibel noise, how many decibels is it?

B

B1. (A) ☺ Logarithms convert multiplication to _____.
  (B) It may be appropriate to graph data on a logarithmic vertical scale. Why?

B2. The amount of light that penetrates water is related to the clarity of the water and the distance the light must penetrate. The model is $I(d) = I(0)\, e^{-kd}$, where d is the distance the light penetrates the water, $I(d)$ is the intensity of the light after penetrating distance d, and k is a parameter that describes how clear the water is.
  (A) If half the light reaches a depth of 10 feet, find k.   [.069]
  (B) How deep will it be where only 1% of the light reaches?   [66]

B3. (See B2) If $\frac{1}{3}$ the light reaches a depth of 20 feet, how much light will reach a depth of 90 feet?   [.0071]

B4. (See B2) (A) If $I(0) = 20$ and $I = 5$ when $d = 10$, find k.   [.14]
  (B) Find I when $d = 15$.   [2.5]

B5. (See B2.) If $I(0) = 100$ and $k = .4$, find d such that $I = 60$.

B6. One type of sheet glass lets through 96% of the light that hits it. Give a mathematical model for the amount of light passing through any number of sheets of this glass.

B7. On Jan. 1, 1989 Sally invested some money which was compounded continuously at a constant rate until Jan. 1, 1999. On that terminal date it was $77,217.82. On Jan. 1, 1995 it was $48,746.34. How much did she invest originally and at what rate?   [$24,000 . . .]

B8. Newton's Law of Cooling yields a model for changes in temperature of small objects moved to cooler or warmer surroundings.

Newton's Law of Cooling Model:
  $T(t) = T_m + (T_0 - T_m)e^{-kt}$, where t is elapsed time, $T(t)$ is the temperature of the small object at time t, T_0 is its initial temperature, and T_m is the temperature of the surrounding medium. The rate of cooling is described by k.
  (A) How many parameters does the model have?
  (B) How many pieces of information will be needed to specify the exact model?
  (C) Suppose T_m is less than T_0. Sketch and label a graph of $T(t)$.

B9. (See B8 for the model) Suppose a small metal ball at 70° is put in boiling water at 212°. Suppose further that $K = 3$ when time is measured in minutes. Find the elapsed time until the ball reaches 150°.   [.28]

B10. (See B8 for the model) A can of soda at 70° is immersed in ice water at 32°. In two minutes it is 55°. Find the formula for its temperature at any time.

B11. pH in chemistry. Acidity is measured in terms of pH, which is defined as the negative of the common logarithm of the molar hydronium-ion concentration. That is, pH $= -\log[H_3O^+]$. Pure water has pH $= 7$, which is considered neutral, neither acidic nor basic. Acids have lower pH numbers.
  (A) Suppose orange juice has $[H_3O^+] = 2.80 \times 10^{-4}$. Give its pH.
  (B) Suppose milk has pH 6.4. What is its ion-concentration?

B12. [See B11] "Normal" rain has a pH of 5.6. If an acid rain has pH 2.6, how much more acidic than normal is it?

B13. [See B11 and B12] If a rain is 50 times as acidic as normal rain, what pH is it?

B14. Star Magnitude. In ancient times people classified stars by how bright they appeared. A star of the first magnitude was distinctly brighter than a star of the second magnitude, etc. Modern astronomers put that intuitive scale on a mathematical basis. They decided that a difference of 5 magnitudes would be equivalent to the brighter star being 100 times as bright as the dimmer star. What does a difference of 1.00 magnitude correspond to?

B15. The Logistic Model. In the real world, exponential growth cannot go on forever. Animal populations run out of food or space. A simple differential equation from calculus yields a growth model with a finite limit to the population. Let b the limiting population and k be a parameter for the growth rate. Then

$$P(t) = \frac{bp}{p + (b - p)e^{-kt}},$$ where p is the population at time 0.

(A) When t is very large, what value is $P(t)$ near?

(B) Let $k = .1$ and $b = 5000$. If $p = 1000$, when will $P(t)$ be 4500?

B16. [Use the logistic model from the previous problem.] If the limiting population is 1,000,000 and $k = .03$ and $p = 2000$, when will the population be 950,000?   [310]

B17. [Use the logistic model from B15.] Let $b = 100$ and $p = 150$ and $k = .02$. Sketch the graph of $P(t)$.

Label both scales. [It should start with population 150 and become population 100.]

B18. [Use the logistic model from B15.] Let $b = 1000$ and $p = 100$ and $k = .05$. Sketch the graph of $P(t)$. Label both scales. [It should start with population 100 and become population 1000.]

B19. Solve for m: $\dfrac{m^2 e^{-m}}{2} = .1$.   [.61 or ...]

B20. The chance of making n baskets in a row is approximately p^n where p is the probability (between 0 and 1) of making one basket. Find the highest value of n such that the chance is still above $\dfrac{1}{2}$.

Reading and Writing Mathematics

B21. One earthquake has waves with amplitude a, and a second, larger one, has waves with amplitude b. Find $\log\left(\dfrac{b}{a}\right)$ in terms of their Richter scale values.

B22. Prove the second half of Theorem 5-3-4.

B23. Do B6. How does the model in B2 compare to the model in B6? Is there a significant difference?

B24. Show why, if a quantity Q grows exponentially with time, then $\ln(Q)$ grows linearly with time.

Section 5-4 MORE APPLICATIONS

Power functions such as $y = x^2$ have curved graphs. However, by graphing $\log y$ above $\log x$ (instead of graphing y above x, as we have always done until this chapter), the graphs of power functions become straight lines. Because we know so much about lines, we sometimes choose to make this change-of-variable.

THE POWER MODEL $y(x) = cx^p$, for some p.

EXAMPLE 1   One example of a power model we have seen is Kepler's Third Law (Example 4-3-14). The time required for a planet to orbit the sun, y, depends upon the mean distance of that planet from the sun, x (these were called by different letters in Example 4-3-14, but that does not matter). Kepler's law is

$$y = cx^{3/2}, \text{ for some } c. \tag{5-4-2}$$

When he started, Kepler did not know the right mathematical description of the relationship between x and y. He did not know a power model was right, and he certainly did not know the right power was $\dfrac{3}{2}$. It took him years to figure out the relationship 5-4-2 from the data he had. Here is how we might do it nowadays with good data and statistical methods that use the upcoming "log-log" idea. The "time" and "distance" columns in the following table give the primary data (Figure 1).

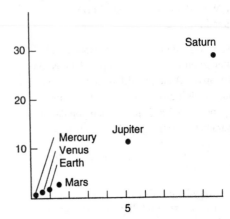

FIGURE 1 Data for Kepler's Third Law. distance [0, 10] by time [0, 40]

Table 5-4-3 Data for Kepler's Law				
Planet	distance (astronomical units)	time	log(distance)	log(time)
	x	y	log x	log y
Mercury	.387	.241	−.412	−.618
Venus	.723	.615	−.141	−.211
Earth	1	1	0	0
Mars	1.524	1.881	.183	.274
Jupiter	5.202	11.862	.716	1.074
Saturn	9.359	29.457	.980	1.469

Distance is measured in "astronomical units" in which the distance of the Earth to the Sun is 1 unit. Plotting that data gives a curve that looks like some power of x, but it is difficult to identify the precise power (Figure 1). Also, too many planets are crammed in the bottom left corner near the origin. The usual uniform scale is awkward.

The table also includes two more columns giving the logs of the distance and time. These are the "change-of-variable" columns.

To see what the two "log" columns can do, take logs of both sides of the Power Model:

$$y = cx^p \qquad\qquad [5\text{-}4\text{-}1]$$

$$\log y = \log(cx^p) \qquad\qquad [\text{taking logs}]$$

$$= \log c + \log(x^p) \qquad\qquad [5\text{-}1\text{-}1\text{L}]$$

$$= \log c + p \log x. \qquad\qquad [5\text{-}1\text{-}3\text{L}]$$

Now change variables. Let $\log y = Y$ and $\log x = X$. This is the change of variable. Log c is a constant; call it, say, b. Rewriting it,

$$Y = pX + b. \tag{5-4-4}$$

You have seen this before. This is the Slope-Intercept form of a line, with "p" denoting the slope. So, plot the X's and Y's (instead of the x's and y's) and a line will fit through the points (if a power model is correct). Figure 2 plots the data about the planets as (X, Y) pairs. The slope is the desired value of p (problem A1). If the log-log data fit a straight line, then the relationship of the original data is given by the Power Model 5-4-1.

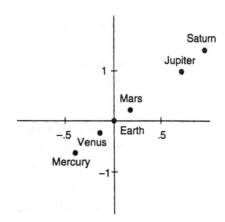

FIGURE 2 The planetary data on a log-log scale

Special graph paper called "log-log" graph paper is used to plot the original values on logarithmic scales directly. Both scales are labeled like the logarithmic vertical scale in Figure 5-3-3. ▬

■ Original Applications

The original application of logarithms was to facilitate multiplication. The values of logs and exponentials were looked up in a table. Now we have calculators which are far more convenient. Logarithms convert multiplication to addition, using properties 5-1-4 and 5-1-1. Here is the theory:

$$a \times b = 10^{\log a} \times 10^{\log b} = 10^{\log a + \log b}. \tag{5-4-5}$$

This is an identity and, like other identities, it expresses an alternative procedure. Therefore, to multiply a times b (expressed on the left of the identity) we may instead do the procedure expressed on the right of the identity:

1. Find $\log a$
2. Find $\log b$
3. *Add* $\log a$ to $\log b$ to obtain the sum. Call it c.
4. Find 10^c, the answer, which is the product ab.

Line 3 exhibits the step where multiplication is converted to addition.

In the old days, the log and exponential functions would have been evaluated with a huge table of values.

Does this seem long? Actually, with practice, people in the pre-calculator age found it a relatively fast way to multiply many-digit numbers by each other. These properties are the principle on which the old engineer's "slide rule" was based (Problem B5). Now the principles are still interesting, but slide rules are obsolete and multiplication by using logarithms is positively medieval. Nevertheless, try the next example.

CALCULATOR EXERCISE 1 Find 56×78 using the above approach, with your calculator replacing a table of logarithms.

> Find log 56. Find log 78. Add them. Find 10 to that power.
> Now check your answer by multiplying the usual way.

Logarithms can be used to evaluate powers. Here is an identity which expresses the theory:

$$b^p = (10^{\log b})^p = 10^{p \log b}. \tag{5-4-6}$$

As with any identity, the right side expresses an alternative procedure to evaluate the left side. To find b^p for any $b > 0$ and any p,

1. Find log b.
2. *Multiply* it by p, to obtain their product. Call it c.
3. Find 10^c, the answer, which is the power b^p.

Prior to the invention of computers this was the only practical way to evaluate non-integer powers.

CALCULATOR EXERCISE 2 Find $2^{4.6}$ using this approach, with your calculator replacing a table of logarithms.

> Find log 2. Multiply it by 4.6. Find 10 to that power. Now check your answer against the usual direct way to evaluate $2^{4.6}$.

> Of course, we do not need logs to multiply, but your calculator does evaluate general powers with a method like the one just expressed (although usually using natural logarithms with base e). These exercises illustrate the remarkable properties of exponential functions and logarithms (Problem B4).

■ Why e?

It takes calculus to fully understand why the preferred base for logarithms is base e. Nevertheless, some reasons can be explained here. For example, the use of base e in the "continuous compounding" model above can be explained.

The usual compound-interest formula is 5-2-3:

$$A = P\left(1 + \frac{r}{k}\right)^{kt},$$

where k is the number of times per year the money is compounded. What happens to the effective yield as k increases?

First consider compounding yearly, then semiannually (twice a year), then quarterly, monthly, daily, and even more frequently. It turns out that the growth factor approaches e^{rt}, the growth factor in the "continuous compounding model."

CALCULATOR EXERCISE 1 Let $t = 1$ and $r = 1$. Evaluate $\left(1 + \dfrac{1}{k}\right)^k$ for various values of k, including large k.

k	1	2	4	12	365	$\rightarrow$
$\left(1 + \dfrac{1}{k}\right)^k$	2	2.25	2.44	2.61	2.714	$\rightarrow 2.718\ldots = e$

As k goes to ∞, $\left(1 + \dfrac{1}{k}\right)^k$ goes to e (This is the definition of e.)

Graph $\left(1 + \dfrac{1}{x}\right)^x$. As x increases, this increases to e (Problem A20).

$$\text{As } k \text{ goes to } \infty, \left(1 + \frac{r}{k}\right)^{kt} \text{ goes to } e^{rt}. \tag{5-4-7}$$

This shows that base e arises naturally as a limit of compounding.

■ Polynomial Approximations

A number like e, $2.718\ldots$, seems like such an unnatural base for a "natural" exponential function and "natural" logs. What could be natural about such a base? For one thing, it yields very simple polynomial approximations.

The image e^x can be closely approximated by "$1 + x$" for small x. No other base yields such a simple approximation.

CALCULATOR EXERCISE 2 Compare "e^x" to "$1 + x$" for various small values of x (Figure 3).

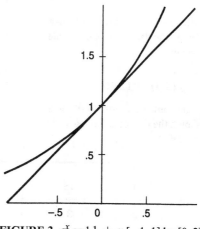

FIGURE 3 e^x and $1 + x$ $[-1, 1]$ by $[0, 2]$

x	e^x	$1+x$	Difference
0	$e^0 = 1$	1	none
.01	$e^{.01} = 1.01005$	1.01	.00005
.1	$e^{.1} = 1.10517$	1.1	.00517

When x is small, there is little difference.

There are better approximations which use higher-degree polynomials (Problem 4-1-B41).

Similarly, natural logarithms have a simple polynomial approximation when the argument is near 1: $\ln(x)$ is near $x - 1$, for x near 1. Graph them both and see (Problem B16).

The point is, polynomial approximations of b^x and $\log_b x$ are particularly simple when $b = e$.

CONCLUSION Taking logs converts a power model into a straight-line model. This can be seen by graphing the logs of the data (or using a log-log scale). The slope of the resulting line is the power.

Terms: power model, change of variable, change of scale.

Exercise 5-4

A

A1. Use the log data from Mercury and Saturn to compute the slope of the line in Figure 2. Is your answer near 1.5?

A2. Use the logarithmic method of Calculator Exercise 1 to multiply 42×36. Show the results of the four steps.

B

B1. The following data come very close to fitting a power model. Graph the data on a log-log scale. Then find the model.

$$f(1.1) = .8, \ f(1.6) = 2.04,$$
$$f(2.1) = 4.0, \ f(3.3) = 12.46.$$

B2. The following data come very close to fitting a power model. Graph the data on a log-log scale. Then find the model.

$$f(.4) = .77, f(1.1) = 2.58,$$
$$f(1.8) = 4.66, f(2.4) = 6.58.$$

B3. Kepler's Third Law gives the periods of planets as a function of their distances from the sun. Assuming circular orbits (not correct, but not far wrong), derive speed in orbit as a function of distance from the sun.

B4. State an identity in the spirit of the ones used in Calculator Exercises 1 and 2 that expresses a way to divide using logarithms.

B5. Suppose two meter sticks are marked with "2" at position log 2, that is, with "2" at 30.1 centimeters from the left end. Similarly, suppose they are labeled "3" at log 3, "4" at log 4, etc. Now, slide one along the other until its left end is above position "2" of the other (see the Figure).

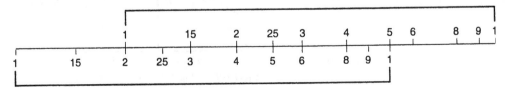

Two similar rules, the upper one slid to the right

Explain how this "slide rule" can facilitate multiplication by any number. Which property is the key? How can you multiply 25 × 30 with it?

B6. (A) Use Kepler's law and the data for Earth to determine "*c*" in formula 5-4-2.
(B) Use the data for Mars to verify that the same value of "*c*" works for Mars.

B7. Assume that Earth is 93,000,000 miles from the sun.
(A) Determine the speed, in miles per hour, of Earth around the sun. [67,000]
(B) Determine the speed, in miles per hour, of Jupiter around the sun. [28,000]

Properties of Exponents and Logs

This table of 2^x is used in the remaining problems.

x	−3	−2	−1	0	1	2	3	4	5	6	7	8	9	10
2^x	$\frac{1}{8}$	$\frac{1}{4}$	$\frac{1}{2}$	1	2	4	8	16	32	64	128	256	512	1024

B8. Multiplying in the bottom row is equivalent to adding in the top row. For example, find 8 and 16 in the bottom row. Multiplying, 8 × 16 = 128. The numbers 8 and 16 correspond to 3 and 4 in the top row. Adding, 3 + 4 = 7. The number 7 in the top row corresponds to the product, 128.

This works because exponents are logs and the log of a product is the sum of the logs. Fill in the blanks in this string of equalities to represent the calculations above.

$$8 \times 16 = \underline{\hspace{1cm}} = \underline{\hspace{1cm}} = 128.$$

B9. Taking powers of numbers in the bottom row is equivalent to multiplying in the top row. For example, find 4 in the bottom row. Cubing it, $4^3 = 64$. The number 4 in the bottom row corresponds to 2 in the top row. Multiplying by 3: 2 × 3 = 6. The number 6 in the top row corresponds to 64, the cube. This works because exponents are logs (so the top row gives logs of the numbers in the bottom row), and

logs convert powers into multiplication (5-1-3 and 5-1-3L). Fill in the blanks in this string of equalities to represent the calculations above.

$$4^3 = \underline{\hspace{1cm}} = \underline{\hspace{1cm}} = 64.$$

B10. Use the table to show how "512 divided by 16" can be computed using subtraction. (Parallel the above problems).

B11. Use the table to show how "$128 \times \frac{1}{8}$" can be computed using addition.

B12. Use the table to show how $16^{1.5}$ can be computed using multiplication.

B13. Use the table to show how $\sqrt{1024}$ can be computed using multiplication. (Treat the square root as the $\frac{1}{2}$ power.)

B14. Consider the table above.
(A) Adding numbers of the top row corresponds to _____ numbers in the bottom row.
(B) Dividing numbers in the bottom row corresponds to _____ numbers in the top row.

B15. Consider the table.
(A) Multiplying numbers in the bottom row corresponds to _____ numbers in the top row.
(B) Multiplying a number in the top row by 3 corresponds to _____ the number in the bottom row.

B16. There is a simple linear approximation to ln *x* if *x* is near 1.
(A) Find ln 1.01, ln 1.02, and ln .97.
(B) When *x* is near 1, ln *x* is near what line?

CHAPTER

6

Trigonometry

Section 6-1   GEOMETRY FOR TRIGONOMETRY

Trigonometry is the branch of mathematics about the relationships of the sides and angles of triangles. Trigonometry also applies to periodic behavior such as the circular motion of rotating engine parts, alternating current electricity, the swing of a pendulum, and waves of all kinds including sound, light, and radio waves. The same trigonometric functions used for triangles are regarded as "circular" functions in the context of motion. The "circular" interpretation of trigonometric functions is important in engineering, physics, and calculus. It will be discussed in Chapter 7. However, as is traditional, we begin with triangle trigonometry.

■ Geometry for Trigonometry

A triangle has six <u>parts</u>—three sides and three angles. To <u>solve</u> a triangle means to find the measures of all the sides and angles.

> We are interested in the *measures* of the angles and sides, not their locations, so we use the term "side" to refer to a length, and "angle" to refer to a measure (at first, in "degrees"). We consider two triangles to be "the same" ("congruent") if they have the same sides and angles, even if they are in different locations.

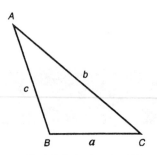

FIGURE 1 A triangle, with typical labeling

See Figure 1. Side *a* is opposite vertex (VER tex) *A*. The plural of vertex is vertices ("VER teh SEES," not "ver TEX is"). Angle *A* is at vertex *A* (using the same letter for two different things, a point and an angle). Lower case letters refer to lengths (sides). Upper case letters refer to angles and points. Angle *A* is opposite side *a* and is formed between sides *b* and c. Side *a* is between angles *B* and *C*.

■ Geometry for Trigonometry

This section reviews geometric constructions that you learned in geometry to remind you **how** and **why** a triangle is, or is not, uniquely determined by three of its parts. Most of all, it is intended to **warn you when not to expect a unique answer** when using trigonometry to solve for unknown parts of a triangle.

Consider a triangle with three parts given. Is it the *only* triangle with those three parts in that arrangement? In many cases the answer is "Yes." But in two cases the answer is "No."

Triangles are said to be <u>similar</u> if they have the same three angles, but not necessarily the same size (Figures 2A and 2B). This case is called Angle-Angle-Angle (AAA).

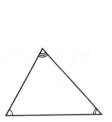

FIGURE 2A A triangle with three given angles

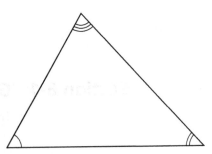

FIGURE 2B A second triangle with the same angles, but twice the size

Given only the three angles of a triangle, the shape is determined, but not the size. Similar triangles have proportional sides. For example, if one side is doubled, the other two sides are also doubled.

The next geometric case is the most interesting. A triangle is not necessarily determined in the Angle-Side-Side case (Figures 3A and 3B).

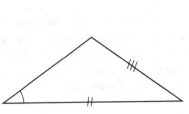

FIGURE 3A A triangle with an angle and two sides given

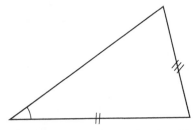

FIGURE 3B A different triangle with the same angle and same sides in the same order

The geometric construction with a compass and straightedge makes it easy to see how this can happen in Angle-Side-Side (Figures 4A and 4B).

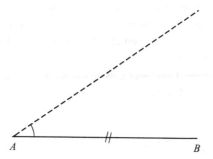

FIGURE 4A The angle and first side of Angle-Side-Side

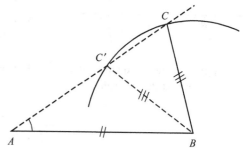

FIGURE 4B The arc from point B uses the length of the second side. It intersects the ray from A twice

Figure 4B shows that side a could be opposite angle A in either of *two* triangles. This is the tricky case in trigonometry. In the upcoming "Law of Sines" *you* must remember that the one solution your calculator finds for angle C is not necessarily the one you want (You might want angle C'). First you must recognize when the solution is not unique. Then you generate the second solution from the one your calculator gives.

All the other geometric cases of a triangle with three given parts yield unique triangles. However, the Angle-Side-Side case has additional complications that we discuss after reviewing the geometric constructions and proving two major theorems:

1. The angles in a triangle sum to 180 degrees (Theorem 6-1-4), and

2. The Pythagorean Theorem (Theorem 6-1-8).

■ Geometric Constructions and the Number of Solutions

In trigonometry, the formula you use depends upon the geometric case. Become familiar with the geometric cases so you can select the proper formula.

SIDE-SIDE-SIDE (SSS) 6-1-1   A triangle is determined by its three sides.

The geometric construction is illuminating. Begin with three given sides (Figure 5). Now construct a triangle with those sides by the following steps.

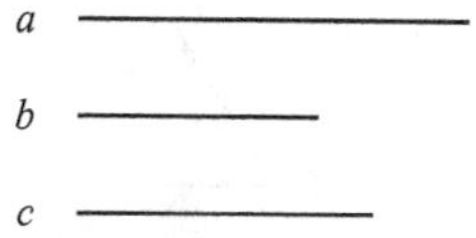

a ─────────

b ──────

c ─────

FIGURE 5 Parts for a SSS construction

1. Lay out a line anywhere (often horizontal) and reproduce length *a* on it. Label the endpoints *B* and *C* (Figure 6).
2. From *B* draw a circle (part of it, an arc, will do) of radius *c* (Figure 7).
3. From *C* draw a circle (an arc will do) of radius *b*.
4. Locate point *A* at the intersection of the two circles. (There will be two candidates for *A*, labeled "*A*" and "*A'*" which is pronounced "*A* prime," Figure 7.)
5. Connect *A* to *B* and *C* with the straightedge, thus forming the sides of the triangle (Figure 8).

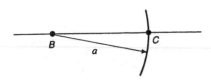

FIGURE 6 Step 1 of SSS. A line segment *BC* of length *a*

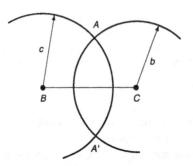

FIGURE 7 Further into the SSS construction. The vertices are located (Steps 2, 3, and 4)

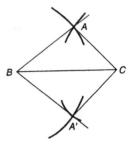

FIGURE 8 The result of the SSS construction. *ABC* and *A'BC* will be regarded as **one** triangle, because their angles and sides are the same—we are not considering location

In Figure 8, the angles and sides in *ABC* have the same measures as the angles and sides in *A'BC*. In trigonometry, different locations do not make for different triangles. In trigonometry they are "the same triangle."

SIDE-ANGLE-SIDE (SAS) 6-1-2 Two sides and the included angle determine a triangle. By "included" angle we mean the angle formed by the two sides. This case is called Side-Angle-Side (SAS) because the angle is between the sides.

Begin with sides *a* and *b* and the angle, *C*, between them (Figure 9).

1. Lay out a line anywhere and reproduce length *a* on it. Label the endpoints *C* and *B* (Figure 10).
2. Reproduce angle *C* at point *C*, creating a ray (half a line) extending from point *C* (How to reproduce an angle is the subject of Problem B18).

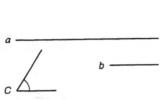

FIGURE 9 SAS. The parts

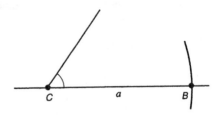

FIGURE 10 Part way through the SAS construction

3. With one end at point C, reproduce length b on the ray. Call the other end A (Figure 11).

4. Draw the line segment through A and B.

Steps 2 and 3 could be interchanged, and, in step 1, length b could have been used instead of length a. This would merely change the orientation of the triangle, but not the measures of the sides or angles.

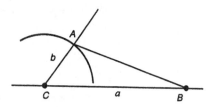

FIGURE 11 The result of the SAS construction

ANGLE-SIDE-ANGLE (ASA) 6-1-3
Two angles and the included side determine a triangle. By " included" we mean the side between the two angles.

Begin with two angles, B and C, and the included side, a (Figure 12).

1. Draw any line and duplicate side a on it. Call the points at the ends B and C.

2. Duplicate angle B at point B, creating a ray from point B (Figure 12).

3. Duplicate angle C at point C, creating a ray from point C (Figure 13).

4. Locate A where the two rays intersect.

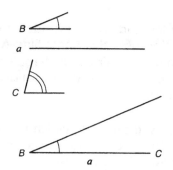

FIGURE 12 Parts for and the beginning of the ASA construction

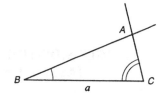

FIGURE 13 The completed ASA construction

■ Any Two Angles Yield the Third

In the Angle-Side-Angle construction the third angle was determined by the two given angles. Two angles of a triangle always determine the third.

THE SUM OF THE ANGLES OF A TRIANGLE 6-1-4
The angles of a triangle sum to a straight angle (180°).

Proof: Given triangle ABC, construct a line through A parallel to BC (Figure 14). Then the angle labeled "1" is congruent to angle B (They are called "alternate interior angles." See Problem B23 for a review of the terms for pairs of angles). The angle labeled "2" is congruent to angle C, for the same reason. Angle 1 plus angle BAC plus angle 2 form a straight angle. Therefore angle B plus angle A plus angle C sum to a straight angle.

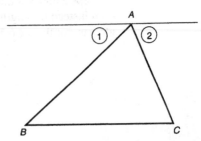

FIGURE 14 The angles of a triangle sum to a straight angle

EXAMPLE 1   Let $\angle A = 32°$ and $\angle C = 16°$. Determine $\angle B$.
   ["$\angle$" is a symbol for "angle".]
   $$\angle A + \angle B + \angle C = 180°.$$
Therefore, $\angle B = 180° - 32° - 16° = 132°$.

   Given two angles and a side, the side is either included (as in ASA, 6-1-3, above) or not included. If the side is not included, we use the previous result to obtain the third angle. Then all three angles will be known and the known side will be included between two of them. Then we can use the above ASA construction.   ▬

ANGLE-ANGLE-SIDE (AAS) 6-1-5   Two angles and a not-included side determine a triangle. The construction uses the Angle-Side-Angle construction (6-1-3) after determining the third angle (Problem B19).
   Combining ASA and AAS, we see that a triangle is determined by a side and any two angles.

ANGLE-SIDE-SIDE (ASS) 6-1-6   Two sides and an angle which is not included *almost* determine a triangle. This is the tricky case. There may be *two* triangles which have the given angle and two sides. In geometry this is easy to see, but, in trigonometry, when your calculator reports a single number (instead of *two* relevant numbers), you may forget that the other number may be the one you really want. Inspect this geometric construction to see how two triangles can result, instead of just one.

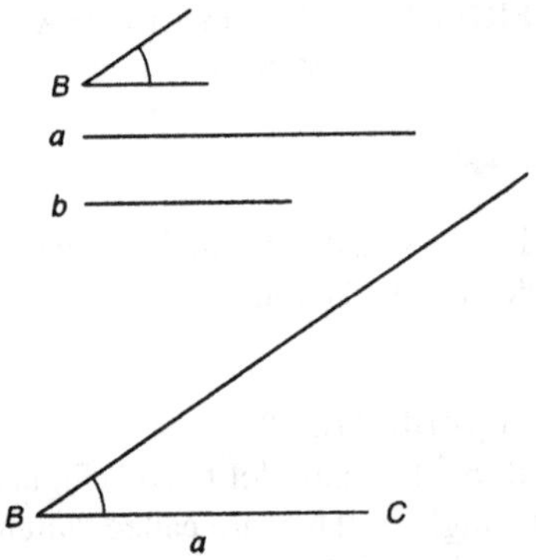

   Begin with an angle, B, and two sides, a and b (Figure 15). One side is opposite the angle, so the angle is not an "included" angle (it is not between the given sides).

   1. Lay out a line anywhere and reproduce length a on it. Call the points at the ends B and C.
   2. Reproduce angle B at point B and create the ray extending from B (Figure 15).
   3. With center at C, draw the circle of radius b (Figure 16).
   4. Label with A and A' the (possibly two) points where the circle intersects the ray (Figure 16).

FIGURE 15 Parts for, and part of, the ASS construction

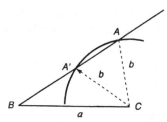

FIGURE 16 The completed ASS construction, with two triangles

This construction uses two sides and an angle that is not included. I use the abbreviation "ASS" for this case is used instead of "SSA" (which would be the same case) because it reflects the order in the picture above.

In ASS, if the side opposite the given angle (side b in Figure 16) is long enough, two triangles fit the conditions, one with an acute ($<90°$) angle at A and the other with an obtuse ($>90°$) angle at A'. The sum of angle A and angle A' is 180 ; (Problem B17). Both angles have the same sine function value. For example, if angle A were 70°, then angle A' would be 110°, and $\sin 70° = \sin 110°$. This will be important when we use the so-called "Law of Sines" to try to determine angle A from the value of its sine. There will be two possible answers, just like there are two possible triangles in Figure 16.

CALCULATOR EXERCISE 1

Evaluate $\sin 70°$. Evaluate $\sin 110°$. Then take inverse sine of each.

Evaluate $\sin 110° = .9397$. Sin 70° is exactly the same. Now, apply $\sin^{-1}$ to the image. ("$\sin^{-1}$" is called "inverse sine" or "arcsine", which is also written "arcsin".)

$$\sin^{-1}(.9397) = 70°.$$

The calculator reports an acute angle that has the given sine value. There is also an obtuse angle (110°) with the proper sine value—but your calculator will not report it. If you want an obtuse angle, you have to think of that yourself. The obtuse angle is 180° minus the given angle.

Watch for this ambiguous case when we get to the Law of Sines.

ANGLE-ANGLE-ANGLE (AAA) 6-1-7

Three angles do *not* determine a triangle, but they do determine its shape. Triangles with the same angles are said to be <u>similar</u> (Figure 17, mirror images are also similar). To describe a triangle, we need at least one side, to fix the size. Three angles alone are not enough.

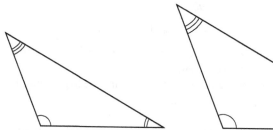
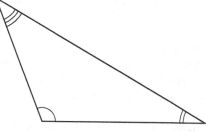
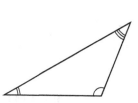

FIGURE 17 AAA. Three similar triangles. Mirror images are also similar

■ The Pythagorean Theorem

The Pythagorean Theorem is a famous measurement theorem of geometry.

THE PYTHAGOREAN THEOREM 6-1-8

Let sides a, b, and c be opposite angles A, B, and C, respectively.

$$a^2 + b^2 = c^2 \text{ if and only if angle } C \text{ is a right angle.}$$

Do *not* think of the Pythagorean Theorem as just "$a^2 + b^2 = c^2$." In most of trigonometry that abbreviation would be false! The connection to a *right* triangle is essential! The Law of Cosines (Theorem 6-3-4) in trigonometry is specifically designed to modify this result to so that non-right triangles can be measured too.

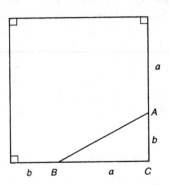

FIGURE 18 A right triangle and a square formed from its extended sides

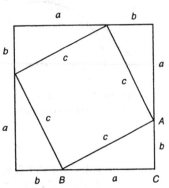

FIGURE 19 The right triangle with the square subdivided into a square and four triangles

Proof: Consider the right triangle ABC with right angle at C as in Figure 18. Extend CA and CB by lengths a and b, respectively, and form the surrounding square. Then area of the square can be computed two ways. On the one hand, its sides are "$a + b$" so its area is "$(a + b)^2$."

On the other hand, the same square can be regarded as the sum of the square on the hypotenuse and four triangles, each congruent to the original triangle (Figure 19). Its area must be the same either way.

The area in Figure 19 is $c^2 + 4\left(\dfrac{1}{2} ab\right)$. Therefore,

$$(a + b)^2 = c^2 + 4\left(\frac{1}{2} ab\right),$$

$$a^2 + 2ab + b^2 = c^2 + 2ab \quad \text{[multiplying each expression out]},$$

$$a^2 + b^2 = c^2 \quad \text{[subtracting } 2ab\text{]}.$$

This proves the "if" half of Theorem 6-1-8. The "only if" half follows, after a clever construction, from SSS (Problem B21).

The Pythagorean Theorem can be used to determine the hypotenuse from two legs, or a leg given the hypotenuse and the other leg.

EXAMPLE 2 Suppose angle C is a right angle and $a = 65$ and $b = 23$. Find c, the hypotenuse.

$$c^2 = a^2 + b^2 = 65^2 + 23^2 = 4754. \text{ So } c = \sqrt{4754} = 68.95. \quad \blacksquare$$

EXAMPLE 3 Suppose angle C is a right angle, $c = 92$ and $b = 37$. Find a.

$$a^2 + b^2 = c^2, \text{ so } a^2 = c^2 - b^2 = 92^2 - 37^2 = 7095.$$

Therefore $a = \sqrt{7095} = 84.23.$ $\quad \blacksquare$

EXAMPLE 4 $a = 24, b = 25,$ and $c = 7$. Is it a right triangle?

A quick glance shows that c is the shortest side, so c^2 cannot be the sum of a^2 and b^2. But that is *not relevant*. The letters have been used differently. Often we use "c"

for the hypotenuse of a right triangle, but that is only when "C" is the right angle. There are many cases when we need to use "a" or some other letter in the position of "c" in the theorem. After all, theorems are stated with placeholders and their results apply to other letters. Of course, in a right triangle the hypotenuse is the longest side. So, if this were a right triangle, "b" would play the role of "c" in the theorem. So we need to know if $a^2 + c^2 = b^2$.

$a^2 + c^2 = 24^2 + 7^2 = 625$, which is $25^2 = b^2$. Therefore, it is a right triangle, with right angle at B.   ▬

EXAMPLE 5 Let $a = 40, b = 50$, and $\angle C = 78°$. Find c.

Do *not* assume "$a^2 + b^2 = c^2$." That is a dangerous abbreviation of the Pythagorean Theorem which holds *only for right triangles*, and this is not a right triangle. We will need a different theorem to do this problem. The so-called Law of Cosines (coming up in Section 6-3) will handle it easily.   ▬

EXAMPLE 6 A rope is tied tight straight between two trees 10 feet apart. If it is pushed sideways at the middle, it stretches one inch. How much will it be deflected from its previous line?

One inch in ten feet doesn't seem like much. Let's determine the deflection as described by the distance CD in Figure 20. Since it is pushed sideways "at the middle," let C be midway between the trees at A and B, so AC is 5 feet. Let D be the midpoint of the stretched rope. $AD = 5$ feet $\frac{1}{2}$ inch. We can determine CD from the Pythagorean Theorem, since triangle ACD is a right triangle.

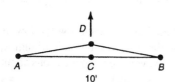

FIGURE 20 A 10-foot rope stretched one inch sideways. (View from above)

To avoid mixing feet and inches, use inches.

$AD^2 = AC^2 + CD^2$. $60.5^2 = 60^2 + CD^2$. $CD = \sqrt{(60.5^2 - 60^2)} = 7.76$ (inches). So the sideways deflection is 7.76 inches (all from a one inch stretch!)   ▬

■ Recognizing the Cases

A triangle has six parts. Sometimes three parts determine the other three.

EXAMPLE 7 A triangle has $a = 12, b = 7$, and $c = 16$. Is the triangle determined?

Yes. This fits SSS (Side-Side-Side). We can construct the triangle using the method in 6-1-1.   ▬

EXAMPLE 8 A triangle has $a = 18, b = 4$, and $\angle C = 134°$. Is it determined?

Angle C is between sides a and b, so this is a case of SAS (Side-Angle-Side). SAS always determines a triangle (construction 6-1-2, Problem A13).   ▬

EXAMPLE 9 A triangle has $\angle A = 39°$, $\angle C = 55°$, and $b = 25$. Is it determined?

Yes, b is the side between angles A and C, so this fits ASA (Angle-Side-Angle), which determines a triangle.   ▬

EXAMPLE 10 Let $\angle A = 120°$, $\angle B = 20°$, and side $a = 12$. Is a triangle determined?

This time the given side is not between the given angles, so it does not yet fit ASA. But the third angle is easy to determine because the angles sum to 180°; it must be 40°. Now the given side *is* between two known angles. So it fits ASA. The triangle is determined (Figure 21).

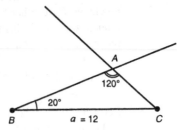

FIGURE 21 $\angle A = 120°$, $\angle B = 20°$, and $a = 12$. Determine $\angle C$ and use ASA ▬

EXAMPLE 11

A triangle has $b = 5$, $a = 10$, and $\angle B = 25°$. Is the triangle determined? Which case is it?

Draw a picture (Figure 22). Angle B is next to side a and opposite side b. So we do not know the included angle. This case is ASS, which could be called SSA. From Figure 22 we can see there will be two possible triangles. The triangle is *not* determined.

In trigonometry, both the Law of Sines and the Law of Cosines will allow us to solve for the other sides and angles. Because there are two possible triangles, the equations they create have two solutions. (I hope you are getting curious about the Laws of Sines and Cosines, but you will have to wait.) ▬

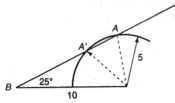

FIGURE 22 $b = 5$, $a = 10$, and angle $B = 25°$

EXAMPLE 12

Let $b = 5$ and $a = 10$ as in the previous example, but let angle $B = 40°$.

See Figure 23. Again, this is ASS. Not only does ASS often give two triangles instead of one, it may not give any triangle because the second side may not be long enough to reach the ray formed by the angle. This happens in Figure 23.

In trigonometry, when we compute the measures of the parts, this behavior would be evident from both the Law of Sines and the Law of Cosines. The equations they create would have no solutions, which would mean there is no such triangle.

Figure 22 illustrates 2 solutions and Figure 23 illustrates no solutions. There is a case in between. If the radius were a bit shorter in Figure 22 or a bit longer in Figure 23, the arc could be made tangent to the ray so they intersected just once—yielding exactly one solution.

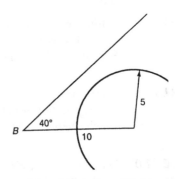

FIGURE 23 $b = 5$, $a = 10$, and angle $B = 40°$. b is not long enough to reach the ray for side c. There is no such triangle ▬

EXAMPLE 13 Let $b = 12$, $a = 10$, and $\angle B = 25°$. Is a triangle determined?

Again, this is ASS. There are similarities with Example 11, which yielded two triangles. But in this example side b is much longer. In fact, it's too long to intersect side a twice on the same side of point B (Figure 24). So one triangle is determined (with vertex A), because the other candidate (with vertex A') would not really have an angle of 25° at B, rather an angle of $180° - 25° = 155°$.

In trigonometry, when we compute the measures of the parts using the Law of Cosines, this case will be evident in that there will be two solutions for a, but one will be negative and therefore extraneous (sides are not negative). Also, the Law of Sines will yield two solutions, but one is extraneous (because the angles sum to more than 180°, which is illegal by Theorem 6-1-4).

Angle-Side-Side is by far the trickiest case. Beware.

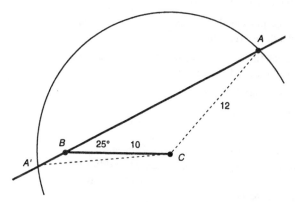

FIGURE 24 ASS. $b = 5$, $c = 12$, and $\angle B = 25°$

CONCLUSION Triangles have six parts. To "solve" a triangle means to find them from given parts. Triangles are determined by SSS, SAS, ASA, and AAS. ASS is the ambiguous case because there may be two solutions (or one, or none).

The angles of a triangle sum to 180 degrees. The Pythagorean Theorem relates the sides of *right* triangles.

Terms: part (of a triangle), solve (a triangle), SSS, SAS, ASA, ASS, AAA, included angle, included side, Pythagorean Theorem.

Exercise 6-1

A

A1.* ☺ (A) What does it mean to "solve a triangle"?

(B) In the context of solving a triangle, how many parts does it have?

A2. ☺ True or false? (Note: Congruent triangles are regarded as the same.)

(A) T F If the case Side-Side-Side yields a triangle, it yields exactly one.

(B) T F If the case Side-Angle-Side yields a triangle, it yields exactly one.

(C) T F If the case Angle-Side-Side yields a triangle, it yields exactly one.

(D) T F If the case Angle-Angle-Angle yields a triangle, it yields exactly one.

(E) T F If the case Angle-Side-Angle yields a triangle, it yields exactly one.

The measures of two angles in a triangle are given. What is the measure of the third angle?

A3. 30° and 85°   **A4.** 60° and 60°

A5. 5° and 75°   **A6.** 130° and 10°

Let a and b be the legs of a right triangle, and c be its hypotenuse. From the two given sides, find the remaining side.

A7. $a = 10$ and $b = 4$ [11]

A8. $a = 10$ and $c = 11$ [4.6]

A9. $c = 20$ and $b = 18$ [8.7]

A10. $c = 100$ and $a = 99$ [14]

A11. If the rope in Example 6 stretched 2 inches, how far would it deflect (in inches)?

A12. How much (in inches) would the rope in Example 6 have to stretch to deflect 1 foot?

A13. Sketch the triangle in Example 8.

A14. Sketch the triangle in Example 9.

B

B1.* Sketch an illuminating picture that clearly shows, geometrically, why ASS might produce two different triangles. Label it so it is clear which parts are given and which are not.

B2.* (A) What are the abbreviations of the cases that determine a unique triangle?
(B) Which case may yield two different triangles?
(C) Which case determines similar, but not necessarily congruent, triangles?

Identify the geometric case (SSS, SAS, etc.) illustrated by the marked parts in the picture and then determine if the marked parts determine a unique triangle, or if there might be another triangle with marked parts with the same measures.

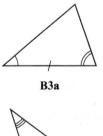

B3a

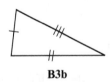

B3b

B3c

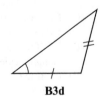

B3d

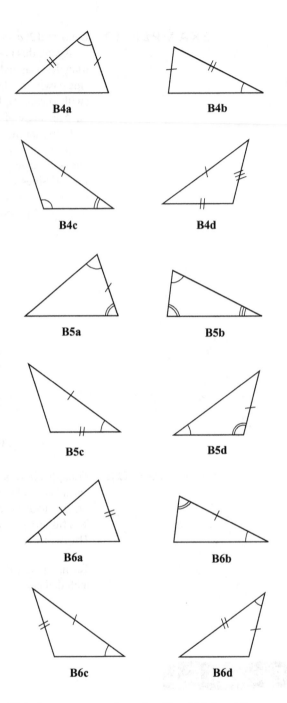

B4a   **B4b**

B4c   **B4d**

B5a   **B5b**

B5c   **B5d**

B6a   **B6b**

B6c   **B6d**

B7.* Suppose two angles and a side which is not the included side are given.
(A) Which geometric case other than AAS is most relevant?
(B) How do these parts fit that case, in spite of the fact that the side is not the included side?

B8. It is possible for ASS to yield exactly one triangle (in two distinctly different ways!). How?

The area of a triangle is one half the base times the height. Find the area of the isosceles triangle with the given sides.

B9. 7, 7 and 5 [16]   **B10.** 10, 10 and 2 [9.9]

B11. 10, 10, and 10 [43] **B12.** *s, s,* and *s*

B13. Suppose you wish to precisely measure the length of a board that is about 8 feet long. If you use a tape measure precisely parallel to the edge of the board, you should get the right measurement.

(A) However, if you measure with the tape one half inch out of parallel and find it to be exactly 8 feet long, how long is it really? (At one end the tape is on the edge but at the other it is one half inch in from the edge.)

(B) How much shorter (in inches) is it than you measured?

(C) What does this tell you about how important that measurements be taken precisely parallel to the edge?

B14. Suppose you intend to measure the length of a rectangular room. You measure BD and find it to be 30 feet, 2 inches long. You did not notice your tape measure was not parallel to the wall. It was 6 inches closer to the wall at one end than at the other (AB is six inches less than CD). How long is the room, really? (How long is AC?)

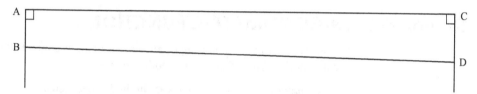

B15. Consider a long thin right triangle with long leg equal to 10 feet and hypotenuse equal to 10 feet, one eighth inch. How long is the short leg (in inches)? This tells you that if you measure across a 10 foot room slightly out of parallel, the measurement will still be pretty close to correct.

Geometry

B16. Outline a proof that "$a^2 + b^2 = c^2$" for a right triangle. Do the algebra.

B17. Explain why, in Figure 16, angle BAC and angle $BA'C$ sum to 180 degrees.

B18. Describe the geometric sequence of steps for reproducing a given angle in a new location, using only a compass and straightedge. Which geometric case is most relevant?

B19. The AAS construction is the most difficult. Given two angles and the side opposite one of them, how can you construct a triangle? [Do not gloss over the non-trivial steps.]

B20. Outline a proof that the angles of any triangle sum to 180°.

B21. Show the "only if" of Theorem 6-1-8. Assume $a^2 + b^2 = c^2$ and show that C must be a right angle. Use SSS (6-1-1) after constructing a triangle with sides a and b and a right angle between them.

B22.* (A) What basic geometrical construction operations are allowed with a straightedge?

(B) What basic geometrical construction operations are allowed with a compass?

Review of Terms about Pairs of Angles.

Intersecting lines form "linear pairs" of adjacent angles. In the figure, angles 1 and 2 form a linear pair (So do 2 and 3, and 3 and 4, and 4 and 1). Intersecting lines form pairs of "vertical angles" on opposite sides of the point of intersection. Angles 1 and 3 are "vertical angles" (So are 2 and 4, 5 and 7, and 6 and 8).

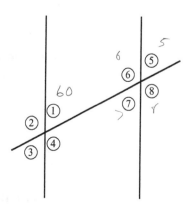

Two parallel lines cut by a transversal

(A) Linear pairs of angles are supplementary. (They sum to 180°.) [For example, 1 and 2 sum to 180°.]

(B) Vertical angles are congruent. [For example, 1 and 3 are congruent.]

A line that intersects two other lines is a "transversal" and forms eight angles as in the figure. Certain pairs have special names. Two angles in corresponding positions are said to be "corresponding angles" (angles 1 and 5, 2 and 6, 3 and 7, and 4 and 8 are corresponding). Two angles between the lines and on opposite sides of the transversal are "alternate interior angles" (1 and 7 are alternate interior angles, as are 4 and 6). Two angles outside the lines and on opposite sides of the transversal are "alternate exterior angles" (2 and 8 are alternate exterior angles, as are 3 and 5).

Let two lines be crossed by a transversal. The two lines are parallel iff

(C) Any two corresponding angles are congruent. [For example, 1 and 5 are congruent.]

(D) Any two alternate interior angles are congruent. [For example, 1 and 7 are congruent.]

(E) Any two alternate exterior angles are congruent. [For example, 2 and 8 are congruent.]

B23. If angle 1 is 60°, find angles 5, 6, 7, and 8.

B24. If angle 6 is 115°, find angles 1, 2, 3 and 4.

Section 6-2   TRIGONOMETRIC FUNCTIONS

Trigonometric functions are important in two distinct contexts–triangles and circular motion. We define them in both contexts:

1. in right triangles where angles are limited to between 0 and 90 degrees, and

2. on the unit circle where angles are unlimited.

The sine function takes an angle and yields a number. You may express the angle in either of two ways: in degrees or in radians. For triangle trigonometry we prefer degrees; for calculus and circular trigonometry we prefer radians. In this chapter we use degrees.

There are 360 degrees in a full circle, 180 degrees in a straight angle, and 90 degrees in a right angle (Figure 1).   (6-2-1)

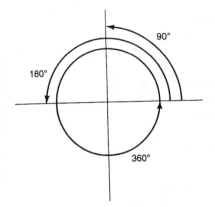

FIGURE 1 360, 180, and 90 degree angles

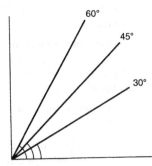

FIGURE 2 30, 45, and 60 degree angles

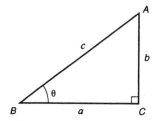

FIGURE 3 A right triangle labeled with letters

Angles between 0° and 90° are <u>acute</u>. Angles between 90° and 180° are <u>obtuse</u>. Right triangles never contain obtuse angles, but many triangles do. Learn to recognize the approximate sizes of angles. There are only three simple acute angles that have convenient trigonometric functions (30°, 45°, and 60°), so memorize their appearance and use them to help you estimate the measures of other angles (Figure 2).

In a *right* triangle the side opposite the right angle is called the <u>hypotenuse</u> and the other two sides are sometimes called the "legs" (Figure 3). When C is the right angle, the hypotenuse is c. Let θ (theta) be the angle at the lower left (also called $\angle B$). So "a" is the side (leg) adjacent to angle θ and "b" is the side (leg) opposite θ.

The three basic trigonometric functions are sine, cosine, and tangent. Their definitions in right triangles apply only to angles between 0 and 90 degrees. They have a second, more general, definition coming up in 6-2-4 that applies to all angles.

For angle θ, $0° < \theta < 90°$, with θ, a, b, and c as in Figure 3,

$$\sin \theta = \frac{opposite}{hypotenuse} = \frac{b}{c} \tag{6-2-2}$$

$$\cos \theta = \frac{adjacent}{hypotenuse} = \frac{a}{c} \quad \text{[in a right triangle]}$$

$$\tan \theta = \frac{opposite}{adjacent} = \frac{b}{a}$$

FIGURE 4 sin θ is opposite over hypotenuse

FIGURE 5 cos θ is adjacent over hypotenuse

FIGURE 6 tan θ is opposite over adjacent

These ratios depend upon the angle, θ, but not on the size of the triangle. Also, these ratios are pure numbers, without units. For example, if the two sides b and c are measured in inches, then their quotient, $\frac{b}{c}$, has "inches/inches" which cancel.

Everyone memorizes these three *right-triangle* interpretations:

> **sine is opposite over hypotenuse.**
> **cosine is adjacent over hypotenuse.**
> **tangent is opposite over adjacent.**

> Some students memorize **soh-cah-toa:**
> <u>s</u>ine is <u>o</u>pposite over <u>h</u>ypotenuse   **soh**
> <u>c</u>osine is <u>a</u>djacent over <u>h</u>ypotenuse   **cah**
> <u>t</u>angent is <u>o</u>pposite over <u>a</u>djacent   **toa**

■ Solving Right Triangles

Trigonometric functions can be used to solve right triangles.

EXAMPLE 1
In Figure 7, if $\theta = 35°$ and $c = 20$, find b and a.

This is a right triangle and the hypotenuse is given. Side b is opposite. In a right triangle, sine is opposite over hypotenuse.

$$\sin \theta = \frac{b}{c}. \quad \sin 35° = \frac{b}{20}.$$

$$b = 20 \sin 35° = 20(.574) = 11.5.$$

Side a is adjacent. Adjacent over hypotenuse is cosine.

$$\cos \theta = \frac{a}{c}. \quad \cos 35° = \frac{a}{20}.$$

$$a = 20 \cos 35° = 20(.819) = 16.4. \quad \blacksquare$$

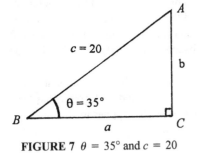

FIGURE 7 $\theta = 35°$ and $c = 20$

We solved for the legs given the hypotenuse. This procedure is so common that you may wish to memorize this result:

In a *right* triangle, "opposite is hypotenuse times sine."
"adjacent is hypotenuse times cosine."   (6-2-3)

EXAMPLE 2
In Figure 8, if $\theta = 32°$ and $b = 12$, find c and a.

The opposite side is given. Side c is the hypotenuse. Sine is opposite over hypotenuse.

$$\sin \theta = \frac{b}{c}. \quad \sin 32° = \frac{12}{c}.$$

$$c = \frac{12}{\sin 32°} = \frac{12}{.530} = 22.6.$$

Side a is adjacent. Tangent is opposite over adjacent.

FIGURE 8 $\theta = 32°$ and $b = 12$

$$\tan \theta = \frac{b}{a}. \quad \tan 32° = \frac{12}{a}. \quad a = \frac{12}{\tan 32°} = \frac{12}{(.625)} = 19.2. \quad \blacksquare$$

EXAMPLE 3
In Figure 9, $\angle B = 23°$ and $a = 480$. Find b and c.

$$\tan \angle B = \frac{b}{a}. \quad \tan 23° = \frac{b}{480}.$$

$$b = 480 \tan 23° = 203.7$$

$$\cos \angle B = \frac{a}{c}. \quad \cos 23° = \frac{480}{c}.$$

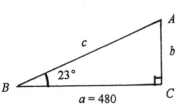

FIGURE 9 $\angle B = 23°$ and $a = 480$

$$c = \frac{480}{\cos 23°} = 521.5 \quad \blacksquare$$

EXAMPLE 4
Consider the triangle in Figure 10. $c = 50$, $\angle B = 20°$, and $a = 50$. Find b.

Side b is opposite the given angle. Can we use "sine is opposite over hypotenuse" to find b?

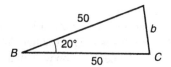

FIGURE 10 $c = 50$, $\angle B = 20°$, and $a = 50$

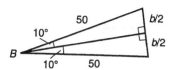

FIGURE 11 Bisecting the angle and opposite side of an isosceles triangle

No. This triangle does not have a hypotenuse. The term "hypotenuse" applies to the side opposite the right angle in a *right* triangle. This is not a right triangle and therefore has no hypotenuse.

However, it is a special type of triangle. Triangles with two equal sides are said to be <u>isosceles</u>. In this example, $a = c = 50$, so the triangle is isosceles. By drawing an auxiliary line we can subdivide it into two congruent right triangles. Bisect angle B. The ray will also be a perpendicular bisector of side b (Figure 24). *Now* we may use sine, because there is a right triangle. The angle is half of $\angle B$, and the opposite side is half of b. Now c is a hypotenuse. "Sine is opposite over hypotenuse."

$$\sin\left(\frac{\angle B}{2}\right) = \frac{\left(\frac{b}{2}\right)}{c}.$$

$$\sin 10° = \frac{\left(\frac{b}{2}\right)}{50} = \frac{b}{100}.$$

$$b = 100 \sin 10° = 17.36. \qquad \rule[0.2em]{1em}{0.3em}$$

EXAMPLE 5   Consider a circle with radius 5 (Figure 12). A chord AB makes a central angle AOB of 72°. How long is the chord (the line segment with ends on the circle)?

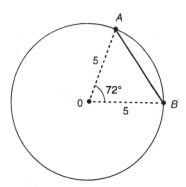

FIGURE 12 A circle of radius 5 and a chord with a 72° central angle

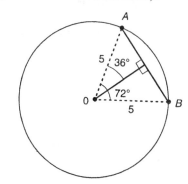

FIGURE 13 Figure 12, with the central angle bisected

Chord AB is opposite angle AOB, but this type of "opposite" is not the "opposite" in the right-triangle interpretation of sine because triangle AOB is not a right triangle. But it is isosceles (the two radii are equal), so it can be split into two right triangles by bisecting the angle and the chord (Figure 13). Now there is a right triangle where half the chord is "opposite" half the angle.

$$\sin\left(\frac{1}{2}\angle AOB\right) = \frac{\left(\frac{1}{2}AB\right)}{OA}.$$

$$\sin 36° = \frac{\left(\frac{1}{2}AB\right)}{5}.$$

$$AB = \frac{(5 \sin 36°)}{\left(\frac{1}{2}\right)} = 5.88. \qquad \rule[0.2em]{1em}{0.3em}$$

EXAMPLE 6   See Figure 14. $\angle A = 13°$, $a = 8.4$, and $b = 35.2$. Find c.

Side a is opposite angle A. Can we use "sine is opposite over hypotenuse"? No. This is a *right* triangle thought, but we are not given this is a right triangle. We don't know $\angle C$. This is a case of Angle-Side-Side. In the next section we will use the "Law of Sines" to find the two possibilities for c. Wait for the Law of Sines.

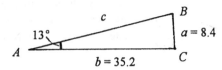

FIGURE 14 $\angle A = 13°$, $a = 8.4$, and $b = 35.2$

EXAMPLE 7   See Figure 15. Find angle θ.

It is a right triangle, so we can use tangent is opposite over adjacent.

$$\tan \theta = \frac{3.2}{9.4}. \quad \theta = \tan^{-1}\left(\frac{3.2}{9.4}\right) = 18.8°.$$

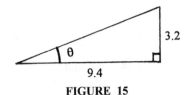

FIGURE 15

EXAMPLE 8   See Figure 16. Find angle θ.

It is a right triangle, so we can use cosine is adjacent over hypotenuse.

$$\cos \theta = \frac{190}{300}. \quad \theta = \cos^{-1}\left(\frac{190}{300}\right) = 50.7°.$$

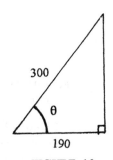

FIGURE 16

EXAMPLE 9   See Figure 17. Find angle θ.

It is a right triangle, so we can use sine is opposite over hypotenuse.

$$\sin \theta = \frac{8}{11}. \quad \theta = \sin^{-1}\left(\frac{8}{11}\right) = 46.7°.$$

In this example we solved "$\sin \theta = \dfrac{8}{11}$" using the inverse-sine function. This is the correct thing to do in a *right* triangle. However, there is actually a second solution to that equation which is important when the triangle has an obtuse angle. Angles larger than $90°$ are common and very important, but do not fit in right triangles.

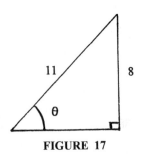

FIGURE 17

■ Unit Circles

There is an easy way to extend these trigonometric functions to handle angles that do not fit in right triangles, including obtuse angles and angles greater than 180°. Put the angle at the center of circle with radius 1.

Draw the unit circle (radius 1, $x^2 + y^2 = 1$) centered at the origin (Figure 18). Let the angle θ be at the origin, with one ray (the "initial side") along the positive x-axis and the other ray (the "terminal side") extending out to cut the circle at point P which depends upon θ. The location of P can be described two ways: by the angle θ, or by its x and y coordinates. Trig functions are, by definition, relationships between these three numbers, θ, x, and y (Figure 19).

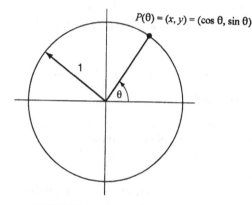

FIGURE 18 A unit circle, angle θ, and.
$P(\theta) = (\cos \theta, \sin \theta) = (x, y)$

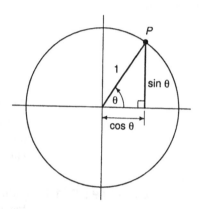

FIGURE 19 A unit circle, θ, $\cos \theta$, and $\sin \theta$.
Memorize this figure

$$\sin \theta = \text{the vertical coordinate, } y \qquad (6\text{-}2\text{-}4)$$

$$\cos \theta = \text{the horizontal coordinate, } x$$

$$\tan \theta = \frac{\sin \theta}{\cos \theta} = \frac{y}{x}, \text{ (for } x \neq 0).$$

For any angle θ, $P(\theta) = (\cos \theta, \sin \theta) = (x, y)$

$\sin^2 \theta + \cos^2 \theta = 1$, a "Pythagorean" identity

---

sin θ is the vertical coordinate of P.
cos θ is the horizontal coordinate of P.

$$\tan \theta = \frac{\sin \theta}{\cos \theta} = \frac{y}{x} = \text{the slope of the line through } P.$$

Inspection of Figure 19 shows that these definitions duplicate the right-triangle interpretations for angles between 0 and 90 degrees (because the hypotenuse is 1). However, these definitions apply to all angles, including those angles greater than 90 degrees and even negative angles. By the way, negative angles are possible; they are just oriented clockwise instead of the usual counterclockwise (Figure 22).

EXAMPLE 10 Picture an angle of 120° and give its sine and cosine (Figure 20).

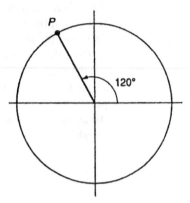

FIGURE 20 A unit circle and **u** = 120°

The x-value of P is negative, so cos 120° is negative. From the picture it appears to be about $-\frac{1}{2}$ (remember the radius is 1). With a good picture you can almost read the sine and cosine of an angle right from the picture. Just estimate the horizontal (cosine) and vertical (sine) coordinates of P. Sin 120° is the y-value, which appears to be a bit less than .9. Check these estimates using a calculator.

EXAMPLE 11 Estimate sin 35° and cos 35° from Figure 21 which has the angle accurately drawn.

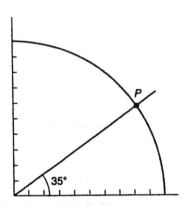

FIGURE 21 $\theta = 35°$ in the first quadrant of a unit circle

It appears that the horizontal coordinate of P is about $\frac{8}{10}$ of the radius. Therefore, cos 35° is about .8. The vertical coordinate appears to be about $\frac{6}{10}$ of the radius. Therefore, sin 35° is about .6. Use your calculator to check the accuracy of these crude estimates.

EXAMPLE 12   Estimate sin 300° and cos 300° from a unit-circle picture (Figure 22).

A 300 degree angle is in the fourth quadrant. A full circle is 360°, so a 300°
angle is 60° short of a full circle. Go backwards 60° from the positive x-axis to
locate the ray (Figure 22). So the angle −60° creates the same P that 300° creates.

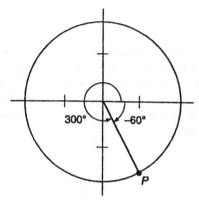

FIGURE 22 A unit circle, θ = 300°, and θ = −60°   **FIGURE 23** A unit circle, θ = 660° = 300° + 360°

From Figure 22, the x-value (cosine) is positive and the y-value (sine) is negative.
The cosine appears to be near $\frac{1}{2}$ and the sine near −.9. Use your calculator to check
the accuracy of these estimates.

Two angles with the same P and therefore the same terminal side are said to be
"coterminal" because they terminate in the same place. **Angles are** <u>coterminal</u> **if
and only if they differ by an integer multiple of 360 degrees (Figure 23).**
Think of P as rotating around the circle. The angle describes how P got where it is.
Take any rotation that yields that position and add or subtract 360 degrees and the
same position results. Coterminal angles have the same trigonometric functions.

■ Sine

The graphs of sine and the other trigonometric functions result from this rotational,
circular, interpretation. Figure 24 shows both a unit circle and a traditional graph
of sin x, side by side.

Imagine the terminal side beginning along the x-axis and rotating counterclockwise. The angle θ and the point P change over time. The graph of sine
exhibits the vertical coordinate of P as time changes on the horizontal axis (called

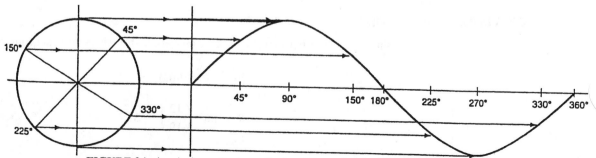

FIGURE 24 sin x is the vertical coordinate of P associated with angle x on the unit circle

the *x*-axis, but it often would be nice if it were the *t*, for time, axis). The height of *P* (sin *x*) goes up from 0 to 1 (at 90°) and then back down to 0 (at 180°). It continues down to −1 (at 270°) and then goes back up to 0 (at 360°). The graph of "sin *x*" plots this up and down motion. This explains why sin *x* repeats every 360 degrees—the same points are retraced.

EXAMPLE 13 Use a unit-circle picture to estimate the solution for θ to sin θ = .4.

On a unit-circle picture, sin θ = *y*. So draw in the line *y* = .4 and see where it intersects the unit circle. It will intersect it twice, at *P* and *P'* (Figure 25), so there will be two solutions between 0° and 360°—one in the first quadrant and one in the second. The first looks to be about 20°, the second about 160°. Check these estimates with a calculator.

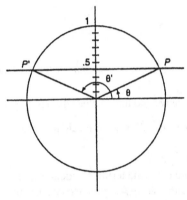

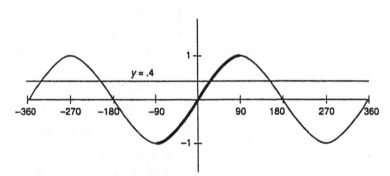

FIGURE 25 sin θ = .4. *P* and *P'* yield two solutions

FIGURE 26 *y* = sin *x* and *y* = .4. sin⁻¹ is defined on the emphasized region. −360° ≤ *x* ≤ 360°

Figure 26 graphs both "*y* = sin *x*" and "*y* = .4". The picture shows there are four solutions to "sin *x* = .4" between −360° and 360°. The inverse sine function will find one of them—the one in the emphasized region. ■

■ Inverse Sine

The inverse sine function is denoted by "sin⁻¹" (read "inverse sine" or "arc sine"), which is sometimes written " arcsin" ("arc sine").

The inverse sine function will find the angle with a given sine value, if you know the angle is in the first quadrant.

EXAMPLE 14 See Figure 27. Find angle θ.

$\sin \theta = \dfrac{12}{19}$. Furthermore, because we are certain that θ is in the first quadrant

$$(0° < \theta < 90°),$$

$$\theta = \sin^{-1}\left(\frac{12}{19}\right) = 39.2°.$$ ■

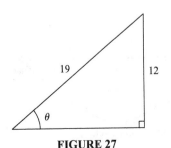

FIGURE 27

EXAMPLE 15   In Figure 28, angle θ satisfies
$\sin \theta = .916$. Find angle θ.
$\sin^{-1}(.916) = 66.3°$, which is obviously not the correct answer. Angle θ is not in the first quadrant. The correct answer is $180° - 66.3° = 113.7°$.   ▬

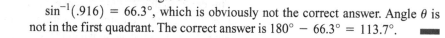

The next theorem explains how to find the right answer when the inverse sine function does not give it (which happens pretty often!).

FIGURE 28

$$\sin x = c \text{ iff } x = \sin^{-1}c, \text{ or} \qquad (6\text{-}2\text{-}5)$$
$$x = 180° - \sin^{-1}c, \text{ or}$$
$$x = \sin^{-1}c \pm 360n°, \text{ for some integer } n, \text{ or}$$
$$x = 180° - \sin^{-1}c \pm 360n°, \text{ for some integer } n.$$

Learn this by memorizing Figure 29.

In Example 15 above, the correct answer was the obtuse angle from the second line, which is not the answer that your calculator gives.

■ Why "$\sin \theta = c$" has Two Solutions

Figure 29 illustrates why "$\sin \theta = c$" has two solutions in triangles (where $0 < c < 1$). Sin θ is the vertical coordinate of P on the unit circle. The line $y = c$ cuts the unit circle twice. There are **two** points with the same vertical coordinate c (points P and P'). There are **two** solutions between $0°$ and $180°$.

Let θ be the first-quadrant solution. The picture shows that $180° - \theta$ is another solution. By Angle-Side-Angle, the picture has two congruent triangles and BP and $B'P'$ are the same length. Therefore, θ and $180° - \theta$ have the same sine.

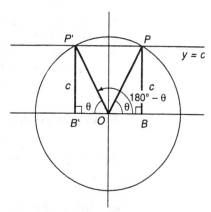

FIGURE 29 Two solutions to $\sin \theta = c$.
θ and $180° - \theta$. $\sin \theta = \sin(180° - \theta)$

$$\sin \theta = \sin(180° - \theta). \qquad (6\text{-}2\text{-}6)$$

This explains line 2 of 6-2-5 above. Therefore, if θ solves "$\sin x = c$" so does $180° - \theta$, because their sine values are the same. Lines 3 and 4 of 6-2-5 follow because P remains the same when the angle is increased or decreased by 360°, or a multiple of 360°. Figure 29 illustrates all of Theorem 6-2-5.

EXAMPLE 16 Solve $\sin x = .8$.

In this problem x is an angle. $x = \sin^{-1}(.8) = 53.13°$. But this is not the only answer. The other answer that could fit in a triangle is $x = 180 - 53.13° = 126.87°$.

Also, all other angles coterminal with these two are also solutions. So $x = 53.13 + 360°$ also solves it, and so do $x = 53.13 - 360°$, $x = 126.87 + 360°$, and $x = 126.36 - 360°$. But these coterminal solutions are not as important as the two solutions between 0° and 180°. ▬

EXAMPLE 17 In Figure 30, $\sin A = .913$. Find $\angle A$.

$\sin^{-1}.913 = 65.9°$. But angle A is obviously not 65.9°, it's obtuse. So you must want the other solution. $\angle A = 180 - 65.9° = 114.1°$.

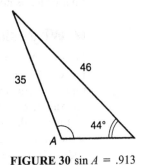

FIGURE 30 $\sin A = .913$ ▬

■ Your Calculator Might Not Give the Right Answer!

There is a parallel here with solving equations like "$x^2 = 30$." You use a keystroke sequence to find one solution ($\sqrt{30} = 5.377$), but you know in your head that there is a second solution (with a negative sign). To solve "$\sin x = c$" you use a keystroke sequence to find one solution ($\sin^{-1}c$), but you know in your head there is second solution (180° minus the first one).

■ What is $\sin^{-1} x$?

"Sine is opposite over hypotenuse." What is a good way to remember what the inverse sine is? In the first quadrant,

---

"$\sin^{-1}x$" is "the angle whose sine is x."

For example, the sentence "$\sin^{-1}\left(\dfrac{1}{2}\right) = 30°$" means "The angle whose sine is $\dfrac{1}{2}$ is thirty degrees," although it is usually read "Inverse sine of $\dfrac{1}{2}$ is thirty

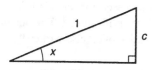

FIGURE 31 $\sin x = c.\ x = \sin^{-1}c$

degrees," or "Arc sine of $\frac{1}{2}$ is thirty degrees." This is the inverse sentence of "sin 30° $= \frac{1}{2}$." For acute angles we can illustrate "$x = \sin^{-1}c$" with a right-triangle picture (Figure 31).

■ Cosine

Cosine, like sine, has both a right-triangle interpretation and a unit-circle definition. Imagine point P rotating around the unit circle. Cosine tracks its horizontal coordinate.

At angle 0, the horizontal value is 1, so cos 0° $= 1$ (Figures 32 and 33). At angle 90°, the horizontal coordinate is 0, so cos 90° $= 0$. Similarly, cos 180° $= -1$, cos 270° $= 0$, and cos 360° repeats cos 0° $= 1$.

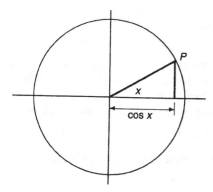

FIGURE 32 A unit circle, angle x, and cos x

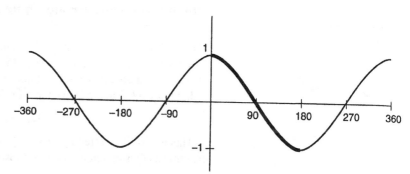

FIGURE 33 cos x. $-360° \le x \le 360°$. The region where cos^{-1} is defined is emphasized

The graph of cosine has exactly the same shape as the graph of sine, but shifted left 90 degrees (compare Figure 33 to Figure 26).

■ Inverse Cosine

Figure 34 illustrates "cos $x = c$." Here x is an angle (x is not the usual horizontal coordinate). The equation "cos $x = c$" says that the horizontal coordinate of P is c.

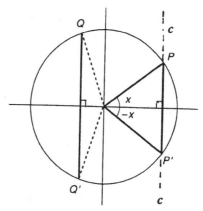

FIGURE 34 cos $x = c$. cos$^{-1}c = x$. Also, cos$(-x) = c$. cos $x = \cos(-x)$

(Draw the usual vertical line $x = c$ and see where it intersects the unit circle.) The vertical line cuts the circle twice, so there are two solutions illustrated. **The one on the top half of the circle is called $\cos^{-1} c$** ("inverse cosine of c," also written " arc-cos c" and called "arc cosine of c").

If $-1 < c < 1$, the vertical line cuts the circle twice, including exactly once on the top half where the angles are between $0°$ and $180°$. If $-1 < c < 1$ there is exactly one triangle-angle solution.

$$\cos x = c \text{ iff } x = \cos^{-1}c \quad \text{[the only triangle-angle}$$
$$\text{solution] or} \tag{6-2-9}$$
$$x = -\cos^{-1}c, \text{ or}$$
$$x = \cos^{-1}c \pm 360n°, \text{ for some integer } n, \text{ or}$$
$$x = -\cos^{-1}c \pm 360n°, \text{ for some integer } n.$$

Learn this by understanding Figure 34.

The first line gives the one and only triangle-angle solution. The second line gives the negative-angle solution visible in Figure 34. Lines 3 and 4 give the solutions for angles that are coterminal to those in lines 1 and 2.

In Figure 34, P and P' have the same horizontal coordinate—the same cosine.

$$\cos(-x) = \cos x \tag{6-2-10}$$

This is also illustrated by Q and Q'. The angle to Q' is clockwise (negative) the same as Q is counterclockwise (positive).

EXAMPLE 18   Solve $\cos x = .2$.

By 6-2-9, $x = \cos^{-1}.2 = 78.46°$ is the only solution which could fit in a triangle. However, there are more solutions which could be rotational angles, including $-78.46°$ (from line 2), and $78.46 + 360° = 438.46°$ and $-78.46 + 360° = 281.54°$. To state them all we can write "$x = \pm78.46° \pm 360n°$, for any integer n." The case $n = 0$ handles lines 1 and 2 of 6-2-9. Thus 6-2-9 could be abbreviated as

(6-2-9, abbreviated) $\cos x = c$ iff $x = \pm\cos^{-1}c \pm 360n°$, for some integer n,

where the "plus or minus" on the "$\pm\cos^{-1}c$" term consolidates lines 1 and 2 of the previous version.   ▬

EXAMPLE 19   In Figure 35, $\cos C = -.394$. Find $\angle C$.

In a triangle, $\cos C = -.394$ iff
$$\angle C = \cos^{-1}(-.394) = 113.2°.$$

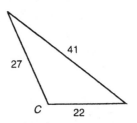

FIGURE 35 $\cos C = -.394$ ▬

E X A M P L E   2 0   In Figure 36, $\angle D = 90°$, $d = 93$, and $m = 38$. Find $\angle K$.
In a right triangle, cosine is adjacent over hypotenuse.

$$\cos K = \frac{38}{93}.\quad K = \cos^{-1}\left(\frac{38}{93}\right) = 65.9°$$

Figure 37 is a right triangle picture that illustrates both "$\cos x = c$" and "$x = \cos^{-1}c$" (for acute angles only).

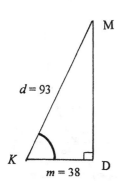

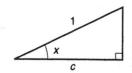

FIGURE 36 $\cos K = \dfrac{38}{93}$

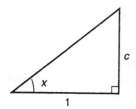

FIGURE 37 $\cos x = c. \cos^{-1}c = x$

"$\cos^{-1}x$" is "the angle whose cosine is x."

■ Tangent

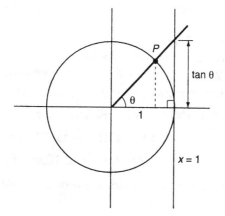

In a right triangle, "Tangent is opposite over adjacent." If the adjacent is 1, tangent is simply the opposite. Figure 38 illustrates both "$\tan x = c$" and "$x = \tan^{-1}c$" ("inverse tangent" or "arc tangent") for acute angles only.

Another way to visualize tangent is to augment the unit circle with a line perpendicular to the x-axis through the point $(1, 0)$ as in Figure 39. Then extend the ray through $P(\theta)$ until it crosses the line. It crosses the line at height "$\tan \theta$."

FIGURE 38 $\tan x = c. x = \tan^{-1}c$

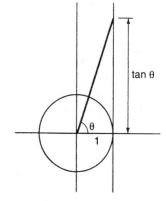

FIGURE 39 A unit circle, angle θ, and $\tan \theta$

FIGURE 40 As θ approaches $90°$, $\tan \theta$ becomes large

As angle θ increases to $90°$, its ray cuts the line $x = 1$ further and further up (Figure 40). If θ is negative in the fourth quadrant, the ray cuts the line below the x-axis and y is negative. Tangent is negative in the fourth quadrant.

This "cut the line" interpretation is valid for angles in all quadrants. In the second quadrant tangent is negative. Extend the ray backwards to cut the line $x = 1$ and the negative y-value there will be $\tan \theta$ (Figure 41).

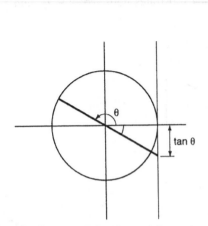

FIGURE 41 $\tan \theta$ for θ in quadrants II and IV

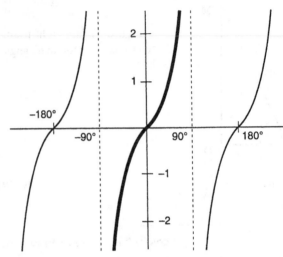

FIGURE 42 $\tan x$. $[-270°, 270°]$ by $[-2.5, 2.5]$. The region where $\tan^{-1}$ is defined is emphasized

The graph of "$\tan x$" is in Figure 42. It repeats every 180 degrees. ▬

EXAMPLE 21 $\angle C = 90°$, $b = 50$, and $\angle A = 80°$. Find a (Figure 43).
From the point of view of A, a is the opposite side and b is the adjacent side. In a right triangle, "tangent is opposite over adjacent."

$$\tan 80° = \frac{a}{50}. \quad a = 50 \tan 80° = 283.6.$$ ▬

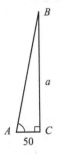

FIGURE 43 $\angle A = 80°$. $b = 50$

Inverse Tangent
The values of tangent repeat every 180 degrees (sine and cosine repeat every 360 degrees). Both 90° and −90° form vertical asymptotes. They are easy to explain: On the unit circle, tangent $= \frac{y}{x}$, therefore there will be a vertical asymptote when $x = 0$, which occurs at 90° and also at −90°.

$$\tan x = c \text{ iff } x = \tan^{-1}c, \text{ or}$$

$$x = \tan^{-1}c \pm 180n°, \text{ for some integer } n. \quad (6\text{-}2\text{-}11)$$

Line 2 actually includes line 1, since n could be 0.

EXAMPLE 22 $\angle C = 90°$, $a = 10$, and $b = 20$. Find angle B (Figure 44).
From the point of view of angle B, b is the opposite side and a is the adjacent side.

$$\tan B = \frac{20}{10} = 2. \ \angle B = \tan^{-1}2 = 63.43°.$$

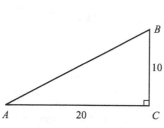

FIGURE 44 $a = 10$ and $b = 20$

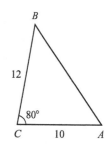

FIGURE 45
$\angle C = 80°. a = 12, b = 10$

EXAMPLE 23   In a triangle, $\angle C = 80°$, $a = 12$, and $b = 10$ (Figure 45). Find $\angle A$.

As always, a is opposite angle A and b is adjacent to angle A. But we cannot use "tangent is opposite over adjacent" here, because that only applies to *right* triangles. Wait for the Law of Cosines.   ▬

■ Basic Facts

There are only a few famous angles. Mathematicians memorize the figures that yield the sines, cosines, and tangents of the three acute angles that work out really nicely: $30°$, $45°$, and $60°$ (Figures 46 and 47. Note that the $30°$ right triangle has the $60°$ angle in it, so *two* pictures suffice for *three* key angles).

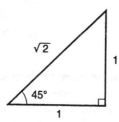

FIGURE 46 A $45°$ right triangle. Angle $45°$. Legs 1 and 1. Deduce the hypotenuse $= \sqrt{2}$. Sides are 1, 1, $\sqrt{2}$. $\tan 45° = 1$

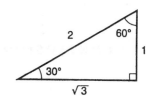

FIGURE 47 A $30°$, $60°$ right triangle. Angle $30°$. Opposite $= 1$. Hypotenuse $= 2$. Deduce the other leg is $\sqrt{3}$. Sides are 1, 2, $\sqrt{3}$. $\sin 30° = \dfrac{1}{2}$

Memorize these two pictures

The values of the trig functions of these famous angles follow from these figures. Memorize the values in **bold**. The rest of the values follow from the Pythagorean Theorem.

θ	sin θ	cos θ	tan θ
TABLE 6-2-12			
0°	0	1	0
30°	$\dfrac{1}{2} = .5$	$\dfrac{\sqrt{3}}{2} = 866$	$\dfrac{1}{\sqrt{3}} = .577$
45°	$\dfrac{1}{\sqrt{2}} = \dfrac{\sqrt{2}}{2} = .707$	$\dfrac{1}{\sqrt{2}} = \dfrac{\sqrt{2}}{2} = .707$	1
60°	$\dfrac{\sqrt{3}}{2} = .866$	$\dfrac{1}{2} = .5$	$\sqrt{3} = 1.732$
90°	1	0	does not exist

The signs of the trig functions in the various quadrants are easy to remember because sine is the vertical coordinate and cosine is the horizontal coordinate.

TABLE 6-2-13

[second quadrant]	$\sin \theta > 0$ $\cos \theta < 0$ $\tan \theta < 0$	$\sin \theta > 0$ $\cos \theta > 0$ $\tan \theta > 0$	[first quadrant]
[third quadrant]	$\sin \theta < 0$ $\cos \theta < 0$ $\tan \theta > 0$	$\sin \theta < 0$ $\cos \theta > 0$ $\tan \theta < 0$	[fourth quadrant]

CONCLUSION Trigonometric functions can be interpreted two ways. Sine is both "opposite over hypotenuse" and the vertical coordinate of a point on the unit circle. Cosine is both "adjacent over hypotenuse" and the horizontal coordinate of a point on the unit circle. Many triangles are not right triangles, so be careful to use the right-triangle interpretations only in right triangles. Learn the unit-circle interpretations.

Terms: acute, obtuse, right, hypotenuse, opposite, adjacent, sine, cosine, tangent, initial side, terminal side, coterminal, inverse sine, inverse cosine, inverse tangent, isosceles.

Exercise 6-2

A

A1.* State and sketch the unit-circle definition of
(A) sine   (B) cosine
(C) tangent
(You may use one picture to illustrate all three.)

A2.* State and sketch the right-triangle interpretation of
(A) sine   (B) cosine
(C) tangent

A3. Sketch and label simple right triangles to illustrate (for acute angles)
(A) $\sin \theta = x$ (B) $\cos \theta = x$
(C) $\tan \theta = x$
(This needs three separate pictures.)

A4. Sketch and label simple right triangles to illustrate (for acute angles)
(A) $\sin x = c$ (B) $\cos x = c$
(C) $\tan x = c$
(This needs three separate pictures.)

See the figure for A5–A14. Sketch it again with the given numerical labels. Then solve.

A5. If $\theta = 24°$ and $a = 16$, find b. [7.1]

A6. If $\theta = 26°$ and $b = 12$, find c. [27]

A7. If $\theta = 29°$ and $c = 8$, find a. [7.0]

A8. If $\theta = 17°$ and $c = 90$, find b. [26]

A9. If $\theta = 21°$ and $b = 50$, find a. [130]

A10. If $\theta = 20°$ and $a = 15$, find c. [16]

A11. If $a = 12$ and $b = 7$, find θ. [30]

A12. If $a = 20$ and $c = 30$, find θ. [48]

A13. If $c = 19$ and $b = 7$, find θ. [22]

A14. If $a = 13$ and $c = 15$, find θ. [30]

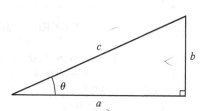

FIGURE For A5–A14 and A25–A33

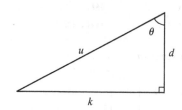

FIGURE For A15–A24 and A34–42

See the figure for A15–A24. Sketch it again with the given numerical labels. Then solve.

A15. If $\theta = 55°$ and $k = 16$, find d. [11]

A16. If $\theta = 50°$ and $d = 12$, find u. [19]

A17. If $\theta = 61°$ and $u = 8$, find k. [7.0]

A18. If $\theta = 73°$ and $u = 90$, find d. [26]

A19. If $\theta = 72°$ and $d = 50$, find k. [150]

A20. If $\theta = 65°$ and $k = 15$, find u. [17]

A21. If $k = 12$ and $d = 7$, find θ. [60]

A22. If $k = 20$ and $u = 30$, find θ. [42]

A23. If $u = 19$ and $d = 7$, find θ. [68]

A24. If $k = 13$ and $u = 15$, find θ. [60]

See the figure for A25–A33.
Set up an equation with sine, cosine, or tangent to solve for the unknown.

A25. Given θ and c, find b.

A26. Given θ and a, find b.

A27. Given θ and b, find c.

A28. Given θ and b, find a.

A29. Given θ and c, find a.

A30. Given θ and a, find c.

A31. Given a and b, find θ.

A32. Given a and c, find θ.

A33. Given b and c, find θ.

See the figure for A34–A42.
Set up an equation with sine, cosine, or tangent to solve for the unknown.

A34. Given θ and u, find d.

A35. Given θ and k, find d.

A36. Given θ and d, find u.

A37. Given θ and d, find k.

A38. Given θ and u, find k.

A39. Given θ and k, find u.

A40. Given k and d, find θ.

A41. Given k and u, find θ.

A42. Given d and u, find θ.

Right-triangle solutions. Let $\angle C = 90°$. (Sketching the figure may help.)

A43. $\angle A = 17°$ and $b = 12$. Find c. [13]

A44. $\angle B = 26°$ and $a = 45$. Find c. [50]

A45. $\angle B = 21°$ and $b = 5$. Find a. [13]

A46. $\angle A = 49°$ and $a = 90$. Find b. [78]

A47. $\angle B = 79°$ and $c = 100$. Find b. [98]

A48. $\angle A = 33°$ and $c = 100$. Find a. [54]

A49. $\angle A = 32°$ and $b = 50$. Find a. [31]

A50. $\angle B = 5°$ and $a = 100$. Find b. [8.7]

A51. $\angle B = 56°$ and $b = 2$. Find c. [2.4]

A52. $\angle A = 89°$ and $a = 100$. Find c. [100]

A53. $\angle A = 4°$ and $b = 100$. Find c. [100]

A54. $\angle B = 6°$ and $a = 10$. Find c. [10]

Solve for θ, $0° < \theta < 180°$.

A55. $\sin \theta = .6$ **A56.** $\sin \theta = .2$

A57. $\sin \theta = .99$ **A58.** $\sin \theta = .4$

A59. $\cos \theta = .6$

A60. $\cos \theta = -.9$

A61. $\cos \theta = -.3$

A62. $\cos \theta = .1$

A63. $\tan \theta = 2$

A64. $\tan \theta = .2$

A65. $\tan \theta = -.9$

A66. $\tan \theta = -5$

Give the quadrant of the angle

A67. (A) $170°$ (B) $57°$
    (C) 57 (D) $-210°$
    (E) $350°$

A68. (A) $-12°$ (B) $370°$
    (C) $-100°$ (D) $100°$
    (E) $210°$

A69. (A) $100°$ (B) $300°$

A70. (A) $210°$ (B) $-60°$

Give the exact value (not a decimal answer)

A71. (A) $\tan 45°$ (B) $\sin 30°$
    (C) $\cos 60°$.

A72. (A) $\sin 0°$ (B) $\cos 0°$
    (C) $\tan 0°$.

A73. (A) $\sin 30°$ (B) $\cos 30°$
    (C) $\tan 30°$.

A74. (A) $\sin 60°$ (B) $\cos 60°$
    (C) $\tan 60°$.

A75. (A) $\sin 90°$ (B) $\cos 90°$
    (C) $\tan 90°$.

A76. (A) $\sin 120°$ (B) $\cos 135°$
    (C) $\cos 150°$.

B

B1.* (A) Draw and label unit-circle picture to illustrate "$\sin x = c$" for $0 < c < 1$ (Your choice of c).
    (B) Use it to explain why that equation has two solutions which could be angles in a triangle.
    (C) Draw and label a right-triangle picture to illustrate "$\sin x = c$."
    (D) Will your picture in part (c) work for second-quadrant angles?

B2.* (A) Draw and label a unit-circle picture to illustrate "$\cos x = c$." for $-1 < c < 1$ (Your choice of c).
    (B) Use it to explain why that equation has two solutions between 0 and 360 degrees, but only one could fit in a triangle.
    (C) Draw and label a right-triangle picture to illustrate "$\cos x = c$."
    (D) Will your picture in part (c) work for second-quadrant angles?

B3.* Sketch the graph of $\sin x$ ($-360° \leq x \leq 360°$) and label multiples of right angles.

B4.* Sketch the graph of $\cos x$ ($-360° \leq x \leq 360°$) and label multiples of right angles.

B5–B8. The figure illustrates θ. Estimate (A) $\sin \theta$. (B) $\cos \theta$.

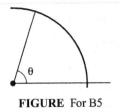 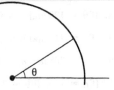

FIGURE For B5   FIGURE For B6

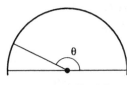 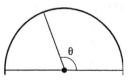

FIGURE For B7   FIGURE For B8

*Sketch a unit-circle picture to illustrate these. Sketch the appropriate line ($y = c$ or $x = c$) **before** sketching the angle(s). Be careful. There may be more than one x.*

B9. $\sin x = .8$ **B10.** $\sin x = .2$

B11. $\sin x = .98$ **B12.** $\sin x = .3$

B13. $\cos x = .2$.   **B14.** $\cos x = .4$

B15. $\cos x = -.9$ **B16.** $\cos x = -.2$

B17. $\tan x = 2$.   **B18.** $\tan x = -.5$

Find two triangle-angle solutions for x.

B19. $\sin x = .8$ $[\ldots, 130°]$

B20. $\sin x = .4$ $[24°, \ldots]$

B21. $\sin x = .2$ $[\ldots, 170°]$

B22. $\sin x = .99$ $[\ldots, 98°]$

Find two solutions for x. Are they both angles that could fit in a triangle?

B23. $\cos x = .5$

B24. $\cos x = .8$ $[37°, \ldots]$

B25. $\cos x = -.9$ $[150°, \ldots]$

B26. $\cos x = -.3$ $[\ldots, -110°]$

B27. If angle A is obtuse and $\sin A = .3$, find $\angle A$.   [160]

B28. If angle A is obtuse and $\sin A = .8$, find $\angle A$.   [130]

B29.* (A) Draw and label unit-circle picture to illustrate "$\tan x = c$" for $0 < c$.
    (B) Draw and label a right-triangle picture to illustrate "$\tan x = c$."

B30.* Draw a unit-circle picture to illustrate "$\sin \theta = \sin(180° - \theta)$." Label both θ and $180° - \theta$.

B31.* In which quadrants are they positive?
    (A) sine   (B) cosine
    (C) tangent

B32. How many degrees does the hour hand of a clock rotate in one hour?

B33. How many degrees does the minute hand of a clock rotate through in one minute?

B34. How many degrees does the hour hand of a clock rotate through in one minute?

B35. How many degrees of longitude does the earth rotate through in one hour?

B36. How many degrees does the minute hand of a clock rotate through in one second?

B37. How many degrees does the second hand of a clock rotate through in one hour?

B38. How many degrees does the second hand of a clock rotate through in one day?

Solve for θ in the second quadrant.

B39. $\sin \theta = .5$ **B40.** $\sin \theta = .9$ [120]
B41. $\tan \theta = -2$ [120]   **B42.** $\tan \theta = -.8$ [140]

Find three solutions to

B43. $\sin x = .2$ [170, ...]
B44. $\sin x = .9$ [420, ...]
B45. $\cos x = .7$ [−46, ...]
B46. $\cos x = -.3$ [250, ...]
B47. $\tan x = 3$ [250, ...]
B48. $\tan x = .45$ [380, ...]
B49. $\sin x = .123$ [170, ...]
B50. $\sin x = -.5$ [210, ...]

Solve for θ in the third quadrant.

B51. $\cos \theta = -.8$ [220]   **B52.** $\cos \theta = -.2$ [260]
B53. $\tan \theta = 1.1$ [230]   **B54.** $\tan \theta = 5$ [260]

Solve for x such that $360° \le x < 720°$.

B55. $\sin x = .6$ [400, 500]
B56. $\sin x = -.88$ [600, 660]
B57. $\cos x = -.4$ [470, 610]
B58. $\cos x = .12$ [440, 640]

Give two solutions for x.

B59. $\tan x = 10$ [..., 260°]
B60. $\tan x = .45$ [24°, ...]
B61. $\tan x = -.3$ [..., 160°]
B62. $\tan x = -5.6$ [..., 100]

Word Problems

B63. A ladder 16 feet long is leaned against a house, with the base 5 feet from the house.
    (A) How high does it reach? [15]
    (B) What angle does it make with the ground? [72]

B64. A rectangle is 30 by 48 inches. What angle does its diagonal make with its longer side?

B65. A board is 8 feet long and 7.25 inches wide. A diagonal line is drawn on it from one corner to the opposite corner. What angle does that line make with the long side?

B66. One side of a rectangle is 48 inches and its diagonal makes a 30° angle with that side. What is the area of the rectangle?

B67. A flagpole of unknown height casts a shadow 28 feet long. At the same time, a yardstick held straight up casts a shadow 15 inches long. How tall is the flagpole? [67]

B68. Hold out your arm and stick up your thumb. If your thumb is 2 feet from your eye and the top part of your thumb is $\frac{7}{8}$ inches wide, what angle does it make with your eye? [Treat your eye as a vertex of a triangle. You can use this to informally measure angles in the real world.] [2.1°]

B69. Get a ruler. Spread your fingers.
    (A) How wide is the spread from tip of thumb to tip of the little finger? Extend your arm straight in front of you and spread your fingers, as if to measure an angle with your eyes.
    (B) Measure how far in front of your eyes your spread fingers are.
    (C) What angle do your outspread fingers make? [Treat your eye as a vertex of a triangle. You can use this to informally measure angles in the real world. I get about 20 degrees.]

B70. In a unit circle,
    (A) How long is a chord associated with central angle 40°? [.69]
    (B) What is the central angle of a chord of length .8? [47°]

B71. In a unit circle,
    (A) How long is a chord associated with central angle 110°? [1.6]
    (B) What is the central angle of a chord of length 1.5? [97°]

B72. In a circle of radius 1000,
    (A) How long is a chord associated with central angle 5°?
    (B) What is the central angle of a chord of length 10? [87, .57°]

B73. In a circle of radius 24,
  (A) How long is a chord associated with central angle 20°?
  (B) What is the central angle of a chord of length 10? [8.3, 24°]

B74. A regular pentagon (5 sides) is inscribed in a unit circle. How long are its sides? [1.2]
[Hint: What is the central angle opposite a side?]

B75. A regular pentagon (5 sides) is circumscribed around a unit circle. How long are its sides? [1.5]
[Hint: What is the central angle opposite a side?]

B76. A right triangle has a 30° angle. Its perimeter is 20 units. How long is the side opposite the 30° angle? [4.2]

B77. A right triangle has a 30° angle. Its area is 50 square units. How long is the side opposite the 30° angle? [7.6]

B78. The diagonal of a rectangle is 30 percent longer than its longest side. Compare its shorter side to the longer side.

Solve for x in $[0°, 360°)$.

B79. $\cos(2x) = .5$ $[\ldots, 150, \ldots]$

B80. $\tan(3x) = 1$ $[\ldots, 75, \ldots]$

B81. $\sin(4x) = .5$ $[\ldots, 130°, \ldots, 310°]$

B82. $\sin(2x) = .5$ $[\ldots, 75°, \ldots]$

B83. $\tan(x - 30°) = 1$ $[75°, \ldots]$

B84. $\cos(4x) = .5$ $[\ldots, 75°, \ldots]$

B85. Solve graphically: $\sin x < \dfrac{1}{2}$ in $[0°, 360°)$.

Illustrations

B86.* If $\sin x = c$, x is not uniquely determined.
  (A) Illustrate this with a unit circle picture.
  (B) Illustrate this with a graph of $\sin x$.

B87. If $\cos x = c$, x is not uniquely determined.
  (A) Illustrate this with a unit-circle picture.
  (B) Illustrate this with a graph of $\cos x$.
  (C) If x were limited to being an angle in a triangle, would it be uniquely determined?

B88. (A) Draw a right triangle and label it to illustrate "$\cos \theta = x$." Use the Pythagorean Theorem to find the other leg. Then give
  (B) $\sin \theta$.   (C) $\tan \theta$.

B89. (A) Draw a right triangle and label it to illustrate "$\tan \theta = x$." Use the Pythagorean Theorem to find the hypotenuse. Then give
  (B) $\sin \theta$ (C) $\cos \theta$.

B90.* (A) Illustrate "$\sin^2 x + \cos^2 x = 1$."
  (B) What is the geometric name of this theorem?

B91.* Draw and label a picture of a triangle that gives the trig functions of 30° and 60°.

B92.* Draw and label a picture of a triangle that gives the trig functions of 45°.

B93. Sketch the graph of $\tan x$ and label multiples of right angles.

Reading and Writing Mathematics

B94. The text draws a parallel between solving "$\sin x = c$" and solving "$x^2 = d$." If you use a calculator, why are they both tricky?

B95.* Some people remember sine as "opposite over hypotenuse." What is "inverse sine"?

B96.* Define "coterminal."

B97. Why do values of sine repeat every 360°?

B98.* In the unit-circle definition, we read $\sin^{-1}$ off the _____ half of the unit circle. We read $\cos^{-1}$ off the _____ half of the unit circle.

B99. (A) Evaluate $\cos 200°$ and then find inverse cosine of that number. Do you get 200°?
  (B) For which values of θ is $\cos^{-1}(\cos \theta) = \theta$?

B100. (A) Evaluate $\sin 160°$.
  (B) Evaluate $\sin^{-1}(\sin 160°)$.
  (C) Why isn't the answer to part b) "160°"?
  (D) For which values of θ is $\sin^{-1}(\sin \theta) = \theta$?

B101. Explain each of these.
  (A) The domain of $\sin^{-1}$ is the interval $[-1, 1]$.
  (B) The range of $\sin^{-1}$ is the interval $[-90°, 90°]$.
  (C) $\sin(\sin^{-1} x) = x$, for all x in the domain.
  (D) $\sin^{-1}(\sin x) = x$ iff $-90° \leq x \leq 90°$.

B102. Explain each of these:
  (A) The domain of $\cos^{-1}$ is the interval $[-1, 1]$.
  (B) The range of $\cos^{-1}$ is the interval $[0°, 180°]$.
  (C) $\cos(\cos^{-1}(x)) = x$, for all x in the domain.
  (D) $\cos^{-1}(\cos x) = x$ iff $0° \leq x \leq 180°$.

Section 6-3 SOLVING TRIANGLES

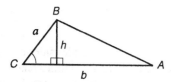

FIGURE 1A Angle C is acute

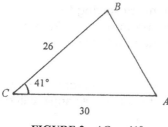

FIGURE 1B Angle C is obtuse

This section is about solving triangles, especially when the triangle is not a right triangle. The Law of Sines and the Law of Cosines use trigonometric functions to determine the remaining angles and sides of a triangle given three parts.

Any side of any triangle can be considered its base. Simply orient the triangle so that side is horizontal. Call its length b. The vertex opposite base b is B. Draw a perpendicular line from B to the base (Figures 1A and 1B).

This perpendicular line is sometimes called an "altitude" and its length is the "height," labeled "h." The area of a triangle in the Side-Angle-Side case follows from this picture and the usual formula for the area of a triangle.

■ Area

The area of any triangle is

$$\text{Area} = \left(\frac{1}{2}\right) \times \text{base} \times \text{height} = \left(\frac{1}{2}\right)bh, \qquad (6\text{-}3\text{-}1)$$

The height h is opposite $\angle C$ in a right triangle with hypotenuse a (Figure 1). So $\sin C = \dfrac{h}{a}$, and $h = a \sin C$. Therefore, substituting for h,

**SIDE-ANGLE-SIDE
AREA FORMULA 6-3-2**

$$\text{Area} = \left(\frac{1}{2}\right)ab \sin C.$$

EXAMPLE 1 Find the area of the triangle in Figure 2 where $a = 26$, $b = 30$, and angle C is $41°$.

$$\text{Area} = \left(\frac{1}{2}\right)ab \sin C$$

$$= \left(\frac{1}{2}\right)26(30)\sin 41° = 255.9.$$

FIGURE 2 $\angle C = 41°$
$a = 26$ and $b = 30$

This area formula is written for the Side-Angle-Side case when the angle is named C. If a different angle is between the two sides, simply switch letters (Problem B34).

EXAMPLE 2 Find the area of the triangle with angle $B = 71°$, $c = 4$, and $a = 9$ (Figure 3).

Now angle B is between the sides. The letters do not match 6-3-2, but that is irrelevant. The SAS Area Formula applies whenever two sides and the included angle are given, regardless of the notation for them.

$$\text{area} = \left(\frac{1}{2}\right)4(9)\sin 71° = 17.02.$$

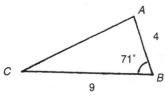

FIGURE 3 $\angle B = 71°$
$c = 4$ and $a = 9$

The SAS Area Formula includes and extends the usual "$A = \left(\dfrac{1}{2}\right)bh$." For the SAS Area Formula you do not need to know h, but you must know angle C. If C

is a right angle, $\sin C = 1$ and a would be the height (Figure 1), so the formula would simplify to "$\left(\dfrac{1}{2}\right)ab$," which would be "$\left(\dfrac{1}{2}\right)bh$."

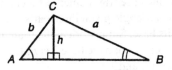

FIGURE 4A Angle A is acute

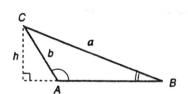

FIGURE 4B Angle A is obtuse

■ The Law of Sines

The Law of Sines is a powerful result with a simple proof. Consider sides a and b and angles A and B in Figures 4A and 4B. How are these four quantities related?

Draw an altitude from C to side c. Now angles A and B are in *right* triangles and we can use sine twice:

$$\sin A = \frac{h}{b}. \quad \sin B = \frac{h}{a}.$$

Now isolate h in each:

$$b \sin A = h. \quad a \sin B = h.$$

Setting these expressions for h equal,

$$b \sin A = a \sin B.$$

Dividing through by ab, $\dfrac{\sin A}{a} = \dfrac{\sin B}{b}$. This is the Law of Sines.

If, instead, you divide through by $(\sin A)(\sin B)$, $\dfrac{b}{\sin B} = \dfrac{a}{\sin A}$.

This is also the Law of Sines. The two versions are equivalent.

Of course, this applies to c and $\sin C$ too (without further proof), because all the letters are placeholders.

THE LAW OF SINES
6-3-3

In any triangle, if side a is opposite angle A, side b opposite angle B, and side c opposite angle C, then

(A) $\dfrac{a}{\sin A} = \dfrac{b}{\sin B} = \dfrac{c}{\sin C}$ [This is convenient for solving for unknown sides, because the unknown side will be in the numerator.]

which is equivalent to

(B) $\dfrac{\sin A}{a} = \dfrac{\sin B}{b} = \dfrac{\sin C}{c}$ [This is convenient for solving for unknown angles, because the unknown angle will be in the numerator.]

You may use the Law of Sines if you know

1. Two angles and an opposite side (Angle-Angle-Side, 6-1-5).
2. Two sides and an opposite angle (Angle-Side-Side, 6-1-6).

EXAMPLE 3   See Figure 5. Let $\angle A = 70°$, $\angle B = 80°$, and side $b = 100$. Find a.
You know two angles and an opposite side.

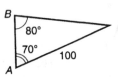

FIGURE 5 $\angle A = 70°$, $\angle B = 80°$, and $b = 100$

$$\frac{a}{\sin A} = \frac{b}{\sin B} \cdot \frac{a}{\sin 70°} = \frac{100}{\sin 80°}.$$

$$a = (\sin 70°)\left(\frac{100}{\sin 80°}\right) = 95.42.$$

If we know two sides and one of the opposite angles, the triangle is "almost" determined. This is the dangerous geometric Angle-Side-Side case (6-1-6) which may yield two answers. ▬

EXAMPLE 4 See Figure 6. Suppose $\angle C = 39°$, $c = 16$, and $a = 20$. Find $\angle A$.

Use the Law of Sines, but beware the possibility of two, one, or no answers because the geometric case is Angle-Side-Side.

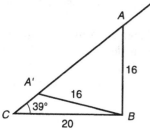

FIGURE 6 $\angle C = 39°$
$a = 20$ and $c = 16$

$$\frac{\sin A}{a} = \frac{\sin C}{c} \cdot \frac{\sin A}{20} = \frac{\sin 39°}{16}.$$

$$\sin A = 20\left(\frac{\sin 39°}{16}\right) = .78665.$$

Here is the tricky part. It is too easy to compute the inverse sine of this number ($\sin^{-1}.78665 = 51.87°$) and feel finished. But there is a second answer—and it is often the one we want. Remember that "$\sin x = c$" has two solutions—the one your calculator provides using the inverse sine function, and 180° minus that one (6-2-5).

So

$$A = \sin^{-1}(.78665) \text{ or } 180° - \sin^{-1}(.78665)$$

$$\angle A = 51.87° \text{ or } 180° - 51.87° = 128.13°.$$

FIGURE 7 $\angle C = 39°$,
$c = 16$, and $a = 20$

Inspect the Figure 6 to see that $\angle A$ is obtuse. The answer must be 128.13°.

Figure 7 shows why the Law of Sines gives two answers. Without the picture, the given information does not determine a triangle! There are two such triangles: CAB (acute A) and $CA'B$ (obtuse A'). ▬

EXAMPLE 5 Find angle A in Figure 8.

This is not given as a right triangle, so do not employ right-triangle thoughts. An angle and opposite side are given, which suggests using the Law of Sines, even though both angle A and its opposite side are unknown. Since b is given, work with angle B first. Later use the fact that the sum of the angles is 180° to find the remaining angle, $\angle A$.

FIGURE 8 Find Angle A

$$\frac{\sin B}{b} = \frac{\sin C}{c} \cdot \frac{\sin B}{50} = \frac{\sin 25°}{22}.$$

$$\sin B = 50\left(\frac{\sin 25°}{22}\right) = .9605.$$

Here is the dangerous part again. Solving for $\angle B$, $\sin^{-1}.9605 = 73.84°$, and the second answer is $180° - 73.84° = 106.16°$. We accept the second answer because the picture shows angle B to be obtuse, not acute. Now

$$\angle A = 180° - \angle C - \angle B = 180° - 25° - 106.16° = 48.84°. \quad ▬$$

EXAMPLE 6   See Figure 9. Let $\angle B = 32°$, $c = 50$, and $b = 70$. Find $\angle C$.
Again, the given parts fit Angle-Side-Side.

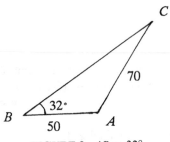

FIGURE 9 $\angle B = 32°$
$c = 50$ and $b = 70$

$$\frac{\sin C}{50} = \frac{\sin 32°}{70}$$

$$\sin C = 50 \left(\frac{\sin 32°}{70} \right) = .3785.$$

Now we are wary. We remember that there are two solutions to "$\sin C = .3785$."
We will not be fooled.

$$\angle C = \sin^{-1} .3785 = 22.24°, \text{ or } 180° - 22.24° = 157.76°.$$

But, if we think there are two answers we will have been fooled after all. Look again.
Figure 9 was good. There is no second triangle (Figure 10). The side opposite the known angle is longer than the adjacent side. This is a case of ASS which has only one solution (Example 6-1-13). The geometric construction has the arc of length b so long that it cuts the ray from angle B only once—the second cut would be on the "backside," where the angle would not really be angle B. Only one triangle meets the given specifications. How can we see that in the Law of Sines?

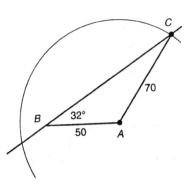

FIGURE 10 $\angle B = 32°$,
$c = 50$, and $b = 70$

The sum of the angles in a triangle is 180 degrees. Angle B is given as 32 degrees, so there is no room for a 157.76 degree angle—the total would be too much. So $157.76°$ is an extraneous solution. The solution must be $22.24°$.

Examples 4, 5, and 6 show that the Law of Sines is tricky because Angle-Side-Side is a tricky geometric result.

■ The Law of Cosines

The Law of Cosines generalizes the Pythagorean Theorem to include non-right triangles. It is derived by using the Pythagorean Theorem twice.

THE LAW OF COSINES

6-3-4

(A) $c^2 = a^2 + b^2 - 2ab \cos C$ [for Side-Angle-Side or

Angle-Side-Side]

(B) $\cos C = \dfrac{c^2 - a^2 - b^2}{-2ab}$, or $C = \cos^{-1}\left(\dfrac{c^2 - a^2 - b^2}{-2ab} \right)$

[for Side-Side-Side]

Part A alone is usually called the Law of Cosines. Part B need not be learned separately; it is what you get when you solve for C in Part A (Problem B37).

Part A resembles the Pythagorean Theorem. The usual "$c^2 = a^2 + b^2$" is there, but with another term in case angle C is not a right angle. When angle C is a right angle, $\cos C = \cos 90° = 0$ and the extra term is zero and disappears. This theorem includes the Pythagorean Theorem!

If angle C is acute, $\cos C$ is positive and the term "$-2ab \cos C$" subtracts some amount from $a^2 + b^2$, so c is shorter than in a right triangle (Figure 11). If angle C is obtuse, $\cos C$ is negative, and then the term "$-2ab \cos C$" *adds* some amount to $a^2 + b^2$, so c is longer than in a right triangle (Figure 12).

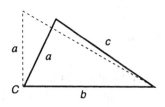

FIGURE 11 Angle C is acute. Side c is shorter than in a right triangle

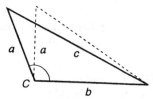

FIGURE 12 Angle *C* is obtuse.
Side *c* is longer than in a right triangle

Use the Law of Cosines when you know

1. two sides and the included angle (Side-Angle-Side), or

2. all the sides (Side-Side-Side).

EXAMPLE 7   See Figure 13. Let $a = 5$, $b = 3$, and $\angle C = 122°$. Find c.

The given angle *C* is between the given sides *a* and *b*, so the geometric case is Side-Angle-Side and the Law of Cosines applies.

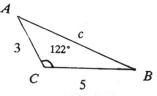

FIGURE 13 $a = 5$,
$b = 3$ $\angle C = 122°$

$$c^2 = a^2 + b^2 - 2ab \cos C$$
$$= 5^2 + 3^2 - 2(5)(3) \cos 122° = 49.9,$$

so *c* is the square root of 49.9. $c = 7.06$. (We do not need to worry about negative solutions, since sides of triangles are never negative.)   ▬

EXAMPLE 8   See Figure 14. Let $a = 9$, $b = 6$, and $c = 4$. Find $\angle C$.

All three sides are given, so the geometric case is Side-Side-Side and Part B of the Law of Cosines applies.

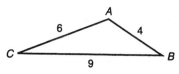

FIGURE 14 $a = 9, b = 6,$
and $c = 4$

$$\cos C = \frac{c^2 - a^2 - b^2}{-2ab}$$
$$= \frac{4^2 - 9^2 - 6^2}{-2(9)(6)} = .9352.$$

This is solved for $\angle C$ by taking the inverse cosine. $\cos^{-1} .9352 = 20.74°$. Angle *C* is 20.74°.

As stated, Part B of the Law of Cosines emphasizes solving for an angle labeled *C*. If you want to solve for angle *B*, change the letters.

Let $a = 9$, $b = 6$, and $c = 4$ (Figure 14, again). Find $\angle A$.

The triangle is the same as in Example 8, but the question is different. To find angle *A*, use the Law of Cosines, rewritten to emphasize angle *A* opposite side *a* (instead of angle *C* opposite side *c*):

$$\cos A = \frac{a^2 - b^2 - c^2}{-2bc} = \frac{9^2 - 6^2 - 4^2}{-2(6)(4)} = -.6042.$$

The cosine of obtuse angles is negative. $\cos^{-1}(-.6042) = 127.17°$. Angle *A* is 127.17°.

Triangles have six parts, which are often determined by three. When you use either the Law of Sines or the Law of Cosines to determine a fourth, there are usually several ways to determine the fifth and sixth parts.

MORE ABOUT
EXAMPLE 8

Suppose $a = 9$, $b = 6$, $c = 4$, and $\angle C = 20.74°$. Find angle A (Figure 15). Should you use the Law of Sines or Law of Cosines?

The Law of Sines is an option (because a given side is opposite a given angle), but the Law of Cosines is a better option. When you know all three sides, the Law of Cosines gives the unique right answer for any angle. The Law of Sines does not. You would need to know if the angle is acute or obtuse. When all three sides are given, the Law of Cosines is preferable. Angle $A = 127.17°$.

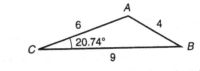

FIGURE 15 $a = 9$, $b = 6$, $c = 4$, and $\angle C = 20.74°$ ▬

EXAMPLE 4
REVISITED

Suppose $\angle C = 39°$, $c = 16$, and $a = 20$. Find side b.

The case is Angle-Side-Side so there are two possible triangles and two answers (Figure 7, again, reproduced as Figure 16). From Part A of the Law of Cosines,

$$c^2 = a^2 + b^2 - 2ab \cos C$$

$$16^2 = 20^2 + b^2 - 2(20)b \cos 39°.$$

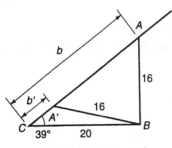

FIGURE 16 $\angle C = 39°$.
$c = 16$, and $a = 20$

This is a quadratic in b. To solve it, put it in "standard" form and treat "b" as the unknown "x."

$$b^2 - (40 \cos 39°)b + (20^2 - 16^2) = 0.$$

Now use the Quadratic Formula by placeholder position, rather than by letter. The "b" you are solving for is not the "b" in "negative b plus or minus . . ." formula.

$$\frac{40\cos 39° \pm \sqrt{(-40\cos 39°)^2 - 4(1)(20^2 - 16^2)}}{2(1)} = 25.42 \text{ or } 5.66.$$

Suppose you also want to find angle A. You could

1. Use the Law of Sines, remembering that there will be two answers, corresponding to angles A and A' in Figure 16 (Example 4).

2. Use the Law of Cosines, 6-3-4B, now that you have already found side b and therefore have the Side-Side-Side case. However, the letters are not the same as originally given in 6-3-4B. To solve for angle A, put A in the place of C. Now a is the opposite side and b and c are the adjacent sides:

$$A = \cos^{-1}\left(\frac{a^2 - b^2 - c^2}{-2bc}\right).$$

Each of the two b's yields an answer. The longer side yields:

$$A = \cos^{-1}\left(\frac{20^2 - 25.42^2 - 16^2}{-2(25.42)(16)}\right) = 51.87°.$$

The calculation for the shorter side is similar.

The Angle-Side-Side case may be ambiguous. You see this when you use the Law of Sines to solve for an angle. There are two answers. You also see the ambiguity when you use the Law of Cosines to solve for a side using the Quadratic Formula, which yields two answers.

Proof of the Law of Cosines

The proof of the Law of Cosines is not hard. It follows from using the Pythagorean Theorem twice on the same sort of picture we used for the SAS Area Formula and the Law of Sines Proof (Figures 17A and 17B).

Label h, the altitude, and d, the directed distance (positive in Figure 17A, negative in Figure 17B) from point C to the place where the altitude intersects the base, b.

Using the two right triangles,

1. $a^2 = h^2 + d^2$.

2. $c^2 = h^2 + (b - d)^2$ [d is negative in Figure 17B.]
   $= h^2 + b^2 - 2bd + d^2$.

Now use (1) to replace $h^2 + d^2$ in (2):

3. $c^2 = a^2 + b^2 - 2bd$.

Now note $\cos C = \dfrac{d}{a}$, so $d = a \cos C$. Substituting this into (3):

$$c^2 = a^2 + b^2 - 2ab \cos C.$$

This is the Law of Cosines (6-3-4).

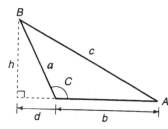

FIGURE 17A Angle C is acute

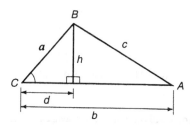

FIGURE 17B Angle C is obtuse

■ The Geometric Cases

Reconsider the geometric cases to determine when the Law of Sines is appropriate and when the Law of Cosines is appropriate.

1. Side-Side-Side (Figure 18)
   For SSS use the Law of Cosines to obtain any desired angle.
   (The Law of Sines has two angles in it. We can solve for one, but not both. The Law of Sines won't work for SSS.)

2. Side-Angle-Side (Figure 19)
   For SAS use the Law of Cosines to find the third side.
   (The Law of Sines has two sides opposite two angles. In SAS neither of the given sides is opposite the given angle. The Law of Sines won't work for SAS.)

3. Angle-Side-Angle (Figure 20)
   or Angle-Angle-Side (Figure 21)
   In ASA the two angles determine the third. Then, and in AAS, the side is opposite a known angle, so the Law of Sines applies.
   (The Law of Cosines uses three sides—knowing one is not enough—it does not apply.)

FIGURE 18 Side-Side-Side

FIGURE 19 Side-Angle-Side

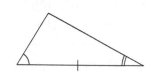

FIGURE 20 Angle-Side-Angle

4. Angle-Side-Side

This case has a known side opposite a known angle. The Law of Sines will work, and so will the Law of Cosines. If the geometric construction is ambiguous, the equations resulting from these laws will be ambiguous too.

FIGURE 21 Angle-Angle-Side

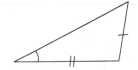

FIGURE 22 Angle-Side-Side

CONCLUSION   If a triangle is geometrically determined, it can be solved using the Laws of Sines and Cosines.

Key Terms: Law of Sines, Law of Cosines.

Exercise 6-3

A

See the Figure.

A1. $b = 33$, $\angle B = 73°$, $\angle A = 57°$. Find a.   [29]

A2. $a = 56$, $b = 64$, $\angle A = 55°$. Find $\angle B$ [69]

A3. $a = 70$, $b = 80$, $\angle C = 51°$. Find c.   [65]

A4. $a = 14$, $b = 16$, $c = 13$. Find $\angle C$.   [51]

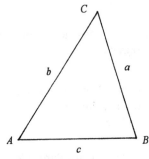

FIGURE For A1–A4

See the Figure.

A5. $b = 9$, $\angle B = 38°$, and $\angle C = 121°$. Find c.   [13]

A6. $a = 5$, $\angle A = 27°$ and $\angle B = 39°$. Find b.   [6.9]

A7. $c = 60$, $\angle C = 119°$, and $b = 42$. Find $\angle B$.   [38]

A8. $c = 50$, $b = 30$, and $\angle B = 35°$. Find $\angle C$.

A9. $c = 100$, $a = 52$, and $\angle A = 23°$. Find $\angle C$.   [130]

A10. $c = 200$, $b = 120$, and $\angle B = 33°$. Find $\angle C$.   [110]

A11. $a = 12$, $b = 18$, $\angle C = 125°$. Find c.   [27]

A12. $a = 50$, $b = 80$, $\angle C = 119°$. Find c.   [110]

A13. $b = 40$, $c = 70$, and $\angle A = 25°$. Find a.   [38]

A14. $a = 6$, $c = 13$, and $\angle B = 32°$. Find b.   [8.5]

A15. $a = 3$, $b = 5$, and $c = 7$. Find $\angle C$.   [120]

A16. $a = 28$, $b = 43$, and $c = 60$. Find $\angle B$.   [41]

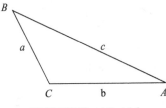

FIGURE For A5–A16

See the Figure for A17–22 on next page.

A17. $a = 78$, $\angle A = 82°$, and $\angle B = 60°$. Find b.   [68]

A18. $a = 20$, $b = 19$, and $c = 7$. Find $\angle C$.   [20]

A19. $a = 10$, $b = 9$, and $c = 4$, Find $\angle C$.   [24]

A20. $a = 20$, $b = 18$, and $c = 10$. Find $\angle B$.   [64]

A21. $a = 49$, $b = 45$, and $c = 22$. Find $\angle A$.   [87]

A22. $\angle B = 71°$, $b = 125$, and $a = 130$. Find $\angle A$.   [80]

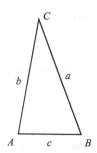

FIGURE for A17–22

Sketch the triangle and find the requested part.

A23. $\angle A = 12°$, $a = 20$, and $\angle B = 75°$. Find b. [93]

A24. $\angle C = 60°$, $c = 56$, and $\angle B = 72°$. Find b. [61]

A25. $\angle A = 40°$, $a = 90$, and $b = 100$. Find $\angle B$. [46, ...]

A26. $\angle B = 18°$, $b = 20$, and $c = 30$. Find $\angle C$. [28, ...]

A27. $\angle C = 55°$, $a = 100$, and $b = 200$. Find c. [160]

A28. $\angle A = 125°$, $c = 4$, and $b = 9$. Find a. [12]

A29. $a = 10$, $b = 12$, and $c = 20$. Find $\angle C$. [130]

A30. $a = 90$, $b = 80$, and $c = 40$. Find $\angle B$. [63]

A31. $a = 100$, $b = 80$, and $\angle A = 37°$. Find c. [150]

A32. $\angle C = 40°$, $c = 10$, and $b = 4$. Find $\angle B$. [15]

A33. $\angle A = 120°$, $a = 7$, and $b = 3$. Find $\angle B$. [22]

A34. $\angle C = 165°$, $b = 200$, and $c = 300$. Find $\angle B$. [9.9]

A35. $\angle A = 70°$, $\angle B = 105°$, and $c = 300$. Find b.
[3300]

A36. $\angle B = 2°$, $\angle C = 5°$, and $a = 1000$. Find c. [720]

A37. Find the area: $\angle C = 67°$, $a = 90$, and $b = 150$.
[6200]

A38. Find the area: $\angle B = 135°$, $a = 3$, and $c = 4$. [4.2]

A39. $a = 10$, $b = 20$, and the area of the triangle is 80. Find
$\angle C$. [53, ...]

A40. $b = 5$, $c = 3$, and the area of the triangle is 3.2.
Find $\angle A$. [25, ...]

B

B1.* (A) When is the Law of Sines applicable?
(B) When does it give unique results?
(C) Which geometric case is dangerous?

B2.* (A) The Law of Cosines generalizes the _____
Theorem.
(B) The Law of Cosines is set up to handle two
geometric cases that yield a unique answer.
Which two?

(C) There is another geometric case it can handle,
although it may not give a unique solution.
Which case?

☺ *For B3–B10. In the figure, the three tick-marked parts are
given. Which of the remaining parts could be found next
and which result would be used to find it? (If more than
one part can be found next, list all of the ones that can be
found next.)*

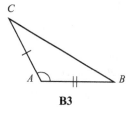

B3

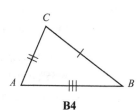

B4

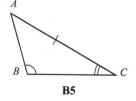

B5

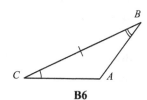

B6

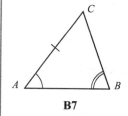

B7

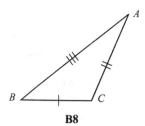

B8

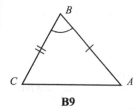

B9

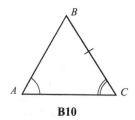

B10

B11. (A) Redo Example 6 (Figure 10) using the Law of
Cosines to solve for side a.
(B) How is it evident in the answer that there is only
one triangle, even though the geometric case
is ASS?

B12.* In each case, which theorem would be most
appropriate for the first step in solving the triangle?
(A) SSS   (B) SAS   (C) ASA
(D) AAS   (E) SSA

Sketch the triangles and find the area.

B13. $\angle A = 20°$, $\angle B = 85°$, and $c = 10$. [18]

B14. $\angle B = 50°$, $\angle C = 110°$, and $b = 20$. [84]

B15. $a = 9$, $b = 10$, $c = 15$. [44]

B16. $a = 8$, $b = 10$, and $\angle A = 40°$. [40, 9.3]

Sketch the triangles and use the Law of Cosines to find b.

B17. $a = 10$, $c = 8$, and $\angle A = 43°$. [14]

B18. $a = 50$, $c = 60$, and $\angle A = 20°$. [100, . . .]

Sketch the triangles and find the requested part.

B19. $\angle A = 55°$, $a = 17$, and $c = 20$. Find $\angle C$. [75,...]

B20. $\angle B = 126°$, $a = 4$, and $c = 6$. Find $\angle A$. [21]

B21. $\angle C = 119°$, $b = 10$, and $\angle A = 20°$. Find c. [13]

B22. $\angle B = 37°$, $\angle A = 109°$, and $c = 100$. Find a. [170]

B23. $\angle A = 22°$, $c = 5$, and $a = 4$. Find $\angle B$. [5.9, . . .]

B24. $\angle B = 59°$, $a = 10$, and $b = 9$. Find $\angle C$. [13, . . .]

Find the angle between the two given sides of the triangle.

B25. $a = 12$, $b = 6$, area $= 30$ [56° or . . .]

B26. $b = 40$, $c = 72$, area $= 200$ [8.0° or . . .]

B27. Here is a problem: "$a = 12$, $b = 10$, and $\angle B = 30°$. Find the Area."
(A) What is tricky about this problem?
(B) Solve it.

B28.* Why can solving for a triangle-angle yield an ambiguous answer in the Law of Sines but not in the Law of Cosines?

B29. $\angle C = 21°$, $c = 5$, and $a = 6$. Given these parts, which part would be easiest to find next using
(A) the Law of Sines and
(B) the Law of Cosines?
(C, D) Do those two computations.
(E) How, in each computation, can you tell that this is a tricky problem?

B30. $\angle B = 25°$, $b = 15$, and $c = 9$. Given these parts, which part would be easiest to find next using
(A) the Law of Sines and
(B) the Law of Cosines?
(C, D) Do those two computations.
(E) How, in each computation, can you tell that this is a tricky problem?

B31. (A) $\angle C = 70°$, $c = 2$, and $b = 5$. Find $\angle B$.
(B) Explain what happened trigonometrically and why it happened, geometrically.

B32. (Heron's Formula for the area of a triangle in the SSS case.) Let a, b, and c be the sides of a triangle (not necessarily a right triangle). Define

$$s = \frac{1}{2}(a + b + c) \text{ [the "semiperimeter"]. Then}$$

the area of the triangle is given by Heron's Formula:

$$A = \sqrt{(s(s - a)(s - b)(s - c))}.$$

(A) Use it to find the area of a 3-4-5 triangle.
(B) Prove this formula using a perpendicular to side c and the Pythagorean Theorem twice on the right triangles formed. [It's quite long, but the mess simplifies very nicely in the end.]

Reading and Writing Mathematics

B33. Suppose a, b, and the area of the triangle are given.
(A) Outline a plan for finding side c.
(B) Is c uniquely determined?

B34.* (A) When is the SAS Area Theorem applicable?
(B) Rewrite it for the case when $\angle B$ is given.

B35.* Rewrite the Law of Cosines to emphasize solving for
(A) Side b given sides a and c and angle B.
(B) angle B.

B36.* Rewrite the Law of Cosines to emphasize solving for
(A) Side a given sides b and c and angle A.
(B) angle A.

B37.* Show the steps in converting the Law of Cosines, Part A, into Part B.

B38. If $\angle B$ is a right angle and a and $\angle A$ are given, then b can be determined from the right-triangle interpretation of sine. Show how b also can be determined by the Law of Sines. [That is, the Law of Sines generalizes the right-triangle interpretation of sine.]

B39. Prove the Side-Angle-Side Area Formula (acute angle case, without looking).

B40. Prove the Law of Sines (acute angle case, without looking).

B41. Prove the Law of Cosines (acute angle case, without looking).

B42. There are no solutions to "$\sin x = c$" if $c > 1$. Here is a problem with no solution: "Let $\angle B = 50°$, $b = 10$, and $a = 15$. Find $\angle A$."
(A) Try to use the Law of Sines to solve it and note why there is no solution.
(B) Sketch a picture to show why there is no solution.

B43.* What must be given to make it easy to compute the area of a triangle? [There are two alternatives.]

Section 6-4 SOLVING FIGURES

Trigonometry can be used to measure distances and angles indirectly. Solving right triangles is relatively easy.

EXAMPLE 1

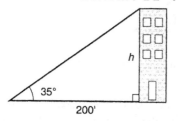

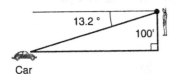

FIGURE 1 A building 200 feet away

Sarah wishes to determine the height of a building. From 200 feet away, the angle to the top from ground level is 35°. How tall is the building (Figure 1)?

The angle (35°) measured upward from level is called the "angle of elevation." In Figure 1 the unknown height is h and opposite the known angle. The known side (200 feet) is adjacent the known angle. Therefore the tangent function applies.

$$\tan 35° = \frac{h}{200} . \; h = 200 \tan 35° = 140.0 \text{ (feet)}.$$

Occasionally angles are measured downward from level.

EXAMPLE 2

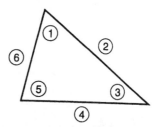

FIGURE 2 Angle of depression 13.2° from a height of 100 feet

A parked car is spotted at an angle of depression of 13.2° from a window at an elevation of 100 feet above the car (Figure 2). How far away is the car from the window?

The term "angle of depression" refers to the angle below level. The hypotenuse is the unknown distance, say d, and the opposite is known. This suggests using sine.

$$\sin 13.2° = \frac{100}{d}. \; d = \frac{100}{(\sin 13.2°)} = 437.9 \text{ (feet)}.$$

■ Solving Triangles

To solve a triangle which is not a right triangle, use the Law of Sines or the Law of Cosines. The upcoming examples illustrate the sequence of steps without actually computing the numbers.

In the next five examples, suppose we know the parts labeled 1, 2, and 3, and we wish to solve for the remaining three parts labeled 4, 5, and 6. The problem is to decide which of parts 4, 5, and 6 to solve for first, and which method to use to solve for it.

EXAMPLE 3

Decide how to solve the triangle in Figure 3, given the first three parts.

FIGURE 3 Parts 1, 2, and 3 are given

The first thing to do is to decide on the geometric case. It is Angle-Side-Angle. Neither the Law of Sines nor the Law of Cosines will work. Use the sum of the angles in a triangle to determine angle 5 first. Then the Law of Sines will work to find either side 4 or side 6.

EXAMPLE 4   Decide how to solve the triangle in Figure 4, given the first three parts.

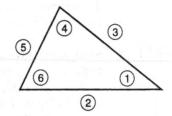

FIGURE 4 Parts 1, 2, and 3 are given

This is a case of Side-Angle-Side. The Law of Cosines makes it possible to find side 5 next. Then use the Law of Cosines to find the remaining angles. The Law of Sines is simpler, but may yield an extraneous solution that can be very hard to distinguish from the actual solution if the angle happens to be near 90°.   ▬

EXAMPLE 5   Decide how to solve the triangle in Figure 5, given the first three parts.

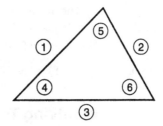

FIGURE 5 Parts 1, 2, and 3 are given

This is a case of Side-Side-Side. The Law of Cosines makes it easy to find any of the three unknown angles next. Then, it could be used again to find another angle. The third would follow easily from the sum of the angles in a triangle. The second angle could also be determined using the Law of Sines, but that law yields an extraneous solution.   ▬

EXAMPLE 6   Decide how to solve the triangle in Figure 6, given the first three parts.

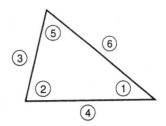

FIGURE 6 Parts 1, 2, and 3 are given

This is a case of Angle-Angle-Side. Whenever two angles are given, the third is easy to obtain using the sum of the angles in a triangle. So find angle 5 first. Then the Law of Sines will yield unambiguous results for the unknown sides.   ▬

EXAMPLE 7   Decide how to solve the triangle in Figure 7, given the first three parts.

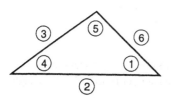

FIGURE 7 Parts 1, 2, and 3
are given

This is a case of Angle-Side-Side. Be careful. In this case you have two options. You may use the Law of Sines to find angle 5 opposite side 2, but there will be two answers to the equation. From the picture, it looks like the obtuse angle is the answer you want.

Then the remaining angle follows from the sum of angles in a triangle, and the remaining side from the Law of Sines or Law of Cosines.

A second option is to use the Law of Cosines to find the third side, side 6. There will be two answers to the resulting quadratic equation because there are two triangles with parts 1, 2, and 3, although Figure 7 only pictures one of them. From the picture, side 6 could be longer and still have side 3 the same length, by having angle 4 larger. Because side 6 could be longer, we must want the shorter of the two solutions for side 6.

Then, with the original three parts and a solution for side 6, we have SSS and may use the Law of Cosines, or, ambiguously, the Law of Sines, to find the remaining angles.

■ More Complex Figures

Complex geometric figures may have more than three sides, but they are still solved by the techniques developed for triangles. Break them into triangular components and solve the triangles successively.

EXAMPLE 8   Figure 8 illustrates three known sides and two known angles of a four-sided geometric figure. Determine the fourth side and the two unknown angles by first subdividing the figure into two triangles as in Figure 9, and then solving each of the triangles.

To find the fourth (dashed) side, we could first find the diagonal (labeled "(1)" in Figure 9) by SAS. Then we could find angle 2 using the Law of Cosines (or the Law of Sines). Angle 3 follows easily by subtracting from 68°. Then side 4 follows from SAS and the Law of Cosines.

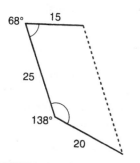

FIGURE 8   A four-sided figure

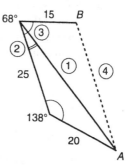

FIGURE 9 The figure subdivided into two triangles, and a solution sequence

CALCULATOR EXERCISE 1   Learn to use your calculator's memories to store and retrieve values. Follow the calculations in Example 8. Side 1 is

$$\sqrt{20^2 + 25^2 - 2(20)(25)\cos 138°}.$$

Result: 42.04931421.

Store this in a memory, say, in C. Now you can retrieve the value of C whenever you need it. You need it to compute angle 2.

Angle 2 follows from the Law of Cosines (or Sines).

$$\cos(\text{angle 2}) = \frac{(20^2 - 25^2 - C^2)}{(-2(25)C)}.$$

Result: .9480034873.

This need not be stored, since we are going to use it only once—immediately. Take the inverse cosine of that to find angle 2 and subtract from 68° to find angle 3. Result: 49.4422741 (degrees).

Store this in, say, T (for no good reason).

Now side 4 follows from SAS, given side 1 (stored as C), angle 3 (stored as T), and the other side of length 15.

Side 4 is computed by:

$$\sqrt{15^2 + C^2 - 2(15)C \cos T}.$$

Result: 34.24782406.

This has more decimal places than we need, so now, and not before, is a good time to "round off." Say, 34.25. People who round off to two decimal places at each step often find they no longer have two accurate decimal places by the time they get to the end. Use your calculator's memory rather than rounding off at each step.

EXAMPLE 8 CONTINUED   Suppose we want to continue Example 8 and find angle B (Figure 9). How would you do it?

There are several ways to find angle B. We could use SSS and the Law of Cosines, now that we have sides 1 and 4 and the side of length 15. We would use the version with side 1 opposite angle B. Also, we could use the Law of Sines, now that we have side 4 opposite angle 3 and side 1 opposite angle B. Still another way would be to find the angle at A first and then subtract the two known angles from 180 degrees. ▬

Trigonometry is useful for indirect measurement.

EXAMPLE 9   Suppose you want to determine the elevation of the top of a mountain above a level plain. Of course, you can not measure from the top to the bottom, since the distance you want is perpendicular to the plain down to a point underneath the mountain. So here are the measurements you take (Figure 10).

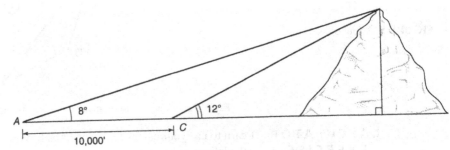

FIGURE 10 The elevation of a mountain. (The vertical scale has been exaggerated to make room for labeling small angles)

You find a point on the plain where you can measure the angle of elevation to the top of the mountain. At point C the angle is 12°. Then you back off 10,000 feet to point A and measure the angle of elevation from there. It is 8°. How high is the mountain above the plain?

There are several ways to do this problem. Before doing any calculations, look at the given parts and plan a sequence of steps that will find h. Labeling the appropriate parts on your picture will be helpful. Figure 11 has useful labels.

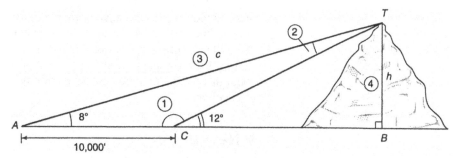

FIGURE 11 Figure 10, with the addition of useful labels

In Figure 11 it is easy to find angle 1 and then angle 2. Then triangle ACT can be solved for side 3 (side c) using the Law of Sines. Then triangle ABT (a right triangle) can be solved for side h, given its hypotenuse c and angle A.

The above plan works. The plan is the key. The steps are then just "plug in" steps. Here they are for this example.

Angle 1 is $180° - 12° = 168°$. Angle 2 is $180° - 168° - 8° = 4°$. By the Law of Sines:

$$\frac{c}{\sin 168°} = \frac{10,000}{\sin 4°}.$$

$$c = \sin 168° \frac{(10,000)}{(\sin 4°)} = 29,805.36 \text{ (feet)}.$$

Store this in C, or, at least, do not erase it yet.

Now

$$\sin 8° = \frac{h}{c}. \quad h = c \sin 8° = 4148.1 \text{ (feet)}.$$

The mountain top is 4148 feet above the level plain. There are other strategies for doing this problem (Problems B46–47). ▬

EXAMPLE 10   Given the marked sides and angles in Figure 12, outline a plan to find x. Illustrate the sequence of steps and name the theorem you intend to apply at each step.

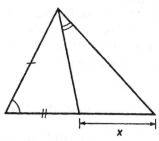

FIGURE 12 Find x, given the marked sides and angles

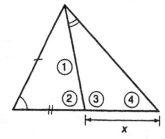

FIGURE 13 The sequence of steps

Figure 13 illustrates the sequence of steps.

Step 1: Find side 1 using the Law of Cosines.

Step 2: Find angle 2 using the Law of Cosines (or the Law of Sines, if you are certain it is acute).

Step 3: Find angle 3 (a straight angle is 180 degrees).

Step 4: Find angle 4 (angles of a triangle sum to 180 degrees).

Step 5: Find x using the Law of Sines.

To solve a complicated problem it is best to outline a plan before worrying about the details. The plan is the interesting part; the steps are just "plug in" calculations. ▬

■ Navigation

Planes and ships can navigate by keeping track of directions and distances. In navigation, directions are related to North, South, East, and West. **One way to give a direction is to state the** <u>bearing</u> **as either one of these four primary directions or as an acute angle away from North or South toward East or West.** For example, "N 75° W" (Read "North seventy five degrees West") would mean 75° away from North toward West. That direction is mostly West. The bearing "S 20° E" would be mostly South, but 20° toward East.

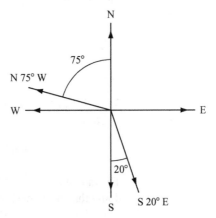

FIGURE 14 Bearings

EXAMPLE 11   Starting from her dock in Key Largo in Florida, Pam navigates S 60° E for 10 miles. How far south of her dock is she? How far east?

Figure 15 illustrates her trip. In circular trigonometry it is common for the hypotenuse to be given, as in this example, and the sides to be determined from the angle. Then it is appropriate to use the following interpretations: In a *right* triangle,

$$\text{"opposite is hypotenuse times sine"} \tag{6-4-1}$$

and

$$\text{"adjacent is hypotenuse times cosine."} \tag{6-4-2}$$

The hypotenuse is 10 and the southward component of her position is the adjacent side. Therefore she has traveled

$$10 \cos 60° = 5 \text{ (miles) south.}$$

The eastward component is the opposite side. She has traveled

$$10 \sin 60° = 8.66 \text{ (miles) east.}$$

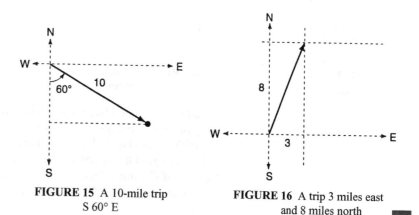

FIGURE 15 A 10-mile trip
S 60° E

FIGURE 16 A trip 3 miles east
and 8 miles north

Given the opposite and adjacent sides, the angle is determined by its tangent.

EXAMPLE 12 The next day Pam piloted straight to a position 3 miles east and 8 miles north of her dock (Figure 16). What bearing did she travel?

In this method of reporting directions we measure the angle away from North. Then 3 is the opposite side and 8 is the adjacent side.

$$\tan \theta = \frac{3}{8}. \quad \theta = \tan^{-1}\left(\frac{3}{8}\right) = 20.56°.$$

She traveled N 20.56° E.

The directional ideas in Examples 11 and 12 can be combined to handle more complicated problems.

EXAMPLE 13 Tamara flew from base N 20° E for 30 miles and then S 55° W for another 22 miles (Figure 17). Where was she then? Which bearing should she fly to go directly back to her base?

There is more than one way to solve this problem. We will give two distinctly different approaches. The key is to plan a sequence of steps to obtain the answer. Pause here and think of a plan.

Both plans given below use the fact from geometry that, when a line cuts two parallel lines, the "alternate interior angles" are equal (Figure 18). We will treat all North-South lines as parallel, although that is not strictly true on the curved surface of the earth.

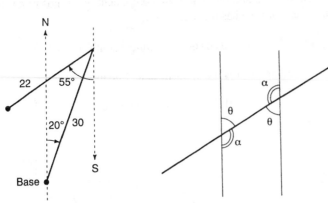

FIGURE 17 Two
legs of a flight

FIGURE 18 Alternate interior angles
are equal

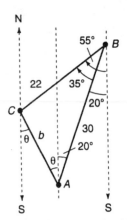

FIGURE 19 Figure 17 with
auxiliary lines

Here is one plan. The leg back to base makes the third side of a triangle. From the given angles in Figure 17, we can determine the angle between the two given legs of the flight (Figure 19). Then, triangle ABC is determined by Side-Angle-Side and we can find the opposite side (b, which is the distance back to base) using the Law of Cosines. Then we can find angle A using the Law of Sines. From angle A, by subtracting 20°, we can find θ, which will give the bearing back to base: S θ° E.

Here are the computations.

Angle B is $55° - 20° = 35°$.

$$b^2 = 30^2 + 22^2 - 2(30)(22)\cos 35°.$$
$$b = 17.40.$$

She is 17.40 miles from base.

Now, by the Law of Sines (or, you could find Angle A using the Law of Cosines),

$$\frac{\sin A}{22} = \frac{\sin 35°}{17.40}.$$

$$\sin A = \frac{22(\sin 35°)}{17.40} = .7252.$$

$$\angle A = 46.49°. \quad \theta = 46.49° - 20° = 26.49°.$$

Therefore, she should fly S 26.49° E to return to base.

EXAMPLE 13 AGAIN
Here is a second plan to find the distance and bearing back to base (Figure 17). As we saw in Example 11, for each leg of her flight we can extract the distance traveled in each primary direction from the bearing and distance of the leg. For this problem, we can do this for each leg and add the contributions to find the total distances in the primary directions. Then use the idea of Example 12 to extract the bearing from the position.

The auxiliary lines and labels in Figure 20 are helpful.

Leg 1: north: $30 \cos 20°$ (*a*)   east: $30 \sin 20°$ (*b*)
Leg 2: north: $-22 \cos 55°$ (*c*)   east: $-22 \sin 55°$ (*d*)

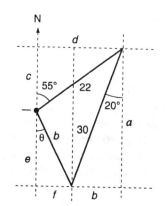

FIGURE 20 Figure 17 with auxiliary lines and labels

[On Leg 2, the minus signs are because South is the negative of North and West is the negative of East.]

Total: north: 30 cos 20° − 22 cos 55° east: 30 sin 20° − 22 sin 55°
= 15.57 (*e*) = −7.76 (*f*)

So to return to base she will have to fly 15.57 miles south and 7.76 miles east. The bearing is S θ° E where θ is in the picture.

$$\tan \theta = \frac{7.76}{15.57} = .498.$$
$$\theta = \tan^{-1}.498 = 26.49°.$$

She should fly S 26.49° E.

The Pythagorean Theorem gives the distance:

$$\sqrt{(15.57^2 + 7.76^2)} = 17.40 \text{ miles.}$$

Both techniques work. They are about equally long. Take your pick. ▬

Bearings from two known positions can be used to locate a third position.

EXAMPLE 14 Sue lives 3 miles Northeast of Amy. When Sue spots a tornado at N 75° W, Amy sees it at N 20° W. At that instant, how far is the tornado from Amy?

This requires a picture (Figure 21). "Northeast" is half way between North and East, so the direction from Amy to Sue is 45° east of North. Plan how you could do this problem. Label the picture with angles and sides you could determine. Fill in auxiliary lines if useful. Your picture should develop from Figure 21 to a more complex picture with important parts labeled, and unimportant parts not labeled (Figure 22).

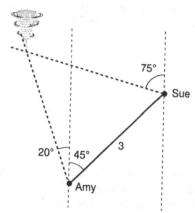

FIGURE 21 A tornado sighting

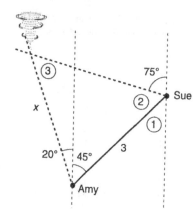

FIGURE 22 The order of solution

By alternate interior angles, angle 1 in Figure 22 is also 45°. Then angle 2 can be found, and then angle 3. Finally the Law of Sines yields side *x*.

$$\text{angle 2} = 180° - 75° - 45° = 60°.$$
$$\text{angle 3} = 180° - 60° - 20° - 45° = 55°.$$

From the Law of Sines

$$\frac{x}{\sin 60°} = \frac{3}{\sin 55°}.$$

Solving for x, $x = 3.17$ miles. The tornado is 3.17 miles from Amy.

There is another way to state a <u>bearing</u> by giving the clockwise (not counterclockwise) angle in degrees away from North. So North is bearing 0, East is bearing 90, South is bearing 180, and West is bearing 270. So, when you watch an old World War II submarine movie and the skipper says "bearing two seven zero" he just means West, and "bearing three zero zero" means 30 degrees north of West, which would be "N 60° W"in the notation we just discussed. ■

In this section we have subdivided complex figures into simpler component figures. Here is a problem about area that utilizes the idea of subdivision in a new way.

EXAMPLE 15

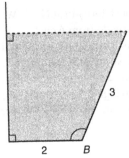

FIGURE 23 Find angle B to maximize the shaded area

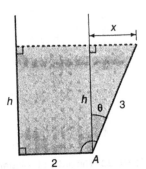

FIGURE 24 Figure 23 with some auxiliary lines and labels

Find angle B such that the area shaded in Figure 23 is its maximum for the given sides and right angles. That is, treat angle B as adjustable, and find how to adjust it to maximize the area.

First build a formula that relates the angle to the area. Given the angle, what is the shaded area? Figure 24 includes a well-chosen auxiliary line and labels. Angle B is $90° + \theta$. The area is the sum of areas of two simple regions, a rectangle and a right triangle:

$$\text{Area} = 2h + \left(\frac{1}{2}\right)xh.$$

We can maximize this graphically as soon as it is expressed in terms of only one variable, x or h or θ. First we write x and h in terms of θ.

$$\cos \theta = \frac{h}{3}. \quad \sin \theta = \frac{x}{3}.$$
$$h = 3 \cos \theta. \quad x = 3 \sin \theta.$$
$$\text{Area} = 2(3 \cos \theta) + \left(\frac{1}{2}\right)(3 \sin \theta)(3 \cos \theta).$$
$$\text{Area} = 6 \cos \theta + 4.5(\sin \theta)(\cos \theta).$$

Now graph this and find the value of θ which maximizes it. Then angle B will be that answer plus 90 degrees. For the numerical value, do Problem A14.

There is a second approach which finds θ last. It is possible to express the area in terms of x (or h) alone, solve for the value of x which maximizes the area, and then solve for the value of θ which produces that x. Of course, this approach yields the same answer. Again,

$$\text{Area} = 2h + \left(\frac{1}{2}\right)xh = \left(2 + \frac{x}{2}\right)h.$$

We want only one variable, so express h in terms of x (or vice versa). Use the Pythagorean Theorem.

$$x^2 + h^2 = 3^2. \quad h = \sqrt{9 - x^2}.$$

$$\text{Area} = \left(2 + \frac{x}{2}\right)\sqrt{9 - x^2}.$$

Now graph this to find the value of x that yields the maximum area. For the numerical value, do Problem A15. Then solve for θ in

$$\sin \theta = \frac{x}{3}$$

to find the value of θ which yields the maximum. Again, angle B is $\theta + 90°$

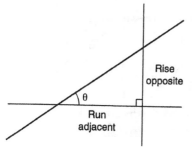

FIGURE 25 The slope and angle of a line

■ Slope

The slope of a line is often denoted by "m", as in the famous slope-intercept equation "$y = mx + b$." The slope describes how steep a line is. Another way to describe the steepness of a line would be to give the angle it makes with the positive x-direction (angle θ in Figure 25). Since the slope is "the rise over the run," which, from the point of view of angle θ is "opposite over adjacent," a simple relationship follows:

If θ is the angle a line "$y = mx + b$" makes with the positive x-direction,

$$\tan \theta = m, \text{ the slope.} \tag{6-4-3}$$

EXAMPLE 16   Find the angle that the line "$x + 2y = 6$" makes with the positive x-direction.

First we need the slope. $x + 2y = 6$ iff $2y = 6 - x$ iff $y = 3 - \dfrac{x}{2}$. The slope is $-\dfrac{1}{2}$. Therefore, $\tan \theta = -\dfrac{1}{2}$.

$$\theta = \tan^{-1}\left(-\frac{1}{2}\right) = -26.56°.$$

The line slopes downward at an angle of -26.56 degrees to the positive x-direction.

The slope of a line is the tangent of θ, the angle it makes with the positive x-axis.

CONCLUSION   Trigonometry is the area of mathematics which assigns measures to angles and sides of geometric figures. Given three parts of a triangle, it is often possible to solve for the remaining three using trigonometry. Given any configuration of known sides and angles, it is appropriate to develop a plan for solving for the unknown sides and angles before worrying about the numbers. If the figure is more complex than a triangle, it may be helpful to subdivide it into triangular components.

Terms: bearing.

Exercise 6-4

A

A1.* Name or state the four most useful trigonometric measurement theorems (that are not simply definitions of trig functions).

A2.* Give the three-letter abbreviations for the geometric cases that
(A) determine a triangle, and
(B) "almost" determine a triangle.

A3. You steer N 82° W for 12 miles. How far north do you go? How far west?   [1.7, ...]

A4. You pilot a plane S 32° E for 30 miles. How far south did you go? How far East?

A5. You end up 6 miles south and 4 miles east of home. Which bearing would lead straight back home? [N 34° ...]

A6. You end up 3 miles north and 4 miles east of home. Which bearing would lead straight back home? [... 53° W]

A7. The moon averages 240,000 miles away and it is 2160 miles in diameter. What is the apparent angle the width of the moon makes, viewed from the earth? (Continued in A8.)   [.52]

A8. (A) The sun averages 93,000,000 miles away and it is 865,000 miles in diameter. What is the apparent angle the width of the sun makes, viewed from the earth?
(B) Compare this to the answer in A7 and note what this has to do with solar eclipses.   [.53]

Find the angle that the line makes with the positive x-direction.

A9. $y = 5x + 7$ [79]   **A10.** $3y - 2x + 7 = 4$ [34]

A11. Give the equation of the line through (1, 8) that makes a 20° counterclockwise angle with the positive x-axis.

A12. Give the equation of the line through (2, 1) that makes a 80° counterclockwise angle with the positive x-axis.

A13. In Example 13, find $\angle C$.

A14. In Example 15, graph the area in terms of θ and find the value of θ which maximizes the area. [θ = ..., maximum area = 7.2]

A15. (A) In Example 15, graph the area in terms of x and find the value of x which maximizes the area.

(B) Then find the associated angle B.
[x = ..., maximum area = 7.2]

A16. You want to measure across the Grand Canyon from observation Deck A on your side to observation deck B on the other side. You pick a third point on your side of the canyon, point C, and measure the distance from A to C, and angles BCA and BAC. How will you figure the distance from A to B?

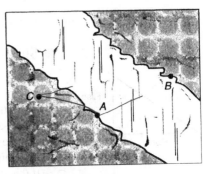

A17. If you stand 100 feet away from a tower than is 1000 feet tall, what angle is it to the top?   [84]

Sketch a figure to illustrate the following information.

A18. B is two miles N 30° W of A.

A19. C is 3 miles S 10° E of D.

A20. K is 4 miles N 70° E of F, and B is 2 miles S 20° E of F.

A21. A is 3 miles N 10° W of B, and C is 2 miles S 65° E of A.

B

☺ *Given parts 1, 2, and 3 of the figure, which part would be easiest to find next, and which theorem would you use? If there is more than one part which can be easily found, mention them all.*

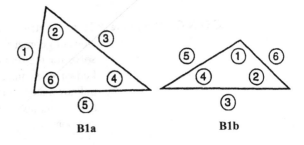

B1a   B1b

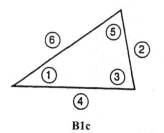

B1c

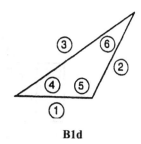

B1d

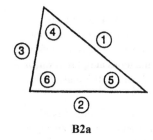

B2a

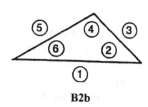

B2b

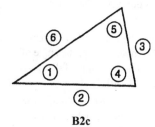

B2c

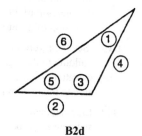

B2d

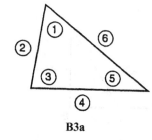

B3a

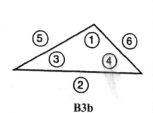

B3b

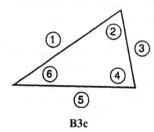

B3c

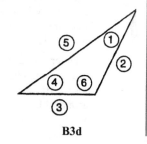

B3d

B4. (See the figure)
$\angle A = 34.633°$, $a = 29$, $b = 41$ *and* $c = 51$.
Find $\angle C$.

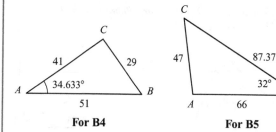

For B4 For B5

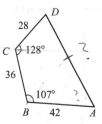 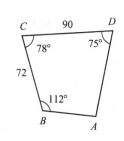

For B6 For B7

B5. (See the figure)
$b = 47$, $\angle B = 32°$, $a = 87.37$, and $c = 66$.
Find $\angle A$.

B6. (See the figure)
$AB = 42$, $BC = 36$, $CD = 28$, $\angle B = 107°$, and
$\angle C = 128°$. Find AD and $\angle A$.

B7. (See the figure)
$\angle B = 112°$, $\angle C = 78°$, $\angle D = 75°$, $BC = 72$, and
$CD = 90$. Find AB and $\angle A$.

B8. A boat heads N 33° E for 2 miles and then N 75° E
for 3 more miles. Sketch a good picture. What
bearing would have led straight from the initial point
to the terminal point? [N 58° . . .]

B9. A boat heads from its dock S 30° E for 20 miles and
then turns around to come back. If it returns 18 miles
at a bearing of N 25° W, how far and in what
direction is the dock? [. . . , N 67° W]

B10. Point B is 25 miles bearing S 63° E from point A.
Point C is 9 miles bearing N 15° W from point B.
How far, and at what bearing, is point A from
point C? [Draw a good picture that shows A slightly
north and mostly west of C.]

B11. Luann flies from base 25 miles S 70° E and then
20 miles S 35° E. How far from base is she and
which bearing heads directly back to base?
[. . . , N 55° W]

B12. Point B is 5 miles southeast of A. From point B, point C is N 75° W. From point A, point C is S 62° W. How far is it from A to C?

B13–16. *Solve for θ in the figure. Reproduce the figure and label your steps on it.*

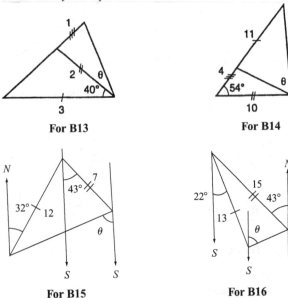

For B13   **For B14**

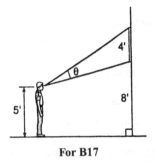

For B15   **For B16**

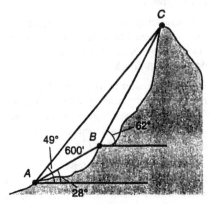

For B17

For B18

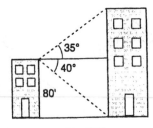

For B19

B17. [See the figure.] A picture that is 4 feet high is mounted on a wall so its bottom edge is 8 feet from the ground. A person whose eye level is 5 feet from the ground stands back and looks at it. From which distance will the angle subtended (the angle θ) be largest?   [$x = \dots, \theta = 24$]

B18. [See the figure.] The elevation of point A is 5100 feet. Point B is 600 feet from point A, at an angle of elevation of 28°. Point C is at an angle of elevation of 62° from point B, and at an angle of elevation of 49° from point A. What is the elevation of point C?

B19. [See the figure.] Compute the height of the taller building in the figure. From 80 feet up, the angle of elevation to the top is 35° and the angle of depression to the bottom is 40°.   [150]

B20. Line segment AB is of length 2 (see the figure). Line segment BC is perpendicular to AB and has length 1. If the sum of the lengths of AD and DC is 2.6, how long is AD?

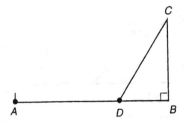

B21. A World War II submarine is 1 mile due south of a ship steaming 20 miles per hour due east when it fires a torpedo which goes straight at 60 miles per hour. At what bearing should the torpedo be aimed?   [19]

B22. A hunter shoots at ducks flying past at 40 miles per hour in a straight line that is 20 yards away at its closest. His shotgun pellets fly at 900 feet per second. If he shoots when the ducks are closest to him, how much should he lead the ducks? [3.9 feet]

Assume the orbits of Venus and Earth are circles about the sun. The radius of the orbit of Venus is 67,000,000 miles and of Earth is 93,000,000 miles.

B23. Viewed from Earth, what is the maximum possible angle between Venus and the Sun?

B24. Viewed from Earth, if the angle between Venus and the Sun is 26°, how far is Venus from Earth?

Sketch a picture to illustrate the relative locations of A, B, and C. Then find the distance and bearing from C to A.

B25. From *A* to *B* it is 5 miles N 30° E. From *B* to *C* it is 2 miles S 50° W.

B26. From *A* to *B* it is 20 miles S 61° E. *C* is 10 miles N 17° W from *B*.

B27. From *A* to *B* is 2000 yards at bearing 300. From *B* to *C* is 400 yards at bearing 20.

B28. From *A* to *B* is 10,000 yards at bearing 80. From *B* to *C* is 8,000 yards at bearing 330.

Plans

Instructions: For all plans, reproduce the figure and label your steps in order: 1, 2, 3, etc. and number the corresponding parts on your picture. For each step, (A) state which part you would find, (B) how (the law or theorem) you would find it, and (C) whether the answer might be ambiguous at that step. [Avoid ambiguous steps, whenever possible.]

B29–B36: Reproduce the given figure. On it, outline a plan to find *x* in the figure, given the marked parts. Label on it the steps (1), (2), etc.

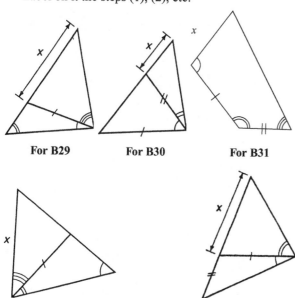

For B29   For B30   For B31

For B32   For B33

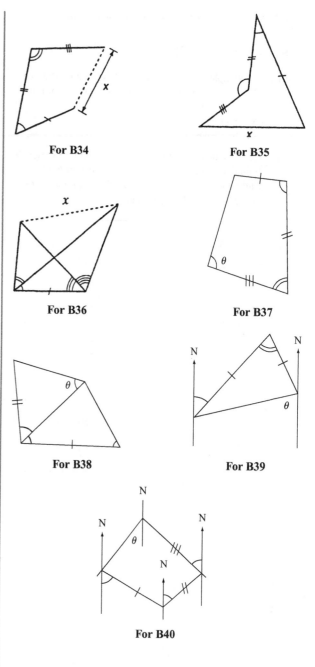

For B34   For B35

For B36   For B37

For B38   For B39

For B40

B37–40. Reproduce the given figure. Outline a plan to find θ, given the marked parts. Label the steps (1), (2), etc. on the figure.

B41. A line crosses the *x*-axis at *x* = 4 and makes a 30° angle (counterclockwise) with the positive *x*-direction. Find the equation of the line.

B42. A line goes through the point (1, 3) and makes a 60° angle (counterclockwise) with the positive *x*-direction. Find the equation of the line.

Find the angle the given line makes with the positive x-direction.

B43. $y = 2x + 7$ [63] **B44.** $y = \dfrac{x}{2} - 5$ [27]

B45. In Example 10, Step 2, you could use either the Law of Cosines or the Law of Sines. What reason might you have to prefer
(A) the Law of Cosines?
(B) the Law of Sines?

B46. Here is a picture of the type of problem in Example 9 (Figures 10 and 11). Outline a plan to find h by finding CT instead of AT. Label the steps and name the law you would use for each.

B47. (A) Set up two equations for finding h in the figure for B46 (again) by using the tangent function twice after labeling the segment CB.
(B) Solve for h with these equations using the data from Example 9 (Figure 10).

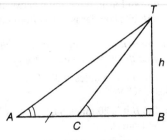

For B46 and B47

CHAPTER

7

Trigonometry for Calculus

Section 7-1   RADIANS

Trigonometry is the branch of mathematics that concerns the relationships of the sides and angles of triangles. It is also the branch of mathematics that concerns circular motion and other types of periodic motion such as waves (including sound, radio, and light waves), the swing of a pendulum, and vibrations of all kinds. In these applications, and in calculus, angles are measured in radians instead of in degrees. A radian is a unit of measure for angles, like a degree but larger (one radian is approximately 57.3 degrees).

■ Radians

By definition, π is the ratio of the circumference to the diameter of a circle: $\pi = \dfrac{C}{d}$, or $C = \pi d$ (Figure 1). Because the diameter is twice the radius, $C = 2\pi r$ (Figure 2)

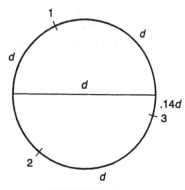

FIGURE 1 $C = \pi d$.
π is approximately 3.14

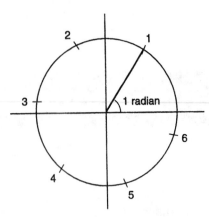

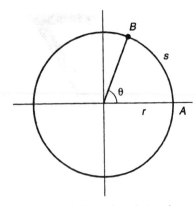

FIGURE 2 $C = 2\pi r$
$2\pi = 6.28$ radians in a full circle

FIGURE 3 Central angle θ and arc
AB of length s

On a circle, **when the arc length is equal to the radius, the central angle is defined to be one radian** (Figure 2). So there are 2π radians in a full circular angle.

Therefore, 2π radians = 360 degrees (Figure 2). The radian measure of any central angle is defined to be the ratio of the arc length to the radius (Figure 3):

$$\theta = \frac{s}{r} \quad \text{(where } s \text{ is arc length and } \theta \qquad\qquad \text{(7-1-1)}$$
$$\text{is expressed in radians).}$$

Equivalently, $s = r\theta$.

On a unit circle the arc length and the central angle have the same numerical value because $\theta = \frac{s}{r} = s$ when $r = 1$.

EXAMPLE 1 An arc of length 5 is part of a circle of radius 10 (Figure 4). What is its central angle?

The arc length is half the radius, so the angle is $\frac{1}{2}$ radian. By Formula 7-1-1, $\theta = \frac{s}{r} = \frac{5}{10} = \frac{1}{2}$. The angle is $\frac{1}{2}$ radian.

It is possible to estimate the radian measure of an angle by eye. To think in radians, compare the arc length to the radius.

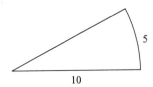

FIGURE 4 Radius 10. Arc length 5. Angle $\frac{1}{2}$ radian

EXAMPLE 2   Estimate the radian measure of the angle in Figure 5.

Compare the length of the arc to the radius. In the picture it looks like the arc length is about $\frac{1}{3}$ the radius. So the angle is about $\frac{1}{3}$ radian.

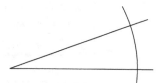

FIGURE 5 An angle. The arc length is $\frac{1}{3}$ the radius.

Therefore, the angle is $\frac{1}{3}$ radian

EXAMPLE 3   A circle has radius 60. How long is the arc opposite a central angle of .3 radian?

$$s = r\theta.\ s = 60(.3) = 18.$$

Of course, radians can be converted to degrees and degrees to radians.

A full circle has arc length $2\pi r$. Its central angle in radians is therefore $\theta = \frac{2\pi r}{r} = 2\pi$. So 2π radians $= 360°$. A straight angle $(180°)$ is π radians, and a right angle $90°$ is $\frac{\pi}{2}$ radians.

Table 7-1-2 Names of Famous Angles							
degrees	360°	180°	90°	60°	45°	30°	0
radians	2π	π	$\frac{\pi}{2}$	$\frac{\pi}{3}$	$\frac{\pi}{4}$	$\frac{\pi}{6}$	0

When these important angles are expressed in radians we almost always carry along the "π" and avoid decimal form. Using the form with π, you can immediately understand an angle as a part of a straight angle. For example, $\frac{\pi}{4}$ radians is one-quarter of a straight angle. In decimal form this would be .785 radians, which is much less illuminating. Memorize the names of these famous angles in radians. Figure 6 illustrates them.

To convert degrees to radians or radians to degrees, use the next result.

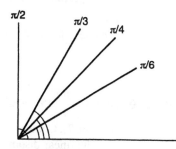

FIGURE 6 Angles $\frac{\pi}{6}, \frac{\pi}{4}, \frac{\pi}{3},$ and $\frac{\pi}{2}$ (in radians)

> The word "radians" need not be used or even abbreviated to give angles in radians. In particular, do *not* use "*r*" to abbreviate "radians" (we need "*r*" for the radius). However, the word "degrees" or the symbol "°" **must** be used to give an angle in degrees. So "sin 1" is the sine of 1 radian. If you want sine of 1 degree, write "sin 1°".
>
> Angles expressed in radian measure frequently have "π" in them, such as "$\frac{\pi}{2}$" or "2π." Then the "π" reminds you that the angles are measured in radians. Be sure to put your calculator in the mode (degree or radian) which matches the measure of the angles you are using.
>
> In the formula "$\theta = \frac{s}{r}$" both *s* and *r* have the same units, say, centimeters. Therefore, in the ratio the units cancel and the angle has no units. Radians do not have units the way measures of length, mass, or time do.

DEGREE-RADIAN CONVERSION

(A) 360 degrees = 2π radians.
180 degrees = π radians.

(B) $\dfrac{360 \text{ degrees}}{2\pi \text{ radians}} = 1.$

(C) $\dfrac{2\pi \text{ radians}}{360 \text{ degrees}} = 1.$

To change units, multiply the given amount by 1.

(7-1-3)

EXAMPLE 4 Change 5 degrees to radians.

$$5 \text{ degrees} = 5 \text{ degrees}\left(\frac{2\pi \text{ radians}}{360 \text{ degrees}}\right) = \frac{5(2\pi)}{360} \text{ radians} = \frac{\pi}{36} \text{ radians}.$$

$\frac{\pi}{36}$ radians is $\frac{1}{36}$ of a straight angle. If we convert this to decimal form it would be .087 radians, which is not very illuminating. Leave it as $\frac{\pi}{36}$. ▬

EXAMPLE 5 Convert 1 radian to degrees.

Multiply by 1 with radians in the denominator and degrees in the numerator.

$$1 \text{ radian} = 1 \text{ radian}\left(\frac{360 \text{ degrees}}{2\pi \text{ radians}}\right) = \frac{360}{2\pi} \text{ degrees} = 57.3 \text{ degrees}.$$

So one radian is about 57.3 degrees. Figure 2 illustrates the size of a radian. ▬

EXAMPLE 6 Form a $\frac{\pi}{4}$ (45°) angle with two yardsticks by putting their left ends together (Figure 7). Compare the length of the line segment *AB* to the length of the arc *AB*.

The linear distance is less than the arc length because "A line is the shortest distance between two points." A yard is 36 inches. The arc length is

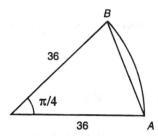

FIGURE 7 A $\frac{\pi}{4}$ (45°) angle.

Compare arc length to linear distance
between points

$s = 36\left(\dfrac{\pi}{4}\right) = 28.27$ inches. On the other hand, the linear distance AB can be calculated by bisecting the central angle (as in Example 5, Section 6-2) or by the Law of Cosines.

$$AB = 2(36)\sin\left(\frac{\pi/4}{2}\right) = 27.55 \text{ (inches)}.$$

This is 97 percent of the arc length. Even when the angle is as great as $\dfrac{\pi}{4}$, the arc length is similar to the length of the corresponding chord.

EXAMPLE 7 See Figure 8. Arc AB is 32 percent longer than BC. What is the central angle?

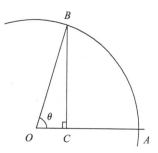

FIGURE 8 An arc

Arc length is easy in radians. Use radians.

$$\text{arc } AB = 1.32\,BC.$$
$$r\theta = 1.32(r\sin\theta).$$
$$\theta = 1.32\sin\theta.$$

This cannot be solved algebraically for $\theta > 0$. By Guess-and-Check, $\theta = 1.25$ radians.

■ Sectors

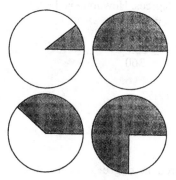

FIGURE 9 Shaded sectors

The part of a circular disk between the center and an arc is known as a <u>sector</u>. A sector is a region shaped like a piece of pie (Figure 9). The area of a sector can be given in terms of the area of the circle and the central angle. The area is proportional to the central angle, just as the arc length is proportional to the central angle.

$$\frac{\text{area of a sector}}{\text{area of a circle}} = \frac{\text{angle of the sector}}{\text{angle of the circle}}.$$

The area of a circle is given by the well-known formula $A = \pi r^2$. If a sector has radius r and central angle θ, then the area of the sector satisfies

$$\frac{\text{area of a sector}}{\pi r^2} = \frac{\theta}{\text{angle of the circle}}$$

where the "angle of the circle" depends upon the units used to express θ. It is either 2π radians or 360°. Solving for area,

$$\text{area} = \left(\frac{\theta \text{ radians}}{2\pi \text{ radians}}\right)\pi r^2 = \frac{\theta r^2}{2}, \text{ if } \theta \text{ is in radians} \qquad (7\text{-}1\text{-}4A)$$

and

$$\text{area} = \left(\frac{\theta \text{ degrees}}{360 \text{ degrees}}\right)\pi r^2 = \frac{\theta \pi r^2}{360}, \text{ if } \theta \text{ is in degrees.} \quad (7\text{-}1\text{-}4B)$$

The π's cancel in the radian version. If θ were 2π or $360°$ (a full circle), then the formulas would yield πr^2, as they should. Note that the formula is simpler when the angle is expressed in radians.

EXAMPLE 8 Find the area of a sector of radius 50 meters and central angle $\frac{\pi}{4}$.

$$\text{Area} = \left(\frac{\pi}{4}\right)\left(\frac{50^2}{2}\right) = 981.7 \text{ (square meters).} \qquad \blacksquare$$

■ Arc Length and Degrees

Arc length is easy to express when the central angle is measured in radians (7-1-1). However, there is also a formula for arc length when the central angle is measured using degrees. It is derived from this proportion:

$$\frac{\text{arc length}}{\text{arc length of the whole circle}} = \frac{\text{central angle of the arc}}{\text{central angle of the whole circle}}$$

If we use degrees to measure the angle,

$$\frac{\text{arc length}}{2\pi r} = \frac{\theta°}{360°}. \qquad (7\text{-}1\text{-}5)$$

Solving this for arc length, s,

$$s = \frac{2\pi r\theta}{360} = \frac{\pi r\theta}{180}, \text{ if } \theta \text{ is expressed in degrees.} \qquad (7\text{-}1\text{-}6)$$

EXAMPLE 9 The central angle of an arc of radius 10 is 72 degrees. How long is the arc? The proportion in formula 7-1-5 is easy to remember:

$$\frac{s}{2\pi r} = \frac{\theta}{360}. \qquad \frac{s}{20\pi} = \frac{72}{360}.$$

$$s = (20\pi)\left(\frac{72}{360}\right) = 12.57. \qquad \blacksquare$$

■ Why Radians?

In school we learn about angles measured in degrees. Protractors are labeled in degrees. The famous angles (such as 90°, 30°, and 45°) are easy to express in degrees. Triangle angles are measured in degrees. So, why switch to radians?

Formulas are simpler in radians. Compare the two arc-length formulas 7-1-1 and 7-1-6. Compare the two area formulas 7-1-4A and B.

Radians are a natural method of measuring angles in circles based on the radius. The famous angles are simple in radians. Angle π is a straight angle. Angle $\dfrac{\pi}{2}$ is a right angle, one half of a straight angle. Angle $\dfrac{\pi}{4}$ is one fourth of a straight angle. Simple.

The trigonometric functions have very nice properties when angles are expressed in radians. A glance at a unit-circle picture (Figure 10) shows that, if the angle is small, x and $\sin x$ are very similar. (We have now switched from angles called "θ" to angles called "x".)

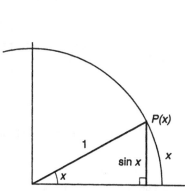

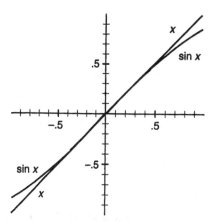

FIGURE 10 x and $\sin x$, where x is both an angle measured in radians and an arc length

FIGURE 11 x and $\sin x$. (in radians) $[-1, 1]$ by $[-1, 1]$

EXAMPLE 10 Put your calculator in radian mode. Evaluate sine at several argument values and compare the images to the arguments.

For small values of x, x itself is a very close approximation to $\sin x$. For example, $\sin(.1) = .0998$ and $\sin(.01) = .0099998$. Figure 10 shows that, on a unit circle, the sine value will be slightly less than the arc length, which is also the measure of the angle.

Figure 11 graphs $y = x$ and $y = \sin x$. They are similar for small x. This simple relationship does not hold for angles measured in degrees. Radian measure yields simple polynomial approximations for the trigonometric functions (Problem B20).

The slope of the line $y = x$ is 1. That line is tangent to the graph of $y = \sin x$ at the origin (Figure 11), so the slope of the sine curve at the origin is 1, which is a very simple result that would be much more complicated if the argument angle were measured in degrees.

This example only hints at the advantages of measuring angles using radians instead of degrees. In calculus, you will discover that the properties of trigonometric functions are distinctly simpler when angles are measured in radians (Problem B31). ▬

■ Why Degrees?

Our method of subdividing the central angle of a whole circle into 360 parts comes from the ancient Babylonians who lived where Iraq is now. We even have a table of a trigonometric function written on a clay tablet which is dated to between 1900 and 1600 BC. The Babylonians used a number system with base 60 (instead of our base 10) and therefore tended to measure things using multiples of 60. A circle is easily divided into 6 parts by swinging a radius from any point on the circle as in Figure 12. The Babylonians divided each of these parts into 60, for a total of 6 times 60 = 360 parts in a circle. Therefore, the central angle of each of the 6 original parts is 60°, and the points in Figure 12 form equilateral triangles with three 60° angles each.

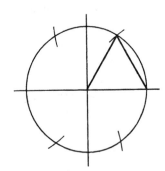

FIGURE 12 A circle divided into six parts

EXAMPLE 11   How big is a degree? For example, if you form an angle of 1° with two yardsticks by putting their left ends together, how far apart would their right ends be?

A yard is 36 inches. The circumference of a circle with radius 36 inches would be $2\pi(36) = 226.2$ inches. A degree is $\frac{1}{360}$ of a circle, so divide the circumference into 360 parts. $\frac{226.2}{360} = .628$ inches, which is about $\frac{5}{8}$ inch. So $\frac{5}{8}$ inch is the arc length of a 1¡ arc with radius 36 inches. Put the other ends $\frac{5}{8}$ inches apart to create a 1° angle.

As Example 9 showed, the arc length is not exactly the straight-line distance between the ends, but it is very close if the angle is small, and 1° is a small angle (Problem B18). ▬

A degree is a small unit of measure, but still smaller units are often necessary. The Babylonians divided degrees into 60 parts called minutes (or "minutes of arc") and minutes are, in turn, divided into 60 parts called seconds (or "seconds of arc"). The abbreviation for minutes is a prime symbol (′) and for seconds is a double prime symbol (″). So 3¡47′ 13″ would be three degrees, 47 minutes, and 13 seconds. Yes, our subdivision of time comes from the Babylonians too.

EXAMPLE 12   How long is a minute of arc on the surface of the Earth?

To answer this we need the size of the Earth. The circumference of the Earth is about 24,856.9 "statute" (usual) miles of 5280 feet each. To find the length of a minute of arc on the surface this would be divided into 360 parts to find a degree and another 60 to find a minute.

$$\frac{24,856.834}{360(60)} = 1.151 \text{ (statute miles)},$$

which, in feet, is

$$1.151(5280) = 6076.1 \text{ feet.}$$

This length is called a "nautical" mile, as opposed to the usual statute mile used on land.

The speed of ships is measured in "knots." One knot is, by definition, one nautical mile per hour. (Speed is not measured in "knots per hour," just "knots.") So a ship that goes 40 knots goes about $40(1.15) = 46$ (statute) miles per hour.

CONCLUSION Radians are a natural way to measure angles in terms of arc length. A central angle is one radian when the arc length equals the radius. In general, the radian measure of the central angle is the arc length divided by the radius: $\theta = \dfrac{s}{r}$. There are 2π radians in a full circle and π radians is a straight angle. Formulas are simpler when expressed in radians.

Trigonometric Functions of Famous Angles (see Section 6-2, Figures 46 and 47)				
Angle	**$\sin\theta$**	**$\cos\theta$**	**$\tan\theta$**	**Angle**
0	0	1	0	0¡
$\dfrac{\pi}{6}$	$\dfrac{1}{2}$	$\dfrac{\sqrt{3}}{2}$	$\dfrac{1}{\sqrt{3}}$	30¡
$\dfrac{\pi}{4}$	$\dfrac{1}{\sqrt{2}}$	$\dfrac{1}{\sqrt{2}}$	1	45¡
$\dfrac{\pi}{3}$	$\dfrac{\sqrt{3}}{2}$	$\dfrac{1}{2}$	$\sqrt{3}$	60¡
$\dfrac{\pi}{2}$	1	0	does not exist	90¡
π	0	-1	0	180¡

Memorize this table. These angles and their trigonometric function values are important and appear frequently.

Terms: central angle, arc length, radian, sector.

Exercise 7-1

A

☺ *Find the central angle in radians if*

A1. The radius is 5 and the arc length is 2.

A2. The radius is 10 and the arc length is 3.

A3. The arc length is 4 and the radius is 20.

A4. The arc length is 3.2 and the radius is 9.6. [.33]

Find the central angle in degrees if

A5. The radius is 10 and the arc length is 5. [29]

A6. The arc length is 5π and the radius is 20. [45]

Find the length of the arc on a circle with the given radius and central angle.

A7. Radius 30 and central angle 10°. [5.2]

A8. Radius 70 and central angle 80°. [98]

A9. Radius 2 and central angle $\dfrac{\pi}{6}$. [1.0]

A10. Radius 25 and central angle $\dfrac{3\pi}{4}$. [59]

A11. An arc length 2 is opposite a central angle 6 radians. What is the radius? [.33]

A12. An arc length of 8 is opposite a central angle of $\dfrac{1}{2}$ radian. What is the radius?

A13. What central angle (in radians) makes an arc of length 2 on a circle of diameter 12?

A14. What central angle (in radians) makes an arc of length 3 on a circle of diameter 2?

What is the area of the sector with the given radius and central angle?

A15. Radius 5 and central angle 20° [4.4]

A16. Radius 30 and central angle 70°. [550]

A17. Radius 2 and central angle $\dfrac{\pi}{6}$. [1.0]

A18. Radius 100 and central angle $\dfrac{\pi}{12}$. [1300]

A19. A sector with radius 10 has area 40. What is the central angle in radians? [.80]

A20. A sector with radius 5 has area 2. What is the central angle in radians? [.16]

A21.* ☺ Give the corresponding radian measure. (Give it exactly, not in decimals.)
(A) 90° (B) 45° (C) 30° (D) 60°

A22.* ☺ Give the corresponding degree measure to these angles given in radians.
(A) $\dfrac{\pi}{3}$ (B) $\dfrac{\pi}{2}$ (C) $\dfrac{\pi}{4}$ (D) $\dfrac{\pi}{6}$

A23. (A) Convert 25° to radian measure.
(B) Convert $\dfrac{\pi}{8}$ radians to degrees. [22]

A24. (A) Convert 10° to radian measure.
(B) Convert $\dfrac{\pi}{12}$ radians to degrees. [15]

A25. (A) Convert 5° 26′ 45″ to decimal degrees. [5.4]
(B) Convert .276° to minutes and seconds of arc. [16′ ...]

A26. (A) Convert 6° 7′ 12″ to decimal degrees. [6.1]
(B) Convert .783° to minutes and seconds of arc. [46′ ...]

☺ *Give the quadrant of the angle.*

A27. (A) $\dfrac{\pi}{3}$ (B) $\dfrac{3\pi}{4}$ (C) $\dfrac{5\pi}{4}$ (D) $\dfrac{2\pi}{3}$

A28. (A) $\dfrac{\pi}{4}$ (B) $\dfrac{\pi}{6}$ (C) $\dfrac{5\pi}{6}$ (D) $\dfrac{-\pi}{4}$

Answer these in radians.

A29. How many radians does the hour hand of a clock sweep out in one hour?

A30. How many radians does the minute hand of a clock sweep out in one minute?

A31. How many radians does the sun move across the sky in one hour?

A32. How many radians does the second hand of a clock sweep out in one hour? [380]

Answer these in degrees.

A33. How many degrees does the hour hand of a clock sweep out in one hour?

A34. How many degrees does the minute hand of a clock sweep out in one minute?

A35. How many degrees the sun move across the sky in one hour?

A36. How many degrees does the second hand of a clock sweep out in one hour?

A37. From 100 yards away, how wide must something be to make an angle of 1°? [1.7]

A38. From 24 inches away, how wide must something be to make an angle of 1°? [.42]

A39. How long is a second of arc on the surface of the earth? [100]

A40. A pulley has a 4 inch diameter. How many radians does it turn when 3 feet of rope are pulled around it? [18]

A41. A bicycle with 26-inch diameter wheels travels at a rate of 30 feet per second. How many radians do the wheels turn in one second? [28]

A42. A wheel of radius 12 inches spins at 15 revolutions per second. What is the speed of the points on the outer edge?

A43. Define "radian." Give the basic equivalence between radians and degrees.

B

B1.* (A) On a circle of radius r, what is the radian measure of the central angle that makes an arc of length s?
  (B) How many radians are there in the central angle of a full circle?

B2.* Give the proportion from which the formula for the length of an arc can be derived if the central angle is given in degrees.

B3.* Give the proportion from which the formula for the area of a sector can be derived if the central angle is given in radians.

B4.* Give the proportion from which the formula for the area of a sector can be derived if the central angle is given in degrees.

B5.* If x is expressed in radians, $\sin x$ is nearly x for small x. Draw and label part of a unit-circle picture to illustrate why.

B6. If x is expressed in radians, $\tan x$ is nearly x for small x.
  (A) Draw part of a unit-circle picture to illustrate why.
  (B) Sketch both $y = x$ and $y = \tan x$ on [0, 1] by [0, 1].
  (C) For small x, $\dfrac{\tan x}{x}$ is nearly one. Find the largest positive x such that $\dfrac{\tan x}{x} < 1.1$.

B7. In the figure, if the radius is 5 and the chord AB is 4, find the arc length AB. [4.1]

B8. In the figure, if the radius is 19 and the chord AB is 11, find the arc length AB. [11]

Hint for B9–12: Guess-and-check is one of the four ways to solve equations.

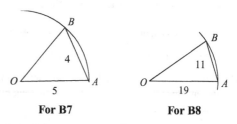

For B7   **For B8**

B9. Find θ (in radians) such that the arc length AC is $1.1BC$ in the figure. [.75]

B10. Find θ (in radians) such that the arc length AC is $4BA$ in the figure. [.51]

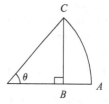

For B9–10

B11. In the figure, the arc length is 10.5 and the chord length is 10. Find the central angle in radians and the radius. [$\theta = 1.1, r = 9.7$]

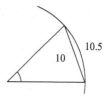

For B11

B12. In the figure, the arc length is 52 and the chord length is 40. Find the central angle and the radius. [$\theta = 2.4, r = 21$]

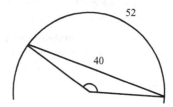

B13. For very small x expressed in radians, $\sin x$ is nearly x.
  (A) Use this to estimate $\sin 1°$ by converting to radians.
  (B) Compare your estimate to the real value of $\sin x$ obtained with your calculator. How far off is it?

B14. The sun is about $\frac{1}{2}^{\circ}$ wide. How long does the sun take to move its own width across the sky?

B15. The moon is about $\frac{1}{2}^{\circ}$ wide. How long does the moon take to move its own width across the sky?

B16. A bike wheel has a 26 inch diameter. How many radians does it go through in a one mile long bike ride? [4900]

B17. A bike is geared so that the pedal gear has 48 teeth and the wheel gear has 18 teeth. When the pedal turns through 1 radian, how many radians will the wheel gear turn through?

B18. In Examples 6 and 11 the linear distance between the ends of the yardsticks was not equal to the arc length, but it was similar. Example 11 found the arc length made by an angle of 1°. Find the linear distance and compare it to the arc length by forming the ratio of the two.

B19. Two points are on a circle. The straight-line distance between them is 90 percent of the arc length between them. What is the central angle of the arc? [1.6 radians]

B20. A better approximation than "x" for sin x (for small values of x) is given by a cubic polynomial "$x - \frac{x^3}{6}$." Compare "sin x" to "$x - \frac{x^3}{6}$" graphically. Which is larger for $x > 0$? Discover when the ratio of the smaller to the larger is at least .99. [$x < 1.0$]

B21. Use graphs of "sin x" and "$x - \frac{x^3}{6} + \frac{x^5}{120}$" (in radians) to find the greatest positive value of x such that they differ by at most .01. [1.7]

B22. Use graphs of "cos x" and "$1 - \frac{x^2}{2} + \frac{x^4}{24}$" (in radians) to find the greatest positive value of x such that they differ by at most .01. [1.4]

B23. An archaeologist finds a broken piece of a pottery cup and asks you to determine the cup's outside diameter. You take the measurements in the picture. What is the diameter? [3.25]

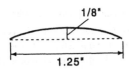

1/8"

1.25"

B24. Suppose you are standing on the seashore and you look out to the horizon. Presume that you can see out to where the tangent line to the earth would come to eye level (exaggerated in the picture). If your eyes are 6 feet above sea level, how far out can you see? (Assume the diameter of the earth is 7900 miles.) [3 miles]

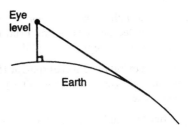

Eye level

Earth

B25. (A) Redo Problem B24 if eye level is h feet above sea level and letting the radius of the Earth be r. Then try to simplify your answer by ignoring any term that is much smaller than the other terms. Plug in for the actual value of r only after you have simplified your expression. [$1.2\sqrt{h}$, where h is in feet. This answer ignores refraction.]

(B) If your eye is 20 feet above sea level, as a boat approaches from over the horizon, when will the top of its 50-foot mast come into view (neglect refraction of light rays). [14]

B26. A belt goes around two wheels, one with diameter of 20 inches and the other with diameter of 6 inches. Their centers are 22 inches apart. How long is the belt?

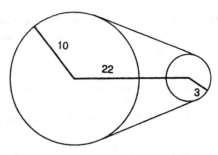

10

22

3

B27. Two points, A and B, are on the unit circle. Determine the central angle when the arc length between them (measured counterclockwise) is twice the straight-line distance between them. [3.8]

B28. Consider two points on a circle, A and B. Determine the central angle when the arc length between them is 10 percent greater than the straight-line distance between them.

B29. Find the angle of the sector with maximum possible area with perimeter 100. [The perimeter of the sector consists of two straight sides and one arc. Neither the radius nor arc length is given.] [2.0]

B30. Suppose one city has latitude 45° and a second is directly north of it at latitude 46°. How far apart are they? [69]

B31. (Why use radians?) Graph sin x (in radians). Estimate the slope of the tangent line at the origin. Use (0, 0) and another nearby point on the graph to estimate the slope. Note how simple the slope of the tangent line is. (This nice number would not result if the angle were measured in degrees.)

B32. An archaeologist finds an ancient coin which was originally circular, of uniform thickness, and 25 millimeters in diameter. Now a piece of the edge is broken off in a straight line 15 millimeters long (forming a chord of the circle). The remaining part weighs 9.56 grams. What did the coin weigh originally?

B33. See the figure for B9–10. If the radius and θ are unknown, $BA = 2$, and arc AC is 5, find θ.

B34. See the figure for B9–10. If the radius and θ are unknown, $BA = 5$, and CB is 11, find θ.

B35. Find the area of the triangle in Figure 7.

B36. A central angle at O of 33° is opposite an arc AB of length 10. Find the area of triangle OAB.

Section 7-2 TRIGONOMETRIC IDENTITIES

This section emphasizes trigonometric identities that follow from two types of pictures:

1. Unit-circle pictures, and
2. Right-triangle pictures.

You have heard that "A picture is worth a thousand words." In this section you will see "A picture is worth several trig identities."

■ Identities from Unit-Circle Pictures

You can "see" many identities just by using the definitions of the functions on the unit circle. The equation of the unit circle is "$x^2 + y^2 = 1$." Let $P(\theta)$ be the point on the unit circle at angle θ counterclockwise from the positive x-axis (Figure 1). Then, by definition,

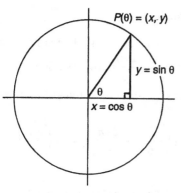

FIGURE 1 The unit circle, $\theta, P(\theta), x$, and y

$$\sin \theta = y \qquad (7\text{-}2\text{-}1)$$
$$\cos \theta = x$$

and

$$\tan \theta = \frac{y}{x} = \frac{\sin \theta}{\cos \theta}.$$

The equation of the unit circle immediately yields the famous

Pythagorean Identity

$$\sin^2 \theta + \cos^2 \theta = 1. \qquad (7\text{-}2\text{-}2)$$

EXAMPLE 1   Let θ be acute and $\cos\theta = .3$. Find $\sin\theta$. Find $\tan\theta$.

Of course, we could use a calculator to solve for θ and then evaluate sine of that angle. But this section is concentrating on theory, not the shortcuts that result from that theory. Sine and cosine are related by the Pythagorean Identity.

$$\sin^2\theta + \cos^2\theta = 1.\ \sin^2\theta + .3^2 = 1.\ \sin^2\theta = 1 - .3^2 = .91.$$

$$\sin\theta = \pm\sqrt{.91} = \pm.954.$$

One solution is extraneous. Since θ is given as acute, the only correct answer is $\sin\theta = .954$.

Tangent is sine over cosine (by 7-2-1). $\tan\theta = \dfrac{\sin\theta}{\cos\theta} = \dfrac{.954}{.3} = 3.18.$ ▬

■ Coterminal Angles

DEFINITION 7-2-3

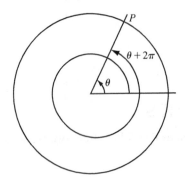

FIGURE 2 Two coterminal angles, θ and $\theta + 2\pi$. Their terminal points are the same. Therefore, their trig function values are the same

Angles, such as θ and $\theta + 2\pi$, which correspond to the same point on the unit circle are said to be <u>coterminal</u> (because they terminate in the same place, Figure 2). Their trigonometric functions are therefore the same.

Coterminal Identities

Let n denote any integer.

$$\sin(\theta \pm 2\pi n) = \sin\theta,\ \text{and} \tag{7-2-4A}$$
$$\cos(\theta \pm 2\pi n) = \cos\theta.$$

$$\sin(\theta \pm 360n°) = \sin\theta,\ \text{and} \tag{7-2-4B}$$
$$\cos(\theta \pm 360n°) = \cos\theta.$$

This identity is used for the first step in evaluating the sine of large angles.

EXAMPLE 2   Find $\sin\left(\dfrac{29\pi}{6}\right)$ using a smaller coterminal angle.

This problem gives you some idea of how calculators work. Here is step that your calculator does (invisibly) inside its guts. Angle $\dfrac{29\pi}{6}$ terminates at the same point as angle $\dfrac{5\pi}{6}$ because they differ by 2 times 2π (two times around the circle).

$$\sin\left(\frac{29\pi}{6}\right) = \sin\left(\frac{29\pi}{6} - 2(2\pi)\right) = \sin\left(\frac{29\pi}{6} - \frac{24\pi}{6}\right) = \sin\left(\frac{5\pi}{6}\right)$$

Instead of finding $\sin\left(\dfrac{29\pi}{6}\right)$, it finds $\sin\left(\dfrac{5\pi}{6}\right)$. The angle is in the second quadrant. We will resume this example shortly by using, instead, a first quadrant angle with the same sine. ▬

"Sine is opposite over hypotenuse" works for acute angles in right triangles. Larger angles do not fit in right triangles. Fortunately, the trigonometric functions of angles which are *not* acute can be given in terms of the trigonometric functions of angles that *are* acute by using the so-called "reference" angles.

■ Reference Angles

Look at Figure 3. Is angle θ positive or negative? Is BP positive or negative?

The answer depends upon whether you are doing geometry or trigonometry. In geometry, angles and sides are never negative, so θ and BP are positive. However, we are doing trigonometry, not geometry. In trigonometry, think of an angle as a rotation from the initial side OA to the terminal position, OP. Then angle θ in Figure 3 is negative and BP is negative.

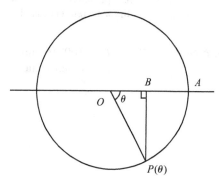

FIGURE 3 The unit circle, θ, $P(\theta)$, and BP

DEFINITION 7-2-5   Let θ be any angle. In a unit-circle picture of θ, the <u>reference angle</u> for θ, is the **positive acute angle** formed by the terminal side and the x-axis.

Figure 4 shows four angles with the same reference angle, α (alpha).

$$\sin \beta = \sin \alpha$$

$$\sin \lambda = -\sin \alpha \quad \text{[negative, because it is in the opposite direction]}$$

$$\sin(-\alpha) = -\sin \alpha.$$

How are β and α related?

$$\beta = \pi - \alpha \text{ and } \alpha = \pi - \beta.$$
$$\lambda = \pi + \alpha \text{ and } \alpha = \lambda - \pi.$$

Also, the cosines of all four angles in Figure 4 are the same, except possibly for a plus or minus sign.

In Figure 5, angle θ is the reference angle for $\pi - \theta$. The congruent right triangles show that

$\sin(\pi - \theta) = \sin \theta$ and

$\cos(\pi - \theta) = -\cos(\theta)$. [The negative sign is because $\cos(\pi - \theta)$ is in the opposite direction.] The sine, cosine, or tangent of any one of the angles in Figure 5 gives the sine, cosine, or tangent of the others, if the correct sign ($\pm$) is attached.

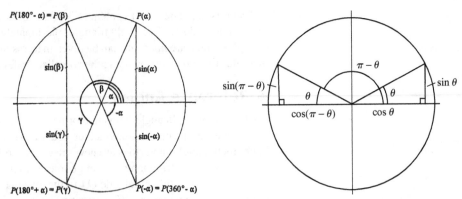

FIGURE 4 Four angles, α (alpha), β (beta), γ [gamma], and $-\alpha$ each with the same reference angle

FIGURE 5 Reference angle θ for angle $\pi - \theta$

EXAMPLE 3 $\angle B = 100$ ¡ (Figure 6). Give its reference angle. Give its trig function values in terms of their values at the reference angle.

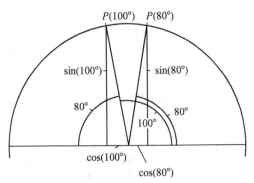

FIGURE 6 100°. sin(100°) = sin(80°). cos(100°) = −cos(80°)

100° is in the second quadrant, 80 degrees back from 180 degrees. The reference angle of 100° is 80°.

sin 100° = sin 80° (sine is positive in the second quadrant).
cos 100° = −cos 80° (cosine is negative in the second quadrant).
tan 100° = −tan 80° (tangent is negative in the second quadrant). ▬

EXAMPLE 4 $\angle D = \dfrac{10\pi}{9}$ (Figure 7). Give its reference angle. Give its trig function values in terms of their values at the reference angle.

Angle D is $\dfrac{\pi}{9}$ into the third quadrant. It is $\dfrac{\pi}{9}$ past π. Its reference angle is $\dfrac{\pi}{9}$.

$\sin\left(\dfrac{10\pi}{9}\right) = -\sin\left(\dfrac{\pi}{9}\right)$ (sine is negative in the third quadrant, since $y < 0$).

$\cos\left(\dfrac{10\pi}{9}\right) = -\cos\left(\dfrac{\pi}{9}\right)$ (cosine is negative in the third quadrant, since $x < 0$).

$$\tan\!\left(\frac{10\pi}{9}\right) = \tan\!\left(\frac{\pi}{9}\right) \text{ (tangent is positive in the third quadrant).}$$ ▬

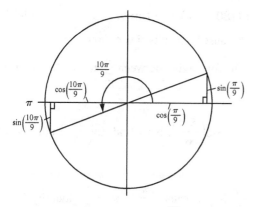

FIGURE 7 $\angle D = \dfrac{10\pi}{9}$

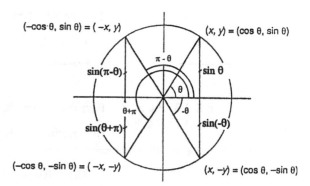

FIGURE 8 Four angles with reference angle θ

Figure 8 is a picture worth at least nine (9!) trigonometric identities. From the coordinates of $P(\theta)$ and $P(-\theta)$ we find

Negatives (Odd-Even Identities)

$$\cos(-\theta) = \cos\theta.$$
$$\sin(-\theta) = -\sin\theta.$$
$$\tan(-\theta) = -\tan\theta. \tag{7-2-6}$$

The result for tangent follows because tangent is sine over cosine.

$$\tan(-\theta) = \frac{\sin(-\theta)}{\cos(-\theta)} = \frac{-\sin\theta}{\cos\theta} = -\tan\theta.$$

These facts say that cosine is an "even function" and sine and tangent are "odd functions" (The terms "even" and "odd" come from even and odd power functions in Section 4-1). The graph of cosine is symmetric about the y-axis (Figure 12), and the graphs of sine and tangent are "point symmetric" about the origin (Figures 11 and 13).

EXAMPLE 5   Find $\sin\!\left(\dfrac{-\pi}{6}\right)$.

$$\sin\!\left(\frac{-\pi}{6}\right) = -\sin\!\left(\frac{\pi}{6}\right) = \frac{-1}{2}.$$ ▬

In Figure 5, from the coordinates of $P(\theta)$ and $P(\pi - \theta)$ we find

Reference-Angle Identities

$$\sin(\pi - \theta) = \sin\theta.$$
$$\cos(\pi - \theta) = -\cos\theta.$$
$$\tan(\pi - \theta) = -\tan\theta. \tag{7-2-7A}$$

$$\sin(180° - \theta) = \sin \theta.$$
$$\cos(180° - \theta) = -\cos \theta.$$
$$\tan(180° - \theta) = -\tan \theta. \tag{7-2-7B}$$

The result for tangent follows because tangent is sine over cosine.

These are particularly useful for angles between $\dfrac{\pi}{2}$ and π (between 90° and 180°).

EXAMPLE 2 CONTINUED   Example 2 showed $\sin\left(\dfrac{29\pi}{6}\right) = \sin\left(\dfrac{5\pi}{6}\right)$. Now find $\sin\left(\dfrac{5\pi}{6}\right)$ in terms of a first quadrant angle.

The reference angle of $\dfrac{5\pi}{6}$ is $\dfrac{\pi}{6}$ because $\dfrac{5\pi}{6} = \pi - \dfrac{\pi}{6}$. Identity 7-2-7A tells us that $\sin\left(\dfrac{5\pi}{6}\right) = \sin\left(\dfrac{\pi}{6}\right)$. $\sin\left(\dfrac{\pi}{6}\right) = \dfrac{1}{2}$,   so $\sin\left(\dfrac{5\pi}{6}\right) = \dfrac{1}{2}$.   ▬

In Figure 8, from the coordinates of $P(\theta)$ and $P(\theta + \pi)$ we find

Reference-Angle Identities

$$\sin(\theta + \pi) = -\sin \theta.$$
$$\cos(\theta + \pi) = -\cos \theta.$$
$$\tan(\theta + \pi) = \tan \theta. \tag{7-2-8A}$$

$$\sin(\theta + 180°) = -\sin \theta.$$
$$\cos(\theta + 180°) = -\cos \theta.$$
$$\tan(\theta + 180°) = \tan \theta. \tag{7-2-8B}$$

This is easy to see on a unit circle because $P(\theta)$ and $P(\theta + \pi)$ are directly opposite one another (Figure 8), so both their horizontal and vertical values are negatives of one another.

EXAMPLE 6   Express "$\tan\left(\dfrac{17\pi}{12}\right)$" in terms of the tangent of an acute angle.

$$\tan\left(\dfrac{17\pi}{12}\right) = \tan\left(\dfrac{5\pi}{12} + \pi\right) = \tan\left(\dfrac{5\pi}{12}\right), \text{ by 7-2-8A. } \dfrac{5\pi}{12} \text{ is acute.}$$

Express "$\tan\left(\dfrac{7\pi}{9}\right)$" in terms of the tangent of an acute angle.

$$\tan\left(\dfrac{7\pi}{9}\right) = \tan\left(\pi - \dfrac{2\pi}{9}\right) = -\tan\left(\dfrac{2\pi}{9}\right), \text{ by 7-2-8A. The reference angle}$$

is $\dfrac{2\pi}{9}$ and the sign is changed because tangent is negative in the second quadrant.

▬

■ Trigonometric Identities

We use unit-circle figures to determine trigonometric identities for combinations of θ, $\pm$, and any of 2π, π, or $\dfrac{\pi}{2}$.

EXAMPLE 7 Find trig identities for $\sin\left(\theta + \dfrac{\pi}{2}\right)$ and $\cos\left(\theta + \dfrac{\pi}{2}\right)$.

Think of angles as rotations, beginning from the positive x-axis. Rotate angle θ (preferably distinctly greater than or less than $\dfrac{\pi}{4} = 45°$, Figure 9). The angle we are interested in is $\theta + \dfrac{\pi}{2}$, which is an additional rotation of $\dfrac{\pi}{2}$ past θ. Locate $\theta + \dfrac{\pi}{2}$ and its terminal point, and then $\sin\left(\theta + \dfrac{\pi}{2}\right)$ and $\cos\left(\theta + \dfrac{\pi}{2}\right)$, which form a triangle (Figure 10). Geometry shows that the triangle is congruent to the triangle formed with $\sin\theta$ and $\cos\theta$. Determine which side is equal to which and whether the new sides are positive or negative. You will have two trig identities:

$$\sin\left(\theta + \frac{\pi}{2}\right) = \cos\theta, \text{ and}$$

$$\cos\left(\theta + \frac{\pi}{2}\right) = -\sin\theta \text{ [negative because the cosine has the opposite sign.]}$$

To find $\tan\left(\theta + \dfrac{\pi}{2}\right)$, use these two and the definition of tangent.

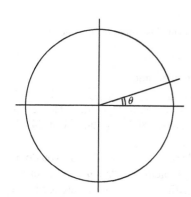

FIGURE 9 θ in the first quadrant

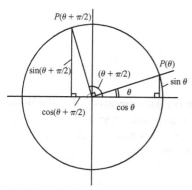

FIGURE 10 The location of

$\theta + \dfrac{\pi}{2}, \sin\left(\theta + \dfrac{\pi}{2}\right)$

and $\cos\left(\theta + \dfrac{\pi}{2}\right)$

$$\tan\left(\theta + \frac{\pi}{2}\right) = \frac{\sin\left(\theta + \dfrac{\pi}{2}\right)}{\cos\left(\theta + \dfrac{\pi}{2}\right)} = \frac{\cos(\theta)}{-\sin(\theta)} = \frac{-1}{\tan(\theta)}.$$

■ Solving "sin x = c"

Identities 7-2-4 and 7-2-6-8 tell us there are many solutions to "$\sin x = c$, "$\cos x = c$," and "$\tan x = c$." Of course, the inverse function is designed to yield one solution. In fact, it yields exactly one, because it is a function. The above identities yield the other solutions. In summary (repeating 6-2-5),

$\sin x = c$ iff $x = \sin^{-1}c \pm 2\pi n$ or $x = \pi - \sin^{-1}c \pm 2\pi n$

(7-2-9A)

$\sin x = c$ iff $x = \sin^{-1}c \pm 360n°$ or

$x = 180° - \sin^{-1}c \pm 360n°$

(7-2-9B)

for some integer n, which may be zero (Figures 2 and 8). This follows from 7-2-4 on coterminal angles which tells us that adding 2π (or 360°) to any solution yields another solution and from 7-2-7 which says that subtracting any solution from π (or 180°) yields another solution. Figures 2 and 8 yield the entire theorem.

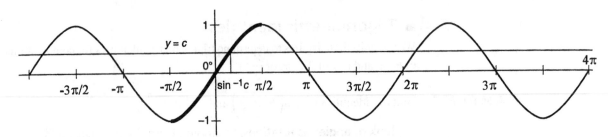

FIGURE 11 $y = \sin x$ and $y = c > 0.$ $-2\pi < x < 4.$ The region where $\sin^{-1}$ is defined is emphasized. Note how the solutions described in 7-2-9 are visible

EXAMPLE 8 Solve $\sin \theta = -.9$ for an angle θ in the third quadrant.

One way is to use your calculator to find $\sin^{-1}(-.9) = -64.2°$ (or -1.12 radians). This is a useful value, but the angle is not in the third quadrant. The second equation in 7-2-9B says another solution is given by $180° - (-64.2°) = 244.2°$, which is in the third quadrant. So $\theta = 244.2°$.

If you understand the basic reference-angle picture (Figure 3 or 8), you do not need to remember the details of the theorem. A picture of "$\sin \theta = -.9$ for θ in the third quadrant" is much like Figure 3 with angle γ. The reference angle, α, has $\sin \alpha = .9$ (positive). Solve that for α.

$\alpha = \sin^{-1}(.9) = 64.2°$. The desired solution is $180° + 64.2° = 244.2°$.

■ Solving "$\cos x = c$"

$$\cos x = c \quad \text{iff} \quad x = \pm\cos^{-1}c \pm 2\pi n \qquad \text{(7-2-10A)}$$

$$\cos x = c \quad \text{iff} \quad x = \pm\cos^{-1}c \pm 360n° \qquad \text{(7-2-10B)}$$

for some integer n, which may be zero (Figure 10). This repeats 6-2-9 and follows from 7-2-4 on coterminal angles (adding 2π to any solution yields another solution) and from 7-2-6 (the negative of any solution is another solution). Figures 2 and 8 labeled for cosines would yield the entire theorem.

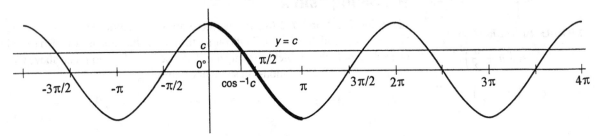

FIGURE 12 $y = \cos x$ and $y = c > 0.$ $-2\pi < x < 4\pi.$ The region where $\cos^{-1}$ is defined is emphasized. Note how the solutions described in 7-2-10 are visible

$$\tan x = c \quad \text{iff} \quad x = \tan^{-1}c \pm \pi n, \qquad \text{(7-2-11A)}$$

$$\tan x = c \quad \text{iff} \quad x = \tan^{-1}c \pm 180n°, \qquad \text{(7-2-11B)}$$

where n is some integer that may be zero (Figure 13). This follows from 7-2-8 which says that adding π to any solution yields another solution.

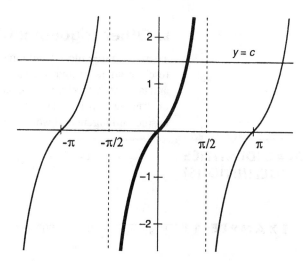

FIGURE 13 $y = \tan x$ and $y = c > 0$. $-\frac{3\pi}{2} < x < \frac{3\pi}{2}$, $-2.5 < y < 2.5$. The region where $\tan^{-1}$ is defined is emphasized. Note how the solutions described in 7-2-11 are visible

A picture is worth a thousand words. To remember these results, simply remember the unit-circle picture in Figure 8.

EXAMPLE 9 Solve $\sin x = \dfrac{1}{2}$.

We know one solution: $x = \dfrac{\pi}{6}$. Theorem 7-2-9 says the others include $\dfrac{\pi}{6} + 2\pi$, $\dfrac{\pi}{6} - 2\pi$, $\pi - \dfrac{\pi}{6} = \dfrac{5\pi}{6}$, and $\dfrac{5\pi}{6} + 2\pi$. We can write all the solutions at once using "n" for an arbitrary integer: "$x = \dfrac{\pi}{6} \pm 2\pi n$ or $x = \dfrac{5\pi}{6} \pm 2\pi n$." ▬

EXAMPLE 10 Solve "$\tan(5x) = 1$" for x between 0 and π.

Use the inverse tangent function first.

$$5x = \tan^{-1}(1) \pm \pi n = \frac{\pi}{4} \pm \pi n.$$

So,

$$x = \frac{\pi}{20} \pm \frac{\pi n}{5} = \frac{\pi}{20} \pm \frac{4n\pi}{20}.$$

The solutions between 0 and π are: $\dfrac{\pi}{20}, \dfrac{5\pi}{20} = \dfrac{\pi}{4}, \dfrac{9\pi}{20}, \dfrac{13\pi}{20}$, and $\dfrac{17\pi}{20}$. ▬

These identities show how **the value of any trigonometric function at any angle can be determined by its value at its acute reference angle.**

■ Other Trigonometric Functions

There are three more trigonometric functions that are less important: cosecant (csc), secant (sec), and cotangent (cot). They are simply reciprocals of sine, cosine, and tangent, respectively. Your calculator does not have keys for cosecant, secant, or cotangent. It is usually appropriate to simply rewrite them in terms of sine, cosine, or tangent and work from there.

RECIPROCAL IDENTITIES (DEFINITIONS)

$$\csc \theta = \frac{1}{\sin \theta}. \quad \sec \theta = \frac{1}{\cos \theta}. \quad \cot \theta = \frac{1}{\tan \theta}. \quad \text{7-2-12}$$

EXAMPLE 11 Evaluate $\csc 30°$, $\sec 80°$, $\cot 12°$, $\sec\left(\dfrac{\pi}{4}\right)$ and $\cot\left(\dfrac{\pi}{6}\right)$.

$$\csc 30° = \frac{1}{\sin 30°} = \frac{1}{\frac{1}{2}} = 2.$$

$$\sec 80° = \frac{1}{\cos 80°} = \frac{1}{.174} = 5.76.$$

$$\cot 12° = \frac{1}{\tan 12°} = \frac{1}{.213} = 4.70.$$

$$\sec\left(\frac{\pi}{4}\right) = \frac{1}{\cos\left(\dfrac{\pi}{4}\right)} = 1.414 \quad \text{[Be sure to use "radian" mode.]}$$

$$\cot\left(\frac{\pi}{6}\right) = \frac{1}{\tan\left(\dfrac{\pi}{6}\right)} = 1.732.$$

To solve equations with these trig functions, convert the equation to one with the familiar trig functions.

EXAMPLE 12 Solve "$\csc x = 2$," "$\sec(2x) = 4$," and "$\cot(x^2) = \dfrac{1}{2}$."

In each case use the reciprocal identities.

$$\csc x = 2 \text{ iff } \frac{1}{\sin x} = 2 \text{ iff } \sin x = \frac{1}{2}. \quad \text{Now solve this.}$$

$$\sec (2x) = 4 \text{ iff } \frac{1}{\cos (2x)} = 4 \text{ iff } \cos (2x) = \frac{1}{4}. \quad \text{Now solve this.}$$

$$\cot(x^2) = \frac{1}{2} \text{ iff } \frac{1}{\tan(x^2)} = \frac{1}{2} \text{ iff } \tan(x^2) = 2. \quad \text{Now solve this.}$$

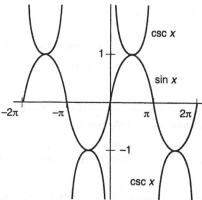

FIGURE 14 $\sin x$ and its reciprocal, $\csc x$

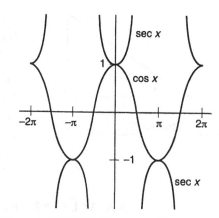

FIGURE 15 $\cos x$ and its reciprocal, $\sec x$ ▬

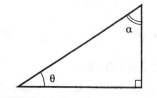

FIGURE 16 Complementary angles θ and α

Cofunction Names

Three trig functions have the names of three others with "co" as a prefix: sine and cosine, tangent and cotangent, secant and cosecant. The "co" refers to complementary angles. Angles are said to be complementary if and only if they sum to a right angle. In a right triangle the two non-right angles are complementary (θ and α in Figure 16). The sine of θ is "opposite over hypotenuse" from the point of view of θ, which is "adjacent over hypotenuse" (cosine α) from the point of view of angle α. Therefore the sine of an angle is the cosine of the complimentary angle. "Cosine" is an abbreviation of the archaic term "complemental sine" (Problems A61-62).

COFUNCTION IDENTITIES

$$\sin\left(\frac{\pi}{2} - \theta\right) = \cos\theta,$$

$$\cos\left(\frac{\pi}{2} - \theta\right) = \sin\theta, \text{ and}$$

$$\tan\left(\frac{\pi}{2} - \theta\right) = \cot\theta = \frac{1}{\tan\theta}. \qquad (7\text{-}2\text{-}13\text{A})$$

$$\sin(90° - \theta) = \cos\theta,$$

$$\cos(90° - \theta) = \sin\theta, \text{ and}$$

$$\tan(90° - \theta) = \cot\theta = \frac{1}{\tan\theta}. \qquad (7\text{-}2\text{-}13\text{B})$$

This result holds even if angle θ does not fit in a right triangle (Problem B56).

PYTHAGOREAN IDENTITIES

$$\sec^2\theta = \tan^2\theta + 1$$

$$\tan^2\theta = \sec^2\theta - 1 \qquad (7\text{-}2\text{-}14)$$

Proof: Divide the Pythagorean identity, $\sin^2\theta + \cos^2\theta = 1$, by $\cos^2\theta$.

$$\frac{\sin^2\theta}{\cos^2\theta} + \frac{\cos^2\theta}{\cos^2\theta} = \frac{1}{\cos^2\theta}$$

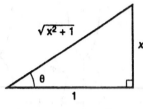

FIGURE 17 $\tan \theta = x$

Now simplify: $\tan^2 \theta + 1 = \sec^2 \theta$.
This is 7-2-14.

The tangent-secant Pythagorean identities (7-2-14) are easy to interpret in a right triangle. Figure 17 illustrates "$\tan \theta = x$" with adjacent side 1. By the Pythagorean Theorem, the hypotenuse is $\sqrt{x^2 + 1}$. Secant, which is the reciprocal of cosine, is "hypotenuse over adjacent." The adjacent is 1, so $\sec \theta = \sqrt{x^2 + 1} = \sqrt{\tan^2 \theta + 1}$.

An effective way to deal with a trigonometric expression with secant, cosecant, or cotangent is to rewrite them in terms of sine and cosine.

EXAMPLE 13 Rewrite $(\sec \theta)(\cot \theta)$ in terms of sine and cosine.

$$(\sec \theta)(\cot \theta) = \left(\frac{1}{\cos \theta} \right) \left(\frac{1}{\tan \theta} \right) = \left(\frac{1}{\cos \theta} \right) \frac{1}{\left(\dfrac{\sin \theta}{\cos \theta} \right)} = \frac{1}{\sin \theta}.$$

If you want to convert the final expression to "$\csc \theta$", you may. ▬

■ Identities from Right-Triangle Pictures

Many trig identities can be derived on demand from a well-labeled picture of a right triangle.

EXAMPLE 14 $\sin \theta = .3$. Find the values of the other trig functions of θ.

One way is to illustrate "$\sin \theta = .3$" with a right triangle (with hypotenuse 1 and opposite side .3) and read the other trig functions from the picture (Figure 18). Calculate the remaining side using the Pythagorean Theorem. From Figure 18,

$$\cos \theta = \sqrt{1^2 - .3^2} = .954.$$

$$\tan \theta = \frac{.3}{\sqrt{1^2 - .3^2}} = .314.$$

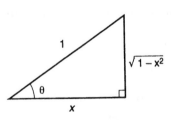

FIGURE 18 $\sin \theta = .3$

Even if angle θ is not acute (and therefore cannot fit in a right triangle), the results hold for its reference angle, and therefore hold in other quadrants too, with at most a change of sign. For example, if θ is in the second quadrant where cosine is negative, then $\cos \theta = -.954$ instead of .954. ▬

EXAMPLE 15 Express tangent in terms of cosine.

Label a good picture. Denote $\cos \theta$ by some letter, say x. Illustrate "$\cos \theta = x$" for an acute angle (Figure 19). Use the Pythagorean Theorem to find the other side. Then read $\tan \theta$ from the picture.

In the picture the opposite side is $\sqrt{1 - x^2}$. Therefore

$$\tan \theta = \frac{\sqrt{1 - x^2}}{x} = \frac{\sqrt{1 - \cos^2 \theta}}{\cos \theta}.$$

FIGURE 19 $\cos \theta = x$

This holds for acute angles—the most important case. If the angle is not acute, this can be interpreted as a picture of the reference angle and the same result holds, possibly requiring a minus sign, depending upon the quadrant of the angle. ▬

Sometimes these problems are disguised as inverse problems.

EXAMPLE 16 Find $\tan(\sin^{-1}x)$.

First draw a right-triangle to illustrate $\sin^{-1}x$.

Since it is an angle, call it θ. $\sin^{-1}x$ is the angle whose sine is x (Figure 20). Make the opposite side x and the hypotenuse 1.

Now, find $\tan(\sin^{-1}x)$.

Use the Pythagorean Theorem to find the other leg. Then read the value of the tangent from the picture.

$$\tan(\sin^{-1}x) = \frac{\pm x}{\sqrt{1-x^2}}.$$

FIGURE 20 $\sin\theta = x$

The most important case is in the first quadrant where the sign is plus. The sign would be minus in the second quadrant. ▬

EXAMPLE 17 Suppose θ is in the first quadrant and $\sec\theta = \dfrac{x}{2}$. Find $\tan\theta$.

Draw a right triangle to illustrate "$\sec\theta = \dfrac{x}{2}$." Use "$\sec\theta = \dfrac{1}{\cos\theta}$" (7-2-12).

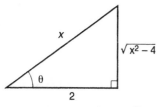

FIGURE 21 $\sec\theta = \dfrac{x}{2}$

$\dfrac{1}{\cos\theta} = \dfrac{x}{2}$ and $\cos\theta = \dfrac{2}{x}$. Label the picture with angle θ, adjacent 2 and hypotenuse x (Figure 21).

Now, use the Pythagorean Theorem to find the remaining side. Because "tangent is opposite over adjacent,"

$$\tan\theta = \frac{\sqrt{x^2-4}}{2}.$$ ▬

■ Trigonometric Substitution

In calculus, the "Pythagorean Identities" are used to rewrite $\sqrt{1-x^2}$, $\sqrt{x^2+1}$, and $\sqrt{x^2-1}$ without the square root signs. Of course, the square root signs do not really go away – they are simply concealed in a new notation which is often convenient. Here is how.

We use the "Pythagorean" identities again. For angles in the first quadrant the trigonometric functions are positive, so we can use the positive sign on the square root.

$$\cos^2\theta = 1 - \sin^2\theta \quad \text{yields} \quad \cos\theta = \sqrt{1-\sin^2\theta} \qquad (7\text{-}2\text{-}2)$$

$$\sec^2\theta = \tan^2\theta + 1 \quad \text{yields} \quad \sec\theta = \sqrt{\tan^2\theta+1} \qquad (7\text{-}2\text{-}15A)$$

$$\tan^2\theta = \sec^2\theta - 1 \quad \text{yields} \quad \tan\theta = \sqrt{\sec^2\theta-1} \qquad (7\text{-}2\text{-}15B)$$

EXAMPLE 18 Rewrite "$\sqrt{x^2 + 1}$" using "trigonometric substitution" so that it is expressed without a square root.

Looking at 7-2-15A, if "tan θ" plays the role of "x," then "$x^2 + 1$" will be "$\tan^2 \theta + 1$" which is "$\sec^2 \theta$." This has a nice square root without a square root sign. So, let $\tan \theta = x$. Then

$$\sqrt{x^2 + 1} = \sqrt{\tan^2 \theta + 1} = \sec \theta$$

This technique is useful in calculus for integrating expressions with square roots. ▬

CONCLUSION A good picture is worth several trig identities. Some identities follow from unit-circle pictures such as 1, 2, 4, 8 and 10. In unit-circle pictures think of angles as **rotations beginning from the positive x-axis**. Other identities follow from right-triangle pictures such as Figures 17, 19, and 20

It is essential to know the definitions of the trigonometric functions (7-2-1) and how to illustrate them on the unit circle and in a right triangle.

Terms: coterminal, reference angle, secant, cosecant, cotangent, complementary angles.

Exercise 7-2

A

A1.* Draw and label one unit-circle picture to illustrate all three of these with θ an acute angle: sin θ, cos θ and tan θ. [Hint: tan θ can be seen along the line $x = 1$.]

A2.* Draw and label one unit-circle picture to illustrate all three of these with θ an obtuse angle: sin θ, cos θ and tan θ. [Hint: tan θ can be seen along the line $x = 1$ (not $x = -1$).]

☺ *Find the reference angle.*

A3. (A) 40° (B) 170° (C) 200° (D) −50°

A4. (A) 100° (B) 300° (C) 19° (D) 260°

A5. (A) $\dfrac{2\pi}{3}$ (B) $\dfrac{5\pi}{6}$ (C) $\dfrac{7\pi}{12}$

A6. (A) $\dfrac{3\pi}{4}$ (B) $\dfrac{9\pi}{4}$ (C) $\dfrac{5\pi}{8}$

A7. (A) $\dfrac{11\pi}{12}$ (B) $\dfrac{-6\pi}{5}$ (C) $\dfrac{9\pi}{7}$

A8. (A) $\dfrac{25\pi}{24}$ (B) $\dfrac{5\pi}{2}$ (C) $\dfrac{3\pi}{5}$

☺ *Find an angle in the first quadrant with the same trig function values, except possibly for a minus sign.*

A9. (A) 100° (B) 345° (C) 500° (D) 175°

A10. (A) 220° (B) 178° (C) −92° (D) 110°

A11. (A) $\dfrac{7\pi}{6}$ (B) $\dfrac{5\pi}{4}$

A12. (A) $\dfrac{-3\pi}{4}$ (B) $\dfrac{7\pi}{12}$

Find the coterminal angle between 0 and 360° or between 0 and 2π.

A13. (A) −210° (B) 1000°

A14. (A) 450° (B) −350°

A15. (A) 13π (B) $\dfrac{17\pi}{4}$

A16. (A) $\dfrac{-3\pi}{4}$ (B) $\dfrac{5\pi}{2}$

Find all solutions satisfying $0° \le x < 360°$.

A17. $\sin x = .85$

A18. $\sin x = .2$

A19. $\cos x = .7$

A20. $\cos x = -.3$

Find x in the given quadrant (use degrees).

A21. $\sin x = .8$, II [130]

A22. $\cos x = .2$, IV [−78]

A23. $\tan x = -.3$, II [160]

A24. $\cos x = -.4$, III [250]

A25. $\sin x = .9$, II [120]

A26. $\tan x = .2$, III [190]

Find x in the given quadrant (use radians).

A27. $\cos x = .8$, IV [−.64]

A28. $\sin x = -.6$, III [3.8]

A29. $\tan x = -5$, II [1.8]

A30. $\sin x = .5$, II [2.6]

A31. $\cos x = .4$, IV [−1.2]

A32. $\tan x = 3$, III [4.4]

☺ *Give the complementary angle:*

A33. (A) $23°$ (B) $80°$

A34. (A) $30°$ (B) $59°$

☺ **secant, cosecant, and cotangent**

A35.* These are the reciprocals of which functions?
   (A) secant   (B) cosecant   (C) cotangent

A36. ☺ Cosine of $40°$ is sine of which angle?

A37. ☺ Cotangent of $80°$ is tangent of which angle?

A38. ☺ Cosecant of $30°$ is secant of which angle?

A39. Evaluate
   (A) $\sec 40°$ [1.3]
   (B) $\csc 20°$ [2.9]
   (C) $\cot 80°$ [.18]

A40. Evaluate
   (A) $\sec 87°$ [19]
   (B) $\csc 5°$ [11]
   (C) $\cot 5°$ [11]

A41. Evaluate
   (A) $\sec\left(\dfrac{\pi}{3}\right)$ [2]
   (B) $\csc\left(\dfrac{\pi}{3}\right)$ [1.2]
   (C) $\cot\left(\dfrac{\pi}{12}\right)$ [3.7]

A42. Evaluate
   (A) $\sec\left(\dfrac{\pi}{4}\right)$ [1.4]
   (B) $\csc\left(\dfrac{\pi}{12}\right)$ [3.9]
   (C) $\cot\left(\dfrac{\pi}{3}\right)$ [.58]

Solve for x in the first quadrant (use degrees).

A43. $\sec x = 3$ [71]

A44. $\csc x = 4$ [14]

A45. $\cot x = \dfrac{1}{2}$ [63]

A46. $\sec x = 1.5$ [48]

A47. $\csc x = 2.5$ [24]

A48. $\cot x = 10$ [5.7]

Solve for x in the first quadrant (use radians).

A49. $\sec x = 2$ [1.0]

A50. $\csc x = 5$ [.20]

A51. $\cot x = .42$ [1.2]

A52. $\sec x = 1.4$ [.78]

A53. $\csc x = 3.2$ [.32]

A54. $\cot x = 6$ [.17]

Reading and Writing Mathematics

A55.* ☺ State the simplest trig identity for
   (A) $\tan x$ (B) $\sec x$
   (C) $\csc x$ (D) $\cot x$

A56.* ☺ Write in terms of $\sin x$, $\cos x$, or $\tan x$, whichever is most convenient.
   (A) $\sec x$ (B) $\csc x$ (C) $\cot x$

A57.* (A) State the sine-cosine Pythagorean identity.
   (B) State the secant-tangent Pythagorean identity.

A58.* (A) Define "coterminal."
    (B) What is important about coterminal angles?

A59. State, using a variable, the trig identity, "The cosine of any angle is the sine of the complementary angle."

A60. State, using a variable, the trig identity, "The sine of any angle is the cosine of the complementary angle."

A61. State, using a variable, the trig identity, "The tangent of any angle is the cotangent of the complementary angle."

A62. State, using a variable, the trig identity, "The secant of any angle is the cosecant of the complementary angle."

Solve for x in [0, 2π):

A63. $\csc x = 2 \quad \left[\ldots, \dfrac{5\pi}{6}\right]$

A64. $\sec(2x) = 4 \quad [.66,\ldots]$

A65. $\cot\left(\dfrac{x}{2}\right) = \dfrac{1}{2} \quad [2.2]$

B

Unit-Circle Pictures

B1. ☺ The picture below is a unit-circle picture with θ in the first quadrant. Give these angles in terms of θ.
    (A) α (B) β (C) θ_2

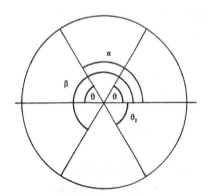

B2. ☺ The picture in the next column is another unit-circle picture with θ in the first quadrant. Give these angles in terms of θ.

    (A) α (B) β (C) θ_2

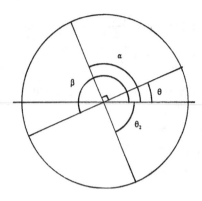

B3. Draw and label a unit-circle picture with angle θ in the first quadrant (Your choice of θ, but make θ distinctly less than or greater than $\pi/4$). Locate and label the other three angles in the other quadrants that have θ as their reference angle. Mark those angles with an arc beginning from the positive x-axis (as in Figure 6).

For each of the next problems, draw and label one unit-circle picture with angle θ and $P(\theta)$ in the first quadrant (Make θ distinctly less than or greater than $\pi/4$). Then, on that same picture, locate and label three things: the angle (beginning from the positive x-axis) requested in (a), the point in (b), and the trig function value in (c).

B4. (A) $\theta + \dfrac{\pi}{2}$ (B) $P\left(\theta + \dfrac{\pi}{2}\right)$ (C) $\sin\left(\theta + \dfrac{\pi}{2}\right)$

B5. (A) $\pi - \theta$ (B) $P(\pi - \theta)$ (C) $\sin(\pi - \theta)$

B6. (A) $\theta - \pi$ (B) $P(\theta - \pi)$ (C) $\sin(\theta - \pi)$

B7. (A) $\theta + \pi$ (B) $P(\theta + \pi)$ (C) $\sin(\theta + \pi)$

B8. (A) $\theta - \dfrac{\pi}{2}$ (B) $P\left(\theta - \dfrac{\pi}{2}\right)$ (C) $\sin\left(\theta - \dfrac{\pi}{2}\right)$

B9. (A) $\dfrac{\pi}{2} - \theta$ (B) $P\left(\dfrac{\pi}{2} - \theta\right)$ (C) $\sin\left(\dfrac{\pi}{2} - \theta\right)$

B10. (A) $\theta + \dfrac{3\pi}{2}$ (B) $P\left(\theta + \dfrac{3\pi}{2}\right)$ (C) $\sin\left(\theta + \dfrac{3\pi}{2}\right)$

See the figure on the next page.

B11. ☺ In the unit-circle figure, give
    (A) a in terms of θ
    (B) c in terms of a
    (C) c in terms of θ

(D) Point (c, d) is the terminal point of which angle?

(E) Complete the identity: $\cos(\pi - \theta) =$

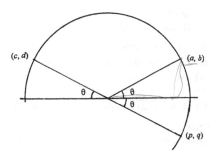

B12. ☺ In the unit circle figure above, give

(A) b in terms of θ

(B) d in terms of b

(C) d in terms of θ

(D) Point (c, d) is the terminal point of which angle?

(E) Complete the identity: $\sin(\pi - \theta) =$

B13. ☺ In the unit circle figure above, give

(A) b in terms of θ

(B) q in terms of b

(C) q in terms of θ

(D) Point (p, q) is the terminal point of which angle?

(E) Complete the identity: $\sin(-\theta) =$

B14. ☺ In the unit-circle figure above, give

(A) a in terms of θ

(B) p in terms of a

(C) p in terms of θ

(D) Point (p, q) is the terminal point of which angle?

(E) Complete the identity: $\cos(-\theta) =$

B15. Let θ be in the second quadrant such that $\cos \theta = -.6$. Draw a unit-circle picture of this. Use it to algebraically find $\sin \theta$ and $\tan \theta$.

B16. Let θ be in the second quadrant such that $\sin \theta = .4$. Draw a unit circle picture of this. Use it to algebraically find $\cos \theta$ and $\tan \theta$.

Trig Identities from the Unit-Circle $\Big[$ *Avoid illustrating θ*

near $\dfrac{\pi}{4}$. Illustrate θ distinctly less than $\dfrac{\pi}{4}$ or distinctly greater

than $\dfrac{\pi}{4}$. $\Big]$

B17. (A) Draw and label a unit-circle picture to illustrate "$\sin \theta$" and "$\cos \theta$" where θ is in the first quadrant. $\Big[$ Avoid illustrating θ near $\dfrac{\pi}{4}$. Illustrate θ

distinctly less than $\dfrac{\pi}{4}$ or distinctly greater than $\dfrac{\pi}{4}$. $\Big]$

Use the picture to illustrate these trig identities:

(B) $\sin(\pi - \theta) = \sin \theta$

(C) $\cos(\pi - \theta) = -\cos \theta$

B18. (A) Draw and label a unit-circle picture to illustrate "$\sin \theta$" and "$\cos \theta$" where θ is in the first quadrant. Use the picture to illustrate these trig identities:

(B) $\sin(-\theta) = -\sin \theta$

(C) $\cos(-\theta) = \cos \theta$

B19. Draw and label a unit-circle picture with an angle θ in the first quadrant. Given your picture of θ, sketch the angle $\pi + \theta$. Use your unit-circle picture to help do this problem. Express the following in terms of the same trigonometric function of θ.

(A) $\sin(\pi + \theta)$

(B) $\cos(\pi + \theta)$

(C) $\tan(\pi + \theta)$

B20. Draw and label a unit-circle picture with an angle θ in the first quadrant. Given your picture of θ, sketch the angle $\theta + \dfrac{\pi}{2}$. Use your unit-circle picture to help do this problem. Express the following in terms of a trigonometric function of θ.

(A) $\sin\left(\theta + \dfrac{\pi}{2}\right)$

(B) $\cos\left(\theta + \dfrac{\pi}{2}\right)$

(C) $\tan\left(\theta + \dfrac{\pi}{2}\right)$

B21. Draw and label a unit-circle picture with an angle θ in the first quadrant. Given your picture of θ, sketch the angle $\dfrac{\pi}{2} - \theta$. Use your unit-circle picture to help do this problem. Express the following in terms of another trigonometric function of θ.

(A) $\sin\left(\dfrac{\pi}{2} - \theta\right)$ (B) $\cos\left(\dfrac{\pi}{2} - \theta\right)$

B22. Draw and label a unit-circle picture with an angle θ in the first quadrant. Given your picture of θ, sketch the angle $\theta - \dfrac{\pi}{2}$. Use your unit-circle picture to help do this problem. Express the following in terms of another trigonometric function of θ.

(A) $\sin\left(\theta - \dfrac{\pi}{2}\right)$ (B) $\cos\left(\theta - \dfrac{\pi}{2}\right)$

B23. Use 7-2, Example 7, to find $\tan\left(\theta + \dfrac{\pi}{2}\right)$ in terms of $\tan\theta$.

B24. See the picture of a unit circle. If $CD = .87$, derive AB.

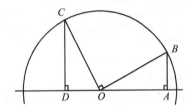

B25. See the picture of a unit circle. If $OD = -.4$, find OA.

B26. See the picture of a unit circle. If $OA = .82$, find OD.

Right-Triangle Pictures

B27. (A) Draw and label a triangle to illustrate "$\sin\theta = .3$."
    (B) Use the Pythagorean Theorem to find the other side and label it too.
    (C) Read your picture to find $\cos\theta$.
    (D) Read the picture to find $\tan\theta$.

B28. (A) Draw and label a triangle to illustrate "$\tan\theta = 1.5$."
    (B) Use the Pythagorean Theorem to find the other side and label it too.
    (C) Read your picture to find $\sin\theta$.
    (D) Read the picture to find $\cos\theta$.

B29. (A) Draw and label a triangle to illustrate "$\cos\theta = .6$."
    (B) Use the Pythagorean Theorem to find the other side and label it too.
    (C) Read your picture to find $\sin\theta$.
    (D) Read the picture to find $\tan\theta$.

B30. (A) Draw and label a triangle to illustrate "$\sin\theta = .9$."
    (B) Use the Pythagorean Theorem to find the other side and label it too.
    (C) Read your picture to find $\cos\theta$.
    (D) Read the picture to find $\tan\theta$.

Trig Identities from Right Triangles

B31. Draw a right triangle to illustrate "$\sin\theta = x$." Fill in the remaining side, and give these in terms of x.
    (A) $\cos\theta$
    (B) $\tan\theta$
    (C) Rewrite (a) and (b) in terms of $\sin\theta$ alone.

B32. Draw a right triangle to illustrate "$\cos\theta = x$." Fill in the remaining side, and give these in terms of x.
    (A) $\sin\theta$
    (B) $\tan\theta$
    (C) Rewrite (a) and (b) in terms of $\cos\theta$ alone.

B33. Draw a right triangle to illustrate "$\tan\theta = x$." Fill in the remaining side, and give these in terms of x.
    (A) $\sin\theta$
    (B) $\cos\theta$
    (C) Rewrite (a) and (b) in terms of $\tan\theta$ alone.

B34. Draw a right triangle to illustrate "$\sec\theta = x$." Fill in the remaining side, and give these in terms of x.
    (A) $\sin\theta$
    (B) $\csc\theta$
    (C) $\tan\theta$
    (D) Rewrite (a), (b), and (c) in terms of $\sec\theta$ alone.

B35. Draw a right triangle to illustrate "$\csc\theta = x$." Fill in the remaining side, and give these in terms of x.
    (A) $\sin\theta$
    (B) $\sec\theta$
    (C) $\cot\theta$
    (D) Rewrite (a), (b), and (c) in terms of $\csc\theta$ alone.

B36. Draw a right triangle to illustrate "$\cot\theta = x$." Fill in the remaining side, and give these in terms of x.
    (A) $\sin\theta$
    (B) $\cos\theta$
    (C) $\csc\theta$
    (D) Rewrite (a), (b), and (c) in terms of $\cot\theta$ alone.

Assume the angles are first quadrant angles. Draw and label a picture to illustrate part (a) and then find part (b).

B37. (A) $\cos^{-1}.4$ (B) Find $\sin(\cos^{-1}.4)$

B38. (A) $\sin^{-1}.9$ (B) Find $\tan(\sin^{-1}.9)$

B39. (A) $\sin^{-1}.7$ (B) Find $\cos(\sin^{-1}.7)$

B40. (A) $\tan^{-1}3$ (B) Find $\sin(\tan^{-1}3)$

Assume the angles are first quadrant angles. Draw and label a picture to illustrate part (a) and then find part (b).

B41. (A) $\sin^{-1}(3x)$ (B) Find $\tan(\sin^{-1}(3x))$

B42. (A) $\tan^{-1}\left(\dfrac{x}{2}\right)$ (B) Find $\cos\left(\tan^{-1}\left(\dfrac{x}{2}\right)\right)$

B43. (A) $\cos^{-1}(2x)$ (B) Find $\sin(\cos^{-1}(2x))$

B44. (A) $\tan^{-1}\left(\dfrac{x}{4}\right)$ (B) Find $\sec\left(\tan^{-1}\left(\dfrac{x}{4}\right)\right)$

Rewrite these in terms of $\sin\theta$ alone.

B45. $(\sec\theta)^2$ **B46.** $(\cot\theta)^2$

B47. $(\sec \theta)(\cot \theta)$ **B48.** $(\tan \theta)^2 + 1$

B49. $\dfrac{(\tan \theta)(\sec \theta)}{\sin \theta}$ **B50.** $\dfrac{\cot \theta}{\sec \theta}$

Use trigonometric substitution (see Example 18) to rewrite the expression as a trigonometric function of θ without a square root symbol. Give the relationship between x and θ.

B51. $\sqrt{x^2 - 1}$ **B52.** $\sqrt{1 - x^2}$

B53. $\sqrt{x^2 + 4}$ **B54.** $\sqrt{x^2 - 9}$

B55. $\sqrt{4 - x^2}$

B56. Draw and label a figure to illustrate that "$\sin\left(\dfrac{\pi}{2} - \theta\right) = \cos \theta$" holds when θ is an angle in the second quadrant.

Reading and Writing Mathematics

B57. If the coordinates of $P(\theta)$ are (a, b), what are the coordinates of $P(90° - \theta)$?

B58.* (A) Define "reference angle."
(B) What is interesting about reference angles?

B59. Find $\sin^{-1}(\sin x)$ for $0 \le x < 2\pi$. Be careful to cover all cases.

B60. Find $\cos^{-1}(\cos x)$ for $0 \le x < 2\pi$. Be careful to cover all cases.

B61. (**Secant, cosecant, and cotangent as lengths in a unit-circle picture**) Draw a tangent line to the circle at point P in Figure 7-2-1.
(A) Where does that tangent line intersect the x-axis?
(B) Where does that tangent line intersect the y-axis?
(C) Find a way to represent cotangent as a length in a unit-circle picture.

B62. Is the function even, odd, or neither?
(A) secant   (B) cosecant   (C) cotangent

B63. Here is one more Pythagorean identity:
$\csc^2 \theta = \cot^2 \theta + 1$. Derive it.

Section 7-3   MORE IDENTITIES

Given the values of the trigonometric functions at α (alpha) and β (beta), what are the values of the trigonometric functions at $\alpha + \beta$ and $\alpha - \beta$? At 2α? At $\dfrac{\alpha}{2}$?

Of course, since order matters, you do not expect $\sin(\alpha + \beta)$ to be $\sin \alpha + \sin \beta$.

■ The Sum Identities 7-3-1

$$\sin(\alpha + \beta) = (\sin \alpha)(\cos \beta) + (\cos \alpha)(\sin \beta) \qquad (7\text{-}3\text{-}1\text{A})$$

$$\cos(\alpha + \beta) = (\cos \alpha)(\cos \beta) - (\sin \alpha)(\sin \beta) \qquad (7\text{-}3\text{-}1\text{B})$$

$$\tan(\alpha + \beta) = \dfrac{\tan \alpha + \tan \beta}{1 - (\tan \alpha)(\tan \beta)} \qquad (7\text{-}3\text{-}1\text{C})$$

EXAMPLE 1   Find $\sin 75°$ exactly using a sum identity.
The values of sine and cosine are known exactly at 45° and 30° (6-2-12).

$$\sin 75° = \sin(45° + 30°) = (\sin 45°)(\cos 30°) + (\cos 45°)(\sin 30°)$$

$$= \left(\dfrac{\sqrt{2}}{2}\right)\left(\dfrac{\sqrt{3}}{2}\right) + \left(\dfrac{\sqrt{2}}{2}\right)\left(\dfrac{1}{2}\right) = \dfrac{\sqrt{6} + \sqrt{2}}{4}.$$

■ Argument for 7-3-1A&B

Here is the most illuminating argument for the sum identities. (An algebraic proof is given with Figures 4 and 5.) The argument determines $\sin(\alpha + \beta)$ and $\cos(\alpha + \beta)$ simultaneously. Following the details will help you learn trigonometry.

Begin with a unit circle. First construct angle α terminating at A (Figure 1). Then construct angle $\alpha + \beta$. Angle $\alpha + \beta$ terminates at some point, P, with angle $POA = \beta$. By definition, $P = (\cos(\alpha + \beta), \sin(\alpha + \beta))$. The coordinates of P are just what we want to know for the sum identities. $\sin(\alpha + \beta) = PE$. $\cos(\alpha + \beta) = OE$.

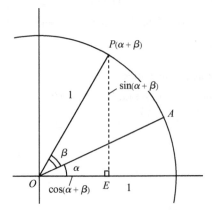

FIGURE 1 Angles α and β, with point P corresponding to angle $\alpha + \beta$

The problem is to express PE and OE in terms of sines and cosines of α and β.

To begin (Figure 2), drop a line from P perpendicular to the line OA, intersecting it at point B. From B (Figure 3) drop a perpendicular to the x-axis at C. Also, draw a horizontal line through B and a vertical line through P and label where they meet D.

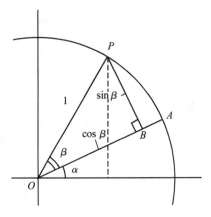

FIGURE 2 Figure 1 again, plus a key line and labels

Then in Figure 3,

$$\sin(\alpha + \beta) = PE = BC + PD.$$ (For 7-3-1A)

$$\cos(\alpha + \beta) = OE = OC - DB.$$ (For 7-3-1B)

We want the lengths of these four line segments.

Begin in Figure 2.

$$OB = \cos \beta, \text{ and}$$
$$PB = \sin \beta.$$

In Figure 3, in the lower triangle OBC,

$$\frac{BC}{OB} = \sin \alpha, \text{ so}$$

$$BC = (\sin \alpha)OB = (\sin \alpha)(\cos \beta).$$ This is the first term in 7-3-1A.

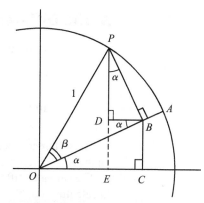

FIGURE 3 Figure 2 again, with more labels

Also, $\dfrac{OC}{OB} = \cos \alpha$, so

$$OC = (\cos \alpha)OB = (\cos \alpha)(\cos \beta).$$ This is the first term in 7-3-1B.

Now consider the triangle PBD. Because DB is parallel to the x-axis and PD is parallel to the y-axis, the angles are α as labeled. Therefore,

$$\frac{PD}{PB} = \cos \alpha, \text{ so}$$

$$PD = (\cos \alpha)PB = (\cos \alpha)(\sin \beta).$$ This is the second term in 7-3-1A.

$$\frac{DB}{PB} = \sin \alpha, \text{ so}$$

$$DB = (\sin \alpha)PB = (\sin \alpha)(\sin \beta).$$ This is the second term in 7-3-1B.

We have the four lengths we needed.

$$\sin(\alpha + \beta) = PE = BC + PD = (\sin \alpha)(\cos \beta) + (\cos \alpha)(\sin \beta).$$ This is 7-3-1A.

$\cos(\alpha + \beta) = OE = OC - DB = (\cos \alpha)(\cos\beta) - (\sin \alpha)(\sin \beta)$. This is 7-3-1B.

The proof of the identity for $\tan(\alpha + \beta)$ is in Example 4 below.

■ The Difference Identities

The identities for $\sin(\alpha - \beta)$ and $\cos(\alpha - \beta)$ follow from the sum identities by treating "$\alpha - \beta$" as "$\alpha + (-\beta)$." Then replace $\cos(-\beta)$ with $\cos \beta$ and $\sin(-\beta)$ with $-\sin(\beta)$, using 7-2-6. Here is the proof of 7-3-2A given 7-3-1A:

$$
\begin{aligned}
\sin(\alpha - \beta) &= \sin(\alpha + (-\beta)) \\
&= (\sin \alpha)(\cos(-\beta)) + (\cos \alpha)(\sin(-\beta)) && \text{[by 7-3-1A]} \\
&= (\sin \alpha)(\cos \beta) + (\cos \alpha)(-\sin \beta) && \text{[by 7-2-6]} \\
&= (\sin \alpha)(\cos \beta) - (\cos \alpha)(\sin \beta) && \text{[This is 7-3-2A]}
\end{aligned}
$$

The proofs for cosine and tangent are similar (Problems B49 and B50).

■ The Difference Identities 7-3-2

$$\sin(\alpha - \beta) = (\sin \alpha)(\cos \beta) - (\cos \alpha)(\sin \beta). \qquad (7\text{-}3\text{-}2A)$$

$$\cos(\alpha - \beta) = (\cos \alpha)(\cos \beta) + (\sin \alpha)(\sin \beta). \qquad (7\text{-}3\text{-}2B)$$

$$\tan(\alpha - \beta) = \frac{\tan \alpha - \tan \beta}{1 + (\tan \alpha)(\tan \beta)} \qquad (7\text{-}3\text{-}2C)$$

The difference identities are very similar to the sum identities—only the signs on the second terms are changed.

The sum and difference identities have numerous applications. They appear in calculus and they even apply to multiplying and dividing complex numbers. It is remarkable that trigonometry appears even in a very algebraic context such as solving quadratic and polynomial equations where complex numbers arise.

EXAMPLE 2   Find $\cos\left(\dfrac{\pi}{12}\right)$ exactly using a difference identity.

Treat $\dfrac{\pi}{12}$ as a difference. $\dfrac{\pi}{12} = \dfrac{\pi}{4} - \dfrac{\pi}{6}$. The trig functions of $\dfrac{\pi}{4}$ and $\dfrac{\pi}{6}$ are known exactly.

$$
\begin{aligned}
\cos\left(\frac{\pi}{12}\right) &= \cos\left(\frac{\pi}{4} - \frac{\pi}{6}\right) \\[2mm]
&= \left(\cos\left(\frac{\pi}{4}\right)\right)\left(\cos\left(\frac{\pi}{6}\right)\right) + \left(\sin\left(\frac{\pi}{4}\right)\right)\left(\sin\left(\frac{\pi}{6}\right)\right) \\[2mm]
&= \left(\frac{\sqrt{2}}{2}\right)\left(\frac{\sqrt{3}}{2}\right) + \left(\frac{\sqrt{2}}{2}\right)\left(\frac{1}{2}\right) = \frac{\sqrt{6} + \sqrt{2}}{4}. \qquad \blacksquare
\end{aligned}
$$

From here it is easy to derive the important "double angle" and "half angle" identities.

■ Double- and Half-Angle Identities

Given the trigonometric functions of θ, what are their values at 2θ? At $\dfrac{\theta}{2}$? Of course, since order matters, you do not expect to simply multiply or divide the values by 2. We begin with "double" angles.

■ Double-Angle Identities 7-3-3

$$\sin 2\theta = 2(\sin \theta)(\cos \theta) \qquad (7\text{-}3\text{-}3\text{A})$$

$$\cos 2\theta = \cos^2\theta - \sin^2\theta \qquad (7\text{-}3\text{-}3\text{B})$$

$$\cos 2\theta = 1 - 2\sin^2\theta \qquad (7\text{-}3\text{-}3\text{C})$$

$$\cos 2\theta = 2\cos^2\theta - 1. \qquad (7\text{-}3\text{-}3\text{D})$$

$$\tan 2\theta = \frac{2\tan\theta}{1 - \tan^2\theta}. \qquad (7\text{-}3\text{-}3\text{E})$$

Proof of 7-3-3A

To find $\sin(2\theta)$, let $\alpha = \beta = \theta$ in 7-3-1A. Then it says:

$$\sin(\theta + \theta) = (\sin \theta)(\cos \theta) + (\cos \theta)(\sin \theta)$$

$$\sin(2\theta) = 2(\sin \theta)(\cos \theta). \text{ [This is 7-3-3A]}$$

Identity 7-3-3B for $\cos(2\theta)$ follows similarly (Problem B51).

Proof of Part C

From Part B, $\cos(2\theta) = \cos^2\theta - \sin^2\theta$. Now replace $\cos^2\theta$ with $1 - \sin^2\theta$ to

obtain $\qquad\qquad \cos(2\theta) = (1 - \sin^2\theta) - \sin^2\theta$

$$= 1 - 2\sin^2\theta.$$

There is another method of proof in Problem B54. The proof of Part D is similar (Problem B52). The proof of the tangent identity is Problem B55.

■ Squared-Function Identities

Two very important identities for calculus follow merely by rearranging the double-angle cosine formulas.

$$\sin^2\theta = \frac{1 - \cos 2\theta}{2}. \qquad (7\text{-}3\text{-}4\text{A})$$

$$\cos^2\theta = \frac{1 + \cos 2\theta}{2}. \qquad (7\text{-}3\text{-}4\text{B})$$

Proof of 7-3-4A

7-3-3C says $\cos(2\theta) = 1 - 2\sin^2\theta$. Therefore, $\cos(2\theta) - 1 = -2\sin^2\theta$. Now divide by -2. This isolates $\sin^2\theta$ and yields 7-3-4A.

In calculus these are used to integrate "$\sin^2 x$" and "$\cos^2 x$."

These, in turn, yield the half-angle formulas by changing placeholders.

Proof of 7-3-5A

θ is a placeholder, so we may, in 7-3-4A, simply replace "θ" with "$\frac{\theta}{2}$".

$$\sin^2\left(\frac{\theta}{2}\right) = \frac{1 - \cos\left(2\left(\frac{\theta}{2}\right)\right)}{2} = \frac{1 - \cos\theta}{2}$$

Now remove the squaring to obtain the half angle identity for sine (Problems B57 and B58).

■ Half-Angle Identities 7-3-5

For parts A and B the right side is either plus or minus, but not both. The sign depends upon the quadrant of $\frac{\theta}{2}$.

$$\sin\left(\frac{\theta}{2}\right) = \pm\sqrt{\frac{1 - \cos\theta}{2}} \qquad\qquad (7\text{-}3\text{-}5A)$$

$$\cos\left(\frac{\theta}{2}\right) = \pm\sqrt{\frac{1 + \cos\theta}{2}} \qquad\qquad (7\text{-}3\text{-}5B)$$

$$\tan\left(\frac{\theta}{2}\right) = \frac{\sin\theta}{1 + \cos\theta} \qquad\qquad (7\text{-}3\text{-}5C)$$

EXAMPLE 3 Find $\cos\left(\frac{\pi}{8}\right)$ exactly.

Treat $\frac{\pi}{8}$ as $\frac{\left(\frac{\pi}{4}\right)}{2}$. We already know that $\cos\left(\frac{\pi}{4}\right) = \frac{\sqrt{2}}{2}$. From the half angle identity,

$$\cos\left(\frac{\pi}{8}\right) = \cos\left(\frac{\frac{\pi}{4}}{2}\right) = \sqrt{\frac{1 + \frac{\sqrt{2}}{2}}{2}} = \sqrt{\frac{2 + \sqrt{2}}{4}} = \frac{\sqrt{2 + \sqrt{2}}}{2}.$$

By determining more and more values we could fill out a trigonometric table for arguments between 0 and $\frac{\pi}{2}$. For example, with this value for $\cos\left(\frac{\pi}{8}\right)$ we could find $\cos\left(\frac{\pi}{16}\right)$ using the half-angle identity again, and we could find $\cos\left(\frac{3\pi}{16}\right)\left(\frac{3\pi}{16} = \frac{\pi}{8} + \frac{\pi}{16}\right)$ using a sum identity. To fill out the entire trigonometric table this way would require a huge amount of work. Fortunately, that work has already been done long ago.

EXAMPLE 4 Prove the sum identity for tangent, 7-3-1C,

$$\tan(\alpha + \beta) = \frac{\tan \alpha + \tan \beta}{1 - (\tan \alpha)(\tan \beta)},$$ (7-3-1C)

given the sum identities for sine and cosine, 7-3-1A and B.

Recall that "tangent is sine over cosine" (7-2-1).

Proof

Begin with the left side of 7-3-1C.

$$\tan(\alpha + \beta) = \frac{\sin(\alpha + \beta)}{\cos(\alpha + \beta)} = \frac{(\sin \alpha)(\cos \beta) + (\cos \alpha)(\sin \beta)}{(\cos \alpha)(\cos \beta) - (\sin \alpha)(\sin \beta)}.$$

Now, divide through both the top and bottom by the first term in the denominator, so that term becomes 1 (as in the finished identity).

$$= \frac{\dfrac{(\sin \alpha)(\cos \beta)}{(\cos \alpha)(\cos \beta)} + \dfrac{(\cos \alpha)(\sin \beta)}{(\cos \alpha)(\cos \beta)}}{1 - \dfrac{(\sin \alpha)(\sin \beta)}{(\cos \alpha)(\cos \beta)}}$$

Some terms cancel to get

$$= \frac{\dfrac{\sin \alpha}{\cos \alpha} + \dfrac{\sin \beta}{\cos \beta}}{1 - \left(\dfrac{\sin \alpha}{\cos \alpha}\right)\left(\dfrac{\sin \beta}{\cos \beta}\right)}$$

Now the sum identity follows immediately by "tangent is sine over cosine":

$$= \frac{\tan \alpha + \tan \beta}{1 - (\tan \alpha)(\tan \beta)}.$$

This is the right side of 7-3-1C. ▬

The difference identity for tangent follows by treating "$\alpha - \beta$" as "$\alpha + (-\beta)$" and using 7-2-6 (Problem B50). The half-angle formula follows using "tangent is sine over cosine" (Problem B58).

EXAMPLE 5 Solve "$\sin \theta = .4 \cos \theta$" for θ in the first quadrant.

$$\frac{\sin \theta}{\cos \theta} = .4 \quad \tan \theta = .4 \quad \theta = \tan^{-1} .4 = 21.8° \qquad ▬$$

EXAMPLE 6 Solve "$\cos^2\theta = \sin \theta + .2$" for θ in the first quadrant.

$1 - \sin^2\theta = \sin \theta + .2$ using $\sin^2\theta + \cos^2\theta = 1$.
                                     This is now a quadratic in "$\sin \theta$."
$0 = \sin^2\theta + \sin \theta - .8$ Use the Quadratic Formula: $a = 1, b = 1$,
                                       and $c = -.8$.
$\sin \theta = .5247$ or -1.525 (extraneous).
$\theta = \sin^{-1} .5247 = 31.6°$. ▬

EXAMPLE 7 Solve "$\cos(2\theta) = .2\cos\theta$" for θ in the first quadrant.
Use the double-angle formula 7-3-3D to eliminate "$\cos(2\theta)$."
$2\cos^2\theta - 1 = .2\cos\theta$. This is a quadratic in "$\cos\theta$."
$2\cos^2\theta - .2\cos\theta - 1 = 0$.
$\cos\theta = .759$ or $-.659$. In the first quadrant, $\theta = \cos^{-1}.759 = 40.6°$. ▬

EXAMPLE 8 Solve "$(\sec\theta)(\tan\theta) = 3$" for θ in the first quadrant.
Switch to sines and cosines.

$$\left(\frac{1}{\cos\theta}\right)\left(\frac{\sin\theta}{\cos\theta}\right) = 3.$$

$\sin\theta = 3(\cos\theta)^2.$

$\sin\theta = 3(1 - \sin^2\theta)$. Now use the Quadratic Formula (Problem A18). ▬

■ Proofs of the Sum and Difference Identities

The proofs of the sum and difference identities (7-3-1 and 7-3-2) can be done by proving 7-3-2B before any of the others.

(7-3-2B, again) $\cos(\alpha - \beta) = (\cos\alpha)(\cos\beta) + (\sin\alpha)(\sin\beta).$

The following proof uses the distance formula (3-3-2), which is basically the Pythagorean Theorem, and the definitions of sine and cosine.

Proof

For convenience, Figure 4 illustrates α greater than β, but unlike the argument for 7-3-1, this proof works the same way regardless of the values of α and β.

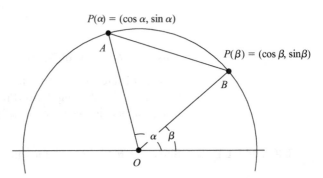

FIGURE 4 α, β, $P(\alpha) = (\cos\alpha, \sin\alpha)$, and $P(\beta) = (\cos\beta, \sin\beta)$

$P(\alpha)$ is the point $(\cos\alpha, \sin\alpha)$ and $P(\beta) = (\cos\beta, \sin\beta)$. Using the distance formula 3-3-2 the distance, d, from $P(\alpha)$ to $P(\beta)$ satisfies:

$$d^2 = (\cos\alpha - \cos\beta)^2 + (\sin\alpha - \sin\beta)^2 \quad \text{[Multiply this out]}$$
$$= (\cos\alpha)^2 - 2(\cos\alpha)(\cos\beta) + (\cos\beta)^2 + (\sin\alpha)^2$$
$$- 2(\sin\alpha)(\sin\beta) + (\sin\beta)^2$$

$$= 2 - 2(\cos \alpha)(\cos \beta) - 2(\sin \alpha)(\sin \beta)$$
$$[\text{because } (\cos \alpha)^2 + (\sin \alpha)^2 = 1]$$

This gives one expression for d^2. We will find a second and set them equal.

In Figure 4 the angle AOB is $\alpha - \beta$. Find $P(\alpha - \beta)$ and $P(0)$ (Figure 5).

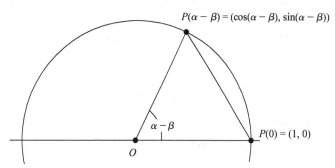

FIGURE 5 Angles $\alpha - \beta$ and 0,
$P(\alpha - \beta) = (\cos(\alpha - \beta), \sin(\alpha - \beta))$,
and $P(0) = (1, 0)$

The distance from $P(0)$ to $P(\alpha - \beta)$ is the same as from $P(\alpha)$ to $P(\beta)$ because they both have angle $\alpha - \beta$ between them.

$$\begin{aligned} d^2 &= (\cos(\alpha - \beta) - 1)^2 + (\sin(\alpha - \beta) - 0)^2 \\ &= (\cos(\alpha - \beta))^2 - 2\cos(\alpha - \beta) + 1 + (\sin(\alpha - \beta))^2. \\ &= 2 - 2\cos(\alpha - \beta). \end{aligned}$$

Setting the two expressions for d^2 equal,

$$2 - 2\cos(\alpha - \beta) = 2 - 2(\cos \alpha)(\cos \beta) - 2(\sin \alpha)(\sin \beta)$$

Now subtract 2 and divide by -2:

$$\cos(\alpha - \beta) = (\cos \alpha)(\cos \beta) + (\sin \alpha)(\sin \beta).$$

This is the desired 7-3-2B.
   The derivation of the remaining parts of 7-3-1 and 7-3-2 is in Problems B59–61.

CONCLUSION   A number of important identities follow from the "sum identities" (7-3-1). They include the difference identities (7-3-2), the double-angle and half-angle identities (7-3-3 and 7-3-5), and the identities for squares of sine and cosine (7-3-4).
   Other useful identities are the sum-to-product identities (Problems B65–B68) and the product-to-sum identities (Problems B69–B72).

Terms: sum identity, difference identity, double-angle identity, half-angle identity.

Exercise 7-3

A

A1.* ☺ True or False?
 (A) $\sin 2x = 2 \sin x$
 (B) $\cos 2x = 2 \cos x$
 (C) $\sin(\alpha + \beta) = \sin \alpha + \sin \beta$
 (D) $\cos(\alpha + \beta) = \cos \alpha + \cos \beta$
 (E) What conclusion can you draw from parts (A) through (D)? [Two words from Section 1-2 will do.]

A2.* ☺ True or False?
 (A) $\sin x = \sin(-x)$ (B) $\cos x = \cos(-x)$
 (C) $\tan x = \tan(-x)$

A3. Use a sum identity to find and simplify
$$\sin\left(\theta + \frac{\pi}{2}\right).$$

A4. Use a sum identity to find and simplify
$$\cos\left(\theta + \frac{\pi}{2}\right).$$

A5. Use a sum identity to find and simplify
$$\tan\left(\theta + \frac{\pi}{4}\right).$$

A6. Use a sum identity to find and simplify $\sin(\theta + \pi)$.

A7. Use a sum identity to find and simplify $\cos(\theta + \pi)$.

A8. Use a sum identity to find and simplify $\tan(\theta + \pi)$.

A9. Use a difference identity to find and simplify
$$\sin\left(\frac{\pi}{2} - \theta\right).$$

A10. Use a difference identity to find and simplify
$$\cos\left(\frac{\pi}{2} - \theta\right).$$

In the unit-circle picture, the two angles are θ.

A11. If $c = .55$, find θ.   [28°]

A12. If $d = .85$, find θ.

A13. If $\dfrac{d}{c} = 1.2$, find θ.   [25°]

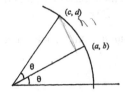

In the unit-circle picture, the two angles are θ.

A14. If $c = -.4$, find θ.

A15. If $d = .9$, find θ.

A16. If $\dfrac{d}{c} = -2$, find θ.

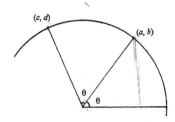

A17. Is the function even, odd, or neither?
 (A) sine   (B) cosine   (C) tangent

A18. Solve the equation in Example 8.   [58°]

A19. In the unit-circle figure, give
 (A) a in terms of θ (B) b in terms of θ
 (C) c in terms of θ (D) d in terms of θ

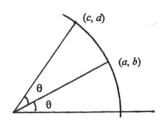

A20. Use a difference identity to find $\sin 15°$ algebraically.

A21. Use a half-angle identity to find $\sin 15°$ algebraically.

B

For B1–3, the circle is a unit-circle.

B1. In the figure on the next page, derive the lengths PB, OB, and BC in terms of α and β.

B2. In the figure on the next page, given $PB = \sin \beta$, derive the lengths PD and DB.

B3. In the figure on the next page, given $OB = \cos \beta$ derive the lengths BC and OC.

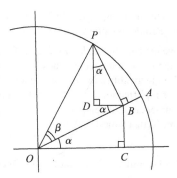

With the help of a trig identity, solve these algebraically for a non-zero answer in the first quadrant, if possible. (Use degrees.)

B4. $(\cos\theta)^2 = \cos\theta - .2$ [74°, . . .]

B5. $(\sin\theta)^2 = \sin\theta - .25$ [30°]

B6. $(\cos\theta)^2 = \sin\theta + .1$

B7. $(\sin\theta)^2 = \cos\theta + .2$ [58°]

B8. $(\cos\theta)^2 = \sin\theta - .13$ [42°]

B9. $(\sin\theta)^2 = \cos\theta - .5$

B10. $\sin(-\theta) = \sin\theta - .5$ [14°]

B11. $\cos(\theta) = 2\sin\theta$ [27°]

B12. $\cos(-\theta) = 3\sin\theta$

B13. $2\cos\theta = 3\sin\theta$ [34°]

B14. $\sin\theta = 3\cos\theta$ [72°]

B15. $\sin\theta = \left(\dfrac{1}{2}\right)\cos\theta$

B16. $\cos(2\theta) = 3\sin(2\theta)$ [9.2°]

B17. $\tan\theta = 2\sin\theta$ [60°]

B18. $\tan\theta = 2\cos\theta$

B19. $\sin(2\theta) = 1.5\cos\theta$ [49°]

B20. $\sin(2\theta) = 1.1\sin\theta$ [57°]

B21. $\cos(2\theta) = \cos\theta - .2$

B22. $\cos(2\theta) = \cos\theta - .6$ [40°]

B23. $\cos(2\theta) = .2\sin\theta$ [41°]

B24. $\cos(2\theta) = \sin\theta$

B25. $\cos\left(\dfrac{\theta}{2}\right) = \cos\theta + .2$ [43°]

B26. $\tan(2\theta) = 3\tan\theta$. [30°]

B27. $\sec^2\theta = \tan\theta + 2$

B28. $\sec\theta = \cos\theta + 2$ [66°]

B29. $\sin\left(\dfrac{\theta}{2}\right) = .6\sin\theta$ [67°]

B30. $\cos\left(\dfrac{\theta}{2}\right) = 1.1\cos\theta$ [28°]

B31. $(\tan\theta)(\cos\theta) = .8$ [53°]

B32. $(\tan\theta)(\sin\theta) = 3$ [72°]

B33. $(\cot\theta)(\sin\theta) = .9$ [26°]

B34. $(\cot\theta)(\cos\theta) = 2.7$ [19°]

B35. In the unit-circle picture both angles are θ. If $OA = .7$, find θ.

B36. In the unit-circle picture both angles are θ. If $AB = .4$, find θ.

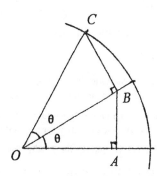

Solve algebraically for θ in the unit-circle figure below.

B37. $AD = 1.7\,BC$ **B38.** $OA = .5\,AD$

B39. $OA = .45\,OB$ **B40.** $OA = BC$

B41. $AD = OB$ **B42.** $1 - OB = OA$

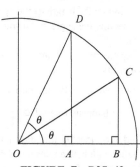

FIGURE For B37–42

Use the figure for Problems B43–48.

Solve algebraically for θ > 0 (use degrees) with the help of trigonometric identities if the following relationships hold.

B43. $d = \dfrac{5b}{3}$ [34]   **B44.** $\dfrac{d}{c} = \dfrac{3b}{a}$ [30]

B45. $a = 2c$ [33]   **B46.** $a = c + .2$ [22]

B47. Solve graphically for θ if $b + .2 = c$. [25]

B48. Solve graphically for θ if $a - .1 = d$.

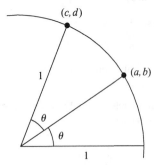

FIGURE For B43–48

Proofs

B49. Start with the sum identity for cosine (7-3-1B) and derive the difference identity for cosine (7-3-2B).

B50. Start with the sum identity for tangent (7-3-1C) and derive the difference identity for tangent (7-3-2C).

B51. Start with 7-3-1B and derive 7-3-3B for $\cos(2\theta)$.

B52. Start with 7-3-3B and derive 7-3-3D for $\cos(2\theta)$.

B53. Begin with a double-angle formula and derive the squared-function identity 7-3-4B for $\cos^2\theta$.

B54. To prove 7-3-3C, "$\cos(2\theta) = 1 - 2\sin^2\theta$" another way, consider an isosceles triangle (or unit circle) with angle 2θ and adjacent sides equal to 1.
(A) Compute the length of the opposite side using the Law of cosines.
(B) Compute the length of the opposite side by bisecting the angle and using the definition of sine.
(C) Equate the results and simplify to get that identity.

B55. (A) Start with 7-3-1C and derive 7-3-3E for $\tan(2\theta)$.
(B) What happens when $\theta = \dfrac{\pi}{4}$?

B56. To find a trig identity for $\sin(2\theta)$, reproduce Figures 1 through 3 with both angles θ and reproduce the steps in the given proof, with the appropriate changes. Which trig identity results?

B57. Derive the half-angle identity 7-3-5B for cosine. [Hint: Parallel the proof for 7-3-5A.]

B58. Derive the half-angle identity 7-3-5C for tangent. Use the fact that tangent is sine over cosine and the half-angle identities for sine and cosine. [Hint: After some simplification you will want to multiply both the numerator and denominator by "$1 + \cos\theta$" to permit further simplification.]

Proofs of 7-3-1 and 7-3-2, the sum and difference identities.

B59. Given the difference-of-angles formula 7-3-2B for cosine, the result 7-3-2A for sine follows from 7-3-2B and a cofunction identity (7-2-13), using $\sin(\alpha - \beta) = \cos\left(\dfrac{\pi}{2} - (\alpha - \beta)\right)$. Complete the proof.

B60. Given the difference-of-angles formula 7-3-2B for cosine, derive the sum-of-angles formula 7-3-1B for cosine.

B61. Given the difference-of-angles formula 7-3-2B for cosine, derive the sum of angles identity 7-3-1A for sine with this approach:

$$\sin(\alpha + \beta) = \cos\left(\dfrac{\pi}{2} - (\alpha + \beta)\right)$$
$$= \cos\left(\left(\dfrac{\pi}{2} - \alpha\right) - \beta\right).$$

Now use the difference identify and simplify.

B62. (This gives a slightly different proof of 7-3-2B before 7-3-1. Then 7-3-2B can lead to a proof of all the sine and cosine sum and difference identities. See Problems B59–61.) On a unit circle locate (arbitrary) angles α and β [For convenience, assume $0 < \beta < \alpha < 2\pi$]. Then locate the points $P(\alpha)$ and $P(\beta)$. Use the Law of Cosines (which works for all angles) to express the distance between $P(\alpha)$ and $P(\beta)$. Also, use the distance formula to express that same distance. Equate the two and simplify to obtain the difference-of-angles formula for cosine.

B63. (A proof of the **Law of Cosines**.) Let θ be the angle between sides a and b, and orient the triangle so that θ is at the origin and a lies along the positive x-axis. Side a ends at $(a, 0)$.
(A) Sketch a picture of the information with θ a second-quadrant angle
(B) What are the coordinates of the end of side b?
(C) Use the distance formula to find the distance between the two endpoints and simplify to obtain the Law of Cosines for θ. [This proof

does not require θ to be a second-quadrant angle.]

B64. Derive this **Pythagorean identity** for cosecant:
$(\csc \theta)^2 = (\cot \theta)^2 + 1$.

*Prove the following important trig identities. You may use lower-numbered ones to prove higher-numbered ones. B65–68 are "**Sum-to-Product**" identities.*

B65. $\sin a + \sin b = 2 \cos\left[\dfrac{a-b}{2}\right]\sin\left[\dfrac{a+b}{2}\right]$.

[Hint for B65: Let $x = \dfrac{a}{2}$ and $y = \dfrac{b}{2}$ and begin with the right side. Hint for B66: Use B65.]

B66. $\sin a - \sin b = 2 \cos\left[\dfrac{a+b}{2}\right]\sin\left[\dfrac{a-b}{2}\right]$

B67. $\cos a + \cos b = 2 \cos\left[\dfrac{a+b}{2}\right]\cos\left[\dfrac{a-b}{2}\right]$

B68. $\cos a - \cos b = 2 \sin\left[\dfrac{a+b}{2}\right]\sin\left[\dfrac{b-a}{2}\right]$

*B69–72 are "**Product-to-Sum**" identities. [Hint: Begin with the right side.]*

[Note: In addition to other important uses, the product-to-sum identities were very valuable when multiplication had to be done by hand. The multiplication on the left, which might be of seven-digit numbers, could be replaced by the simpler additions and subtractions on the right.]

B69. $(\sin a)(\sin b) = \left(\dfrac{1}{2}\right)[\cos(a-b) - \cos(a+b)]$

B70. $(\sin a)(\cos b) = \left(\dfrac{1}{2}\right)[\sin(a+b) + \sin(a-b)]$

B71. $(\cos a)(\sin b) = \left(\dfrac{1}{2}\right)[\sin(a+b) - \sin(a-b)]$

B72. $(\cos a)(\cos b) = \left(\dfrac{1}{2}\right)[\cos(a+b) + \cos(a-b)]$

B73. (Trig identities) Consider an isosceles triangle with base between the two equal angles and angle θ opposite the base. The base can be computed in terms of the two equal sides and θ using the Law of Cosines or, alternatively, by bisecting angle θ to create two right triangles.

(A) Do it both ways and equate the two answers to find a trigonometric identity for $\sin\left(\dfrac{\theta}{2}\right)$.

(B) Rewrite it to give an identity for $\cos(2\alpha)$.

B74. Set up equations for θ in the picture. Solve them algebraically, using the double-angle identity for tangent. [18]

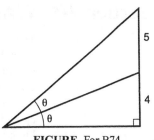

FIGURE For B74

B75. Consider the graph of x^2 to be a reflecting surface. Let a ray of light come in from above, parallel to the y-axis and reflect off the curve (see the figure). Where does the ray intersect the y-axis? To do this, use the fact from physics that "the angle of incidence is equal to the angle of reflection" and the fact from calculus that the slope of the curve x^2 at any point (x, x^2) is $2x$. You might first try the particular case of a ray parallel to the y-axis which reflects off the curve at $(1, 1)$.

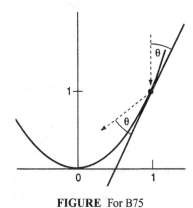

FIGURE For B75

B76. The area of the parallelogram in the figure is given by "$A = ad - bc$." Prove this result using the SAS Area Formula (6-3-2) and a difference of angles identity.

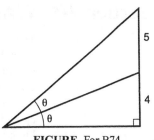

FIGURE For B76

Section 7-4 WAVES

The shape of the graph of "sin t" (Figure 1) is called a "sine wave." (In this section we will use the letter "t" for time, instead of "x".) By definition, a sine wave describes one type of periodic motion—the vertical motion of a point rotating uniformly about a circle. Circular motion has two components as time elapses: the point goes both up-and-down and side-to-side. The sine wave isolates the up-and-down part and ignores the side-to-side part. Think of the horizontal axis in Figure 1 as time and the vertical axis as the vertical position of a rotating or oscillating point.

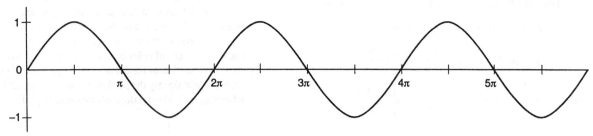

FIGURE 1 sin t. $0 \le t \le 6\pi$

Modifications of sine waves are well-suited to describing other types of periodic behavior—behavior that repeats after a constant time interval, such as the position of a swinging pendulum.

The <u>period</u>, p, of a periodic function f is the time between repetitions. Then the graphs of $f(t)$ and $f(t \pm p)$ are the same. The period of "sin t" is 2π. (7-4-1)

A function can be periodic without being exactly a sine wave. For example, a regular heartbeat can produce a repeating up and down wave that is not as simple as a sine wave.

EXAMPLE 1 Sound waves are sine waves. That is, when you hear a single note at constant volume, the pressure at your eardrum goes up and down according to a sine wave.

Pluck a string on a guitar and watch it vibrate at the middle. The position of the string changes very rapidly back and forth. The position is described by a sine wave with a short period.

Electricity coming out of a wall plug in the U.S. has a frequency of 60 cycles per second (in England it is 50 cycles per second). That is, it repeats 60 times a second. It can be described by a sine wave with a period of $\dfrac{1}{60}$ second. ▬

Other electrical phenomena, such as radio and television waves, x-rays, gamma rays, light waves (as in lasers that read music off of compact disks), microwaves (for cooking) are described by sine waves. Radio waves repeat from thousands to millions of times a second. Visible light repeats on the order of about 10^{15} times a second X-rays repeat on the order of 10^{20} times per second.

Even if a periodic motion is not exactly described by a sine wave, sine waves are often involved in combination with other functions.

EXAMPLE 2

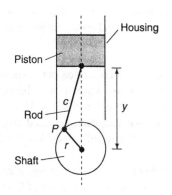

Piston

Housing

Rod — c

P

r

y

Shaft

FIGURE 2 A piston connected to a rotating shaft

The position of a piston in an engine running at constant speed is periodic (Figure 2, position y). The piston is connected to a shaft which is going around at a constant angular speed, so the connection (P) moves up and down in a sine wave. That is, the vertical component of P moves in a sine wave. But the piston itself does not move the same vertical distance that P does because it is connected to the rotating point with a piston rod which is not constantly vertical (Problem B9). So, in Figure 2, y is closely related to a sine wave but it is not exactly a sine wave. ▬

Sine waves are important. Any one of the applications in Examples 1 and 2 could be studied for a long time (in another course). We will just introduce some of the key terms and indicate some fascinating properties of sine waves.

Again, imagine the argument of a sine wave to be time. When $P(t)$ goes around the circle once, one <u>revolution</u> (or <u>cycle</u>) has been completed. The amount of time one revolution takes is the <u>period</u>. The <u>frequency</u> of a particular sine wave is the number of periods (cycles, revolutions) per unit of time, for example, "60 cycles per second" or "60 revolutions per second."

EXAMPLE 3

The period of "sin t" is 2π, because its values repeat after 2π. Its frequency is $\frac{1}{2\pi}$, because that is the number of periods (revolutions) that occur in one unit of time.

Sometimes sine waves are written with the "2π" included so the period and frequency will be simpler numbers. The period of "sin($2\pi t$)" is 1. As t goes from 0 to 1, the function goes though a full cycle (2π). Its frequency is $\frac{1}{1} = 1$.

The period of "sin($5(2\pi t)$)" is $\frac{1}{5}$. Scale changes were discussed in Section 2-2 on composition of functions. Multiplying by 5 before applying sine makes the graph $\frac{1}{5}$ as wide and the period $\frac{1}{5}$ as much. Therefore the frequency is 5 times as much. For instance, as t goes from 0 to 1, the argument of sine goes from 0 to 10π, 5 periods. Its frequency is 5 and its period is $\frac{1}{5}$. ▬

Period p corresponds to frequency $\frac{1}{p}$. "sin($k(2\pi t)$)" has period $\frac{1}{k}$ and frequency k. "sin(kt)" has period $\frac{2\pi}{k}$ and frequency $\frac{k}{2\pi}$. (7-4-2)

If the frequency is described in "cycles per second," the period is described in "seconds (per cycle)."

EXAMPLE 4 The musical note called "concert A" has frequency 440 cycles per second. Give a sine wave with its frequency. What is its period?

The frequency of "sin(440(2πt))" is 440 (cycles per second). Its period is

$\dfrac{1}{440}$ second. ▄▄

So many types of real-world waves have such short periods that we often describe sine waves in terms of frequency rather than period so the numbers are big rather than tiny.

EXAMPLE 5 The frequency of a local FM radio station is 93.7 million cycles per second. A "cycle per second" in the context of electricity is also known as a "hertz" after an important scientist, Heinrich Hertz, who worked in the late 1800's. (Think of something important and you might have something named after you, too.) The prefix "mega" is used for million, so this would be called "93.7 megahertz."

The distance that a wave travels in one period is called a <u>wavelength</u>. Electromagnetic waves (including radio and light waves) travel at the speed of light, about 300,000 kilometers per second (186,000 miles per second). That's fast.

What is the wavelength of a 93.7 megahertz radio wave?

The wavelength is distance per cycle. Divide "distance per second" by "cycles per second" to obtain "distance per cycle."

$$\frac{300{,}000 \text{ kilometers per second}}{93{,}700{,}000 \text{ cycles per second}} = .0032 \text{ kilometers per cycle}$$

$$= 3.2 \text{ meters per cycle.}$$

The wavelength is 3.2 meters. ▄▄

EXAMPLE 6 Visible light has frequency on the order of 10^{15} cycles per second. What is the wavelength?

$$\frac{300{,}000 \text{ kilometers per second}}{10^{15} \text{ cycles per second}} = 3 \times 10^{-10} \text{ kilometers per cycle}$$

$$= 3 \times 10^{-7} \text{ meters per cycle.}$$

This is on the order of $\dfrac{1}{10{,}000}$ inch. A short wavelength! ▄▄

EXAMPLE 7 What is the wavelength of the musical note concert A?

For this we need to know the frequency of concert A (440 cycles per second) and the speed of sound. The speed of sound is about 1100 feet per second in air at 0° Celsius at sea level. (It's slower at warmer temperatures and at higher altitudes.)

$$\frac{1100 \text{ feet per second}}{440 \text{ cycles per second}} = 2.5 \text{ feet per cycle.}$$

The wavelength is about $2\dfrac{1}{2}$ feet. ▄▄

The previous facts about the speed of sound and the speed of light can be combined to give a quick way to tell how far away lightning is. If lightning is, say 5000 feet away, the light flash will arrive in almost no time, but the sound will travel only about 1100 feet per second. The sound will arrive about $\dfrac{5000}{1100} = 4.5$ seconds later than the light. This type of computation can be converted into a simple formula. Since a mile is 5280 feet, sound takes close to 5 seconds to travel a mile, 10 seconds to travel two miles, etc. So the method to tell how far away lightning is: Count the time in seconds from the times you see the flash until you hear the thunder. Divide by 5 to determine how far away the lightning was in miles.

EXAMPLE 8   Rose sees a lightning bolt and hears the thunder 7 seconds later. How far away was the bolt?

$\dfrac{7}{5} = 1\dfrac{2}{5}$. It was about one and two fifths miles away.   ▬

■ Describing Sine Waves

Sine waves are described by three features: amplitude, period, and phase shift. The frequency determines the period, so it can be given in place of the period. These features produce scale and location changes of the sort discussed in Section 2-2 on composition of functions.

Let the basic function of the following discussion be the fundamental sine wave (in radians), "sin t," with period 2π. It has amplitude (maximum value) 1. "A sin t" differs from "sin t" only in the vertical scale. If $A > 0$, its maximum value is A which is called its <u>amplitude</u> (Figure 3).

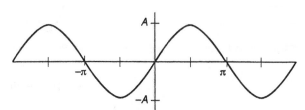

FIGURE 3 A sin t. amplitude $= A$. $-2\pi \le t \le 2\pi$

"Sin(Bt)" differs from "sin t" only in horizontal scale. Since sine repeats every 2π, sin(Bt) repeats every $\dfrac{2\pi}{B}$. If B is greater than 1 the period of "sin(Bt)" is shorter than the period of "sin t" and the frequency is greater (Figure 4).

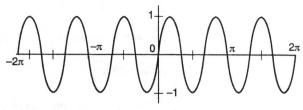

FIGURE 4   sin($3t$). period $= \dfrac{2\pi}{3}$. $-2\pi \le t \le 2\pi$

If the graph is shifted horizontally, the value of t that makes the argument of sine zero is called the <u>phase shift</u>. Sin$(t - c)$ has argument zero when $t = c$; its phase shift is c. When a sine wave has phase shift c, it begins at c the way the usual sine graph begins at 0 (Figure 5).

The function $\sin(Bt - C) = \sin\left(B\left(t - \dfrac{C}{B}\right)\right)$ combines a period change and

a phase shift (Figure 6). The period is $\dfrac{2\pi}{B}$. The phase shift is $\dfrac{C}{B}$, because the argument

of the sine function is zero when $t = \dfrac{C}{B}$. If the phase shift is an integer multiple

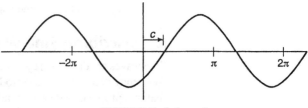

FIGURE 5 $\sin(t - c)$

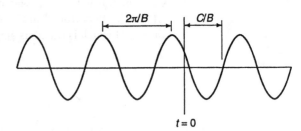

FIGURE 6 $\sin(Bt - C). = \sin\left(B\left(t - \dfrac{C}{B}\right)\right)$

of the period the two waves "$\sin(Bt)$" and "$\sin(Bt - C)$" have the same graph and

are said to be "in phase". If $\dfrac{C}{B}$ is $\dfrac{1}{2}$ the period, or $\dfrac{3}{2}$ the period, etc., they are "out of phase" because the high points of one correspond to the low points of the other (Figure 9).

These three types of scale changes and location shifts can be expressed in one expression. Let $B > 0$. Then

$y = A \sin(Bt - C)$ has amplitude |A| (absolute value, in case A is

negative), period $\dfrac{2\pi}{B}$ $\left[\text{which is equivalent to frequency } \dfrac{B}{2\pi}\right]$, and

phase shift $\dfrac{C}{B}$. (7-4-3)

Any curve with this equation is called a "sinusoid," or a "sinusoidal" curve. Any motion satisfying this equation is said to be "simple harmonic motion."

EXAMPLE 9 Give the amplitude, period, and phase shift of

$$y = 500 \sin(20(t - 5)).$$

The amplitude is 500, the period is $\dfrac{2\pi}{20} = \dfrac{\pi}{10}$, and the phase shift is 5. ▬

As you have seen, sine and cosine waves have the same shape and the same period, 2π. Actually, you don't hear the term "cosine waves" much because they are just shifted sine waves.

$$\cos \theta = \sin\left(\theta + \frac{\pi}{2} \right) \text{ [in radians].}$$
$$\cos \theta = \sin(\theta + 90\,\text{º}) \text{ [in degrees]}$$

(7-4-4)

This tells us that the graph of cosine is the graph of sine shifted *left* $\dfrac{\pi}{2}$.

EXAMPLE 10 When a spring is stretched and let go it vibrates. The motion can be modeled by $y(t) = A \cos(Bt)$.

The initial stretch at time 0 is $y(0) = A$, because $\cos(0) = 1$. B determines the frequency of the vibration.

Suppose a spring is stretched 4 inches and let go and it vibrates up and down 5 times a second. Give the formula for the position at time t.

We are given $y(0) = 4$ (inches). B is the other parameter to be determined. The problem says that five cycles occur in 1 second. By 7-4-2 (which works for cosines exactly as for sines), using $k = 5$, the formula is $y(t) = 4 \cos(5(2\pi t)) = 4 \cos(10\pi t)$. ▬

■ Adding Sine Waves

Some of the most fascinating features of sine waves are their additive properties. We will use sound waves to illustrate them. For this subsection you will want to get out your graphing calculator and play around with these ideas.

Most sound you hear consists of a mixture of frequencies, but to begin discussion, consider a single frequency. We will consider combining two sound waves of the same frequency, but with possibly different amplitudes (volumes) and different phases. To make things easy, we will not bother to use a realistic scale; we will use "sin t" as the basic sinusoid.

Any sinusoid of the form "$A \sin(t - c)$" has the same frequency as "sin t." What do you hear when you hear two sources of sound producing the same note? What happens when you add two such curves together?

Of course, if two people hit two tuning forks and both tuning forks produce middle C, you will still hear middle C, but probably louder. That is, you expect the amplitude to change, but not the frequency. Mathematically, this real-life result corresponds to examples such as the following.

EXAMPLE 11 Use radians to graph "$\sin t + \sin(t - 1)$".

Figure 7 graphs three graphs. You recognize "sin t." The graph of "$\sin(t - 1)$" is similar, but shifted right 1 unit $\left(1 \text{ radian is about } \dfrac{1}{6.28} \text{ period} \right)$. The graph of their sum appears to have the same period, but be higher (which corresponds to louder). It does not cross the t-axis at the same place as either, so it has a different phase.

In this example, the amplitude of the sum is greater than 1, but not as great as 2, which would be the sum of the amplitudes. The amplitude is about 1.755.

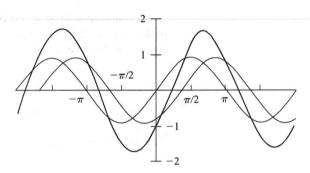

FIGURE 7 $\sin t$, $\sin(t - 1)$ and their sum (bold). $[-2\pi, 2\pi]$

EXAMPLE 12 Use radians to graph "$\sin t + \sin(t - 3)$."

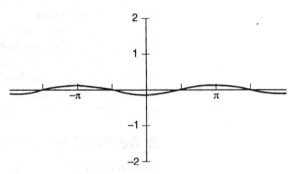

FIGURE 8 $\sin t + \sin(t - 3)$ $[-2\pi, 2\pi]$

Figure 8 shows that the sum is *less* than either. That is because the components are almost completely "out of phase" (Figure 9). The components, rather than working together to produce a larger sum, work against one another and tend to cancel each other. The amplitude of the sum is about .141, far less than the sum of the amplitudes and even less than the amplitude of either component.

Example 12 illustrates some very high-tech ideas. If an out-of-phase wave of the same frequency can be added to a wave, it can cancel the original wave. This "out-of-phase" idea will eventually have applications in the car and in the

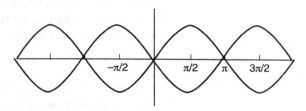

FIGURE 9 $\sin t$, $\sin(t - 3)$. $[-2\pi, 2\pi]$. They are out of phase

home. There will be "silence-generating machines." The machine will analyze incoming sound for amplitude, frequency, and phase, and in an instant generate canceling waves from a speaker. If you are in the right position (say, the car driver's seat), you will hear very little. The machine will have turned sound into silence. You can already buy headphones that use this principle to *cancel* (not just muffle) the stressful monotonous drone of jet-engines during air travel. ▬

EXAMPLE 13 Graph "sin t + 2 sin(t − 4)."

Even if the components do not have the same amplitudes, the sum is again sinusoidal, as long as the frequencies of the components are the same. Figure 10 shows that, again, the sum has the same frequency as the components. The amplitude is about 1.55.

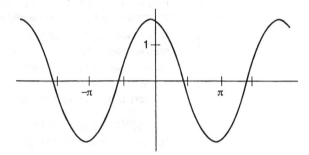

FIGURE 10 sin t + 2 sin(t − 4). [−2π, 2π] ▬

The previous examples concerned waves that are not "in phase." That is, the components of the sum had different phases. If two sines waves are in phase, the amplitude of the sum is the sum of the amplitudes. For example, "sin t + 2 sin t" is clearly just "3 sin t," which has amplitude 3, which is the sum of the amplitudes.

When waves with different frequencies are added together, the result can be complicated and no longer simply sinusoidal.

EXAMPLE 14 Graph sin t + sin(2t).

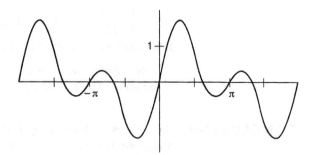

FIGURE 11 sin t + sin(2t). [−2π, 2π]

The graph is more complicated and not simply sinusoidal (Figure 11). It is a mathematical challenge to figure out how to take such a curve and decompose it into a sum of sinusoidal curves. Imagine you are a sound engineer helping design sound recording and analyzing machines. For example, you may wish to filter out "noise" of certain frequencies. But the music already consists of numerous frequencies at various amplitudes and all phases. Sound hitting a microphone does not come labeled by its component frequencies. It's just sound. How can your machine, which "hears" it all at once, find the components at the frequencies it wants to filter out?

In the ocean there are many sources of sound. How can sonar (say, in a submarine) be analyzed to isolate the components of interest from the "background noise." Tough question. If you want to know, you will have to take more math!

Things are mathematically much simpler if all waves are of the same frequency. Normal light, say, from a flashlight, consists of many waves of many frequencies added together. Even if they were all of the same frequency, they would not add very efficiently because they would still be of different phases. If fact, some components would cancel others, as in Example 12 (Figure 8). The reason laser light is so powerful is that lasers emit light waves *of the same frequency* that are *in phase*, so they add together efficiently and do not cancel each other.

Here is a theorem about the phase shift that will cancel a wave.   ▬

$$\sin(B(t - p)) + \sin(B(t - p) \pm \pi) = 0,$$
for all t, for any B and p.   (7-4-5)

EXAMPLE 15   What wave, when added to "$2\sin(t - 1)$" will cancel it?

We need a wave of the same amplitude, but out of phase.

The wave "$2\sin(t - 1 - \pi)$" will do it. I'd show a graph of the sum, but there wouldn't be much to see. The height would always be zero.   ▬

EXAMPLE 16   What wave, when added to $\sin(12t - 4) = \sin\left(12\left(\dfrac{t - 4}{12}\right)\right)$ will cancel it?

An answer (one of many) is simply $\sin(12t - 4 - \pi) = \sin\left(12\left(\dfrac{t - 4}{12} - \dfrac{\pi}{12}\right)\right)$. Adding or subtracting π to the argument of a sine function makes a new function completely out of phase with the original.   ▬

Waves that reinforce or cancel one another play an important role in physics in the study of sound and light (Problems B16-21).

CONCLUSION   Sines waves are described by their amplitude, period (or frequency or wavelength), and phase. If waves of the same frequency are out of phase, they can cancel one another.

Terms: sine wave, periodic, period, cycle, revolution, frequency, wavelength, amplitude, sinusoid, phase, in phase, out of phase.

Exercise 7-4

A

A1. Use the speed of 1100 feet per second to compute how many miles per hour sound travels.

A2. How far away is lightning if you hear the thunder
(A) 15 seconds after you see the flash?
(B) 2 seconds after you see the flash?

A3. Some microwaves have a frequency of 5×10^{10} cycles per second. What is the corresponding wavelength?

A4. Some x-rays have a frequency of 10^{18} cycles per second. What is the corresponding wavelength?

Identify the amplitude, period, and phase shift.

A5. (A) $2 \sin(3t - 4)$

(B) $5 \sin\left(\dfrac{t - 7}{6}\right)$.

A6. (A) $8 \sin\left(\dfrac{t}{9} + 10\right)$

(B) $11 \sin(12(t - 13))$.

A7. When a typical car engine is idling, approximately what is its frequency?

A8. What is the frequency of household electricity?

B

B1.* Give the relationships between period, frequency, and wavelength.

B2.* Explain how *adding* a wave to a given wave can cancel the given wave. (Subtracting would be easy to understand.)

B3. Find a wave, which when added to the given wave, would cancel it.
(A) $3 \sin(t - 2)$
(B) $2 \sin(1000t)$

B4. Find a wave, which when added to the given curve, would cancel it.
(A) $\sin t + 2 \sin(t + 4)$
(B) $\sin 2t + 2 \sin t$

B5. Use the model in Example 10. A guitar string pulled aside 2 millimeters vibrates 400 times per second. Give the position at time t.

B6. Use the model in Example 10. A spring stretched $\dfrac{1}{2}$ centimeter vibrates 20 times per second. Give the position at time t.

B7. (A) Convert c cycles per second to radians per second. "Radians per second" serves as other units for measuring <u>angular speed</u>.
(B) If a wheel spins around a fixed shaft at d radians per second, what is the linear speed of a point on its circumference?

B8. True or false? Explain. If $f(x)$ has maximum value M an $g(x)$ has maximum value m, then $h(x) = f(x) + g(x)$ has maximum value $M + m$.

Graph the expression (in radians) and use it approximate the amplitude and phase.

B9. $\sin x + \sin(x - 2)$ $[\ldots, 1.0]$

B10. $\sin x + 2 \sin(x + 1)$ $[2.7, \ldots]$

B11. In Figure 2, determine y in terms of time if the shaft is rotating at 50 cycles per second, $r = 2$, and $c = 6$. Begin time when P is at the top.

Damped Waves

B12. Let $f(x) = e^{-x/5}\cos(x)$. The exponential decay factor makes a "damped" wave.
(A) Graph it for $0 \le x \le 2\pi$.
(B) Find the ratio of the height at 2π to the height at 0.

B13. Let $f(x) = e^{-kx}\cos(x)$. The exponential decay factor makes a "damped" wave. Approximate the parameter "k" such that $e^{-kx}\cos(x)$ has the height at each local maximum about $\dfrac{9}{10}$ the height at the previous local maximum. [Hint: Graph some possibilities until you understand the curve.]

B14. Let $f(x) = e^{-kx}\cos(5x)$. The exponential decay factor makes a "damped" wave. Approximate the parameter "k" such that the height of $e^{-kx}\cos(5x)$ at each local maximum is about $\dfrac{3}{4}$ the height at the previous local maximum. [Hint: Graph some possibilities until you understand the curve.]

Artifacts

B15. Many calculators produce artifacts with one or both of these graphs: $\sin(50x)$, $\sin(66x)$.
(A) Use theory to decide what the graphs should look like on $[-2\pi, 2\pi]$.
(B) Graph them with a graphing calculator and see if it does look like what it should. Do you find

an artifact? If so, why is there one? [Calculators have trouble with "$\sin(nx)$" on $[-2\pi, 2\pi]$ when $2n$ is close to the number of columns of pixels on the screen.]

Phase

B16. (A) Use the identity in exercise 7-3-B65 to prove this result about the combination of two sine waves of the same frequency and amplitude but different phases. $\sin(Bt + k) + \sin(Bt + j)$

$$= 2\cos\left(\frac{k - j}{2}\right)\sin\left(Bt + \frac{k + j}{2}\right).$$

   (B) What is the amplitude of the sum? How must the phases be related to reach the maximum possible amplitude?

B17. When two sound waves are very similar, but not identical, in frequency, they will reinforce each other at times and nearly cancel each other at other times. For example, if one note is played at 440 cycles per second and a another note is played at 438 cycles per second, there will be an audible increase and decrease in amplitude twice a second known as a "beat." This can be illustrated with a very wide graph of the sum of two sine waves. However, your calculator does not have enough columns of pixels to display such graphs. Near $x = 0$ the graph of "$\sin x + \sin(1.01x)$" displays reinforcement (graph it and see). Here is the problem: Find, very roughly, the smallest positive c such that this graph displays nearly complete canceling on the interval $[c, c + 10]$.

B18. (See B17). Suppose 2 sine waves of amplitude 1 begin in phase, one at 440 cycles per second and the other at 438 cycles per second.
   (A) Give the expression for their sum.
   (B) Find a window on your graphics calculator that clearly displays both non-canceling and the near-canceling which occurs periodically.
   (C) At which value of t does the first such instance of that near-canceling occur?
   (D) When is the next instance of that near-canceling?

B19. Imagine a wave coming from the left and passing through two slits, A and B, in phase (See the figure, which is not to scale). From each slit the waves spread out to reach the points on the line CM, which is parallel to the line containing the slits. Since the distance from A to C is not the same as the distance from B to C (except when C is at M, right in the middle), the waves which were in phase when they passed through the slits may not be in phase when they reach C. When the waves are light waves, the difference in distance causes a pattern of alternating light and dark bands along the line CM because of the reinforcing or canceling of the two waves. Suppose the slits are .03 mm (millimeter) apart,

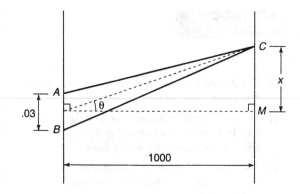

and the line through C and M is 1000 mm from the slits, and the wavelength of the light passing through the slits in phase is .00051 mm. Find the smallest distance, x, from M to C such that the waves are completely out of phase when they reach C. Hint: If you treat AC and BC as parallel, and x as small relative to 1000, the approximation will be good enough and substantial simplifications will occur. Remember, the picture is not to scale.

B20. (See B19.) If the distance between the slits is d, the wavelength is w, and the distance to the line is L, find a simple formula for the distance, x, from M to C such that the waves are completely out of phase when they reach C. If you treat AC and BC as parallel, and x as small relative to L, the approximation will be good enough and substantial simplifications will occur. Remember, the picture is not to scale.

B21. Here is a **trigonometric identity** for adding waves of the same frequency:
$c \sin(ax) + d \cos(ax) = A \sin(ax + b)$, where

$$(7\text{-}4\text{-}6)$$

$A = \sqrt{c^2 + d^2}$ and $\cos b = \dfrac{c}{A}$ and $\sin b = \dfrac{d}{A}$.

   (A) Find an expression equivalent to $\sqrt{3}\sin x + \cos x$ that involves only the sine function.
   (B) Graph the original expression and the result of part (a) and see if the amplitude and phase you found are right.

B22. [Use the data about Earth and Venus in Table 5-4-3.] Suppose Earth and Venus are on the same side of the Sun and all three are in a line. Because Venus goes around the Sun faster than Earth, they will not stay in line. How long will it be until they are in line again after being in line? [1.6 years]

B23. [Use the data about Earth and Mars in Table 5-4-3.] Suppose Earth and Mars are on the same side of the Sun and all three are in a line. Because Earth goes around the Sun faster than Mars, they will not stay in line. How long will it be until they are in line again after being in line? [2.1 years]

B24. [Recall the lessons of Section 2-2 on composition of functions and shifting and re-scaling curves.] The number of daylight hours depends upon the time of year. For example, you might discover that on the longest day, June 21, there are 16 daylight hours and on the shortest day, December 21, only 8 daylight hours. The data almost fits a sine curve. Fit a curve of the form "$a \sin(b(t - c)) + d$" to those extreme points, adjusting the parameters appropriately. Measure time in months, beginning on the 21st. [It would be more interesting for you to find out the number of daylight hours for some days of the year in your own area and determine the parameters.]

B25. Tourist guidebooks sometimes publish the mean monthly temperatures for the twelve months of the year. For example, you might find the coldest month is February with an average temperature of 30 degrees, and the warmest is August with an average temperature of 70 degrees. The data almost fit a sine curve. Determine the parameters of a sine curve of the form "$a \sin(b(t - c)) + d$" that fits that data. Measure t in months. [It would be more interesting to find data for your own home town, fit the curve, and then see how well it fits.]

B26. (A) There is an x-interval $[c, c + 10]$ such that the graph of $\sin(1.02x) + \sin(x)$ will be very close to the x-axis. Find such a c (in radians).

(B) Explain what this has to do with 7-4-5.

B27. (A) There an x-interval $[c, c + 10]$ such that the graph of $\sin(.97x) + \sin(x)$ will be very close to the x-axis. Find such a c (in radians).

(B) Explain what this has to do with 7-4-5.

INDEX